Wen-Lian Hsu Ming-Yang Kao (Eds.)

Computing and Combinatorics

4th Annual International Conference
COCOON'98
Taipei, Taiwan, R.o.C., August 12-14, 1998
Proceedings

Series Editors

Gerhard Goos, Karlsruhe University, Germany
Juris Hartmanis, Cornell University, NY, USA
Jan van Leeuwen, Utrecht University, The Netherlands

Volume Editors

Wen-Lian Hsu
Academia Sinica, Institute of Information Science
Nankang 11529, Taipei, Taiwan, R.o.C.
E-mail: hsu@iis.sinica.edu.tw

Ming-Yang Kao
Yale University, Department of Computer Science
51 Prospect Street, New Haven, CT 06520, USA
E-mail: kao-ming-yang@cs.yale.edu

Cataloging-in-Publication data applied for

Die Deutsche Bibliothek - CIP-Einheitsaufnahme

Computing and combinatorics : 4th annual international conference ; proceedings / COCOON '98, Taipei, Taiwan, August 12 - 14, 1998. Wen-Lian Hsu ; Ming-Yang Kao (ed.). - Berlin ; Heidelberg ; New York ; Barcelona ; Budapest ; Hong Kong ; London ; Milan ; Paris ; Santa Clara ; Singapore ; Tokyo : Springer, 1998
(Lecture notes in computer science ; Vol. 1449)
ISBN 3-540-64824-0

CR Subject Classification (1991): F.2, G.2.1-2, I.3.5, C.2.3-4, E.1

ISSN 0302-9743
ISBN 3-540-64824-0 Springer-Verlag Berlin Heidelberg New York

Typesetting: Camera-ready by author
SPIN 10639047 06/3142 – 5 4 3 2 1 0 Printed on acid-free paper

Preface

The papers in this volume were selected for presentation at the *Fourth Annual International Computing and Combinatorics Conference* (COCOON'98), held on August 12–14, 1998, in Taipei. The topics cover most aspects of theoretical computer science and combinatorics related to computing.

Submissions to the conference this year were only conducted electronically. Thanks to the excellent software developed by the system team of the Institute of Information Science, we were able to make virtually all communications through the World Wide Web.

A total of 69 papers was submitted in time to be considered, of which 36 papers were accepted for presentation at the conference. In addition to these contributed papers, the conference also included four invited presentations by Christo Papadimitriou, Michael Fishcher, Fan Chung Graham, and Rao Kosaraju. It is expected that most of the accepted papers will appear in a more complete form in scientific journals. Moreover, selected papers will appear in a special issue of Theoretical Computer Science.

We thank all program committee members, their support staff and referees for excellent work within demanding time constraints. We thank all authors who submitted papers for consideration. We are especially grateful to our colleagues who worked hard and offered widely differing talents to make the conference both possible and enjoyable.

August 1998

Wen-Lian Hsu and Ming-Yang Kao
Program Co-chairs
COCOON'98

Organization

COCOON'98 is organized by the Institute of Information Science, Academia Sinica, Taipei, Taiwan, ROC and in cooperation with Institute of Information and Computing Machinery (IICM), Taiwan, ROC.

Program Committee

Dana Angluin (Yale U., USA)
Ricardo Baeza-Yates (U. Chile, Chile)
Allan Borodin (U. Toronto, Canada)
Francis Chin (U. Hong Kong, PRC)
Anne Condon (U. Wisconsin, USA)
Joan Feigenbaum (AT&T Labs, USA)
Mordecai J. Golin (Hong Kong U. Science and Technology, PRC)
Xin He (SUNY Buffalo, USA)
Jan-Ming Ho (Academia Sinica, ROC)
Michael Houle (U. Newcastle, Australia)
Wen-Lian Hsu (Academia Sinica, ROC, co-chair)
Xiaodong Hu (Academy of Science, PRC)
Frank Hwang (Chiao Tung U., ROC)
Toshihide Ibaraki (Kyoto U., Japan)
Hiroshi Imai (Tokyo U., Tokyo)
Ravi Janardan (U. Minnesota, USA)
Ming-Yang Kao (Yale U., USA, co-chair)
Philip Klein (Brown U., USA)
Andrzej Lingas (Lund U., Sweden)
Richard J. Lipton (Princeton U., USA)
Rolf Möehring (TU Berlin, Germany)
Maurice Nivat (U. Paris, France)
Sartaj Sahni (U. Florida, USA)
Uwe Schöning (U. Ulm, Germany)
Paul Spirakis (U. Patras, Greece)
Tandy Warnow (U. Pennsylvania, USA)
Louxin Zhang (U. Singapore, Singapore)

Executive Committee

Organizing:	Wen-Lian Hsu (Academia Sinica, ROC)
	Ming-Tat Ko (Academia Sinica, ROC)
Local arrangement:	Jan-Ming Ho (Academia Sinica, ROC)
	Ming-Tat Ko (Academia Sinica, ROC)
	Ya-Chiung Lee (Academia Sinica, ROC)
	Tze-Heng Ma (Academia Sinica, ROC)
	Kuo-Hui Tsai (Academia Sinica, ROC)
Web:	Wen-Jyi Sen (Academia Sinica, ROC)
	Chiu-Feng Wang (Academia Sinica, ROC)
	Da-Wei Wang (Academia Sinica, ROC)
Finance:	Ting-Yi Sung (Academia Sinica, ROC)
Publicity and Publication:	Tsan-sheng Hsu (Academia Sinica, ROC)

Sponsoring Institutions

ACM Taipei/Taiwan Chapter.
National Science Council, Taiwan, ROC.

Table of Contents

Invited Presentations

Algorithmic Approaches to Information Retrieval and Data Mining 1
Christos H. Papadimitriou

Combinatorial Problems Arising in Massive Data Sets 2
Fan Chung Graham

Estimating Parameters of Monotone Boolean Functions 3
Michael J. Fischer

De-amortization of Algorithms 4
S. Rao Kosaraju and Mihai Pop

Computational Geometry

On Computing New Classes of Optimal Triangulations with Angular Constraints 15
Yang Dai and Naoki Katoh

Maximum Weight Triangulation and Its Application on Graph Drawing . 25
Cao An Wang, Francis Y. Chin and Bo Ting Yang

The Colored Sector Search Tree: A Dynamic Data Structure for Efficient High Dimensional Nearest-Foreign-Neighbor Queries 35
T. Graf, V. Kamakoti, N. S. Janaki Latha and C. Pandu Rangan

Space-Efficient Algorithms for Approximating Polygonal Curves in Two Dimensional Space 45
Danny Z. Chen and Ovidiu Daescu

Parallel Geometric Algorithms in Coarse-Grain Network Models 55
Mikhail J. Atallah and Danny Z. Chen

Algorithms and Data Structures

On the Bahncard Problem 65
Rudolf Fleischer

The Ultimate Strategy to Search on m Rays? 75
Alejandro López-Ortiz and Sven Schuierer

Better Approximation of Diagonal-Flip Transformation and Rotation Transformation 85
Ming Li and Louxin Zhang

A Cost Optimal Parallel Algorithm for Computing Force Field in N-Body Simulations 95
Guoliang Xue

Efficient 2 and 3-Flip Neighborhood Search Algorithms for the MAX SAT 105
Mutsunori Yagiura and Toshihide Ibaraki

Algorithms for Almost-uniform Generation with an Unbiased Binary Source 117
Ömer Eğecioğlu and Marcus Peinado

Improved Algorithms for Chemical Threshold Testing Problems 127
Annalisa De Bonis, Luisa Gargano and Ugo Vaccaro

Min-Max-Boundary Domain Decomposition 137
Marcos Kiwi, Daniel A. Spielman and Shang-Hua Teng

Computational Complexity

On Boolean Lowness and Boolean Highness 147
Steffen Reith and Klaus W. Wagner

The Inherent Dimension of Bounded Counting Classes 157
Ulrich Hertrampf

An Exact Characterization of Symmetric Functions in $qAC^0[2]$ 167
Chi-Jen Lu

Robust Reductions 174
Jin-Yi Cai, Lane A. Hemaspaandra and Gerd Wechsung

Approaches to Effective Semi-continuity of Real Functions 184
Vasco Brattka, Klaus Weihrauch and Xizhong Zheng

On the Power of Additive Combinatorial Search Model 194
Vladimir Grebinski

Parallel and Distributed Processing, and Switching Networks

The Number of Rearrangements in a 3-stage Clos Network Using an Auxiliary Switch 204
Frank K. Hwang and Wen-Dar Lin

Lower Bounds for Wide-Sense Non-blocking Clos Network 213
Kuo-Hui Tsai and Da-Wei Wang

Multirate Multicast Switching Networks 219
Dongsoo S. Kim and Ding-Zhu Du

Efficient Randomized Routing Algorithms on the Two-Dimensional Mesh of Buses 229
Kazuo Iwama, Eiji Miyano, Satoshi Tajima and Hisao Tamaki

Executing Divisible Jobs on a Network with a Fixed Number of Processors 241
Tsan-sheng Hsu and Dian Rae Lopez

Graph Theory

On the Ádám Conjecture on Circulant Graphs 251
Bernard Mans, Francesco Pappalardi and Igor Shparlinski

Proof of Toft's Conjecture: Every Graph Containing No Fully Odd K_4 Is 3-Colorable 261
Wenan Zang

A New Family of Optimal 1-Hamiltonian Graphs with Small Diameter .. 269
Jeng-Jung Wang, Ting-Yi Sung, Lih-Hsing Hsu and Men-Yang Lin

A Linear-Time Algorithm for Constructing an Optimal Node-Search Strategy of a Tree 279
Sheng-Lung Peng, Chin-Wen Ho, Tsan-sheng Hsu, Ming-Tat Ko and Chuan Yi Tang

A Graph Optimization Problem in Virtual Colonoscopy 289
Jie Wang and Yaorong Ge

Approximation and Exact Algorithms for Constructing Minimum Ultrametric Trees from Distance Matrices 299
Bang Ye Wu, Kun-Mao Chao and Chuan Yi Tang

An Optimal Algorithm for Finding the Minimum Cardinality Dominating Set on Permutation Graphs 309
H. S. Chao, F. R. Hsu and R. C. T. Lee

Combinatorics and Cryptography

Similarity in Two-Dimensional Strings 319
Ricardo A. Baeza-Yates

On Multi-dimensional Hilbert Indexings 329
Jochen Alber and Rolf Niedermeier

Combinatorial Properties of Classes of Functions Hard to Compute in Constant Depth 339
Anna Bernasconi

Eulerian Secret Key Exchange 349
Takaaki Mizuki, Hiroki Shizuya and Takao Nishizeki

One-Time Tables for Two-Party Computation 361
Donald Beaver

Author Index 371

Algorithmic Approaches to Information Retrieval and Data Mining

Christos H. Papadimitriou
University of California Berkeley, USA

Abstract

The emerging globalized information environment, with its unprecedented volume and diversity of information, is creating novel computational problems and is transforming established areas of research such as information retrieval. I believe that many of these problems are susceptible to rigorous modeling and principled analysis. In this talk I will focus on recent research which exemplifies the value of theoretical tools and approaches to these challenges.

Researchers in information retrieval have recently shown the applicability of spectral techniques to resolving such stubborn problems as polysemy and synonymy. The value of these techniques has more recently been demonstrated rigorously in a PODS 98 paper co-authored with Raghavan, Tanaki, and Vempala, by utilizing a formal probabilistic model of the corpus. Also in the same paper, a rigorous randomized simplification of the singular value decomposition process was proposed. In a paper in SODA 98, Kleinberg shows how spectral methods can extract in a striking way the semantics of a hypertext corpus, such as the world-wide web.

Although data mining has been promising the extraction of interesting patterns from massive data, there has been very little theoretical discussion of what "interesting" means in this context. In a STOC 98 paper co-authored with Kleinberg and Raghavan, we argue that such a theory must necessarily take into account the optimization problem faced by the organization that is doing the data mining. This point of view leads quickly to many interesting and novel combinatorial problems, and some promising approximation algorithms, while leaving many challenging algorithmic problems wide open.

Combinatorial Problems Arising in Massive Data Sets

Fan Chung Graham

University of Pennsylvania
Philadelphia, PA 19104

Abstract. We will discuss several combinatorial problems which are motivated by computational issues arising from the study of large graphs (having sizes far exceeding the size of the main memory). A variety of areas in graph theory are involved, including graph labeling, graph embedding, graph decomposition, and spectral graph theory as well as concepts from probabilistic methods and dynamic location/scheduling.

Estimating Parameters of Monotone Boolean Functions

(abstract)

Michael J. Fischer

Department of Computer Science, P.O. Box 208285,
Yale University, New Haven CT 06520–8285, USA,
fischer-michael@cs.yale.edu,
WWW home page: http://www.cs.yale.edu/~fischer/

Let $F: \{0,1\}^n \to \{0,1\}$ be a monotone Boolean function. For a vector $\mathbf{x} \in \{0,1\}^n$, let $\#(\mathbf{x})$ be the number of 1's in $\mathbf{x}$, and let $S_k = \{\mathbf{x} \in \{0,1\}^n \mid \#(\mathbf{x}) = k\}$. For a multiset set $R \subseteq \{0,1\}^n$, define the *F-density* of R to be

$$D(R) = \frac{|\{\mathbf{x} \in R \mid F(\mathbf{x}) = 1\}|}{|R|} .$$

Thus, $D(R) = \text{prob}[F(X) = 1]$, where X is uniformly distributed over R.

Let R_k be a random multiset consisting of m independent samples drawn uniformly from S_k. Then the random variable $Y_k = D(R_k)$ is a sufficient estimator for $D(S_k)$. A naive algorithm to compute $Y_0, \ldots, Y_n$ evaluates F on each of the $m(n+1)$ random samples in $\bigcup_k R_k$ and then computes each $D(R_k)$.

The following theorem shows that the number of evaluations of F can be greatly reduced.

Main Theorem. *There is a randomized algorithm for computing $Y_0, \ldots, Y_n$ that performs at most $m\lceil \log_2(n+2) \rceil$ evaluations of F.*

When n is large and F takes a considerable amount of time to evaluate, this theorem permits a dramatic decrease in the time to compute the Y_k's and hence to estimate the parameters $D(S_k)$ to a given degree of accuracy.

The problem of estimating the parameters $D(S_k)$ arises in the study of error-correcting codes. The vector $\mathbf{x} \in \{0,1\}^n$ represents a particular error pattern, and $F(\mathbf{x}) = 1$ iff the error-correction algorithm fails to correct all errors. Many error correction algorithms are monotone in the sense that additional errors can never turn an uncorrectable error pattern into a correctable one. In this context, $D(S_k)$ is the probability that a corrupted code word containing k independent bit errors is uncorrectable by the algorithm. Regarded as a function of k, it is a useful measure of the overall error-correction ability of an algorithm.

De-amortization of Algorithms*

Preliminary version

S. Rao Kosaraju and Mihai Pop

Department of Computer Science
Johns Hopkins University
Baltimore, Maryland 21218

Abstract. De-amortization aims to convert algorithms with excellent overall speed, $f(n)$ for performing n operations, into algorithms that take no more than $O(f(n)/n)$ steps for each operation. The paper reviews several existing techniques for de-amortization of algorithms.

1 Introduction

The worst case performance of an algorithm is measured by the maximum number of steps performed by the algorithm in response to a single operation, while the amortized performance is given by the total number of steps performed in response to n operations. De-amortization seeks to convert an algorithm with $f(n)$ amortized speed to another with close to $O(\frac{f(n)}{n})$ worst case speed.

Consider, for example, the standard pushdown in which the allowable single step primitives are *push*, which pushes a single item on top of the pushdown, and *pop*, which removes the topmost item of the pushdown. The pushdown is initially empty, and it is not allowed to pop when it is empty. We want to implement the pushdown automaton with *push*, specified above, and *pop** which performs a sequence of pops until a specified condition is satisfied. Note that at any stage, the total number of pops performed cannot be more than the total number of pushes performed. Hence, if we implement *pop** by a sequence of pops, the realization of a sequence of n operations composed of *pushes* and *pop**s can have at most $n-1$ pushes and $n-1$ pops, resulting in an amortized speed of $2(n-1)$. However, all the $n-1$ pops will be performed on the last *pop** if the sequence of operations is $n-1$ *pushes* followed by a *pop**. Thus, the worst case speed for implementing a single operation is $n-1$, while the average number of steps performed by this algorithm is $2-2/n$. The average amortized speed of this algorithm is $2-2/n$. In de-amortization we seek an algorithm whose worst case single operation speed is $O(2-2/n)$. It can easily be shown that pushdowns cannot be de-amortized. However, when the pushdowns are allowed to have a limited "jumping" capability, de-amortization can be achieved ([13, 18]).

Many techniques were developed for achieving de-amortization of algorithms, and we review some of them. Even though we present them as distinct techniques, they are quite inter-related.

---

* Supported by NSF Grant CCR9508545 and ARO Grant DAAH04-96-1-0013

In section 2 we present how data duplication can be used to achieve de-amortization. Section 3 discusses techniques that are based on partially or fully rebuilding a data-structure. Approaches that maintain global invariants in order to achieve worst-case performance guarantees are presented in section 4. In section 5 we examine the relationship between a certain class of pebble games and de-amortization.

2 Data Duplication

The earliest application of de-amortization is for converting a Turing tape with multiple heads, M, into Turing tapes with one head per tape, M' [11]. Each tape of M' tries to maintain a neighborhood of one of the heads of M, such that the neighborhoods are non-overlapping and together cover the complete tape of M. However, this process runs into difficulty when some head h of M tries to move out of the neighborhood maintained by the corresponding tape of M'. An amortized multi-tape single head per tape simulator of M, denoted by $\hat{M}$, will stop the simulation and will redistribute the tape between h and its corresponding neighbor evenly between the two tapes and then will resume the simulation. Such a simulator $\hat{M}$ guarantees an amortized speed of $O(n)$ for simulating n steps of M. Fischer, et al. ([11]) de-amortize this simulator by keeping essentially two copies of $\hat{M}$, one in the background and the other in the foreground. While the current simulation is being performed on the foreground copy, redistribution of data will be performed on the background copy. By the time the foreground copy needs redistribution, the background copy will be ready to come to the foreground, and the current foreground copy becomes the background copy. Fischer, et al. showed that such a de-amortization strategy achieves a worst case performance of $O(1)$ steps per operation of M — matching the average amortized speed.

Baker [2] applies a similar data duplication technique for implementing a real-time mechanism for performing garbage collection in a list processing system. Continuous garbage collection is achieved by performing (worst case) constant number of steps after each CONS operation.

The main idea is to keep two list spaces, one which is active (the *from* space), and another one into which garbage collection will put the accessible nodes (the *to* space). During each CONS operation the algorithm performs a constant number of garbage collection steps, moving accessible nodes into the *to* space and keeping pointers to them in the *from* space. By correctly choosing the number of garbage collection operations performed during each CONS, it is guaranteed that eventually the *to* space will contain all accessible nodes, while the *from* space will contain contain garbage. At this point the roles of the two spaces can be exchanged.

Gajewska and Tarjan ([12]) present a nice adaptation of the data duplication technique while implementing double ended queues (deques) with heap order.

This technique of maintaining duplicate copies of the data structure plays a significant role in many other strategies that will be covered in the subsequent sections.

3 Rebuilding a Data Structure

Overmars [28] applies the data duplication technique to achieving dynamization of data structures for multi-dimensional search problems. The main idea is to start constructing in the background a new version of the data-structure once the original structure has reached a certain level of unbalance. The new structure is slowly constructed over a series of updates. A certain restriction on updates, designated as *weak updates*, guarantees that the new structure is fully built before the original structure has become too unbalanced. Then the new structure becomes active and the old one is removed. An update is *weak* iff there exist constants α and k such that after αn updates (n = size of data-structure) the query time doesn't deteriorate by more than a k factor. For data-structures that allow weak updates, Overmars establishes the following worst case result.

Theorem 1. *Any data structure S for some searching problem, that permits weak deletions and weak insertions, can be dynamized into a structure S' such that:*

$$Q_{S'} = O(Q_S(n))$$
$$U_{S'} = O(U_S(n) + P_S(n)/n)$$

where $Q_S(n)$, $U_S(n)$, and $P_S(n)$ are the average amortized speeds for performing queries, updates, and preprocessing in S, respectively, and $Q_{S'}$, and $U_{S'}$ are the worst case speeds for performing queries and updates in S', respectively.

This result is applied to quad trees and k-d trees obtaining $O(\log^2 n)$ insertion time and $O(\log n)$ deletion time, and to d-dimensional super B-trees yielding an insertion time of $O(\log^d n)$ and a deletion time of $O(\log^{d-1} n)$.

Earlier, Bentley applied a similar technique for decomposable search problems [3]. A search problem is called *decomposable* iff for any partition of the data set and any query object, the answer can be obtained in constant time from the answers to the queries on the blocks of the partition. The idea is to split the structure into a logarithmic number of bags of blocks of increasing sizes. Insertions and deletions may cause blocks to over- or underflow, in which case the block is either moved up in the bag list, by merging with another block of the same size, or is split into two smaller blocks that are inserted at a lower level in the list. It is shown that this technique yields amortized time bounds for the updates. Amortization is then removed by spreading the splitting and merging procedures over a sequence of updates. The increase in block sizes guarantees that at any time, in any bag, only one block is in the process of being built.

The result is summarized in the following theorem:

Theorem 2. *Let $g(n)$ be a smooth, nondecreasing integer function with $0 < g(n) \leq n$. Given a data structure S for a decomposable searching problem PR,* *there exists a structure S' for solving PR dynamically such that:*

$$Q_{S'}(n) = O(g(n))Q_S(n)$$

$$I_{S'}(n) = \begin{cases} O(\log n/\log(g(n)/\log n))P_S(n)/n & \text{when } g(n) = \Omega(\log n) \\ O(g(n)n^{1/g(n)})P_S(n)/n & \text{otherwise} \end{cases}$$

This result is used to solve *nearest neighbor searching* and *common intersection of halfspaces* in $Q_S(n) = O(\sqrt{n \log n})$, $I_S(n) = O(\log n)$, $D_S(n) = O(\sqrt{n \log n})$; *d-dimensional range counting* and *rectangle intersection counting* in $Q_S(n) = O(\log^{d+1} n)$, $I_S(n) = O(\log^d n)$, $D_S(n) = O(\log^d n)$.

4 Global Invariants

In one instance of this technique, the data structure that needs to be de-amortized is first partitioned into many (non-constant) sub-structures which permit a restricted set of operations. These sub-structures are of varying sizes, and the size of any sub-structure will be upperbounded by an appropriate function of the sizes of the smaller sub-structures.

The first problem that is solved by this technique is the step-for-step simulation of concatenable deques by deques (without concatenation) [19]. A *concatenable deque* is a deque machine, with a constant number of deques, in which one allows as a primitive the operation of concatenation of any two deques. The main result of the paper is that every operation of a k concatenable deque machine can be simulated within $O(1)$ steps by an $O(k)$ deque.

Leong and Seiferas ([23]) earlier established that a restricted stack of deque machine (with a non-constant number of deques) can be simulated step-for-step by a deque machine. Kosaraju [19] first simulates step-for-step any concatenable deque machine by a stack of deques. This is achieved by partitioning each deque into many subdeques, then viewing all the resulting subdeques as a stack of deques, and finally specifying a global upperbound constraint for the size of any subdeque in the stack as a function of the sizes of the smaller subdeques in the stack. The global constraints on deque lengths guarantee that during the simulation access to the stack of deques is limited to the top. It is then shown that a step-for-step simulation of the concatenable deque machine can be performed by the stack of deques while maintaining the global constraints. Finally, the invocation of the result of [23] establishes the main result.

A second type of global constraints is explored in [20, 21]. The paper [20] shows how to perform search-tree operations in $O(\log d)$ steps, where d is the difference in ranks of the previous searched element and the currently sought element. The idea is to keep the keys in a set of balanced search trees of increasing heights. Then a global constraint on the number of search trees of height no more than h is imposed for every h. It is shown that the maintenance of these constraints assures $O(\log d)$ step worst case performance, and that the constraints can be maintained dynamically.

The paper [21] translates this technique into a pebble game on a set of bins numbered 1 through n. A move can remove two pebbles from bin i and add a pebble in bin $i+1$, or remove one pebble from bin i and add two pebbles into bin $i-1$. The goal of the game is to allow *insert pebble* and *delete pebble* operations on any bin. If a bin is empty then a sequence of moves needs to be performed in order to allow a delete from that bin. The complexity is measured by the number of moves necessary. The paper shows how to perform any insert or delete operation in bin k with at most $2k+1$ moves.

Another type of global invariants is explored in [22]. The paper contains an implementation of generalized deques that allow the normal deque operations, concatenation and `findmin`. The implementation is on a random access machine (RAM). The implementation achieves all the above deque operations in $O(1)$ RAM steps in the worst case. The idea is to maintain a tree of deques which satisfies certain global invariants. It is shown that updates can be spread over a series of subsequent operations while preserving the invariants.

A similar technique is applied in [9] to obtain worst case performance guarantees in relaxed heaps. Their data-structure is a priority queue that supports the operations `decrease_key` and `delete_min` in $O(1)$ and $O(\log n)$ time respectively. Insertions are supported in constant time. The authors first show how they can obtain these bounds in amortized time by relaxing the heap order constraint at a logarithmic number of nodes (called "active nodes"), none of which are siblings. They further relax this invariant by allowing the active nodes to be siblings, in order to obtain the worst case bound. For both cases the paper presents transformations that preserve the invariant and yield the specified time bounds.

A similar technique, called *data structural bootstrapping* was introduced in [4, 5]. The main idea is to allow elements of a data-structure to be smaller versions of the same data-structure. In [4], the authors create concatenable heap ordered deques by using heap ordered deques. Concatenation is realized by allowing elements of a deque to point to other deques. The resulting tree-like structure can perform all operations in $O(1)$ worst case if only a constant number of concatenations are allowed. This variation of data structural bootstrapping is called by the authors *structural abstraction.*

In [5], the authors introduce a new variant of data structural bootstrapping, called *structural decomposition.* They address the problem of making deques confluently persistent. They represent deques as balanced trees. The trees are, in turn, decomposed into *spines* (i.e. extremal paths from a node to a leaf). The spines are recursively represented as deques. This approach yields a $O(\log^* n)$ worst case bound for the n-th operation on the data-structure. This is due to the fact that each operation needs to traverse the whole recursive structure. An improvement is presented in [16] by using a technique called *recursive slow-down.* This technique avoids the need to recurse at each step. The authors insure that two operations need to be performed at a level of the recursive structure before the next level structure needs to be accessed. They achieve this, in a manner similar to binary counting, by assigning each level a color (yellow, green, or red) which represents its unbalancedness. Green levels are good, red levels are bad, and yellow levels are intermediate. A red level can be made green with the cost of having to change the color of the next level from green to yellow, or yellow to red. They maintain as invariant that red and green colors alternate among levels. This technique allows the authors to improve the results of [5] so that each operation can be done in $O(1)$ worst case time.

In [10], Driscoll et al. examine the problem of making data-structures persistent. Relative to de-amortization they show how balanced search trees can be made fully persistent with $O(\log n)$ insertion and deletion time and $O(1)$ space in the worst case. Their approach is also based on maintaining a global invariant. Change records for the nodes of the tree are not necessarily located at the modified node, but somewhere on an access path from the root to the node, called a *displacement path.* As invariants, they require that all change records have displacement paths, for any change record, the nodes on its displacement path are not its version-tree descendents, and that the displacement paths for the version-tree ancestors of a node, that are also on its displacement path, are disjoint from the node's displacement path. The authors show how to maintain these invariants during update operations.

Willard and Lueker [32] show how to extend dynamic data-structures to allow range restriction capabilities with only a logarithmic blowup in space, update time and query time. Their structure is a balanced binary tree on the range components, augmented at each node v with a version of the original structure ($AUX(v)$) for the elements stored in the subtree rooted at v. The amortized analysis requires rebuilding of the $AUX(v)$ structures whenever the range tree needs rebalancing. For the worst case analysis they allow incomplete AUX fields to exist within the tree while being slowly completed over a series of updates. They preserve the invariants that each incomplete node has complete children and the tree is balanced. It is shown that even though the rebalancing procedure cannot be performed at certain nodes in order to preserve the first invariant, the tree does not become unbalanced and a constant number of operations per update are sufficient to maintain the invariants.

5 Pebble Games and Applications

A set of pebble games have been used to design worst case algorithms for a series of problems. The pebble games involve a set of piles of pebbles. Two players take turns at adding or removing pebbles from the piles. The increaser (player I) can add pebbles to several piles while the decreaser (player D) can remove pebbles from the piles.

A first game [8] allows player I to split a pebble into fractional pebbles and to add them to some of the piles. Player D can remove all the contents of a single pile. It is shown that if D always removes all the pebbles in the largest pile, the size of any pile is bounded by the harmonic number H_n where n is the number of piles.

This result is applied to the following problem.

Design a data-structure that supports the following operations in constant time.

- insert(x, y) - insert record y after record x
- delete(x) - delete record x
- order(x, y) - return true if x is before y and false otherwise.

The data-structure proposed in [8] is a 4 level tree in which each internal node has a fan-out of $O(\log n)$ elements. The root has $O(n/\log^3 n)$ children. The children of the root are stored in a data-structure due to Willard [30, 31] that allows insertion, deletion, and order operations to be performed in $O(\log^2 n)$. The order of two nodes is determined by finding their nearest common ancestor (nca) and comparing the order of the children of nca that are ancestors of the two nodes.

Whenever insertions cause an internal node to have too many descendants (more than $\log^h n$ where h is the height of the node) the node is split into two nodes which are inserted in the parent. This operation is performed over $\log n$ insertions so that the time per update is $O(1)$.

The pebble game is used to guarantee that at the root, insertions in the data-structure can be done slowly over $\log^2 n$ updates, without the need to split a depth 1 node (child of the root) before a previous split has been completed. The algorithm anticipates the split of a level 1 node by inserting overflow nodes into the root structure whenever the node is close to becoming full (has almost $\log^3 n$ descendants). We define the *fullness* of the node to be the difference between the size of the subtree rooted at the node and $2/3 \log^3 n$. The connection with the pebble game becomes apparent. The chosen node is always the one with the largest value for fullness. Due to the result of the game, the fullness increase of any node is bounded by $O(\log n)$ for each insertion. Thus, over $\log^2 n$ operations, the fullness of any node cannot increase by more than $O(\log^3 n)$, implying that the overflow node has been inserted into the root by the time a node needs to be split.

Raman [29] uses the same game to insure worst case performance in dynamic two-level data-structures. His approach starts with a data-structure organized as a collection of $O(n/s(n))$ buckets of size $O(s(n))$. If a bucket of size k can be created(deleted) in $T_t(k)$, partitioned(fused with another bucket) in $T_p(k)$ and an element can be inserted in $T_b(k)$, he shows how the structure can be dynamized to allow updates to be performed in $O(\frac{T_t(n/(s \log n)) + T_p(s \log n)}{s} + T_b(s \log n))$ worst case time where s is the bucket size of the static structure.

The main idea is to split or fuse buckets that are either too big or too small. The *criticality* of buckets is defined as a value that reflects the largeness or smallness of the bucket. The value of the criticality of a bucket plays the role of the pebbles in the above zeroing game. This pebble game insures that buckets will not exceed $O(s \log n)$ in size when splits and fuses are spread over $O(s)$ operations, thus obtaining the time bounds claimed above. As an application of this result, the author shows how to construct a finger search tree with constant update time. Previous results showed only amortized constant time bounds [15].

As another application of the zeroing pebble game, Raman [29] presents a data-structure for dynamic fractional cascading [6, 7, 25] that supports insertions and deletions in worst case $O(\log d + \log\log s)$, where s is the space requirement and d is the degree. Queries are supported in $O(t(\log d + \log\log s) + \log s)$, where t is the path length. The pebble game is used to guarantee that blocks are of bounded size in the context of updates, thus allowing efficient transitions across the edges of the graph.

Raman [29] introduces a pebble game played on bounded in-degree digraphs. In this game player I can add a pebble to some vertex or modify the connectivity of the graph, while preserving the degree constraints. Player D can remove all pebbles from a constant number of vertices, at the same time adding one pebble to each predecessor of each processed vertex. It is shown that player D has a strategy which guarantees that no vertex can have more than $O(\log n)$ pebbles for any constant degree bound. Moreover, the strategy can be implemented in constant time.

This result is applied to the problem of making bounded in-degree data-structures partially persistent. The main result is

Theorem 3 ([29]). *Data structures where the in-degree of any node is at most $\Theta(\log^c n)$ after n operations on an initially empty data structure, for some constant c, can be made persistent with $O(1)$ worst case slowdown on a RAM.*

The idea of the persistent data-structure is to keep at each node a set of versions of the data in the node. Nodes that are too large are slowly split over a series of updates, however the size of the largest node will remain bounded due to the result shown for the pebble game.

As an application, the author shows a data-structure that performs partially persistent set union in $O(\log n / \log\log n)$ worst case time and $O(n\alpha(m, n))$ space on a RAM.

Raman [29] also introduces a variant of the zeroing game in which player D is allowed to remove at most a constant number of pebbles from a single pile. He shows that if D always chooses the largest pile and removes as many pebbles as possible from it, the size of the largest pile will be bounded by $\ln n + 1$, even if only one pebble can be removed at a time.

This result is applied to making bounded in-degree data-structures fully persistent. The author shows how a data structure with degree bound d can be made fully persistent with $O(\log\log m + \log d)$ worst case slowdown for access and update steps, and $O(1)$ worst case space per update.

Unlike the partial persistence case, in the case of full persistence it is not possible to always split a node whenever it becomes too large. The algorithm picks the largest node and transfers $O(d)$ version records to a new node and then restores pointer invariants. It is now clear that the decrementing game will be used, instead of the zeroing game, to guarantee that the number of versions stored at each node is bounded by $O(d \log m)$ after m update steps.

Levcopulos and Overmars [24] use a version of a pebble game in which player I can add a total of k pebbles to some piles, and player D can split a pile in half.

If D always splits the largest pile, the authors show that the size of any pile is bounded by $4k \log n$.

This game is applied to a data-structure for answering member and neighbor queries in $O(\log n)$ time and that allows for $O(1)$ time updates once the position of the element is known. Previous solutions could achieve only amortized constant time bounds for updates [14, 15, 27].

The data-structure is a search tree that stores at most $O(\log^2 n)$ elements in each leaf. After every $\log n$ insertions, the largest leaf is split into two equal halves. Due to the result proved for the game, the size of the largest leaf is bounded by $O(\log^2 n)$ ($\log n$ over $\log n$ updates). The split and the insertion of the new leaf in the tree are performed over $\log n$ updates during which time no leaves are split (splits occur only every $\log n$ updates). Thus the size invariant is preserved and each insertion is performed in constant time. Deletions are handled by "global rebuilding" as described in [28].

6 Conclusions

The paper reviewed several known techniques for de-amortization of algorithms. The techniques ranged from simple data duplication to elaborate pebble games played on graphs. In spite of the existence of these techniques, it is extremely hard to de-amortize new algorithms. A systematic study of de-amortization will be of great utility.

References

1. M. Ajtai, M. Fredman, and J. Komlós. Hash functions for priority queues. *Information and Control*, 63:217–225, 1984.
2. H. G. Baker, jr. List processing in real time on a serial computer. *Communications of the ACM*, 21(4):280–294, April 1978.
3. J. L. Bentley. Decomposable searching problems. *Information Processing Letters*, 8:244–251, 1979.
4. A. L. Buchsbaum, R. Sundar, and R. E. Tarjan. Data structural bootstrapping, linear path compression, and catenable heap ordered double ended queues. In *Proceedings of the 33rd Symposium on Theory of Computing*, pages 40–49, 1992.
5. A. L. Buchsbaum and R. E. Tarjan. Confluently persistent deques via data structural bootstrapping. In *Proceedings of the 4th ACM/SIGACT-SIAM Symposium on Discrete Algorithms*, pages 155–164, 1993.
6. B. Chazelle and L. J. Guibas. Fractional cascading: I. A data structuring technique. *Algorithmica*, 1:133–162, 1986.
7. B. Chazelle and L. J. Guibas. Fractional cascading: II. Applications. *Algorithmica*, 1:163–191, 1986.
8. P. F. Dietz and D. D. Sleator. Two algorithms for maintaining order in a list. In *Proceedings of the 19th ACM Symposium on Theory of Computing*, pages 365–371, 1987.
9. J. R. Driscoll, H. N. Gabow, R. Shrairman, and R. E. Tarjan. Relaxed heaps: An alternative to Fibonacci heaps with applications to parallel computation. *Communications of the ACM*, 31(11):1343–1454, November 1988.

10. J. R. Driscoll, N. Sarnak, D. D. Sleator, and R. E. Tarjan. Making data structures persistent. *Journal of Computer and System Sciences*, 38:86–124, 1989.
11. P.C. Fischer, A. R. Meyer, and A. L. Rosenberg. Real-time simulation of multihead tape units. *Journal of the ACM*, pages 590–907, 1972.
12. H. Gajewska and R. E. Tarjan. Deques with heap order. *Information Processing Letters*, pages 197–200, April 1986.
13. S. A. Greibach. Jump pda's, deterministic context-free languages: Principal afdls and polynomial time recognition. In *Proceedings of 5th annual ACM Symposium on Theory of Computing*, pages 20–28, April-May 1973.
14. L. Guibas, E. McCreight, M. Plass, and J. Roberts. A new representation for linear lists. In *Proceedings of the 9th ACM Symposium on Theory of Computing*, pages 49–60, 1977.
15. S. Huddleston and K. Mehlhorn. A new data structure for representing sorted lists. *Acta Informatica*, 17:157–184, 1982.
16. H. Kaplan and R. E. Tarjan. Persistent lists with catenation via recursive slow-down. In *Proceedings of the 27th ACM Symposium on Theory of Computing*, pages 93–102, May 1995.
17. H. Kaplan and R. E. Tarjan. Purely functional representations of catenable sorted lists. In *Proceedings of the 28th ACM Symposium on Theory of Computing*, pages 202–211, May 1996.
18. S. R. Kosaraju. 1-way stack automaton with jumps. *Journal of Computer and System Sciences*, 9(2):164–176, October 1974.
19. S. R. Kosaraju. Real-time simulation of concatenable double-ended queues by double-ended queues. In *Proceedings of the 11th ACM Symposium on Theory of Computing*, pages 346–351, 1979.
20. S. R. Kosaraju. Localized search in sorted lists. In *Proceedings of the 13th ACM Symposium on Theory of Computing*, pages 62–69, 1981.
21. S. R. Kosaraju. Redistribution of computations for improving the worst-case. In *Proceedings of the First Conference on Foundations of Software Technology and Theoretical Computer Science*, pages 3–8, December 1981.
22. S. R. Kosaraju. An optimal RAM implementation of catenable min double-ended queues. In *Proceedings of the 5th ACM/SIGACT-SIAM Symposium on Discrete Algorithms*, pages 195–203, 1994.
23. B. Leong and J. Seiferas. New real-time simulations of multihead tape units. In *Proceedings of the 9th ACM Symposium on Theory of Computing*, pages 239–248, 1977.
24. C. Levcopoulos and M. H. Overmars. A balanced search tree with $O(1)$ worst-case update. *Acta Informatica*, 26:269–277, 1988.
25. K. Mehlhorn and S. Näher. Dynamic fractional cascading. *Algorithmica*, 5:215–241, 1990.
26. C. Okasaki. Amortization, lazy evaluation and purely functional catanable lists. In *Proceedings of the 36th Symposium on Foundations of Computer Science*, pages 646–654, 1995.
27. M. H. Overmars. A $O(1)$ average time update scheme for balanced search trees. *Bull. EATCS*, 18:27–29, 1982.
28. M. H. Overmars. *The Design of Dynamic Data Structures.* Lecture Notes in Computer Science. Springer Verlag, 1983.
29. R. Raman. *Eliminating Amortization: On Data Structures with Guaranteed Response Time.* PhD thesis, University of Rochester, 1992.

30. D. E. Willard. Maintaining dense sequential files in a dynamic environment. In *Proceedings of the 14th ACM Symposium on Theory of Computing*, pages 114–121, May 1982.
31. D. E. Willard. Good worst-case algorithms for inserting and deleting records in dense sequential files. In *Proceedings of the ACM SIGMOD Conference on Management of Data*, pages 251–260, May 1986.
32. D. E. Willard and G. S. Lueker. Adding range restriction capability to dynamic data structures. *Journal of the Association for Computing Machinery*, 32(3):597–617, July 1985.

On Computing New Classes of Optimal Triangulations with Angular Constraints *

Yang Dai[1] and Naoki Katoh[2]

[1] Tokyo Institute of Technology (dai@is.titech.ac.jp)
[2] Kyoto University (naoki@archi.kyoto-u.ac.jp)

1 Introduction

Given a planar point set S, a triangulation of S is a maximal set of non-intersecting edges connecting points in S. Triangulating a point set has many applications in computational geometry and other related fields. Specifically, in numerical solutions for scientific and engineering applications, poorly shaped triangles can cause serious difficulty. Traditionally, triangulations which minimize the maximum angle, maximize the minimum angle, minimize the maximum edge length, and maximize the minimum hight are considered. For example, if angles of triangles become too large, the discretization error in the finite element solution is increased and, if the angles become too small, the condition number of the element matrix is increased [1, 10, 11]. Polynomial time algorithms have been developed in determining those triangulations [2, 7, 8, 15]. In computational geometry another important research object is to compute the minimum weight triangulation. The weight of a triangulation is defined to be the sum of the Euclidean lengths of the edges in the triangulation. Despite the intensive study made during the lase two decades, it remains unknown that whether the minimum weight triangulation problem is NP-complete or polynomially solvable. In this paper we consider two new classes of optimal triangulations :

Problem (1) the minimum weight triangulation with the minimum (resp. maximum) angle in the triangulation not smaller (resp. greater) than a given value α (resp. γ);

Problem (2) the triangulation which minimizes the sum over all triangles of ratios defined by the values of the maximum angles to the minimum angles in the triangles.

If the value of α is zero and γ is equal to π, then Problem (1) is reduced to the minimum weight triangulation problem. If α is defined as the maximum value of the minimum angles among all possible triangulations, the solution of Problem (1) may give the Delaunay triangulation, although this case is not equivalent to the Delaunay triangulation problem. Therefore, Problem (1) contains the minimum weight triangulation as a special instance. We identify Problem (1) as the minimum weight triangulation problem with angular constraints and Problem (2) as the angular balanced triangulation problem. In Problem (1) we proposed somewhat more general criteria of the minimum weight triangulation and in

* This research is supported by the Grant-in-Aid of Ministry of Science, Culture and Education of Japan.

Problem (2) a new criterion which is different from the minmax angle criterion. Although no evidences of applications of these new criteria have been found, we believe that they should be potentially useful, since the angular conditions on the angles in the minimum weight triangulation allows one to control the quality of the triangulation generated, and the sum of the ratios of the value over all triangles in the angular balanced triangulation contains more information than the minmax or the maxmin angle criterion does.

The main purpose of the paper is to provide solution methods for computing the optimal triangulations defined above. The difficulty of determining the optimal triangulation depends on the position of the points in the given set. If the points are vertices of a simple polygon then the problems are easily solvable. Actually the dynamic programming approach can be applied to both classes and provides polynomial time algorithms. For general point sets, the apparent difficult to compute the minimum weight triangulation problem means that it is unlikely that we can design a polynomial algorithm for the min-sum type problems at the moment. Moreover due to the angular conditions we can not expect the heuristic methods such as edge-flipping and greedy methods work well for these new classes.

On the other hand, recent research has revealed promising ways to determine large subgraphs, namely, the β-skeleton [12, 13] and the LMT-skeleton [4, 5, 6] of the minimum weight triangulation. The experimental results show that these subgraphs are well connected for most of the point sets having relatively small sizes that are generated from uniform random distributions. Therefore they are useful for the design of algorithms to compute the exact minimum weight triangulation of point sets having small sizes. When the number of connected components is small, a complete exact minimum weight triangulation can be produced by using the $O(n^3)$ dynamic algorithm [14] on each possible polygon which is reduced by the subgraph.

Unfortunately, the definition of the β-skeleton relies heavily on the distances of pairs of the points, it is not applicable to the new problems which involve angular conditions. However, since the main idea of the LMT-skeleton for the minimum weight triangulation is the determination of the edges in every locally minimal triangulation, it suggests that there is room left to generalize the concept to the new classes of optimal triangulations through an appropriate definition of local optimality. This motivates us to design a generalized unifying method for the computation of subgraphs for other classes of optimal triangulations. Our new results are as follows.

- $O(n^3)$ time and $O(n^2)$ space algorithms for the problems in each class with the point set being a vertex set of a simple polygon;
- $O(n^4)$ time and $O(n^2)$ space algorithms for computing the subgraphs of the minimum weight triangulation with angular constraints and the angular balanced triangulation;
- the computational results for the two algorithms which demonstrate their usefulness.

The organization of this paper is as follows. Section 2 gives polynomial time algorithm based on dynamic programming approach for computing the optimal triangulations defined above with the point set being a vertex set of a simple polygon. Section 3 presents the algorithm for the computation of the subgraph of the minimum weight triangulation with angular constraints. Section 4 introduces the algorithm for the determination of the subgraph of the angular balanced triangulation. Section 5 states the conclusions.

2 Polynomial Time Algorithms

In this section we confine ourselves to the point set which is a vertex set of a simple polygon. We give polynomial time algorithms based on the dynamic programming approach for determining the minimum weight triangulation with angular constraints and the angular balanced triangulation.

In Bern and Eppstein [3] they discussed a class of optimal triangulation problems which admit efficient solutions. The class possesses so called decomposable measures which allow one to compute the measure of the entire triangulation quickly from the measures of two pieces of the triangulation, along with the knowledge of how the pieces are put together. The decomposable measures include the minimum (maximum) angle in the triangulation, the minimum (maximum) circumcircle of a triangle, the minimum (maximum) length of an edge in the triangulation, the minimum (maximum) area of a triangle, and the sum of edge lengths in the triangulation. They also presented polynomial time algorithms which use the dynamic programming approach attributed to Klincsek [14]. We will add the problems of determining the minimum weight triangulations with angular constraints and the angular balanced triangulation to the decomposable class and present polynomial time algorithms for solving these problems.

The Minimum Weight Triangulation with Angular Constraints:

Denote the minimum weight triangulation of a point set S with respect to α and γ by $MWT(S, \alpha, \gamma)$. A triangle is defined admissible if it satisfies the angular conditions, otherwise it is inadmissible. Label the vertices $p_1, p_2, \cdots, p_n$ of the simple polygon in the clockwise order. The polygon defined by the point set S is called P. An edge p_ip_j $(j > i+1)$ is said to be interior to P if the line segment connecting p_i and p_j splits P into two polygons whose union is P. An interior edge is called a diagonal edge of P. Let $w(i,j)$ be the weight of the minimum weight triangulation of the polygon involving the points $p_i, p_{i+1}, \ldots, p_j$. However, if p_ip_j is not a diagonal edge of the polygon S define $w(i,j) = +\infty$. The algorithm for computing the minimum weight triangulation $MWT(S, \alpha, \gamma)$ is as follows.

Algorithm $MWT(S, \alpha, \gamma)$**:**

Step 1. For $k = 1, i = 1, 2, \ldots, n-1$ and $j = i+k$, let $w(i,j) = d(p_i, p_j)$, where $d(p_i, p_j)$ is the length of edge p_ip_j.

Step 2. Let $k = k+1$. For $i = 1, 2 \ldots, n$ and $j = i+k \leq n$, if the edge p_ip_j is not interior to P let $w(i,j) = +\infty$. Otherwise let

$$M = \{m \mid i < m < j, p_ip_mp_j \text{ is admissible and both the edge } p_ip_m \text{ and the edge } p_mp_j \text{ are diagonal}\}$$

Compute

$$w(i,j) = \begin{cases} d(p_i, p_j) + \min_{m \in M}\{w(i,m) + w(m,j)\} & \text{for } M \neq \emptyset \\ +\infty & \text{otherwise.} \end{cases} \quad (1)$$

For each pair (i,j) such that $w(i,j) < \infty$, let $m^*(i,j)$ be the index where the minimum $w(i,j)$ in (1) is achieved.

Step 3. If $k < n-1$, go to Step 2. Otherwise $w(1,n)$ is the minimum weight.

Step 4. If $w(1,n) < \infty$, then backtrace along the pointers m^* to determine the edges of the minimum weight triangulation. Otherwise no triangulation satisfying the angular conditions exists.

For each pair (i,j) whether the edge p_ip_j is interior to P can be decided by the calculation of the intersection of the edge p_ip_j with $O(n)$ boundary edges of P. More precisely, if there is intersection then the edge p_ip_j is not interior to P, otherwise it is interior to P. We may calculate $d(p_i,p_j)$ for each pair (i,j) and store the information at the begining of the algorithm. $O(n)$ testing time for each edge shows that the interior test for all edges p_ip_j needs $O(n^3)$ time and $O(n^2)$ space. Since the admissibility test of a triangle takes constant time, Step 2 requres $O(kn)$ time for each $k = 2,3,...,n-2$. Therefore Steps 1-3 take $O(n^3)$ time. In Step 4 the determination of the edges in minimum weight triangulation can be done in $O(n)$ time, therefore the algorithm runs in a total $O(n^3)$ time and $O(n^2)$ space.

The Angular Balanced Triangulation:

Let t be an arbitrary triangle in some triangulation T. Let θ_{large} (resp. θ_{small}) be the largest (resp. smallest) angle in the triangle. Recall that the measure function $f(t)$ of the triangle t is defined by $f(t) = \theta_{\text{large}}/\theta_{\text{small}}$. Specifically for a degenerate t, i.e., t is an edge we define $f(t) = +\infty$. For any two nondegenerate triangles t and t', let $f(t,t') = f(t) + f(t')$. The sum of the ratios of the triangulation T, denoted by $f(T)$, is defined as $f(T) = \sum_{t\in T} f(t)$. Therefore, the angular balanced triangulation is the triangulation which minimizes the value $f(T)$ over all triangulations T.

We denote by $P(i,j)$ the polygon formed by points $p_i, p_{i+1}, ..., p_j$. Let $F(i,j)$ be the minimum value of $f(T_{ij})$ over all triangulations T_{ij} of $P(i,j)$. Define $F(i,j) = +\infty$ for each boundary and non-diagonal edge p_ip_j of the polygon P. We compute $F(1,n)$ by the dynamic programming method. Suppose that an diagonal edge p_ap_b splits $P(i,j)$ into two polygons P_1 and P_2. Let T_1 and T_2 be the triangulations of P_1 and P_2, respectively and $T_{ij} = T_1 \cup T_2$. We define a function g as follows.

$$g(f(T_1), f(T_2), a, b) = f(T_1) + f(T_2) = f(T_{ij}).$$

If the edge p_ap_b is on the boundary of the polygon P, then p_ap_b is an edge of the polygon $P(i,j)$. We define in this case

$$g(f(T_{ij}), f(\{a,b\}), a, b) = f(T_{ij}).$$

Note that in any triangulation of $P(i,j)$, p_ip_j must be a side of a triangle, say $p_ip_jp_m$, with $i < m < j$. We can compute the value of the angular balanced triangulation of $P(i,j)$ by trying all choices of m. The algorithm for computing the angular balanced triangulation can be obtained by replacing $w(i,j)$ with $F(i,j)$ and Step 2 with the following Step 2' in **Algorithm** $MWT(S,\alpha,\gamma)$.

Sept 2'. let $k = k+1$. For $i = 1,2\ldots,n$ and $j = i+k \le n$, Compute

$$F(i,j) = \min_{i<m<j} g(g(f(p_ip_jp_m), F(i,m), p_i, p_m), F(m,j), p_m, p_j)) \tag{2}$$

For each pair (i,j) let $m^*(i,j)$ be the index where $F(i,j)$ in (2) is achieved.

Note that each evaluation of g takes constant time per possible value of m, therefore a total $O(kn)$ time in Sept 2'. Following a similar analysis of **Algorithm** $MWT(S,\alpha,\gamma)$ one can show that this algorithm also takes $O(n^3)$ time and $O(n^2)$ space.

3 The Subgraph of the Minimum Weight Triangulation with Angular Constraints

In this section we present the algorithm for computing the subgraph of the minimum weight triangulation with angular constraints for point set in general position. Designate the set of all possible edges connecting two points in S by $E(S)$. We assume that the values of α and γ are given so that there are always triangulations satisfying the angular constraints. In order to determine the minimum weight triangulation $MWT(S,\alpha,\gamma)$, we only need to consider triangulations which contain no inadmissible triangles. For the entirety of our discussion in this section, the term triangulation will mean a triangulation which consists solely of admissible triangles.

Let e be an edge in an arbitrary triangulation. If e is not on the boundary of the convex hull of the set S, then there exist two admissible triangles t_1 and t_2 such that $t_1 \cap t_2 = e$. If the quadrilateral $t_1 \cup t_2$ is convex, then it contains another diagonal, e'. Denote by t_1' and t_2' as the two triangles formed by connecting the edge e'. The edge e is defined to be locally minimal with respect to (α,γ) if either one of the following two cases holds: Case 1: $t_1 \cup t_2$ is not convex; Case 2: $t_1 \cup t_2$ is convex and either (i) $|e| \leq |e'|$, or (ii) at least one of the triangles t_1' and t_2' is inadmissible. A triangulation is called locally minimal with respect to (α,γ) if each edge e in the triangulation is locally minimal with respect to the two triangles containing e and (α,γ). From the definition, it follows that $MWT(S,\alpha,\gamma)$ is a locally minimal triangulation with respect to (α,γ). The intersection of all locally minimal triangulations must be a subgraph of $MWT(S,\alpha,\gamma)$. Our algorithm intends to find a subgraph of the intersection.

We define a triangle as empty if it contains no points of S except the vertices. For each edge and empty triangle if they are known not to be contained in any locally minimal triangulation we call them dead. Therefore all inadmissible triangles are dead initially. When each edge is examined by the algorithm, its status will be determined as active, inactive, or dead as follows. Let $\mathcal{T}$ be the set of pairs $\{axb, ayb\}$ of empty active triangles, one from each side of edge ab such that $axb \cap ayb = ab$. The edge ab is labeled active if it lies on the boundary of the convex hull of S, or if there exists $\{axb, ayb\} \in \mathcal{T}$ such that $ab \cap xy \neq \emptyset$ and e is locally minimal with respect to $\{axb, ayb\}$ and (α,γ).

Suppose that ab is not labeled active. Then, if $\mathcal{T} = \emptyset$, or $ab \cap xy \neq \emptyset$ for all $\{axb, ayb\} \in \mathcal{T}$, we label ab dead. Otherwise, we label ab inactive. If an edge ab becomes inactive or dead, we label some of the admissible triangles bounding ab as dead. More precisely, define the set $\mathcal{A}$ as the collection of the empty admissible active triangles axb which satisfy:

> $axb \cap ayb = ab$, $ab \cap xy \neq \emptyset$ for all empty admissible active triangles ayb such that x, y are on different sides of ab, respectively.

We label all triangles in the set $\mathcal{A}$ dead.

We present below the algorithm. It produces a set of edges which is called the LMT-skeleton of the minimum weight triangulation with angular constraints and denote it by LMT-skeleton(S, α, γ). Three edge sets are used in the algorithm; they are *candEdges*, *edgeIn* and *deadEdges*. All edges in $E(S)$ are initially active. We note additionally that (1) all edges in $E(S)$, except the convex hull edges, are in *candEdges*, (2) *edgeIn* contains the convex hull edges and (3) *deadEdges* is empty.

The Algorithm LMT(S, α, γ):
Input: point set S.
Output: edge set *edgeIn*.

Step 0. Set all edges of *candEdges* unexamined.
Step 1. If there are no unexamined edges, go to Step 4. Otherwise choose an unexamined edge $e \in$*candEdges*, check all empty triangles on both sides of e. If they are all inadmissible then delete e from *candEdges* and move it to *deadEdges*. Otherwise,
Step 2. Find all combinations of empty admissible triangles t_i and t_j on the two sides of e such that t_i and t_j are not bordered by an edge in *deadEdges*.
Step 3. For each combination of t_i and t_j, test if e is locally minimal with respect to t_i, t_j and (α, γ). If e is not locally minimal to any such pair t_i and t_j, then move e to *deadEdges*. Otherwise, mark e active or inactive according to the definitions. Go to **Step 1**.
Step 4. For each edge marked active or inactive, if it intersects no other active edges then move it to *edgeIn*.

The algorithm iterates Steps 1-3 for $O(n^2)$ times, once for each candidate edge. Computing the empty triangles for each edge requires $O(n \log n)$ time by using the method of [4]. We can also preprocess the data in $O(n^2)$ time and $O(n^2)$ space by the algorithm in [9] so that all empty triangles sharing an edge can be computed in linear time. The admissibility for each empty triangle can be tested in $O(1)$ time. So Step 1 needs at most $O(n^2)$ time. To test if an edge is locally angular balanced can be done in $O(1)$ time. Since e might have a linear number of triangles on each side, we may test $O(n^2)$ combinations of adjacent triangles. Step 4 tests an edge against at most $O(n^2)$ other edges and there are at most $O(n^2)$ edges. Thus, the Algorithm LMT(S, α, γ) runs in $O(n^4)$ time and $O(n^2)$ space.

The following lemmas guarantee the correctness of the algorithm.

Lemma 1. *If an empty admissible triangle t is labeled dead, then $t \notin MWT(S, \alpha, \gamma)$.*

Proof. We prove the lemma by contradiction. Let axb be the first triangle in $MWT(S, \alpha, \gamma)$ that it is labeled dead because ab becomes either inactive or dead. Since ab is not an edge of the convex hull, there exists a triangle $ayb \in MWT(S, \alpha, \gamma)$ such that $axb \cap ayb = ab$. Since axb is the first triangle becoming dead, both axb, ayb are active immediately after labeling ab inactive or dead. Labeling axb dead implies that $ab \cap xy \neq \emptyset$. Moreover, in order that ab is labeled inactive or dead, we must have $|xy| < |ab|$ and that neither of the triangles xay and xby is inadmissible. Therefore, in the convex quadrilateral $axby$ in $MWT(S, \alpha, \gamma)$, we can replace the diagonal ab by xy to decrease the total length without destroying the angular condition. This leads to a firm contradiction. □

Lemma 2. *If $ab \notin MWT(S, \alpha, \gamma)$, then ab intersects some active edge.*

Proof. Let ab be some edge not in $MWT(S, \alpha, \gamma)$. Since $ab \notin MWT(S, \alpha, \gamma)$, ab must intersect some edge in $MWT(S, \alpha, \gamma)$. Let x_0y_0 be such an edge that $ab \cap x_0y_0$ is closest to a and the triangle $x_0ay_0 \in MWT(S, \alpha, \gamma)$. Since a and b lie on opposite sides of x_0y_0, x_0y_0 is not on the convex hull. Therefore, there is another triangle x_0zy_0 in $MWT(S, \alpha, \gamma)$ adjacent to x_0ay_0. For convenience, we rename a as v_0. Let C_0 be the cone bounded by the two rays originating from v_0 through x_0 and y_0, respectively. By construction, $b \in C_0$. Note that both triangles x_0ay_0 and x_0zy_0 are admissible. See Figure 1. Assume to the contrary

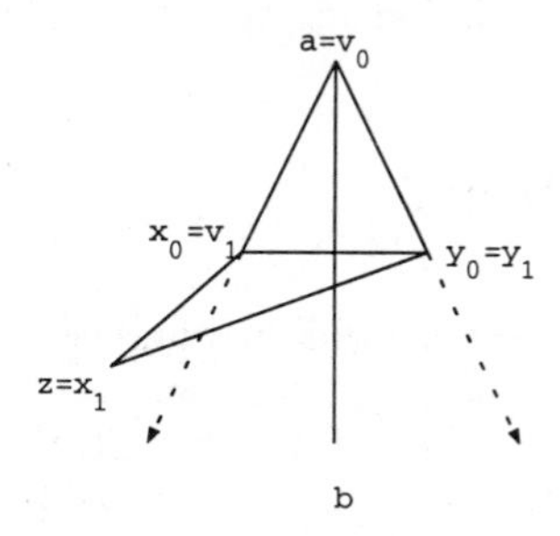

Fig. 1.

that ab does not intersect any active edge. By Lemma 1, $x_0v_0y_0$ and $x_0z_0y_0$ are not labelled dead, even if x_0y_0 is labeled dead or inactive. This means that $v_0z \cap x_0y_0 = \emptyset$. Therefore, $z \notin C_0$ and $z \neq b$. Suppose z lies on the left side of v_0b. We rename x_0 as v_1 , z as x_1 and y as y_1. By construction, we discover a new edge x_1y_1 in $MWT(S, \alpha, \gamma)$ that intersects ab and b lies inside of the cone C_1 bounded by the two rays from v_1 through x_1 and y_1. Thus, by application of the previous argument to $x_1v_1y_1$ and x_1y_1, we obtain a new edge x_2y_2 that intersects ab and the triangle $x_2x_1y_2$ conserves satisfaction of the angular condition. Infinite iteration of this argument gives an infinite sequence of edges x_iy_i, $i > 1$ that intersect v_0b. This contradicts to the finiteness of $MWT(S, \alpha, \gamma)$. □

4 The Subgraph of the Angular Balanced Triangulation

Denote the angular balanced triangulation of a general position point set S by ABT(S). Initially, we define the concept the local optimality of the angular balanced triangulation, and subsequently describe the algorithm which can be used to determine some of the edges in angular balanced triangulation.

Let e be an arbitrary edge in some triangulation but not on the convex hull. As discussed in Section 3, there exist two triangles t_1 and t_2 such that $t_1 \cap t_2 = e$ and a corresponding pair of triangles t_1' and t_2' such that $t_1' \cap t_2' = e'$. The edge e is defined to be locally angular balanced if $t_1 \cup t_2$ is not convex or if $t_1 \cup t_2$ is convex and $f(t_1, t_2) \leq f(t_1', t_2')$. A triangulation is called a locally angular balanced triangulation (LABT) if each of its edge is locally angular balanced. Obviously, the angular balanced triangulation must be locally angular balanced; otherwise the exchange of the diagonals in a convex quadrilateral containing a non-balanced edge will reduce the sum of the values.

Naturally, a concept similar to the LMT-skeleton for the minimum weight triangulation can be considered. Correspondingly, we name the subgraph which contains a set of edges that must be in every locally angular balanced triangulation the LABT-skeleton and denote it by LABT(S).

Obviousely the framework of Algorithm LMT(S, α, γ) works for for determination of LABT(S) if a replacement of the locally minimal test by the locally angular balanced test and removal of the admissibility test are made. Since the locally angular balanced test runs in $O(n)$ time, the computation of the LABT-skeleton needs the same time and space bounds as the that of LMT(S, α, γ).

5 Computational Results

Since for a uniformly distributed random point set there almost always exist a triangle with the smallest angle as well as the largest angle which should be included in every triangulation, it has very little meaning to consider the minimum weight triangulation with angular constraints in the random data set. The data we used for computing LMT(S, α, γ) are designed as follows. First, grid points in a fixed-size square are generated and then each point inside of the square is purtabated within a circle of small radius. By changing the value of the radius, different types of data sets can be produced. Figures 2-6 show results of LMT-Skeletons LMT(S, α, γ) by testing a set of 64 points. In this example the angle α changes between 0 and the value of the minimum angle α_D of the Delaunay triangulation, i.e., $0 = \alpha_1 < \alpha_2 < \alpha_3 < \alpha_4 < \alpha_D$. We set $\gamma = \pi$ and set the radius to be half of the grid size in our test. We tested for point sets of size up to 100. The obsevation is that for this type of point sets the effectiveness of the algorithm is influenced heavily by the value of the radius. The smaller the radius is the larger LMT-Skeletons can be found. Small purtabation results in small number of connected components in the subgraph.

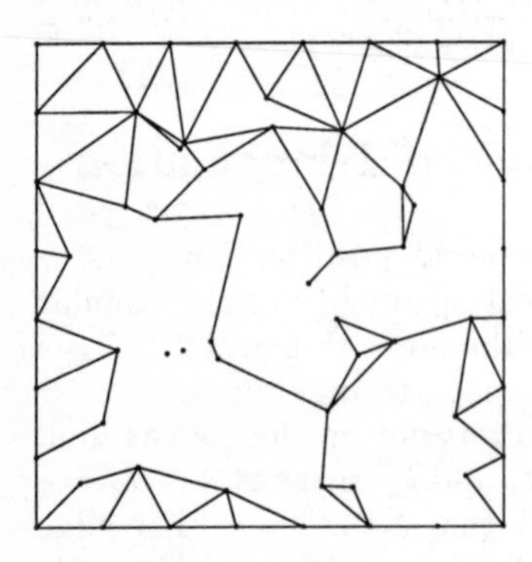

Fig. 2. LMT-skeleton(S, α_1, γ)

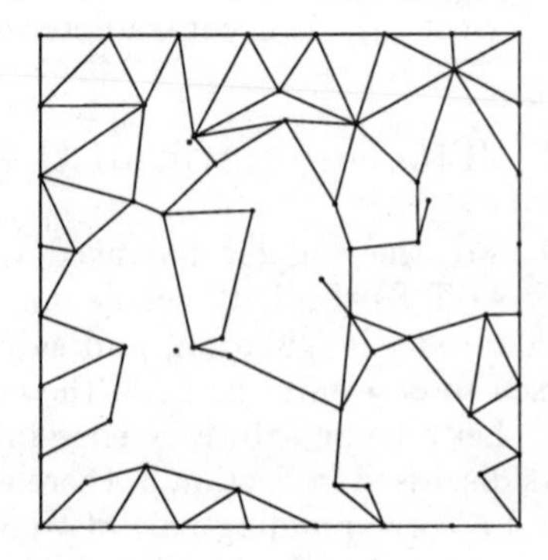

Fig. 3. LMT-skeleton(S, α_2, γ)

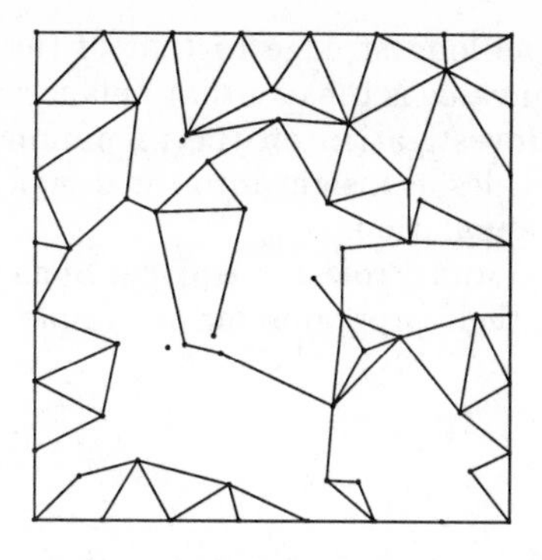

Fig. 4. LMT-skeleton(S, α_3, γ)

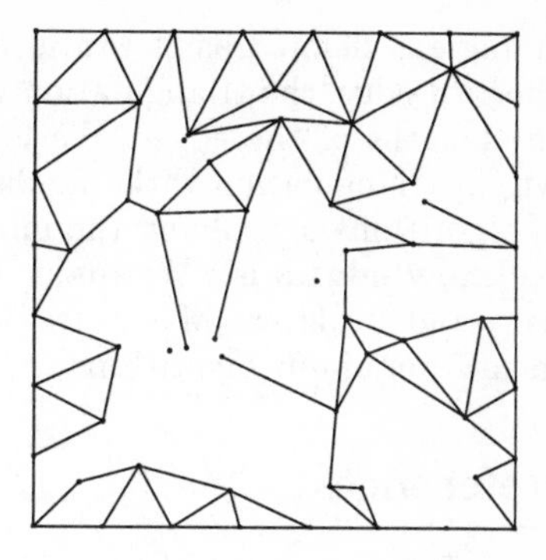

Fig. 5. LMT-skeleton(S, α_4, γ)

We use uniformly random distribute point sets for computing the LABT-skeleton. Point sets of size range from 50 to 500 are tested. The results in Table 1 are average of ten runs for each size. We show the number of connected components(#comp), the percentage of the identified edges(% MWT) in angular balanced traingulation, the numbers of active and dead edges(its ratio) and CPU time in the table. The programs are coded in C and the experiments are conducted on a Ultra SPARC (143Mhz.) workstation. From our result shows that the algorithm works quite well for the point set with small size. Note we could improve the result by repeating the algorithm until no more edges can be identified as the edge in angular balance triangulation.

6 Conclusions

We have shown that the identification of the appropriate definition of local optimality leads to a simple unified method for the computation of the subgraphs of the optimum triangulations. It is worth noticing that this unified method also works for determining the subgraph of other optimal triangulations with min-sum type quality measurement. For example, if the measure function f for each

#nodes	#comp.	% MWT	# active/dead edges(%)	CPU time (sec.)
50	1.0	0.75	182/1031(0.177)	0.04
100	1.5	0.69	430/4501 (0.096)	0.30
200	4.3	0.62	1021/18838(0.054)	2.37
300	10.3	0.59	1633/43129(0.038)	8.66
400	24.0	0.55	2380/77288(0.031)	21.45
500	40.2	0.52	3135/121406(0.026)	429.87

Table 1. The computational results for LABT-skeleton.

triangle is defined as the ratio of the length of the longest edge to that of the shortest edge, the triangulation with the minimum value is the one that balances the lengths of the edges. We hope that further investigation on the minimum wright triangulation with angular constraints provides us insight into the design of algorithms for solving the minimum weight triangulation.
Acknowledgment: We thank Mr. Manabu Sugai and Professor Kokichi Sugihara and Professor Masao Iri [16] for sharing us their programs for the implementation of our algorithms.

References

1. I. Babuska and A. Aziz, "On the angle condition in the finite element method", *SIAM Journal on Numerical Analysis* 13, pp.214-226, 1976.
2. M. Bern, H. Edelsbrunner, D. Eppstein, S. Mitchell, and T. S. Tan, "Edge insertion for optimal triangulations", *Discrete Comput. Geom.*, 10, pp.47-65, 1993.
3. M. Bern and D. Eppstein, "Mesh generation and optimal triangulation", in D.-Z. Du eds. *Computing in Euclidean Geometry*, World Scientific Publishing, pp.23-90, 1992.
4. S.-W. Cheng, N. Katoh and M. Sugai, "A study of the LMT-Skeleton", *Proceedings of ISAAC'96*, Lecture Notes in Computing Science 1178, pp.339-359, 1996.
5. S.-W. Cheng and Y. Xu, "Approaching the largest $\beta-$ skeleton within a minimum weight triangulation", *Proceedings of the 12th Annual ACM Symposium on Computational Geometry*, 1996.
6. M. T. Dickerson and M. H. Montague, "The exact minimum weight triangulation", *Proceedings of the 12th Annual ACM Symposium on Computational Geometry*, 1996.
7. H. Edelsbrunner and T. S. Tan, "A quadratic time algorithm for the minmax length triangulation", *SIAM J. Comput.* 22, pp. 527-551, 1993.
8. H. Edelsbrunner, T. S. Tan and R. Waupotitsch, "A $O(n \log n)$ time algorithm for the minmax angle triangulation", *SIAM J. Sci. Statist. Comput.* 13, pp. 994-1008, 1992.
9. D. Eppstein, M. Overmars, G. Rote and G. Woeginger, "Finding Minimum Area k-gons", *Discrete Comput. Geom.* 7, pp. 45-58, 1992.
10. L. Freitag, M. Jones and P. Plassmann, "An efficient parallel algorithm for mesh smoothing", *Proceedings of the 5th International Mesh Roundtable*, 1996.
11. I. Fried, "Condition of finite element matrices generated from nonuniform meshes", *AIAA Journal* 10, pp.219-221, 1972.
12. J. M. Keil, "Computing a subgraph of the minimum weight triangulation", *Computational Geometry: Theory and Applications*, Vol. 4, pp.13-26, 1994.
13. D. Kirkpatrick and J. Radke, "A framework for computational morphology", in G. Toussaint ed, *Computational Geometry* , pp. 217-248, Elsevier Science Publishers, 1985.
14. G. Klincsek, "Minimum triangulations of polygonal domains", *Annual Discrete Mathematics* 9, pp.121-123, 1980.
15. R. Sibson, "Locally equiangular triangulations", *Comput. J*, 21, pp.243-245, 1978.
16. K. Sugihara and M. Iri, "VORONOI2 reference manual, Research Memorandum RMI-89-04, Department of Mathematical Engineering and Information Physics, Faculty of Engineering, University of Tokyo, 1993.

Maximum Weight Triangulation and Its Application on Graph Drawing[1]

Cao An Wang [2], Francis Y. Chin [3], and Bo Ting Yang.[2]

Abstract

In this paper, we investigate the maximum weight triangulation of a polygon inscribed in a circle (simply inscribed polygon). A complete characterization of maximum weight triangulation of such polygons has been obtained. As a consequence of this characterization, an $O(n^2)$ algorithm for finding the maximum weight triangulation of an inscribed n-gon is designed. In case of a regular polygon, the complexity of this algorithm can be reduced to $O(n)$. We also show that a tree admits a maximum weight drawing if its internal node connects at most 2 non-leaf nodes. The drawing can be done in $O(n)$ time. Furthermore, we prove a property of maximum planar graphs which do not admit a maximum weight drawing on any set of convex points.

1 Introduction

Triangulation of a set of points is a fundamental structure in computational geometry. Among different triangulations, the *minimum weight triangulation*, $MinWT$, of a set of points in the plane attracts special attention [3,6,9,10]. The construction of the $MinWT$ of a point set is still an outstanding open problem. When the given point set is the set of vertices of a convex polygon (so-called *convex point set*), then the corresponding $MinWT$ can be found in $O(n^3)$ time by dynamic programming [5,8].

According to the authors' best knowledge, there is not much research done on *maximum weight triangulation*, $MaxWT$. From the theoretical viewpoint, the maximum weight triangulation problem and the minimum weight triangulation problem are equally important, and one seems not to be easier than the other. The study of maximum weight triangulation will help us to understand the nature of optimal triangulations.

An *inscribed polygon* is one all whose vertices lie on a circle. In this paper, we show that the maximum weight triangulation of an inscribed polygon can be found in quadratic time, and the graph extracted from its maximum weight triangulation by omitting the boundary edges must form a special tree. Furthermore, if the polygon is

[1]This work is supported by NSERC grant OPG0041629 and RGC grant HKU 541/96E.

[2]Department of Computer Science, Memorial University of Newfoundland, St. John's, Newfoundland, Canada A1B 3X5 (email: wang@garfield.cs.mun.ca)

[3]Department of Computer Science, University of Hong Kong, Hong Kong (email: chin@cs.hku.hk)

regular, i.e., all its edges are of the same length and all its inner angles are equal, then its maximum weight triangulation can be found in linear time.

Straight-line drawing is a field of growing interest [2]. A special type of straight-line drawings is minimum weight drawings. Let C be a class of graphs, and S be a set of points in the plane. Let $G = (V, E)$ be a graph of C such that $V(G) = S$, E is a set of non-crossing straight-line segments connecting pairs of points of S, and the sum of the lengths of the edges in E is minimized over all straight-line graphs of class C on S, G is called a ***minimum weight representative of*** C with respect to S. A graph $G \epsilon C$ is said to admit a ***minimum weigh drawing*** if G is a minimum weight representative of C with respect to some point set S. In particular, if C is the class of trees, a tree G admits a minimum weight drawing if there exists a set S of points in the plane such that G is isomorphic to an Euclidean minimum weight spanning tree of S. For example, tree T in the Figure 1a has a minimum weight drawing as T is isomorphic to an Euclidean minimum weight spanning tree as given in Figure 1b.

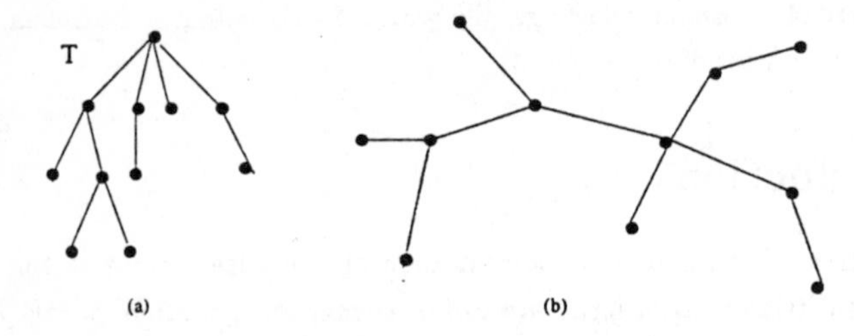

Figure 1: An illustration of minimum weight drawing.

In the area of minimum weight tree drawing, it is proved that every tree with maximum node degree of at most five admits a minimum weight drawing, i.e., it can be drawn as a minimum weight spanning tree of some set of points. On the other hand, a tree with maximum node degree of more than six cannot be drawn as a minimum weight spanning tree [13]. Interestingly, deciding whether a tree with maximum degree of six has a minimum weight drawing is NP-Hard [4].

The problem of constructing a minimum weight drawing for a planar triangulation was first studied by Lenhart and Liotta [11]. They showed that the greedy triangulation of a regular polygon is of minimum weight. Furthermore, they investigated the drawing of *maximum outerplanar graphs*. A graph G is *outerplanar* if it has a planar embedding such that all its nodes lie on a single face; an outerplanar graph is *maximum* if all the other faces of G are bounded by exactly three edges. They devised a linear-time algorithm that takes a maximum outerplanar graph G as input and constructs its straight-line drawing G' as output such that G' is a minimum weight triangulation of the set of points representing the nodes of G.

In this paper, we say graph G is a *maximum weight drawing* $MaxWD$ if G is isomorphic to a straight-line graph in the plane with maximum weight. In particular, we show that any tree whose internal nodes connect to at most 2 non-leaf nodes can be realised as a maximum weight drawing in linear time. We further show that any graph with a special forbidden property does not admit a maximum weight drawing, i.e., there does not exist a convex point set whose maximum weight triangulation is isomorphic to this graph.

In Section 2, we present some definitions on $MaxWT$. In Section 3, the $MaxWT$ of inscribed and regular polygons will be discussed. The maximum weight drawing of a special type of tree on a convex point set will be described in Section 4. We shall discuss in Section 5 that some graphs with a certain property do not have a $MaxWD$. Section 6 is the conclusion.

2 Preliminaries

Let P be a set of points in the plane. A *triangulation* of P, denoted by $T(P)$, is a maximal set of non-crossing line segments with their endpoints in P. It follows that the interior of the convex hull of P is partitioned into non-overlapping triangles. The weight of a triangulation $T(P)$ is given by

$$\omega(T(P)) = \sum_{\overline{p_i p_j} \in T(P)} \omega(\overline{p_i p_j}),$$

where $\omega(\overline{p_i p_j})$ is the Euclidean length of line segment $\overline{p_i p_j}$.

A *maximum weight triangulation* of P ($MaxWT(P)$) is defined as for all possible $T(P)$, $\omega(MaxWT(P)) = max\{\omega(T(P))\}$.

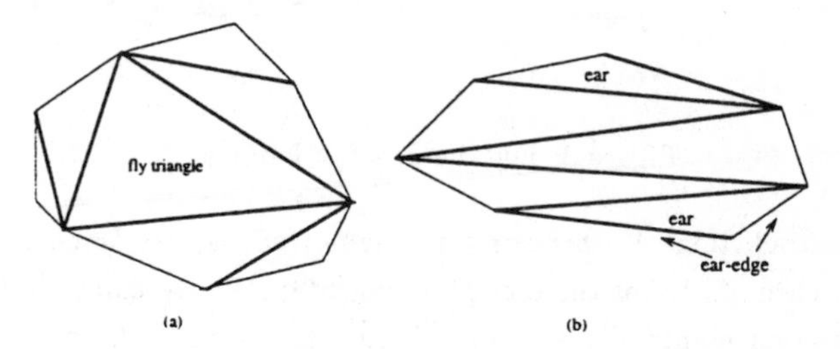

Figure 2: An illustration of the definitions.

Let P be a convex polygon (whose vertices are a *convex point set*) and $T(P)$ be its triangulation. A *fly triangle* of $T(P)$ is one consists of three diagonals (Figure 2a). An *ear* of $T(P)$ is a triangle containing two consecutive boundary edges of P, which are called *ear edges*. An *inner-spanning tree* of the vertices of P is a subgraph of $T(P)$ whose nodes are those vertices of P and whose edges are the internal edges of $T(P)$ plus two ear edges, one per ear (Figure 2b).

For simplicity, in the proofs of the lemmas, we use '$\leq, =, <$' etc. to denote the comparison of the lengths of arcs or line segments, $\overline{xy}^2$ denotes the square of the length of line segment $\overline{xy}$, i.e., $\overline{ab} < \overline{cd}$ means $\omega(\overline{ab}) < \omega(\overline{cd})$.

3 Maximum weight triangulation of inscribed polygons

The following lemma shows an important property of the maximum weight triangulation of an inscribed polygon.

Lemma 1 *Let P be an inscribed polygon. Then $MaxWT(P)$ cannot contain any fly triangle.*

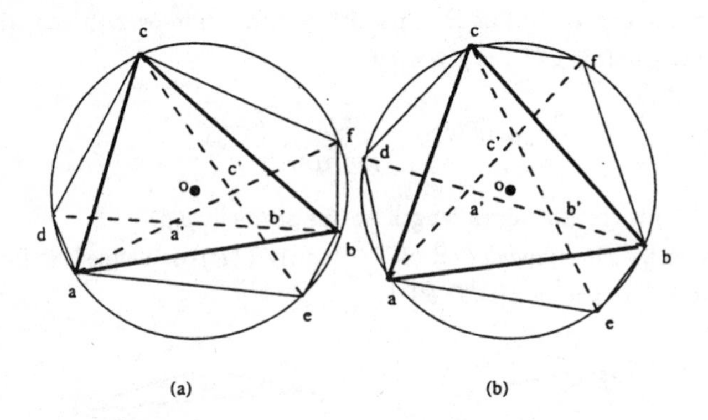

Figure 3: For the proof of Lemma 1.

Proof: By contradiction. With respect to Figure reff2, let Δabc be a fly triangle of $MaxWT(P)$. Then, Δabc has three neighboring triangles, say Δaeb, Δbfc, and Δcda. Let the intersection points of diagonals $\overline{af}$, $\overline{bd}$, and $\overline{ce}$ be a', b', and c', respectively. There are two distinct cases, depending on whether center o of the circumcircle lies inside $\Delta a'b'c'$ or not.

(1) Let o lie outside the triangle $\Delta a'b'c'$ (Figure 3a). Then, o must lie inside one of the areas bounded by ($\overline{cc'}$, $\overline{c'a}$, $\widehat{adc}$) or ($\overline{aa'}$, $\overline{ba'}$, $\widehat{aeb}$) or ($\overline{bb'}$, $\overline{cb'}$, $\widehat{bfc}$, where $\widehat{abc}$ denotes arc abc. W.l.o.g., let o lie inside the area bounded by $\overline{cc'}$, $\overline{c'a}$, and $\widehat{adc}$. Then, arcs

$\widehat{fbea} < \pi$ and $\widehat{cfbe} < \pi$. In quadrilateral $\square aebc$ of $MaxWT(P)$, given $\overline{ce} \leq \overline{ab}$, then $\widehat{cfbe} \leq \widehat{bea}$. Similarly, in $\square abfc$ of $MaxWT(P)$, as $\overline{af} \leq \overline{bc}$, then $\widehat{fbea} \leq \widehat{cfb}$. Thus, we have $\widehat{cfbe} + \widehat{fbea} \leq \widehat{bea} + \widehat{cfb}$, or $\widehat{fbe} \leq 0$, a contradiction.

(2) Let o lie inside $\Delta a'b'c'$ (Figure 3b). In $\square aebc$ of $MaxWT(P)$, $\overline{ce} \leq \overline{ab}$, then we have $\widehat{cfbe} \leq \widehat{bea}$. Similarly, we have $\widehat{adcf} \leq \widehat{cfb}$ as $\overline{af} \leq \overline{bc}$, and we have $\widehat{bead} \leq \widehat{adc}$ as $\overline{db} \leq \overline{ac}$. Adding them accordingly, we have $(\widehat{bead} + \widehat{adcf} + \widehat{cfbe}) \leq (\widehat{adc} + \widehat{cfb} + \widehat{bea})$. Then, $(\widehat{ad} + \widehat{cf} + \widehat{be}) \leq 0$, a contradiction.

Both of the above cases assume center o lies inside fly triangle Δabc. If center o lies outside Δabc, one of the angles of Δabc must be larger than $\frac{\pi}{2}$. By a lemma to be proved later, Δabc cannot be a fly triangle. □

Corollary 1 *The $MaxWT(P)$ of an inscribed polygon P contains an inner-spanning tree which is the maximum weight spanning tree of the vertices of P.*

Proof: As $MaxWT(P)$ does not contain any fly triangle, except for the two ear triangles, each triangle of $MaxWT(P)$ contains exactly one boundary edge of P. Moreover, one of the two boundary edges in an ear triangle must belong to any spanning tree. Thus, $MaxWT(P)$ contains an inner-spanning tree of P. By the property of $MaxWT$, the inner-spanning tree of $MaxWT(P)$ is the maximum weight spanning tree of the vertices of P. □

Theorem 1 *The $MaxWT(P)$ for an inscribed n-gon P can be found in $O(n^2)$ time.*

Proof: Assume $P = (0, 1, ..., n-1)$ and all the vertex indices are modulo n. Let $W_{i,j}$ with $i < j$ denote the weight of $MaxWT(P_{i,j})$, where $P_{i,j}$ is the convex subpolygon of P. By Lemma 1, $MaxWT(P)$ does not contain any fly triangle. Thus, for each internal edge $\overline{ij}$ in $MaxWT(P)$, the triangle in P_{ij} associated with edge $\overline{ij}$ must involve with either boundary edge $\overline{i(i+1)}$ and diagonal $\overline{(i+1)j}$ or boundary edge $\overline{(j-1)j}$ and diagonal $\overline{i(j-1)}$.

Thus, we have the following recursion for W_{ij},

$W_{i,j} = \omega(\overline{ij})$ if $j = i+1$,

$W_{i,j} = max(W_{i,j-1} + \omega(\overline{(j-1)j}) W_{i+1,j} + \omega(\overline{i(i+1)})) + \omega(\overline{ij})$, if $i+1 < j$.

Since the recursion indices i and j range from 0 to $n-1$ and each evaluation of W_{ij} takes constant time, all W_{ij} for $0 \leq i, j \leq n-1$ can be evaluated in $O(n^2)$ time. Finally, $\omega(MaxWT(P)) = max\{W_{(i+1)i} \mid 0 \leq i \leq n-1\}$ which takes another $O(n)$ time. □

When P is a regular polygon, the following theorem shows that $MaxWT(P)$ is not unique.

Theorem 2 *Any inner-spanning tree of a regular n-gon P together with the boundary edges of P form a $MaxWT(P)$.*

Proof: Corollary 1 implies that the $MaxWT(P)$ does not contain any cycle formed by diagonals. Moreover, as every boundary edge of P is shorter than any internal edge of $MaxWT(P)$, all the internal edges of $MaxWT(P)$ and two edges of P form a maximum weight spanning tree. We say in P that a diagonal *bridging* a piece of the boundary with k edges if they form a cycle of length $k+1$. For every inner-spanning tree of P, it must consist of two boundary edges and a set of diagonals in which a diagonal bridging two boundary edges, a diagonal bridging three boundary edges, ..., a diagonal bridging $(n-2)$ boundary edges. As P is regular, all diagonals bridging the same number of boundary edges must be of the same length. Thus, all the inner-spanning tree must be of the same weight. □

4 Maximum weight drawing of Caterpillar graphs

Let C be a class of graphs and S be a set of points in the plane. Let $G = (V, E)$ be a graph of C such that $V(G) = S$, E is a set of straight-line segments connecting pairs of points of S, and the sum of the lengths of the edges in E is maximized over all graphs of class C on S. G is called a ***maximum weight representative of*** C with respect to S. A graph $G(\epsilon C)$ is said to admit a ***maximum weight drawing*** ($MaxWD$) if G is a maximum weight representative of C with respect to some point set S.

A *caterpillar* is a tree such that all internal nodes connect to at most 2 non-leaf nodes. Figure 4 gives an example of caterpillar.

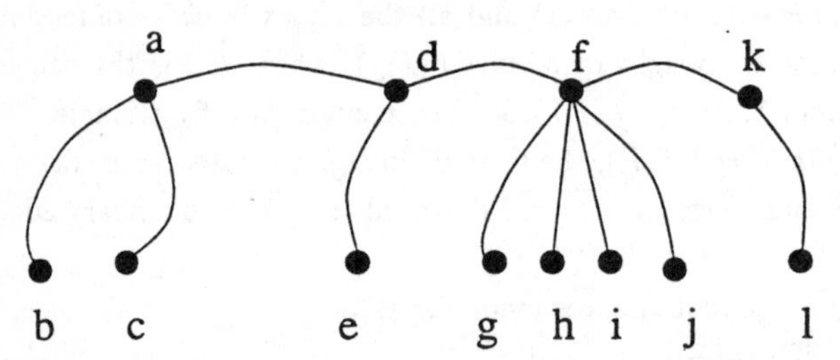

Figure 4: An illustration of the definition of caterpillar.

Now let C be the class of caterpillars. We say a caterpillar G_c has a maximum weight drawing if there exists a convex point set S in the plane such that G_c is isomorphic to an Euclidean maximum weight spanning tree of S.

In this section, we present a linear-time algorithm for the $MaxWD$ of caterpillars through the inner-spanning trees of the vertex set of a regular polygons. Given a caterpillar of n nodes, we construct a *regular point set*, i.e., the vertex set of a regular n-gon, $(0, 1, ..., n-1)$. The drawing starts from a *head* of the caterpillar G_c, i.e., an internal node with exactly one other internal node as its neighbor. For example, nodes a and k are heads in the caterpillar given in Figure 5a. The next step is to select a vertex, say $(n-1)$, in the regular n-gon to represent the head, and to act as the center of a fan to vertices $0, 1, \dots$ to represent edges adjacent to the head (Figure 5b). The drawing of the spanning tree will continue with the head's neighboring internal node to be represented by the last vertex in the fan (Figure 5b). The drawing will proceed along the chain of internal nodes of G_c and the detailed algorithm is given below.

Algorithm MaxWDRAW
Input: Caterpillar graph G_c
Output: Maximum weight spanning tree isomorphic to G_c
Method:

1 $n \leftarrow |V(G_c)|$; Draw a regular point set $(0, 1, \dots, n-1)$.

2 Let V_I be the chain of internal nodes starting from a head of G_c.
  $s \leftarrow n-1$; $t \leftarrow 0$; Draw $\overline{st}$

3 *While* $V_I \neq \emptyset$ *do*
  $v_I \leftarrow Extract(V_I)$; $k \leftarrow degree(v_I)$;
  Draw $\overline{sj}$ for $j = t+1, t+2, \dots, t+k-1$;
  $t \leftarrow t+k-1$;
  if $V_I \neq \emptyset$ *then*
    $v_I \leftarrow Extract(V_I)$; $k \leftarrow degree(v_I)$;
    Draw $\overline{tj}$ for $j = s-1, s-2, \dots, s-k+1$;
    $s \leftarrow s-k+1$;

EndDo

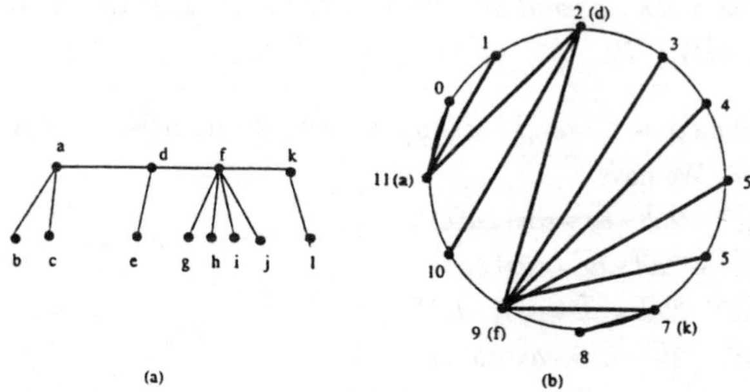

Figure 5: (a) caterpillar graph (b) the corresponding maximum weight drawing.

Theorem 3 *Any caterpillar graph has a straight-line maximum weight drawing and which can be found in linear time.*

Proof: Apply algorithm *MaxWDRAW* to G_c. The output is a spanning tree over a regular point set, which is isomorphic to G_c. By Theorem 2, this spanning tree gives the maximum weight triangulation of the regular polygon formed by the point set. Corollary 1 implies that this spanning tree is of the maximum weight. Finally, it is easy to see that *MaxWDRAW* takes $O(n)$ time. □

5 Forbidden graphs for maximum weight drawing on a convex point set

In this section, we shall prove that some maximum outerplanar graphs do not admit an *MaxWD*, these graphs are called *forbidden graphs*.

Lemma 2 *If P is a convex point set, then each interior angle of any fly triangle of the $MaxWT(P)$ must be no less than $\frac{\pi}{4}$.*

Proof: By contradiction. W.o.l.g., assume $\triangle abd$ is a fly triangle in the $MaxWT(P)$ with $\angle a < \frac{\pi}{4}$ as the smallest angle and $\overline{ah}$ be the line segment perpendicular to $\overline{bd}$ from a (Figure 6). As $\angle a = (\alpha + \beta) < \frac{\pi}{4}$, we have $\overline{bd} = \overline{ah} * (tan\alpha + tan\beta) < \overline{ah} * tan(\alpha + \beta)$ $< \overline{ah}$. Replacing $\overline{bd}$ by $\overline{ac}$ (as $\overline{ac} > \overline{ah} > \overline{bd}$) would arrive at another triangulation whose weight is larger than the weight of $MaxWT(P)$. This leads to a contradiction. □

Corollary 2 *If P is a convex point set, then no interior angle of any fly triangle of $MaxWT(P)$ is larger than $\frac{\pi}{2}$.*

Lemma 3 *If P is a convex point set, then there cannot exist two fly triangles sharing an edge in the $MaxWT(P)$.*

Proof: By contradiction. W.o.l.g., assume the two fly triangles are $\triangle abd$ and $\triangle bcd$ as shown in Figure 6. We have

$$\begin{aligned}\overline{ac}^2 &= \overline{ab}^2 + \overline{bc}^2 - 2\overline{ab} * \overline{bc} * cos(\angle abc)\\ &= \overline{ad}^2 + \overline{dc}^2 - 2\overline{ad} * \overline{dc} * cos(\angle cda).\\ \overline{bd}^2 &= \overline{ab}^2 + \overline{ad}^2 - 2\overline{ab} * \overline{ad} * cos(\angle dab)\\ &= \overline{bc}^2 + \overline{dc}^2 - 2\overline{bc} * \overline{dc} * cos(\angle bcd).\end{aligned}$$

From Lemma 2, since all angles of the fly triangles are larger than $\frac{\pi}{4}$, $\angle abc$ and $\angle cda$ are larger than $\frac{\pi}{2}$, i.e., $cos(\angle abc)$ and $cos(\angle cda)$ are negative. Thus, we have $2\overline{ac}^2 - 2\overline{bd}^2 > 0$ or $\overline{ac} > \overline{bd}$. This contradicts that $\overline{bd}$ is an edge in $MaxWT(P)$. □

Let C be the class of all maximum outerplanar graphs. A maximum outerplanar graph G has a maximum weight drawing if there exists a convex point set S in the plane such that G is isomorphic to an Euclidean maximum weight triangulation of S.

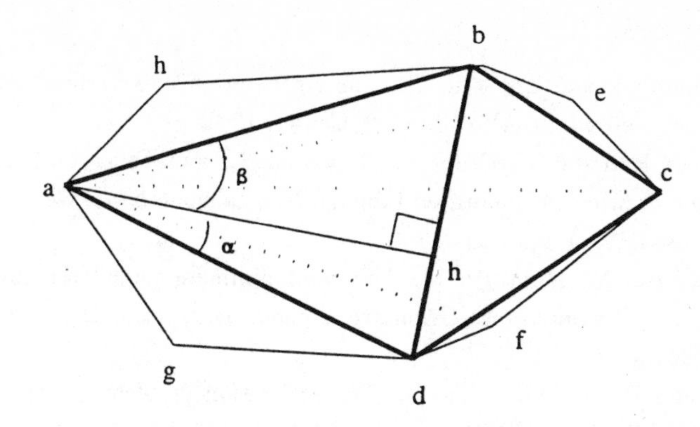

Figure 6: Fly triangles in the $MaxWT(P)$.

Based on Lemma 3, the following theorem shows that some maximum outerplanar graphs do not have maximum weight drawings. Figure 6 illustrates such an example.

Theorem 4 *If $G(V, E)$ is a maximum outerplanar graph containing a simple cycle C with four nonconsecutive nodes which form two triangles sharing a common edge, then G cannot have a maximum weight drawing.*

Proof: Figure 6 shows a maximum outerplanar graph which does not have a maximum weight drawing, cycle $C = ahbecfdga$ and the four nonconsecutive nodes a, b, c, d, as specified in the theorem. In fact, as long as nodes a, b, c, d are nonconsecutive, any ear edges in the triangulation (Figure 6) can be replaced by chains of nodes (note that many edges are needed to connect these nodes to make the graph maximum). The proof follows directly from Lemma 3 as any triangulation of a convex polygon isomorphic to a maximum outerplanar graph having the property specified in the Theorem would imply the existence of two fly triangles sharing a common edge. □

6 Concluding Remarks

Maximum weight triangulation problem is the counterpart of and no easier than the minimum weight triangulation problem. In this paper, we study the maximum weight triangulation of inscribed polygons, in particular regular polygons. Some properties of the triangulation have been obtained. We utilized this property to design a linear-time algorithm for maximum weight drawing of caterpillar graphs. We also showed the property of some forbidden maximum outerplanar graphs which do not admit maximum weight drawings.

References

[BeEp92] Bern M. and Eppstein D., Mesh generation and optimal triangulation, *Technical Report*, Xerox Palo Alto Research Center, 1992.

[DETT94] Di Battista G., Eades P., Tamassia R., and Tollos I., Algorithms for automatic graph drawing: A annotated bibliography, *Computational Geometry: Theory and Applications*, 4(1994), pp. 235-282.

[DM96] Dickerson M., Montague M., The exact minimum weight triangulation, *Proc. 12th Ann. Symp. Computational Geometry, Philadelphia, Association for Computing Machinery*, 1996, pp. .

[EW94] Eades P. and Whitesides S., The realization problem for Euclidean minimum spanning tree is NP-hard, *Proceedings of 10th ACM Symposium on Computational Geometry*, Stony Brook, NY (1994), pp. 49-56.

[Gilb79] Gilbert P., New results on planar triangulations, Tech. Rep. ACT-15 (1979), Coord. Sci. Lab., University of Illinois at Urbana.

[Keil94] Keil M., Computing a subgraph of the minimum weight triangulation, *Computational Geometry: Theory and Applications*, 4(1994), pp.13-26.

[Kirk80] Kirkpatrick D., A note on Delaunay and optimal triangulations, *Information Processing Letters* 10 (3) (1980), pp. 127-128.

[Kl80] Klincsek G., Minimal triangulations of polygonal domains, *Annual Discrete Mathematics* 9 (1980) pp. 121-123.

[LeLi87] Levcopoulos C. and Lingas A., On approximation behavior of the greedy triangulation for convex polygon, *Algorithmica* 2 (1987), pp. 175-193.

[LeLi90] Lingas A., A new heuristic for minimum weight triangulation, *SIAM J. Algebraic Discrete Methods*, (1987), pp. 646-658.

[LeLi96] Lenhart W. and Liotta G., Drawing outerplanar minimum weight triangulations, *Information Processing Letters*, 57 (1996) pp.253-260.

[MR92] Meijeri H. and Rappaport D., Computing the minimum weight triangulation of a set of linearly ordered points, *Information Processing Letters*, 42 (1992), pp. 35-38.

[MS91] Monna C. and Suri S., Transitions in geometric minimum spanning trees, *Proceedings of 7th ACM Symposium on Computational Geometry*, North Conway, NH (1991), pp.239-249.

[PrSh85] Preparata F. and Shamos M., *Computational Geometry* (1985), Springer-Verlag.

[WCX97] Wang C., Chin F., and Xu Y., A new subgraph of minimum weight triangulations, *Journal of Combinatorial Optimization*, Volume 1, No. 2 (1997), pp. 115-127.

The Colored Sector Search Tree: A Dynamic Data Structure for Efficient High Dimensional Nearest-Foreign-Neighbor Queries

T. Graf[1], V. Kamakoti[2], N. S. Janaki Latha[3] and C. Pandu Rangan[3]

[1] Research Center Jülich,
52425 Jülich,
Germany,
t.graf@kfa-juelich.de

[2] Institute of Mathematical Sciences,
CIT Campus, Tharamani,
Chennai - 600 113, India,
kama@imsc.ernet.in

[3] Department of Computer Science and Engineering,
Indian Institute of Technology, Madras
Chennai - 600 036, India
rangan@iitm.ernet.in

Abstract. In this paper we present the new data structure *Colored Sector Search Tree* (*CSST*) for solving the *Nearest-Foreign-Neighbor Query Problem* (*NFNQP*): Given a set S of n colored points in $\mathbb{R}^D$, where $D \geq 2$ is a constant, and a subset $S' \subset S$ stored in a *CSST*, for any colored query point $q \in \mathbb{R}^D$ a nearest foreign neighbor in S', i.e. a closest point with a different color, can be reported in $O(\log n (\log \log n)^{D-1})$ time w.r.t. a polyhedral distance function that is defined by a star-shaped polyhedron with $O(1)$ vertices; note that this includes the Minkowski metrics d_1 and d_∞. It takes a preprocessing time of $O(n(\log n)^{D-1})$ to construct the *CSST*. Points from S can be inserted into the set S' and removed from S' in $O(\log n(\log \log n)^{D-1})$ time. The *CSST* uses $O(n(\log n)^{D-1})$ space. We present an application of the data structure in the parallel simulation of solute transport in aquifer systems by particle tracking. Other applications may be found in GIS (geo information systems) and in CAD (computer aided design). To our knowledge the *CSST* is the first data structure to be reported for the *NFNQP*.

1 Introduction

1.1 Related work

The *Closest Pair Problem* (*CPP*) is one of the fundamental problems in Computational Geometry [13]. Given a set S of points in $\mathbb{R}^D$, $D \geq 2$ and constant, the *CPP* is to find a Closest Pair of points in S, where the distances are measured w.r.t. an L^t-metric d_t, $(1 \leq t \leq \infty)$. The *Closest Foreign Pair Problem* (*CFPP*) is a generalization of the *CPP*, where the input S is a set of colored points in $\mathbb{R}^D$. The *CFPP* asks for a pair of points realizing the minimal bichromatic distance in S, i.e. a pair of points with minimal foreign distance. In the *All-Nearest-Foreign-Neighbors Problem* (*ANFNP*) one has to compute for each point p of a fixed colored point set S a nearest foreign neighbor in S w.r.t. a given metric, e.g. an L^t-metric d_t, $1 \leq t \leq \infty$ ([1]).

Optimal $\Theta(n \log n)$ algorithms for the *CFPP* and the *ANFNP* in two dimensions are given in [1,6,7] . Efficient randomized algorithms for the *CFPP* in $D \geq 2$ dimensions are presented in [10]. Efficient randomized algorithms for *On-line CFPP* where the point set S is modified by insertions have been given in [9]. Recently, an optimal $\Theta(n \log n)$ algorithm for the *ANFNP* in arbitrary dimensions for a fixed number of colors has been given in [8].

In this paper we study the problem of reporting nearest foreign neighbors inside a given set S of colored points in $\mathbb{R}^D$ (or inside a subset $S' \subset S$) for any colored query point $q \in \mathbb{R}^D$.

1.2 Applications

Our new data structure *CSST* meets the requirements of one of the parallelization strategies that are supported by our massively-parallel 3D particle tracking system that is used for the simulation of reactive solute transport in aquifer systems [3,5]:

The strategy uses a distribution of the particles and the parameters given for the domain, e.g. the velocity field, once for all at the beginning of the simulation; during the simulation the particles inside a processor have to query for the parameters at their specific location which involves network communication between the processors. Each processor stores its particles in a *CSST*. The particles come out of different molecule classes which are represented by different colors in the terminology of our data structure. For the computation of sorption and chemical reactions we compute in each time step for each particle p the L^∞-closest particle q with a molecule type different from p's. In the query, only the unadsorbed particles are of relevance, which correspond to the points in the subset S'.

To reduce the communication overhead, the processors use packers that collect queries to inject them into the network as a single package. Since in production runs of the simulation, the number of processors (about 50) is very small compared with the total number of particles (about 500 millions) used, the communication overhead is acceptable.

Other applications of the *CSST* may be found in GIS (geo information systems) ([13,14,4,11]) for $D \geq 2$ dimensions, and in 3D CAD (computer aided design) for specialized vertex selections. It is worthwhile to notice that from the $\Omega(n \log n)$ lower bound for the *ANFNP* in the algebraic decision tree model of computation ([1]) we obtain an $\Omega(\log n)$ lower bound for the amortized complexity of a nearest-foreign-neighbor query.

1.3 Polyhedral distances

Let S be a set of n points in $\mathbb{R}^D$, $D \geq 2$. A *polyhedral distance function* is defined by a polyhedron P given by its $O(1)$ vertices such that P is star-shaped w.r.t. the origin. For $p \in \mathbb{R}^D$ and $0 < \delta \leq \infty$ we denote by $P_\delta(p)$ the open polyhedron which we obtain by scaling P w.r.t. the origin by a factor of δ and translating the resulting polyhedron by the vector p. The (oriented) *polyhedral distance* from $p = (p.1, \ldots, p.D) \in \mathbb{R}^D$ to $q = (q.1, \ldots, q.D) \in \mathbb{R}^D$ is defined by

$$d(p,q) := \begin{cases} \max\{\delta \mid q \notin P_\delta(p)\} & \text{if } p.i \neq q.i \text{ for at least one } i \in \{1,\ldots,D\} \\ 0 & \text{otherwise} \end{cases}$$

Well-known polyhedral distances are given by $d_\infty(p,q) := \max_{i=1,\ldots,D}\{|p.i - q.i|\}$ defined by $P = \{p \in \mathbb{R}^D \mid p.i \in \{-1,1\}\,\forall i = 1,\ldots,D\}$ and $d_1(p,q) := \sum_{i=1,\ldots,D} |p.i - q.i|$ defined by $P = \{p \in \mathbb{R}^D \mid p.i \in \{1,0\}\,\forall i = 1,\ldots,D \wedge \sum_{i=1}^{D} p.i = 1\}$, also known as Minkowski-metrics ([13])[1].

1.4 Nearest-foreign-neighbor query problem

We assign each point $p \in S$ a color $c(p) \in \mathbb{N}$. For $c(p)$, $p \in S$, we denote by $S_{c(p)}$ the subset of S containing all points with color $c(p)$. The (oriented) *polyhedral foreign distance* $d_f(p,q)$ from p to q is defined by

$$d_f(p,q) := \begin{cases} d(p,q) & \text{if } p \in S \setminus S_{c(q)} \\ \infty & \text{otherwise} \end{cases}$$

The *nearest-foreign-neighbor distance* from p to S w.r.t. d_f is defined by $\delta(p,S) := \min\{d_f(p,q) \mid q \in S\}$. The *nearest foreign neighbors* of p in S w.r.t. d_f are the points in $S \setminus S_{c(p)}$ lying on the boundary of $P_{\delta(p,S)}(p)$. We formulate the *nearest-foreign-neighbor query problem* ($NFNQP$) as follows:

Definition 1. *Given a set S of n colored points in $\mathbb{R}^D$, a possibly empty subset $S' \subset S$, and a polyhedral foreign distance function d_f defined by a polyhedral P which is star-shaped w.r.t. the origin with $O(1)$ vertices. Preprocess S in such a way that*

a. points from S can be inserted into S' and removed from S' efficiently, and
b. for any colored query point $q \in \mathbb{R}^D$, a nearest foreign neighbor in the current set S' w.r.t. d_f can be reported efficiently.

1.5 Contribution of this paper

We introduce the new data structure *Colored Sector Search Tree* ($CSST$) for answering nearest-foreign-neighbor queries w.r.t. polyhedral foreign distance functions efficiently. The $CSST$ is obtained by "cascading" *complete binary skeleton*

---

[1] For $1 < t < \infty$ the Minkowski metric d_t is defined by $d_t(p,q) := \left(\sum_{i=1}^{D} |p.i - q.i|^t\right)^{1/t}$

search trees (*CBSST*) [12] and *colored quadrant priority skeleton search trees* (*CQPSST*) ([2]). The main result of the paper is as follows:

Given a set S of n colored points in $\mathbb{R}^D$, a subset $S' \subset S$, and a polyhedral foreign distance function d_f defined by a polyhedron P with $O(1)$ vertices which is star-shaped w.r.t. the origin. A *CSST* storing S' using $\Theta(n(\log n)^{D-1})$ space can be computed in $O(n(\log n)^{D-1})$ time such that

a. In $O(\log n(\log\log n)^{D-1})$ time, a point $s \in S$ can be inserted into S' and removed from S'.
b. In $O(\log n(\log\log n)^{D-1})$ time, for any colored query point $q \in \mathbb{R}^D$, a nearest foreign neighbor in $S' \subset S$ w.r.t. the polyhedral distance function d_f can be reported.

The result differs from the $\Omega(\log n)$ lower bound by a factor of $(\log\log n)^{D-1}$. For fixed $S' = S$ and S as query point set the *CSST* improves the result in [8] to handle not only $O(1)$ but $\Theta(n)$ colors in the *ANFNP* for the cost of an additional $(\log\log n)^{D-1}$ factor in the time complexities. To the best of our knowledge the *CSST* is the first data structure to be reported for the *NFNQP*.

2 The colored sector search tree

Let S be a set of colored points in $\mathbb{R}^D$ and Δ be a $(D+1)$-faced polyhedron with one vertex in the origin and supporting hyperplanes $h_1, \ldots, h_{D+1}$ (the polyhedron P from section 1.3 can be divided into $O(1)$ such elementary polyhedrons Δ). W.l.o.g. we assume that the hyperplanes $h_1, \ldots, h_D$ contain the origin. Let h_i^+ denote the (closed) halfspace on that side of the hyperplane h_i which contains Δ. By $h_i(p)$ (resp. $h_i^+(p)$) we denote the hyperplane (resp. the halfspace) which we obtain by translating h_i (resp. h_i^+) by the vector p. By $\Delta_\delta(p)$ we denote the polyhedron which we obtain by scaling Δ by a factor of δ w.r.t. the origin and translating the resulting polyhedron by the vector p. Hence $\Delta_\infty(p)$ denotes the entire *sector* $\bigcap_{i\in\{1,\ldots,D\}} h_i^+(p)$. Given the polyhedron Δ, the *foreign sector neighbors* in S of a colored query point $q \in \mathbb{R}^D$ are those points in $(S \setminus S_{c(q)}) \cap \Delta_\infty(q)$, such that the foreign distance from q to these points is minimal among all points in $(S \setminus S_{c(q)}) \cap \Delta_\infty(q)$.

The *colored sector search tree* (*CSST*) supports the following operations in $O(\log n(\log\log n)^{D-1})$ time each:

a. *insert*(p): Insert the point $p \in S$ into the subset $S' \subset S$ stored in the *CSST*.
b. *delete*(p): Delete the point p from the subset $S' \subset S$ stored in the *CSST*.
c. *nfsn*(q): Report a foreign sector neighbor of q inside the subset $S' \subset S$ stored in the *CSST*.

In the following we assume that the points in S are in general position, i.e. no two points of S lie on a common hyperplane $h_i(r)$, $i \in \{1, \ldots, D+1\}$, $r \in \mathbb{R}^D$. We will show in section 2.5 how to overcome this restriction.

2.1 Basic structures of the CSST

A *base tree* for key values $\mathcal{K}_1, \ldots, \mathcal{K}_n$ is a 0-2 binary tree ([12]), i.e. each inner node has exactly two sons, and a leaf search tree for the key values, i.e. for every key value $\mathcal{K}_i$, $i \in \{1, \ldots, n\}$, there exists one leaf in the tree. For a node k we denote by $LST(k)$ and $RST(k)$ the left and the right subtrees of the tree rooted by k, and by $ST(k)$ the subtree consisting of k, $LST(k)$ and $RST(k)$. Each inner node k of a base tree stores a split value which is the minimal key value stored in a leaf of $RST(k)$. The depth $\eta(k)$ of a node k in a base tree, say T, is defined as the number of nodes on the path from the root of T to k. The base tree is almost balanced in the sense that for any two leaves l_1 and l_2 we have $| \eta(l_1) - \eta(l_2) | \leq 1$. In standard terms, the base tree is an almost balanced binary skeleton search tree.

We assign each point $p \in S$ a unique key value $\mathcal{K}(p)$. From a base tree for the key values $\mathcal{K}(p)$, $p \in S$, we obtain a *colored binary skeleton search tree* ($CBSST$) by inserting into each node enough space to store a colored point in $\mathbb{R}^D$ which is possibly the *nil*-point. In the $CBSST$ we store points according to the following conditions:

a. Each point $p \in S$ lies on the path from the root to the leaf with value $\mathcal{K}(p)$.
b. If a node stores a point then, its father also stores a point.

Denote by S' the subset of S that is currently stored in the $CBSST$. A point $p \in S$ is *inserted* into S' as follows: We sift the point p down the tree along the path from the root to the leaf whose value is $\mathcal{K}(p)$. We store the point p in the first node of this path in which no point is stored so far. Since all key values $\mathcal{K}(p)$ are unique by our restriction, condition (a.) ensures that such a node exists. A point $p \in S'$ is *deleted* from S' as follows: After a binary search of the node storing p we remove p from this node. We fill the gap by sifting successor points up the tree. It is easy to show that inserting a point $p \in S$ into S' and deleting a point $p \in S'$ from the set S' can be done in $O(\log n)$ time.

From the properties of the $CBSST$ we easily obtain the following Theorem:

Theorem 1. *Given a colored query point $q \in \mathbb{R}^D$, let $k_1, \ldots, k_m$ denote the nodes on the root-to-leaf path which we traverse in a binary search in the CBSST for the value $\mathcal{K}(q)$. All points in S' with key value smaller that $\mathcal{K}(q)$ are either stored in a node k_i or in a subtree $LST(k_i)$ for $i \in \{1, \ldots, m\}$. All points in S' with key value equal to or greater than $\mathcal{K}(q)$ are either stored in a node k_i or in a subtree $RST(k_i)$ for $i \in \{1, \ldots, m\}$.*

Assume that we are given for the points $p \in S$ a *primary key* $\mathcal{K}_1(p)$ and a *secondary key* $\mathcal{K}_2(p)$. As described in [2] we may base the *colored quadrant priority search tree* ($CQPSST$) on a $CBSST$ for the key values $\mathcal{K}_1(p)$, $p \in S$. Additionally each node k of the $CQPSST$ stores the maximal $\mathcal{K}_2$-value of the points stored in $ST(k)$ with a color different from the color of the point stored in k. For storing points in the $CQPSST$ we have the following additional condition:

c. The $\mathcal{K}_2$-values of the points stored along an arbitrary root-to-leaf path are in increasing order.

Again denote by S' the subset of S that is currently stored in the *CQPSST*. Inserting a point $p \in S$ into S' and deleting a point $p \in S'$ from the set S' can be done in $O(\log n)$ time ([2]). From [2] we obtain the following Theorem:

Theorem 2. *For any colored query point $q \in \mathbb{R}^D$, a point $q' \in S' \cap (S \setminus S_{c(q)})$ for which $\mathcal{K}_1(q) \leq \mathcal{K}_1(q')$, $\mathcal{K}_2(q) \leq \mathcal{K}_2(q')$ and $\mathcal{K}_2(q')$ is minimal among all points in $S' \cap (S \setminus S_{c(q)})$ that satisfy the first two conditions, can be reported by the CQPSST in $O(\log n)$ time; if such a point does not exist the query returns the nil-point.*

It is easy to show that both data structures *CBSST* and *CQPSST* use $O(n)$ space to store a set S of n colored points.

We obtain the *CSST* by "cascading" *CBSSTs* and *CQPSSTs*. The reason why the trees are based on the base tree and not on an arbitrary binary search tree is to avoid rebalancing in the *CSST*.

2.2 The cascading mechanism

Denote by v_i, $i \in \{1, \ldots, D+1\}$, the vector perpendicular on h_i starting in the origin. Let v_i be oriented inside h_i^+ for $i \in \{1, \ldots, D\}$ and v_{D+1} be oriented outside h_{D+1}^+. For $i \in \{1, \ldots, D+1\}$ we define the key-value functions $\mathcal{K}_i : \mathbb{R}^D \to \mathbb{R}$, $i \in \{1, \ldots, D+1\}$, by

$$\mathcal{K}_i(q) := t \quad \text{such that } q \in h_i(tv_i)$$

which are injective on S by our assumption of general position. In level L_1 of the *CSST* we store the values $\mathcal{K}_1(p)$, $p \in S$, in a *CBSST*. In level L_i, $i \in \{2 \ldots, D-1\}$, of the *CSST* we additionally store for each inner node k in a tree of level L_{i-1} the values in $\{\mathcal{K}_i(p) | p \in ST(k)\}$ in a separate *CBSST*. In level L_D of the *CSST* we generate for each inner node k in a tree of level L_{D-1} a separate *CQPSST* with primary key values $\{\mathcal{K}_D(p) | p \in ST(k)\}$ and secondary key values $\{\mathcal{K}_{D+1}(p) | p \in ST(k)\}$ for the points stored in $ST(k)$. Again we denote by S' the subset of S that is currently stored in the *CSST*.

Theorem 3. *A CSST for a set S of n colored points in $\mathbb{R}^D$ can be stored using $O(n(\log n)^{D-1})$ space.*

Proof. By construction, level L_1 requires $O(n)$ space. Level L_2 contains at most 2^j trees of size bounded by $\frac{2^{\lceil \log n \rceil}}{2^{j-1}} - 1$ for each $j \in \{0, \ldots, \lceil \log n \rceil\}$. The levels $L_2, \ldots, L_D$ can be interpreted as a collection of *CSSTs* with $D-1$ levels for the trees in level L_2 (see above). It can be seen easily that we may apply the Theorem recursively, and we obtain that the space needed to store all D levels of the *CSST* is bounded by

$$\sum_{j=0}^{\lceil \log n \rceil} O\left(2^j \left(\frac{2^{\lceil \log n \rceil}}{2^{j-1}} - 1\right) \left(\log \left(\frac{2^{\lceil \log n \rceil}}{2^{j-1}} - 1\right)\right)^{D-2}\right) \leq n \sum_{j=0}^{\lceil \log n \rceil} O\left((\log n)^{D-2}\right)$$
$$= O(n(\log n)^{D-1})$$

which proves the Theorem. □

2.3 The operations insert and delete

Inserting (resp. deleting) a point $p \in S$ into (resp. from) the set $S' \subset S$ currently stored in the *CSST* is performed in D steps where p gets inserted (resp. removed) into (resp. from) some trees of level L_i in the ith step.
In the first step we insert (resp. remove) p into (resp. from) the *CBSST* of level L_1. Denote by N_1 the set of nodes thus considered. In the second step we insert (resp. remove) p into (resp. from) those *CBSST*s of level L_2 which are rooted by nodes in N_1. We denote by N_2 the set of nodes in the trees of level L_2 thus traversed. Generally, in the ith step, $2 < i < D$, we insert (resp. remove) p into (resp. from) those *CBSST*s of level L_i which are rooted by a node in N_{i-1}; by N_i we denote the set of nodes on level L_i thus traversed. In the Dth step we insert (resp. remove) p into (resp. from) those *CQPSST*s of level L_D which are rooted by a node in N_{D-1} (see [2]).

Theorem 4. *A point $p \in S$ can be inserted into and removed from the set $S' \subset S$ currently stored in a CSST in $O(\log n(\log\log n)^{D-1})$ time.*

Proof. By construction we have $N_1 = O(\log n)$. Level L_2 contains at most 2^j trees of size bounded by $\frac{2^{\lceil \log n \rceil}}{2^{j-1}} - 1$ for $j \in \{0, \ldots, \lceil \log n \rceil\}$. The levels $L_2, \ldots, L_D$ can be interpreted as a collection of *CSST*s with $D-1$ levels for the trees in level L_2 (see section 2.2). For a *CSST* in level L_2 with m nodes inserting and deleting can be done in $O(\log m(\log\log m)^{D-2})$ time by an inductive argument. Hence the total time for inserting p into the *CSST*s on level L_2 corresponding to the nodes in N_1 is bounded by

$$O\left(\sum_{j=0}^{\lceil \log n \rceil} \log(\frac{2^{\lceil \log n \rceil}}{2^{j-1}} - 1) \left(\log\log(\frac{2^{\lceil \log n \rceil}}{2^{j-1}} - 1)\right)^{D-2}\right)$$
$$\leq O\left((\log\log n)^{D-2} \log \prod_{j=0}^{\lceil \log n \rceil} \frac{2^{\lceil \log n \rceil}}{2^{j-1}}\right) = O\left(\log n(\log\log n)^{D-1}\right)$$

which completes the proof. □

2.4 The operation nearest foreign sector neighbor (nfsn)

The operation *nfns* reports for a query point q a foreign sector neighbor of q inside the subset $S' \subset S$ stored in the *CSST*. Let $q \in \mathbb{R}^D$ be a colored query point. A nearest foreign neighbor of q in the subset $S' \subset S$ of points currently stored in the *CSST* is computed in D steps as follows:

In the first step we perform a binary search for the value $\mathcal{K}_1(q)$ in the *CBSST* of level L_1. Denote by N_1 the set of nodes contained in the root-to-leaf path thus obtained. Obviously $|N_1| = O(\log n)$. In the second step we perform a binary search for the value $\mathcal{K}_2(q)$ in those *CBSST*s of level L_2 which are rooted by a node in N_1. We denote by N_2 the set of nodes in the trees of level L_2 thus traversed. Generally, in the ith step, $2 < i < D$, we perform a binary search for the value $\mathcal{K}_i(q)$ in those *CBSST*s of level L_i which are rooted by a node in N_{i-1}; by N_i we denote the set of nodes on level L_i thus traversed. In the Dth step we perform a range query as stated in Theorem 2 for the *CQPSST*s of level L_D which are rooted by a node in N_{D-1}.

Note that Theorem 1 ensures in the ith step that all points of $S' \cap_{j=1}^{i} h_j^+(q)$ are either stored in nodes of $\cup_{j=1}^{i} N_j$ or in the right subtrees of these nodes.

Theorem 5. *For any colored query point $q \in \mathbb{R}^D$ a nearest foreign sector neighbor in the set $S' \subset S$ currently stored in the CSST w.r.t. a polyhedral foreign distance function can be reported in $O(\log n (\log\log n)^{D-1})$ time.*

Proof. A computation similar to the computation in the proof of Theorem 4 shows that $\sum_{i=1}^{D} |N_i| = O(\log n(\log\log n)^{D-1})$ and that performing the D steps given above requires $O(\log n(\log\log n)^{D-1})$ time. Hence, computing a foreign sector neighbor of q among the points in S' can be done in $O(\log n(\log\log n)^{D-1})$ time. □

2.5 Overcoming the restriction of general position

So far we assumed that $|h_i(r) \cap S| \leq 1$, for all $i \in \{1, \ldots, D+1\}$ and for all $r \in \mathbb{R}^D$, i.e. no two points of S lie on a common hyperplane $h_i(r)$. This restriction can be removed without any additional overhead in the run time by assuming the lexicographic order among the points in $\mathbb{R}^D$ in addition to the linear order existing among their key values while performing the operations insert, delete and query for nearest foreign neighbors in the *CSST*. Since D is assumed to be a constant the upper time bounds for insertion, deletion and nearest-foreign-neighbor query do not change.

3 Nearest-foreign-neighbor queries

Again, denote by S a set of colored points in $\mathbb{R}^D$ and $S' \subset S$ be a subset currently stored in a *CSST*. Denote by d_f a polyhedral foreign distance function that is generated by a polyhedron P which is starshaped w.r.t. the origin given by its $O(1)$ vertices. Applying the results of section 2 to the *NFNQP* is straightforward now:

Theorem 6. *For any colored query point $q \in \mathbb{R}^D$ a nearest foreign neighbor in the set S' w.r.t. the polyhedral distance function d_f can be reported in $O(\log n (\log\log n)^{D-1})$ time.*

Proof. It is easy to show that the surface of the starshaped polyhedral P can be triangulated in $O(1)$ time. Therefore P can be subdivided into a constant number of $D+1$-faced subpolyhedrals $P^1, \ldots, P^m$ such that each polyhedral P^j, $j \in \{1, \ldots, m\}$, has the origin as one vertex. The supporting hyperplanes of the polyhedral P^j are denoted by $h^+_{j,i}$, $j \in \{1, \ldots, m\}$, $i \in \{1, \ldots, D+1\}$, analogously to section 2. By $v_{j,i}$, $j \in \{1, \ldots, m\}$, $i \in \{1, \ldots, D+1\}$, we denote the vector perpendicular on $h_{j,i}$ starting in the origin. Assume that $v_{j,i}$ is oriented inside $h^+_{j,i}$ and that $v_{j,D+1}$ is oriented outside $h^+_{j,i}$. We generate a *CSST* for each polyhedral P_i with key value functions $\mathcal{K}_{j,i} : \mathbb{R}^D \to \mathbb{R}$ by

$$\mathcal{K}_{j,i}(q) = t \quad \text{such that } q \in h_{j,i}(t\, v_{j,i})$$

By Theorem 5 foreign sector neighbors can be reported in the sectors $P^j_\infty(q)$, $j \in \{1, \ldots, m\}$, in total $O(\log n (\log\log n)^{D-1})$ time since m is treated as a constant. The foreign neighbor of q is the closest among the $O(m)$ foreign sector neighbors thus found and that is reported. This completes the proof. □

4 Conclusions

We have presented the *colored sector search tree* (*CSST*). After a preprocessing time of $O(n(\log n)^{D-1})$ the *CSST* reports for any colored query point $q \in \mathbb{R}^D$ a nearest foreign neighbor in a subset $S' \subset S$ of a fixed set $S \subset \mathbb{R}^D$ containing n colored points in $O(\log n(\log\log n)^{D-1})$ time using $O(n(\log n)^{D-1})$ space. Points from S can be inserted into the set S' and removed from S' in $O(\log n(\log\log n)^{D-1})$ time. Distances are measured w.r.t. a polyhedral foreign distance function defined by a star-shaped polyhedron with $O(1)$ vertices, e.g. one of the Minkowski distance functions d_1 and d_∞. We have presented an application of the *CSST* for parallelly simulating reactive solute transport in aquifer systems by particle tracking.

References

1. A. Aggarwal, H. Edelsbrunner, P. Raghavan, and P. Tiwari. Optimal time bounds for some proximity problems in the plane. *Inform. Process. Lett.*, 42(1):55–60, 1992.
2. A. Brinkmann, T. Graf, and K. Hinrichs. The colored quadrant priority search tree with an application to the all-nearest-foreign-neighbors problem. In *Proc. 6th Canad. Conf. Comput. Geom.*, pages 69–74, 1994.
3. U. Doering, O. Neuendorf, R. Seidemann, U. Jaekel, and H. Vereecken. Transport of reactive solutes in a heterogeneous aquifer : Simulations with a stochastic approach. In *Groundwater in the Urban Environment: Problems, Processes and Management, Chilton et al. (eds). Balkema, Rotterdam*, pages 121–126, 1997.

4. M.F. Goodchild. A spatial analytic perspective on geographical information systems. *Intern. Journ. Geograph. Inform. Syst.*, 1:327–334, 1987.
5. T. Graf, U. Hashagen, and H. Vereecken. 3d massive-parallel computation of solute transport in heterogeneous porous media. In *Fourth SIAM Conference on Mathematical and Computational Issues in the Geosciences*, 1997.
6. T. Graf and K. Hinrichs. Distribution algorithms for the all-nearest-foreign-neighbors problem in arbitrary L^t-metrics. In *Proc. 6th Canad. Conf. Comput. Geom.*, pages 69–74, 1994.
7. T. Graf and K. Hinrichs. Simple search for the nearest foreign neighbors in arbitrary L^t-metrics. In *Proceedings of the 10th European Workshop on Computational Geometry*, pages 25–28, 1994.
8. T. Graf, N.S. Janaki Latha, V. Kamakoti, and C. Pandu Rangan. An optimal algorithm for the all-nearest-foreign neighbors problem in arbitrary dimensions (submitted). 1996.
9. T. Graf, V. Kamakoti, N.S. Janaki Latha, and C. Pandu Rangan. Efficient randomized algorithms for the on-line closest-foreign-pair problem (submitted). 1996.
10. V. Kamakoti, K. Krithivasan, and C. Pandu Rangan. Efficient randomized algorithms for the closest pair problem on colored point sets. *Nordic Journal of Computing*, 2:28–40, 1995.
11. I. Masser and Blakemore M., editors. *Handling geographical information: methodology and potential applications.* Long Scientific & Technical, 1991.
12. K. Mehlhorn. *Data structures and algorithms, Vol.1-3.* Springer, New York, 1984.
13. F.P. Preparata and Shamos M.I. *Computational geometry: An introduction.* Springer, New York, 1985.
14. Jonathan F. Raper, editor. *Three dimensional applications in Geographic Information Systems.* Springer-Verlag, 1989.

Space-Efficient Algorithms for Approximating Polygonal Curves in Two Dimensional Space*

Danny Z. Chen and Ovidiu Daescu**

Department of Computer Science and Engineering, University of Notre Dame
Notre Dame, IN 46556, USA
{chen,odaescu}@cse.nd.edu

Abstract. Given an n-vertex polygonal curve $P = [p_1, p_2, \ldots, p_n]$ in the 2-dimensional space R^2, we consider the problem of approximating P by finding another polygonal curve $P' = [p'_1, p'_2, \ldots, p'_m]$ of m vertices in R^2 such that the vertex sequence of P' is an ordered subsequence of the vertices along P. The goal is to either minimize the size m of P' for a given error tolerance ϵ (called the *min-# problem*), or minimize the deviation error ϵ between P and P' for a given size m of P' (called the *min-ϵ problem*). We present useful techniques and develop a number of efficient algorithms for solving the 2-D min-# and min-ϵ problems under two commonly-used error criteria for curve approximations. Our algorithms improve substantially the space bounds of the previously best known results on the same problems while maintain the same time bounds as those of the best known algorithms.

1 Introduction

In this paper, we consider the problem of approximating an arbitrary n-vertex polygonal curve P in the 2-dimensional space R^2 by another polygonal curve P' whose vertices form an ordered subset of the vertices along the original curve P. An n-vertex polygonal curve in R^2 is specified by an ordered set $[p_1, p_2, \ldots, p_n]$ of vertex points in R^2 such that any two consecutive vertices p_i, p_{i+1} are connected by the line segment $\overline{p_i p_{i+1}}$, $1 \leq i < n$. It is possible that such a polygonal curve has self-intersections. Specifically, the problem is, given a polygonal curve $P = [p_1, p_2, \ldots, p_n]$ in R^2, to determine another polygonal curve $P' = [p'_1, p'_2, \ldots, p'_m]$ of m vertices in R^2 such that:

1. $m \leq n$ (desirably, m is much smaller than n),
2. the vertex sequence of P' is a subsequence of the vertex sequence of P, with $p'_1 = p_1$ and $p'_m = p_n$, and
3. each edge $\overline{p'_i p'_{i+1}}$ of P' is an *approximating line segment* of the subcurve $[p_j, p_{j+1}, \ldots, p_k]$ of P, where $p'_i = p_j$, $p'_{i+1} = p_k$, and $j < k$. That is, for every point p of the subcurve $[p_j, p_{j+1}, \ldots, p_k]$ of P, the error incurred by using $\overline{p'_i p'_{i+1}}$ ($= \overline{p_j p_k}$) to approximate p, based on a given error criterion, is no

* This research was supported in part by the National Science Foundation under Grant CCR-9623585.

** This author was supported in part by a fellowship of the Center for Applied Mathematics of the University of Notre Dame.

bigger than a specified error tolerance ϵ. Such a line segment $\overline{p'_i p'_{i+1}}$ is called the *approximating line segment* of the corresponding subcurve $[p_j, p_{j+1}, \ldots, p_k]$ of P.

The parameter m specifies the size (i.e., the number of vertices) of the "compressed" version P' of P, and the parameter ϵ controls the "closeness" of P' to P (under a certain error criterion). Actually, there is a trade-off between the two parameters m and ϵ: The smaller ϵ is, the larger m tends to be, and *vice versa*. Based on this relation between m and ϵ, Imai and Iri [8,9] considered two versions of optimization problems on approximating polygonal curves in the plane: (i) Given ϵ, minimize m (called the *min-# problem*), and (ii) given m, minimize ϵ (called the *min-ϵ problem*). In this paper, we study both the min-# and min-ϵ problems in 2-D space.

Curve approximation problems appear in many applications, such as image processing, computer graphics, cartography, and data compression. In these applications, it is often desirable to approximate a complex graphical or geometric object (possibly specified by a polygonal curve in R^2) by a simpler object that captures the essence of the complex one yet achieves a certain degree of data compression [11].

An error criterion defines the goodness of fit in terms of the deviations between the approximated and approximating objects. Different error criteria have been used in solving various polygonal curve approximation problems (e.g., see [2, 4–6, 8–10, 12–14]). In this paper, we will use two commonly-used error criteria for studying polygonal curve approximations: The error criterion used in [2, 8–10], which we call the *tolerance zone criterion*, and the criterion used in [4, 9, 12], which we call the *infinite beam criterion* (the infinite beam criterion is also called the *parallel-strip* criterion in [4, 9, 12]).

Under the *tolerance zone criterion*, the *approximation error* between a segment $\overline{p_j p_k}$ and the corresponding subcurve $S = [p_j, p_{j+1}, \ldots, p_k]$ of P is defined as the maximum distance in an L_h metric between $\overline{p_j p_k}$ and each point on the subcurve S (we consider $h \in \{1, 2, \infty\}$). Because P is a polygonal curve, the maximum L_h distance between $\overline{p_j p_k}$ and the points of S can be computed by simply finding the maximum L_h distance between $\overline{p_j p_k}$ and each vertex p_l of S (with $j \leq l \leq k$). We denote by $dist_h(\overline{p_j p_k}, p_l)$ the L_h distance between the segment $\overline{p_j p_k}$ and the vertex p_l. Under the *infinite beam criterion*, the *approximation error* between a segment $\overline{p_j p_k}$ and the corresponding subcurve $S = [p_j, p_{j+1}, \ldots, p_k]$ of P is defined as the maximum L_h distance between the line $L(\overline{p_j p_k})$ that contains $\overline{p_j p_k}$ and each point of the subcurve S. We denote by $dist_h(L(\overline{p_j p_k}), p_l)$ the L_h distance between the line $L(\overline{p_j p_k})$ and a vertex p_l of S. According to these error criteria, the *approximation error* incurred by using a curve $P' = [p'_1, p'_2, \ldots, p'_m]$ to approximate P is defined as the maximum error among those of the edges of P' with respect to their corresponding subcurves of P (e.g., $\max_{i=1}^{m-1}\{\max\{dist_h(\overline{p'_i p'_{i+1}}, p_l) \mid p'_i = p_j, p'_{i+1} = p_k, \text{ and } j \leq l \leq k\}\}$). The parameter ϵ specifies the upper bound of the approximation error of P' with respect to P.

Using the tolerance zone criterion, Imai and Iri [8, 9] and Melkman and O'Rourke [10] studied the 2-D min-# and min-ϵ problems; their algorithms for the 2-D min-# (resp., min-ϵ) problem take $O(n^2 \log n)$ (resp., $O(n^2 \log^2 n)$) time and $O(n^2)$ space. Recently, Chan and Chin [2] reduced the time complexity of the 2-D min-# (resp., min-ϵ) problem under the tolerance zone criterion to $O(n^2)$ (resp., $O(n^2 \log n)$). Based on the infinite beam criterion, Toussaint [12] solved the 2-D min-# problem in $O(n^2 \log n)$ time and $O(n^2)$ space, and Imai and Iri [9] gave an $O(n^2 \log^2 n)$ time and $O(n^2)$ space algorithm for the 2-D min-ϵ problem. Eu and Toussaint [4] then published another algorithm for the 2-D min-# (resp., min-ϵ) problem under the infinite beam criterion that was claimed to take $O(n^2)$ (resp., $O(n^2 \log n)$) time and $O(n^2)$ space. They also used the infinite beam criterion based on the L_1 and L_∞ metrics. Very recently, Barequet *et al.* presented efficient algorithms for approximating polygonal curves in 3-D and higher dimensional spaces under the tolerance zone criterion [1]. Varadarajan [13] studied the min-# and min-ε problems for 2-D monotone polygonal paths, using the uniform measure of error. He gave $O(n^{4/3+\delta})$ time and space algorithms for both problems, where $\delta > 0$ is an arbitrarily small constant. However, those algorithms cannot be extended for the tolerance zone and infinite beam criteria.

In this paper, we present a number of efficient algorithms for the 2-D min-# and min-ϵ problems [4, 2]. In particular, we solve the min-# and min-ϵ problems under both the tolerance zone and infinite beam criteria in the same time bounds as [4, 2], but using only $O(n)$ space, in comparison with the $O(n^2)$ space used in [4, 2]. Our algorithms are based on several new ideas and techniques. In [4, 2, 8–10, 12], the min-# problem is solved by first constructing a directed acyclic graph $G = (V, E)$ for the curve approximation (where V is the vertex set of the curve P), and then finding a p_1-to-p_n shortest path in G. G has an arc $e_{ij} = (p_i, p_j)$, $i < j$, if and only if (iff) $\overline{p_i p_j}$ is the approximating line segment for the chain $[p_i, p_{i+1}, \ldots, p_j]$ of P. For the 2-D case, the number of edges in G, $|E|$, is $O(n^2)$, and the time complexity of the min-# algorithms in [4, 2, 8–10, 12] is in general dominated by the time for constructing G. For the min-ϵ problem, the algorithms in [4, 2, 8–10, 12] first compute and store the $O(n^2)$ approximation errors for all the segments $\overline{p_i p_j}$ defined on P, and then perform a binary search on these errors for the sought error ϵ, at each step of the search applying a min-# algorithm. In comparison, we are able to compute a p_1-to-p_n shortest path in G without having to maintain G explicitly for the min-# problems, and to store only a fraction of the $O(n^2)$ approximation errors for the binary search process for the min-ϵ problems. Thus, we reduce the space bound of the 2-D min-# and min-ϵ problems to $O(n)$. Our results can be further generalized to several other cases of the 2-D polygonal curve approximation problems, as well as to some 3-D cases of those problems (our 3-D results will be given in the full paper).

2 Useful Structures

Let $P = [p_1, p_2, \ldots, p_n]$ be an arbitrary polygonal curve in 2-D that is to be approximated using an L_h metric, where $h \in \{1, 2, \infty\}$. Due to the similarity of our algorithms for these three metrics, we will first illustrate the main ideas and steps of our algorithms with the L_2 metric; the differences between the algorithms for the L_2 metric and those for the L_1 and L_∞ metrics will be discussed later.

Henceforth, unless otherwise specified, the metric we use is L_2, and the subscript 2 is omitted in all notations related to L_2 distances (e.g., $dist(\overline{p_i p_j}, p_k)$ is used instead of $dist_2(\overline{p_i p_j}, p_k)$).

Based on the tolerance zone criterion, a vertex p_k of P is within distance ϵ from a line segment $\overline{p_i p_j}$, where $i \leq k \leq j$, if the following conditions are all satisfied [2]:

Condition 1: $dist(L(\overline{p_i p_j}), p_k) \leq \epsilon$.

Condition 2: If the convex angle defined by $\overline{p_i p_k}$ and $\overline{p_i p_j}$ is greater than $\pi/2$, then $d(p_k, p_i) \leq \epsilon$, where $d(p_k, p_i)$ denotes the L_2 distance between p_k and p_i, and an angle defined by two line segments is said to be *convex* if the angle is no larger than π.

Condition 3: If the convex angle defined by $\overline{p_j p_k}$ and $\overline{p_j p_i}$ is greater than $\pi/2$, then $d(p_k, p_j) \leq \epsilon$.

These three conditions together define a region called the *error tolerance region* of the line segment $\overline{p_i p_j}$. Let $ray(p_i, p_j)$ denote the ray emanating from p_i and passing through p_j. Since the line segment $\overline{p_i p_j} = ray(p_i, p_j) \cap ray(p_j, p_i)$, the error tolerance region of $\overline{p_i p_j}$ is the intersection of the error tolerance regions of $ray(p_i, p_j)$ and $ray(p_j, p_i)$.

In comparison, for the infinite beam criterion, only Condition 1 needs to be satisfied, and hence the shape of the error tolerance region of a segment $\overline{p_i p_j}$ is an infinite "strip" of width 2ϵ in 2-D.

Let p_i, p_j, and p_k be vertices of P with $i < k \leq j$. If $dist(ray(p_i, p_j), p_k) \leq \epsilon$, then $ray(p_i, p_j)$ is said to be an *approximating ray* of p_k. If $dist(ray(p_i, p_j), p_k) \leq \epsilon$ for each k with $i < k \leq j$, then $ray(p_i, p_j)$ is an *approximating ray* of the chain $[p_i, p_{i+1}, \ldots, p_j]$ of P (an approximating ray, for short). Thus, under the tolerance zone criterion, $dist(\overline{p_i p_j}, p_k) \leq \epsilon$ iff $dist(ray(p_i, p_j), p_k) \leq \epsilon$ and $dist(ray(p_j, p_i), p_k) \leq \epsilon$ [2]. In other words, $\overline{p_i p_j}$ is an approximating line segment of p_k iff $ray(p_i, p_j)$ and $ray(p_j, p_i)$ are both approximating rays of p_k. Therefore, for a given ϵ, one can first compute all approximating rays $ray(p_i, p_j)$, then all approximating rays $ray(p_j, p_i)$, with $1 \leq i < j \leq n$, and finally find the set of approximating line segments from the set of approximating rays [2].

Consider the tolerance zone criterion. In 2-D, for two vertices p_i and p_k, let r_a and r_b be two rays emanating from p_i such that the distance between p_k and each of r_a and r_b is exactly ϵ. Let D_{ik} be the whole plane if $d(p_i, p_k) \leq \epsilon$, and let D_{ik} be the convex region bounded by r_a and r_b otherwise. Then, by Conditions 1–3, $dist(ray(p_i, p_j), p_k) \leq \epsilon$ iff $p_j \in D_{ik}$ [2].

3 Min-# and Min-ϵ Problems with the L_2 Metric

The algorithms in [2, 4] for solving the 2-D versions of the min-# and min-ϵ problems all use $O(n^2)$ space. In this section, we present several algorithms for the 2-D min-# and min-ϵ problems. The time bounds of our algorithms match those of their corresponding solutions in [2, 4], but our space bounds are $O(n)$ instead of $O(n^2)$. We will mainly describe the algorithms under the tolerance zone criterion. The different aspects of the algorithms under the infinite beam criterion will be pointed out at appropriate places.

3.1 2-D min-# algorithms

The basic approach for solving the 2-D min-# problem on an arbitrary polygonal curve $P = [p_1, p_2, \ldots, p_n]$ is as discussed in Section 1: Construct a directed acyclic graph $G = (V, E)$ for the curve P (where $V = \{p_1, p_2, \ldots, p_n\}$ and every edge e_{ij} of E, $1 \leq i < j \leq n$, represents an approximating segment $\overline{p_i p_j}$ of P), and find a shortest p_1-to-p_n path in G. Note that most of the previous 2-D min-# algorithms [4, 2, 8–10, 12] separate the stage of constructing the graph G from the stage of computing a shortest path in G, thus having to store G explicitly, which requires $O(n^2)$ space. Our idea for solving the 2-D min-# problem is to mix these two stages together. In particular, we present a technique for finding a shortest path in G incrementally. This technique computes the edges of G only as they are needed by the shortest path computation. Hence it avoids maintaining G explicitly, while is still able to find a shortest path in G in a certain topological-sort fashion. As it turns out, only $O(n)$ space is needed by our 2-D min-# algorithm.

For $1 \leq i < j \leq n$, let $SD(p_i, p_j)$ denote the length of a shortest path from p_i to p_j in G, and let $w(e_{ij})$ denote the weight of an edge $e_{ij} \in E$. For this problem, $w(e_{ij}) = 1$ for every $e_{ij} \in E$ (but our algorithm still works even if $w(e_{ij})$ is of any fixed value). Since the graph G is directed acyclic, it is clear that the inductive relation $SD(p_1, p_j) = \min\{SD(p_1, p_k) + w(e_{kj}) \mid 1 \leq k < j \leq n \text{ and } e_{kj} \in E\}$, where $SD(p_1, p_1) = 0$, holds. This immediately suggests an incremental algorithm for computing $SD(p_1, p_j)$ (from the $SD(p_1, p_k)$'s, with $1 \leq k < j$). As in the rest of this section, we WLOG describe only our algorithm for computing the length $SD(p_1, p_n)$ of a shortest p_1-to-p_n path in G (the algorithm can be easily modified to produce, in addition to $SD(p_1, p_n)$, a shortest path tree of G rooted at p_1).

Let D_{ik} be defined as in Section 2, for $1 \leq i < k \leq n$. That is, D_{ik} is either the whole plane if $d(p_i, p_k) \leq \epsilon$ or is otherwise the planar cone (or wedge) that is bounded by the two rays emanating from p_i and tangent to the disc of radius ϵ centered at p_k. For $1 \leq i < j \leq n$, let $F_{ij} = \cap_{k=i+1}^{j} D_{ik}$ and let $B_{ij} = \cap_{k=i}^{j-1} D_{jk}$. Observe that, if F_{ij} (resp., B_{ij}) is not empty, then every ray $r \in F_{ij}$ (resp., B_{ij}) that starts from p_i (resp., p_j) has nonempty intersection with the disc of radius ϵ centered at p_k, for each $i \leq k \leq j$. Thus, under the tolerance zone criterion, $\overline{p_i p_j}$ is an approximating line segment of P (and hence $e_{ij} \in E$) iff $p_j \in F_{ij}$ and $p_i \in B_{ij}$. Note that F_{ij} (resp., B_{ij}) always consists of one (possibly empty) cone. Hence each F_{ij} (resp., B_{ij}) takes $O(1)$ space to store. Also note that if $\overline{p_i p_j}$ is an approximating line segment of P, then $\overline{p_i p_j} \in F_{ij} \cap B_{ij}$.

Our incremental algorithm, based on the above inductive relation among the $SD(p_1, p_k)$'s, is described as follows. We start with the basis $SD(p_1, p_1) = 0$. Now suppose that for a vertex p_j such that $2 \leq j \leq n$, we have obtained $SD(p_1, p_k)$ for every k with $1 \leq k < j$. We must show how to obtain $SD(p_1, p_j)$. Clearly, the key to computing $SD(p_1, p_j)$ from the $SD(p_1, p_k)$'s is to identify all the edges $e_{kj} \in E$ such that $1 \leq k < j$. To find out whether an edge $e_{kj} \in E$, we need to compute F_{kj} and B_{kj} and to test whether both $p_j \in F_{kj}$ and $p_k \in B_{kj}$. Testing whether $p_j \in F_{kj}$ and $p_k \in B_{kj}$ can be easily done in $O(1)$ time for each k if F_{kj} and B_{kj} are already available, so we focus on computing the F_{kj}'s and

B_{kj}'s. Assume that we have computed and maintained $F_{k,j-1}$ for every k with $1 \leq k \leq j-1$ (WLOG, let $F_{k,k}$ be the whole plane). Then by definition, $F_{kj} = F_{k,j-1} \cap D_{kj}$. So computing an F_{kj} from an already available $F_{k,j-1}$ takes $O(1)$ time. In contrast, we need not maintain the B_{kj}'s. Simply, B_{kj}'s can be computed by definition: $B_{kj} = B_{k+1,j} \cap D_{jk}$. Hence each B_{kj} can be obtained from $B_{k+1,j}$ in $O(1)$ time.

It is now clear that, once the $SD(p_1, p_k)$'s and $F_{k,j-1}$'s are available for every k with $1 \leq k < j$, the edges $e_{kj} \in E$ can all be identified in $O(j)$ time. Thus $SD(p_1, p_j)$ can also be computed in $O(j)$ time. Maintaining the $SD(p_1, p_k)$'s and $F_{k,j-1}$'s uses $O(n)$ space. Therefore, our 2-D min-# algorithm under the tolerance zone criterion takes altogether $O(n^2)$ time and $O(n)$ space.

Lemma 1. *The 2-D min-# problem on an arbitrary polygonal curve under the tolerance zone criterion can be solved in $O(n^2)$ time and $O(n)$ space.*

Our 2-D min-# algorithm under the infinite beam criterion, although is likewise based on the inductive relation among the $SD(p_1, p_k)$'s, is somewhat different from the algorithm under the tolerance zone criterion presented above. Due to the space limit, we leave the description of this algorithm to the full paper.

3.2 2-D min-ϵ algorithms

The basic approach for solving the 2-D min-ϵ problems is as follows: Perform binary search on a sorted set of the $O(n^2)$ approximation errors of the polygonal curve P, at each step of the search using the corresponding min-# algorithm on P and a certain approximation error ϵ'. Since our 2-D min-# algorithms in Subsection 3.1 all use $O(n)$ space, the main difficulty to improving the $O(n^2)$ space bound of the 2-D min-ϵ algorithms in [2, 4] is now at the binary search process of this approach.

The previous 2-D min-ϵ algorithms [4, 2, 8–10, 12] explicitly store the $O(n^2)$ approximation errors of P in a sorted array for the binary search process. Our 2-D min-ϵ results are based on the observation that binary search on a set A can be performed in multiple stages: In the first stage, perform binary search on a sorted subset of the set A (and hence only this subset needs to be stored), and then recursively perform the remaining stages on an appropriate subset of A. Our technique actually stores only $O(n)$ judiciously chosen *sample* approximation errors of P out of the total $O(n^2)$, while still enables us to perform binary search on all the $O(n^2)$ errors of P, without increasing the time bound of the binary search process. The space bound for all our 2-D min-ϵ algorithms is hence $O(n)$. This technique could be applicable to other problems involving similar binary search processes.

Our technique consists of $O(1)$ stages, each of which performs the following computation. (1) Select $O(n)$ samples from the set S of all the elements that are currently active for the binary search; these samples are such that between any two consecutive samples, there is a provably "small" subset of S. (2) Perform binary search on the $O(n)$ samples. (3) At the end of this binary search, reduce the problem to one such "small" subset of the set S (thus only the elements in this subset continue to be currently active). The details of this technique are unfolded as follows.

First, we need to organize the $O(n^2)$ approximation errors into n sets, each of which is of size $O(n)$. Let $err(\overline{p_ip_j}) = \max_{k=i}^{j}\{dist(\overline{p_ip_j}, p_k)\}$ (resp., $err(L(\overline{p_ip_j}))$ $= \max_{k=i}^{j}\{dist(L(\overline{p_ip_j}), p_k)\}$) denote the approximation error of the segment (resp., line) $\overline{p_ip_j}$ (resp., $L(\overline{p_ip_j})$).

Since $err(\overline{p_ip_j}) = \max\{err(L(\overline{p_ip_j})), \max_{k=i}^{j}\{d(p_i,p_k), d(p_j,p_k)\}\}$, for all the segments $\overline{p_ip_j}$ such that $1 \leq i < j \leq n$, we have $\{err(\overline{p_ip_j}) \mid 1 \leq i < j \leq n\} \subseteq \{err(L(\overline{p_ip_j})) \mid 1 \leq i < j \leq n\} \cup \{d(p_i,p_j) \mid 1 \leq i \leq j \leq n\}$. Let $ERR(P) = \{err(\overline{p_ip_j}) \mid 1 \leq i < j \leq n\}$, $L\text{-}ERR(P) = \{err(L(\overline{p_ip_j})) \mid 1 \leq i < j \leq n\}$, and $V\text{-}ERR(P) = \{d(p_i,p_j) \mid 1 \leq i \leq j \leq n\}$. Since $L\text{-}ERR(P) \cup V\text{-}ERR(P)$ is a superset of $ERR(P)$, the sought error $\epsilon \in ERR(P)$ that is the solution for the 2-D min-ϵ problem is contained in $L\text{-}ERR(P) \cup V\text{-}ERR(P)$. Hence we search for the $\epsilon \in ERR(P)$ by performing binary search on $L\text{-}ERR(P) \cup V\text{-}ERR(P)$. Note that $|L\text{-}ERR(P) \cup V\text{-}ERR(P)| = O(n^2)$.

For every i with $1 \leq i < n$, let $L\text{-}ERR_i$ be the set of the approximation errors $err(L(\overline{p_ip_j}))$ for the lines $L(\overline{p_ip_j})$, and $V\text{-}ERR_i$ be the set of the $d(p_i,p_j)$'s, for all j such that $i \leq j \leq n$ (assume that $err(L(\overline{p_ip_i})) = 0$). Note that all the errors in $L\text{-}ERR_i$ can be computed in $O(n\log n)$ time and $O(n)$ space by using Toussaint's approach based on maintaining on-line convex hulls of planar points [12] (in fact, $L\text{-}ERR_i$ can also be obtained in the same complexity bounds by using a tree-guided scheme). Also, it is easy to compute $V\text{-}ERR_i$ in $O(n)$ time and space. Let $ERR_i = L\text{-}ERR_i \cup V\text{-}ERR_i$. Then $|ERR_i| \leq 2n$ and ERR_i can be obtained in $O(n\log n)$ time and $O(n)$ space. Since $L\text{-}ERR(P) \cup V\text{-}ERR(P) = \cup_{i=1}^{n-1} ERR_i$, we organize the $O(n^2)$ approximation errors of $L\text{-}ERR(P) \cup V\text{-}ERR(P)$ into the sets ERR_1, ERR_2, ..., ERR_{n-1}. WLOG, we assume that the errors in $L\text{-}ERR(P) \cup V\text{-}ERR(P)$ are distinct (ties can be easily broken in a systematic way).

The following lemma, which has been used in various selection algorithms before, is a key to our $O(n)$ space binary search algorithm.

Lemma 2. *Suppose that a set S of r distinct elements is organized as m sorted sets C_i of size $O(r/m)$ each. For every $i = 1, 2, \ldots, m$, let C_i' be the subset of C_i that consists of every s-th element of C_i (i.e., the s-th, $(2s)$-th, $(3s)$-th, ..., elements of C_i). Let $S' = \bigcup_{i=1}^{m} C_i'$. If w (resp., z) is the α-th (resp., β-th) smallest element of S', with $w < z$, then there are at most $s(\beta - \alpha + m - 1)$ elements of S that are between w and z (i.e., these elements are $\geq w$ but $\leq z$).*

Proof. See the proof of Lemma 1 in [3].

Now from the $n-1$ sets ERR_1, ERR_2, ..., ERR_{n-1} of size $O(n)$ each, we need to choose $O(n)$ sample elements for the binary search process. Our basic idea for the sampling is as follows:

> Partition the $n-1$ sets ERR_k into $\sqrt{n}$ groups G_i of (roughly) $\sqrt{n}$ sets each. Treat each group G_i (of size $O(n^{1.5})$) as one single sorted set and select $O(\sqrt{n})$ sample elements from G_i, such that there are $O(n)$ elements of G_i between every two consecutive samples from G_i. The total number of samples so selected from all the $\sqrt{n}$ groups is $O(n)$.

Note that, since we use only $O(n)$ space, we cannot explicitly store a group G_i for the sampling process. Instead, we sample from the $\sqrt{n}$ sets of G_i with the following procedure:

Procedure Sampling(G_i)

1 For every set ERR_k of G_i, first compute ERR_k and sort it; then select every $(\sqrt{n})$-th element from the sorted set ERR_k, and put the selected elements in the set S_i for G_i.

2 Sort the set S_i, and choose every $(\sqrt{n})$-th element from S_i. These chosen elements form the samples of G_i, and are put into the set $Sample(G_i)$.

The total time of the above procedure is $O(n^{1.5}\log n)$, since we need to compute and sort each of the $\sqrt{n}$ sets ERR_k of G_i. The space is $O(n)$, because we only need to store each ERR_k once in Step 1, and store the set S_i of size $O(n)$. The size of $Sample(G_i)$ is clearly $O(\sqrt{n})$. The quality of the samples in $Sample(G_i)$ is ensured by the following lemma.

Lemma 3. *There are $O(n)$ elements of G_i between every two consecutive samples in $Sample(G_i)$.*

To avoid unnecessary repetitions on describing our solutions to various 2-D min-ϵ problems, we assume below that we will be solving an "abstract" 2-D min-ϵ problem. Let $T_{\min-\#}$ denote the time bound of an appropriate 2-D min-# algorithm (note that each of our 2-D min-# algorithms uses $O(n)$ space). This "abstract" 2-D min-ϵ algorithm consists of three stages.

The First Stage

In the first stage, we perform **Procedure Sampling** on each of the $\sqrt{n}$ groups G_i and obtain $O(n)$ samples, in altogether $O(n^2\log n)$ time and $O(n)$ space. Let $Sample = \cup_{i=1}^{\sqrt{n}} Sample(G_i)$. We then perform binary search on the sorted set $Sample$, at each step of the search using the min-# algorithm on P and a certain approximation error ϵ' of $Sample$. This binary search takes $O(T_{\min-\#}\log n)$ time. Therefore, this stage uses $O(n^2\log n + T_{\min-\#}\log n)$ time and $O(n)$ space.

At the end of the binary search of the first stage on the set $Sample$, we obtain two values a and b that are two consecutive errors in $Sample$, such that the sought error ϵ which is the solution to the min-ϵ problem satisfies $a \le \epsilon \le b$. Then, the elements of $L\text{-}ERR(P) \cup V\text{-}ERR(P)$ between a and b are regarded as *currently active*, and all other elements are not. The set of the currently active elements (i.e., between a and b) is characterized by the following lemma.

Lemma 4. *There are $O(n^{1.5})$ elements of $L\text{-}ERR(P) \cup V\text{-}ERR(P)$ between a and b.*

The Second Stage

In the second stage, we first partition the $O(n^{1.5})$ currently active elements into $\sqrt{n}$ subsets G'_i of size $O(n)$ each. The following steps carry out this partitioning. (1) Compute ERR_k, for every k with $1 \le k < n$, and find the number of the currently active elements in ERR_k (by comparing the elements of ERR_k with a and b). (2) Partition the currently active elements of the ERR_k's into $\sqrt{n}$ subsets G'_i, by associating each G'_i with several appropriate ERR_k's; this is done by a simple prefix sum computation on the numbers of the currently active elements in ERR_1, ERR_2, ..., ERR_{n-1}. This partitioning process takes $O(n^2 \log n)$ time and $O(n)$ space (the costly time is on re-computing the ERR_k's).

Next, we compute each set G'_i from its associated ERR_k's, sort G'_i, and select as a sample every $(\sqrt{n})$-th element from G'_i. Note that $|G'_i| = O(n)$ and there are $\sqrt{n}+1$ elements of G'_i between every two consecutive samples from G'_i. Let $Sample'$ be the set of such selected samples from all the $\sqrt{n}$ G'_i's. Then $|Sample'| = O(n)$. Performing binary search on $Sample'$ again gives two consecutive errors a' and b' in $Sample'$ such that $a' \leq \epsilon \leq b'$. The binary search of the second stage has the same complexity bounds as the first stage. Based on Lemma 2, it is now not hard to see that between a' and b', there are $O(n)$ (currently active) elements of $L\text{-}ERR(P) \cup V\text{-}ERR(P)$.

The Third Stage

The third stage computes the $O(n)$ currently active elements from the ERR_k's, and simply performs binary search on these $O(n)$ elements. This binary search obtains the sought error ϵ. The third stage is again carried out in the same complexity bounds as the first stage.

In summary, the above "abstract" 2-D min-ϵ algorithm runs in $O(n^2 \log n + T_{\min-\#} \log n)$ time and uses $O(n)$ space.

Lemma 5. *The 2-D min-ϵ problem under the tolerance zone criterion can be solved in $O(n^2 \log n)$ time and $O(n)$ space.*

A similar result holds under the infinite beam criterion, which we leave to the full paper.

4 Min-# and Min-ϵ Algorithms with L_1 and L_∞ Metrics

In this section, we sketch our algorithms for the 2-D min-# and min-ϵ problems under the tolerance zone and infinite beam criteria that are based on the L_1 and L_∞ metrics. On one hand, the algorithms with the L_1 and L_∞ metrics are quite similar to those with the L_2 metric that were described in Sections 2, and 3. On the other hand, the L_1 and L_∞ versions of the problems have different geometric structures that can be exploited by our algorithms. In fact, some of these structures enable us to obtain more efficient solutions for several L_1 and L_∞ problems than their L_2 counterparts in the previous sections. For these reasons, our discussions will be focusing on the differences between the L_1 and L_∞ algorithms and their L_2 counterparts.

The 2-D min-# and min-ϵ problems under the infinite beam criterion using the L_1 and L_∞ metrics were considered in [4]. It has been shown in [4] that a main difference between the L_1 and L_∞ problems and the L_2 ones is at the different *shapes* of the error tolerance regions. While the 2-D L_2 error tolerance region of a point is a disc in the plane, the 2-D L_∞ (resp., L_1) error tolerance region of a point is a square (resp., diamond). Our algorithms solve these 2-D L_1 and L_∞ problems in the same time bounds as those in [4], but using only $O(n)$ space. Our 2-D L_1 and L_∞ algorithms are much like their L_2 counterparts in Section 3.

Similarly, we solve the 2-D min-# (resp., min-ϵ) problems under the tolerance zone criterion using the L_1 and L_∞ metrics in $O(n^2)$ (resp., $O(n^2 \log n)$) time and $O(n)$ space. Although we are not aware of any previous algorithms specifically for these 2-D L_1 and L_∞ problems, one could certainly use the 2-D min-# (resp., min-ϵ) techniques with the L_2 metric in [2] to solve these problems in $O(n^2)$ (resp., $O(n^2 \log n)$) time and $O(n^2)$ space.

References

1. Barequet, G., Chen, D.Z., Daescu, O., Goodrich, M.T., Snoeyink, J.: Efficiently Approximating Polygonal Paths in Three and Higher Dimensions. Proc. of the 14th ACM Symp. on Comp. Geometry (1998, to appear)
2. Chan, W.S., Chin, F.: Approximation of polygonal curves with minimum number of line segments or minimum error. International Journal of Computational Geometry and Applications **6** (1996) 59–77
3. Chen, D.Z., Chen, W., Wada, K., Kawaguchi, K.: Parallel algorithms for partitioning sorted sets and related problems. Proc. of the 4th Annual European Symp. on Algorithms (1996) 234–245
4. Eu, D., Toussaint, G.T.: On approximation polygonal curves in two and three dimensions. CVGIP: Graphical Models and Image Processing **56** (1994) 231-246
5. Goodrich, M.T.: Efficient piecewise-linear function approximation using the uniform metric. Proc. 9th Annual ACM Symp. on Computational Geometry (1994) 322–331
6. Guibas, L.J., Hershberger, J.E., Mitchell, J.S.B., Snoeyink, J.S.: Approximating polygons and subdivisions with minimum link paths. International Journal of Computational Geometry and Applications **3** (1993) 383–415
7. Hakimi, S.L., Schmeichel, E.F.: Fitting polygonal functions to a set of points in the plane. CVGIP: Graphical Models Image Process. **53** (1991) 132–136
8. Imai, H., Iri, M.: Computational-geometric methods for polygonal approximations of a curve. Computer Vision, Graphics, and Image Processing **36** (1986) 31–41
9. Imai, H., Iri, M.: Polygonal approximations of a curve-Formulations and algorithms. Computational Morphology (1988) 71–86
10. Melkman, A., O'Rourke, J.: On polygonal chain approximation. Computational Morphology (1988) 87–95
11. Natarajan, B.K., Ruppert, J.: On sparse approximations of curves and functions. Proc. 4th Canadian Conference on Computational Geometry (1992) 250–256
12. Toussaint, G.T.: On the complexity of approximating polygonal curves in the plane. Proc. IASTED International Symp. on Robotics and Automation (1985)
13. Varadarajan, K.R.: Approximating monotone polygonal curves using the uniform metric. Proc. 12th Annual ACM Symp. on Computational Geometry (1996) 311–318
14. Wang, D.P., Huang, N.F., Chao, H.S., Lee, R.C.T.: Plane sweeping algorithms for polygonal approximation problems with applications. Proc. of the 4th Annual International Symp. on Algorithms and Computation (1993) 323–332

Parallel Geometric Algorithms in Coarse-Grain Network Models

Mikhail J. Atallah* and Danny Z. Chen**

Abstract. We present efficient deterministic parallel algorithmic techniques for solving geometric problems in BSP like coarse-grain network models. Our coarse-grain network techniques seek to achieve scalability and minimization of both the communication time and local computation time. These techniques enable us to solve a number of geometric problems in the plane, such as computing the visibility of non-intersecting line segments, computing the convex hull, visibility, and dominating maxima of a simple polygon, two-variable linear programming, determination of the monotonicity of a simple polygon, computing the kernel of a simple polygon, etc. Our coarse-grain algorithms represent theoretical improvement over previously known results, and take into consideration additional practical features of coarse-grain network computation.

1 Introduction

A *coarse-grain* parallel computer has a relatively small number of processors (the number of processors may range from several to a few thousands). Each processor of a coarse-grain parallel computer, usually a state-of-the-art processor in itself, has fairly sophisticated computing power and a quite large local memory; hence the processor can store a large amount of data and perform considerably complicated computation by itself. The processors are connected together with an interconnection network (e.g., a hypercube, mesh, or fat tree architecture). A processor can access data stored in a non-local memory by communicating with other processors via the network. This class of machines represents a main stream of today's general-purpose parallel computers that are marketed commercially (e.g., nCUBE, KSR, Intel Paragon, Intel iPSC/860, CM-5, Cray T3E, SP-1 and SP-2), and have been used in various applications.

Clearly, the coarse-grain network models are more practical than the fine-grain models (it is assumed that a fine-grain processor has only $O(1)$ local memory but the number of such processors in a parallel machine can be as large as needed). However, there are additional obstacles to designing efficient algorithms on coarse-grain networks. A very crucial one is the communication bottleneck. This is because more data items are likely to be exchanged among processors but there are less communication links available in a (moderate-size) coarse-grain

* Department of Computer Sciences, Purdue University, West Lafayette, IN 47907, USA. E-mail: mja@@cs.purdue.edu. Portions of this work were supported by sponsors of the COAST Laboratory.

** Department of Computer Science and Engineering, University of Notre Dame, Notre Dame, IN 46556, USA. E-mail: chen@@cse.nd.edu. The research of this author was supported in part by the National Science Foundation under Grant CCR-9623585.

network. Besides, the theoretical assumption that both the inter-processor communication operations and local computation operations take $O(1)$ time does not hold very well for current coarse-grain computers. On many commercial coarse-grain parallel machines today, communication operations are in general much more time-consuming than local computation operations (typically, the time is an order of magnitude more). For example, on a 64-processor nCUBE 2 supercomputer, an addition operation performed on local data takes 0.5 microseconds, while an operation for transferring a data item between two neighboring processors takes about 200 microseconds [16], 400 times of that for the addition operation. Therefore, it makes sense to distinguish two kinds of time complexity, one for local computation and the other for communication, and to seek to minimize each of them.

Not only we target minimization of both the local computation time and communication time, also we aim to obtain coarse-grain network algorithms that are scalable (i.e., the algorithmic efficiency is achievable on a wide range of coarse-grain machines with various ratios of problem size to number of processors). The 1992 "Grand Challenges" report [13] listed as a major goal in research the design of scalable parallel algorithms for application problems. In this paper, we focus on developing efficient deterministic algorithmic techniques for solving computational geometry problems in coarse-grain network models.

In the last several years, there has been considerable work on developing coarse-grain algorithms for geometric problems. These coarse-grain geometric algorithms are typically based on Valiant's "bulk synchronous" processing (BSP) model [19, 20] or some of its variations. In the BSP model, a problem of size n is stored evenly in a p-processor parallel computer (each processor contains $O(n/p)$ data). A parallel algorithm based on this model consists of a sequence of *supersteps* (or *communication rounds*). In each superstep or round, every processor can send/receive messages of size $O(n/p)$ to other processors, and perform computation on its local data. Typical communication operations performed in each round are **global sort** [11], **all-to-all broadcast** [7], **personalized all-to-all broadcast** [7], **partial sum (scan)** [7]. The goal is to minimize the number of communication rounds as well as the local computation time taken by the algorithm. If the best known sequential algorithm for a problem of size n takes $T_s(n)$ time, then ideally one would like to obtain a parallel algorithm in the BSP model using $O(1)$ communication rounds and $O(T_s(n)/p)$ local computation time.

For the problems studied in this paper, there have been interesting parallel geometric algorithms in BSP like models. In particular, efficient algorithms have been known for computing 2-D convex hull [4, 6, 8–10, 21] and 3-D convex hull [5, 6, 12], dominating maxima [4, 6, 7, 12, 18], and visibility of planar non-intersecting line segments from a point [6, 7, 18]. All these algorithms rely on sorting $O(n)$ input items and using messages of size $O(n/p)$. Note that recently, Goodrich [11] developed an efficient sorting algorithm in the BSP model with optimal local computation time and communication rounds.

We use a variation of the BSP model in which the p processors of a parallel computer are connected by a communication network (e.g., mesh, hypercube, or fat tree). This model is called the *coarse grained multicomputer* (CGM) in [7]. We assume that each processor stores $O(n/p)$ data items for a problem of size n and that n/p is sufficiently large. In contrast to previous algorithms in BSP like models which usually use messages of size $O(n/p)$, we further assume that it takes less time to send "short" messages of size $O((n/p)^\alpha)$, for some constant α with $0 < \alpha < 1$, than "long" messages of size $O(n/p)$. This assumption is reasonable because on current parallel computers, the time for sending a message typically consists of a startup overhead time and a time depending on the message length. When n/p is very large, sending messages of size $O(n/p)$ can become expensive because of their lengths. On the other hand, if one only sends/receives short messages of size $o(n/p)$, then massive global data movement operations such as global sorting would require many communication rounds (instead of the desirable $O(1)$ rounds). Therefore, an efficient algorithm in the model we use should try to solve the target problem by avoiding as much as possible using long messages while in the same time using as few communication rounds as possible. Specifically, this means that one should try to avoid performing massive global data movement operations such as sorting $O(n)$ data items unless necessary. Also, this means that one should try to solve problems by exchanging only $o(n)$ (instead of $O(n)$) data among the p processors.

We present efficient deterministic coarse-grain network algorithms for solving several important geometric problems in the plane, such as computing the convex hull, visibility, and dominating maxima of a simple polygon, two-variable linear programming, determination of the monotonicity of a simple polygon, computing the kernel of a simple polygon, etc. Most of these algorithms take $O(1)$ communication rounds and optimal local computation time, yet are able to avoid global sorting and to use only "short" messages of size $O((n/p)^\alpha)$, where α is a constant with $0 < \alpha < 1$. Note that using the algorithms in [7] to solve some of the problems we study would be less efficient (requiring massive data movement). As far as the scalability is concerned, our algorithms require that $n/p \geq p^\epsilon$ for any constant ϵ with $0 < \epsilon < 1$, in comparison to that of $n/p \geq p$ in [7]. We also give an efficient algorithm for computing the visibility of planar non-intersecting line segments from a point. This algorithm improves the scalability of the previously best known algorithm in [7] from $n/p \geq p$ to $n/p \geq p^\epsilon$, while still using $O(1)$ communication rounds and optimal local computation time (but, by using long messages as in [7]). Our visibility algorithm is in fact quite different from that in [7].

2 Visibility of a Simple Polygon from a Point

An important case of the visibility problem studied in Section 3 is that when the n line segments in S form the boundary of a simple polygon P (i.e., every one of the n segments shares each of its two endpoints with exactly one other segment and has no other intersection with any other segments). In his book on art gallery problems and algorithms [17], O'Rourke argues that this case is perhaps the most fundamental problem in visibility.

We are not aware of any previous BSP like algorithm specifically for computing the visibility of a simple polygon. Although one could use the CGM

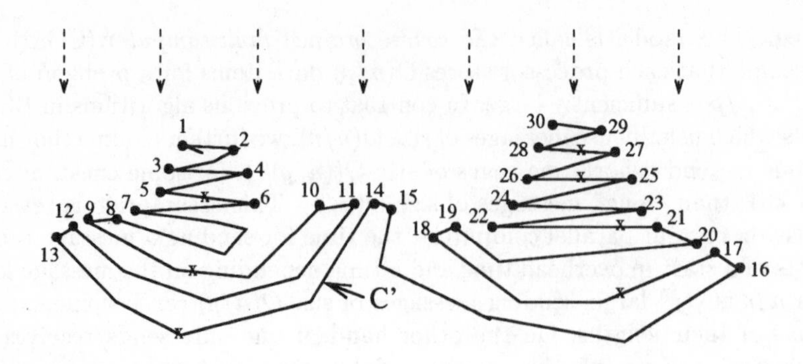

Fig. 1. The $\prec$ order of a visibility chain need not be consistent with the chain order $\prec_c$ along C.

visibility algorithm in [7] to solve the polygon case, doing so would require massive data movement such as global sorting. In this section, we give an efficient algorithm for this problem that uses $O(1)$ communication rounds and optimal local computation time. Furthermore, this algorithm only uses messages of size $O((n/p)^\alpha)$, where α is a fixed constant with $0 < \alpha < 1$, thus avoiding expensive communication operations like global sorting.

Let P be a simple polygon bounded by a closed polygonal chain C of n vertices $p_1, p_2, \ldots, p_n$, in the clockwise order. WLOG, let the source point q be at $(0, \infty)$ (the general case can be handled in a similar fashion, as shown in [2]).

We assume that the input chain C is given sorted by the *chain order* $\prec_c$, i.e., the order in which the vertices appear along C. The order $\prec_c$ is described implicitly by the way in which the elements (e.g., the vertices and edges) of C are initially stored in the processors $PE_1, PE_2, \ldots, PE_p$:

- Every element in processor PE_i is $\prec_c$ every element in PE_{i+1}.
- The n/p elements in each PE_i are in the sorted order of $\prec_c$.

The output to be produced is a *visibility chain* of C, which we denote by $VIS(C)$, consisting of a subset of C based on some sorted order $\prec$ that is different from $\prec_c$. E.g., for a finite source point q, $\prec$ is the sorted order of the vertices of $VIS(C)$ according to their polar angles with respect to q. When q is at $(0, \infty)$, $\prec$ is simply the $<$ order of the x-coordinates of $VIS(C)$. Hence, the desired output is the set of $n' \leq n$ vertices of $VIS(C)$, stored in processors $PE_1, PE_2, \ldots, PE_p$ with each PE_i containing $O(n/p)$ such vertices, in the sorted order of $\prec$.

One may point out that along the *final* visibility chain $VIS(C)$, the $\prec$ and $\prec_c$ orders are consistent with each other. This is indeed true for $VIS(C)$ if C is a *closed* chain. However, since our algorithm uses a divide and conquer strategy, we must cut the closed chain C into many *open* chains, and the $\prec$ and $\prec_c$ orders on the visibility chain of an open chain C' need not be consistent with each other. Figure 1 illustrates that $<$ is quite different from $\prec_c$. The vertices of $VIS(C')$ in this example are in the following order which is clearly different from $\prec_c$: p_{13}, $p_{12}, p_9, p_8, p_7, p_5, p_3, p_1, p_2, p_4, p_6, p_{10}, p_{11}, p_{14}, p_{15}, p_{18}, p_{19}, p_{22}, p_{24}, p_{26}, p_{28}$, $p_{30}, p_{29}, p_{27}, p_{25}, p_{23}, p_{21}, p_{20}, p_{17}, p_{16}$.

The known optimal parallel fine-grain algorithms for this problem (e.g., on the PRAM [2] or hypercube [1]) explicitly maintain the vertices of $VIS(C')$, for every open chain C' used in such an algorithm, in the sorted order of $\prec$. This sorted order of $VIS(C')$ is needed because some of the key operations in such algorithms are parallel k-ary searches on $VIS(C')$ (where $k > 2$ is an integer parameter chosen by the algorithms). Although maintaining $VIS(C')$ in the sorted order of $\prec$ can be done efficiently in the fine-grain models (by exploiting some useful properties of this visibility problem and certain network structures), doing so in our coarse-grain model would mean massive data movement, which we seek to avoid as much as possible. In fact, we want to send only short messages of size $O((n/p)^\alpha)$, for a fixed positive constant $\alpha < 1$. Therefore, a difficulty to our coarse-grain algorithm is to avoid, especially at recursion levels below the top one, explicitly maintaining visibility chains in the sorted order of $\prec$, while still being able to perform parallel k-ary searches correctly and efficiently. We will make crucial use of the geometry of this problem, especially the fact that knowing the order $\prec_c$ enables one to obtain significant information about $VIS(C')$ in an *almost sorted* $\prec$ order.

Let $g = \min\{(n/p)^{\alpha/3}, p\}$ be a control parameter of our divide and conquer algorithm. Let the p processors $PE_1, PE_2, \ldots, PE_p$ form a group G, with each PE_i storing an $(\frac{n}{p})$-vertex contiguous subchain of C. We henceforth let the order $\prec$ be the order $<$ of the x-coordinates.

Algorithm Polygon-Visibility(G)

1. If the group G consists of one processor PE: Let PE compute $VIS(C')$, where C' is the subchain of C stored in PE. This computation is done by PE by simply using the sequential algorithm in [15]. After this computation is finished, return.
2. Otherwise: Let the group G consist of $m > 1$ processors $Q_1, Q_2, \ldots, Q_m$. The following steps are carried out.
   (2.1) Partition G into g subgroups $G_1, G_2, \ldots, G_g$, with each group G_i consisting of processors $Q_{\frac{m(i-1)}{g}+1}, Q_{\frac{m(i-1)}{g}+2}, \ldots, Q_{\frac{mi}{g}}$. Let C_i be the polygonal chain which is the concatenation of the chains stored in the processors of G_i.
   (2.2) In parallel for i = 1, 2, ..., g, recursively call **Algorithm Polygon-Visibility**(G_i) to compute $VIS(C_i)$.
   (2.3) Combine the subsolutions from the g recursive calls to obtain $VIS(C_G)$, where C_G is the chain which is the concatenation of all the C_i's.

The keys are on resolving appropriately two related issues: (1) How to represent $VIS(C_i)$ without having to explicitly maintaining $VIS(C_i)$ in the sorted order of $<$, and (2) how to perform efficiently parallel searches on $VIS(C_i)$ using such a representation. Our ideas for handling these two issues are as follows.

1. Partition the portions of $VIS(C_i)$ which are stored in a processor PE of G_i into $O(1)$ connected components (this is possible by Lemma 4.1 of [2]).
2. Select a sorted set of $O(m/g)$ vertices of $VIS(C_i)$ in the $<$ order to represent the sorted order of $VIS(C_i)$ and to guide the parallel searches into appropriate portions of individual processors.

Our algorithm makes use of many geometric structures of this polygon visibility problem, some of which were given in [1, 2]. We must omit many details due to the space limit.

3 Visibility of Non-intersecting Line Segments from a Point

Given a set S of n non-intersecting "opaque" line segments $s_1, s_2, \ldots, s_n$ and a *source point* q in the plane, the problem is to compute the (possibly unbounded) region of the plane that is visible from the point q. Without loss of generality (WLOG), we assume that the point q is at $(0, \infty)$ (the algorithm for the general case is similar). This case of the visibility problem is also called the *upper envelope* problem. This problem is solvable sequentially in $O(n \log n)$ time [14].

In [7], an efficient CGM algorithm was presented for the upper envelope problem that uses $O(1)$ communication rounds and $O(\frac{n \log n}{p})$ optimal local computation time, provided that $n/p \geq p$, with each round sending/receiving messages of size $O(n/p)$. It was posed in [7] as an open problem to improve the scalability from $n/p \geq p$. In this section, we present a different algorithm than that in [7], improving the scalability from $n/p \geq p$ to $n/p \geq p^{\epsilon}$ for any small constant $\epsilon > 0$, while still using $O(1)$ communication rounds and $O(\frac{n \log n}{p})$ local computation time, in the same CGM model as [7]. Our algorithm is based on the observations of Bertolazzi, Salza, and Guerra [3] for solving this visibility problem.

Let $Vert(S)$ denote the end vertices of the segments in S. Then the size of $Vert(S)$, $|Vert(S)|$, is $2n$. WLOG, we assume that no two points in $Vert(S)$ have the same x-coordinate. The upper envelope $UE(S)$ of S, consisting of the visible portions of the segments of S from the point $q = (0, \infty)$, is specified by a sequence of its vertices, in sorted order of their x-coordinates. Each vertex p of $UE(S)$ can be characterized as one of the following three types.

1. $p \in Vert(S)$ and p is visible from $(0, \infty)$. That is, there is a vertical ray going downwards from $(0, \infty)$ that hits p before it intersects any other points on the segments of S.
2. p is on a segment $s(p)$ of S, $p \notin Vert(S)$, and there is a vertical ray going downwards from $(0, \infty)$ that hits p right after it passes through for the first time an end vertex v of another segment of S such that v is visible from $(0, \infty)$ (i.e., the interior of the line segment $\overline{vp}$ intersects no segments of S). We call in this case the segment $s(p)$ the *projection segment* of v (since a light source at $(0, \infty)$ projects v onto $s(p)$ at p).
3. p is not on any segment of S but is on a horizontal line $y = -\infty$, and there is a vertical ray going downwards from $(0, \infty)$ that hits p after it passes through for the first time an end vertex v of another segment of S such that v is visible from $(0, \infty)$. In this case, we assume that p is on a special line segment s^* and s^* is the *default* projection segment of v.

For a segment set $B \subseteq S$, we say that a point p is *visible from q with respect to B* if there is a vertical ray going downwards from $(0, \infty)$ that intersects no other segment of B before hitting p. For segment sets A and B with $A \subseteq B \subseteq S$, we denote by $VIS_B(A)$ the set of vertices in $Vert(A)$ that are visible from q with respect to B. Note that $UE(S)$ is completely described by the vertices in $VIS_S(S)$ and their projection segments, in sorted order of the x-coordinates. Hence, our problem becomes one of computing $VIS_S(S)$, as well as their projection segments, in sorted order,

Let R be the vertical region bounded by two vertical lines $L_l(R)$ and $L_r(R)$, with $L_l(R)$ to the left and $L_r(R)$ to the right. Then a segment of S, if it intersects R, may or may not have an end vertex contained in R. We call the segments of S that have an end vertex in R the *internal segments* of R, denoted by $S_I(R)$, and the segments of S that intersect R but do not have any end vertices in R the *crossing segments* of R, denoted by $S_C(R)$. Since the segments of S do not cross each other at an interior point, only the *highest crossing segment* of R, which has the highest intersection point with the bounding lines $L_l(R)$ and $L_r(R)$ of R, may have portions in R that are visible from q. Clearly, all the visible vertices of $Vert(S)$ in R, $VIS_S(S) \cap R$, are vertices of $S_I(R)$, and the projection segments of these visible vertices in R are either some of those in $S_I(R)$ or the highest crossing segment in $S_C(R)$. For a vertex v of $Vert(S_I(R))$ such that v is in R and v is visible from q with respect to $S_I(R)$ (i.e., $v \in VIS_{S_I(R)}(S_I(R))$), we define the *projection segment of v in R* as the first segment among those in $S_I(R)$ that is hit by a downward vertical ray from v.

The following observations, given in [3], are a key to our algorithm.

Lemma 1. *Let R and R' be two vertical regions with $R \subseteq R'$. Let v be a vertex of $Vert(S_I(R))$ such that v is in R and v is visible from q with respect to $S_I(R)$. Then the following are true:*

- *The vertex v is visible from q with respect to $S_I(R')$ if and only if v is visible from q with respect to the highest crossing segment of R in $S_I(R')$.*
- *The projection segment of v in R' is either (1) the projection segment of v in R, (2) the highest crossing segment of R in $S_I(R')$, or (3) the default projection segment s^*.*

We are now ready to present our divide and conquer algorithm. Let $g = \min\{n/p, p\}$ (g is a control parameter for our divide and conquer strategy).

Algorithm Upper-Envelope-Main(S, n)

(1) Sort the $2n$ end vertices of $Vert(S)$ by increasing x-coordinates. Use $p-1$ vertical lines $L_1, L_2, \ldots, L_{p-1}$ to partition the plane into p vertical regions $R_1, R_2, \ldots, R_p$, each of which contains $2n/p$ vertices of $Vert(S)$. For $i = 1, 2, \ldots, p$, let processor PE_i store the $2n/p$ vertices of $Vert(S) \cap R_i$ as well as all the internal segments of R_i. (Note that a segment of S can be stored in two processors, but this does not matter.)

(2) Let the p processors $PE_1, PE_2, \ldots, PE_p$ form a group G, and call **Procedure Upper-Envelope**(G) to compute the upper envelope of S.

Procedure Upper-Envelope(G)

1. If the group G consists of one processor PE: Let PE compute the upper envelope in the region R with respect to $S_I(R)$ (i.e., the set of the internal segments of R), where R is the vertical region with which the processor PE is associated. That is, compute the vertices in $VIS_{S_I(R)}(S_I(R)) \cap R$ and their projection segments in R. This computation is done by PE by simply using the sequential algorithm in [14]. After this computation is finished, return.

2. Otherwise: Let the group G consist of $m > 1$ processors $Q_1, Q_2, \ldots, Q_m$. The following steps are carried out.
   (2.1) Partition G into g subgroups $G_1, G_2, \ldots, G_g$, with each group G_i consisting of processors $Q_{\frac{m(i-1)}{g}+1}, Q_{\frac{m(i-1)}{g}+2}, \ldots, Q_{\frac{mi}{g}}$. Let $R(G_i)$ be the vertical region which is the union of all the vertical regions associated with the processors of G_i.
   (2.2) In parallel for $i = 1, 2, \ldots, g$, recursively call **Procedure Upper-Envelope**(G_i) to compute the upper envelope in the region $R(G_i)$ with respect to the set $S_I(R(G_i))$ of the internal segments of $R(G_i)$.
   (2.3) Combine the subsolutions from the g recursive calls to obtain the upper envelope in the region $R(G) = \cup_{i=1}^{g} R(G_i)$ with respect to the set of the internal segments of $R(G)$.

Suppose that **Procedure Upper-Envelope**(G) performs its computation correctly (particularly in step (2.3)), then at the top level of the recursion, it finds the upper envelope of S. This is because at the top recursion level, the region associated with all the p processors is the whole plane and certainly all the segments of S are internal segments of the whole plane.

In the rest of this section, we discuss the computation of **Procedure Upper-Envelope**(G). (In fact, we only need to discuss step (2.3), which we call the **Combining Step**.) We also analyze the number of communication rounds and local computation time taken by **Algorithm Upper-Envelope-Main**(S, n).

Note that based on Lemma 1, to obtain visibility information for a vertical region R' from that of another vertical region R with $R \subseteq R'$, it is sufficient to identify the highest crossing segment of R in $S_I(R')$. In the case of step (2.3), once the highest crossing segment h_i of $R(G_i)$ in $S_I(R(G))$ is identified for every $R(G_i)$, we are done. This is because we then only need to check every vertex $v \in VIS_{S_I(R(G_i))}(S_I(R(G_i))) \cap R(G_i)$ against h_i to see whether v belongs to $VIS_{S_I(R(G))}(S_I(R(G))) \cap R(G)$ and (if so) whether the projection segment of v in $R(G)$ should become h_i. Hence, we only need to show how to compute the highest crossing segment h_i of $R(G_i)$ in $S_I(R(G))$ for every $R(G_i)$.

The **Combining Step** consists of the following substeps, performed in parallel for every G_i.

Substep (i) Let $L_l(R(G_i))$ (resp., $L_r(R(G_i))$) be the left (resp., right) vertical bounding line of the region $R(G_i)$. Let LS_i (resp., RS_i) be the segments of S that intersect $L_l(R(G_i))$ (resp., $L_r(R(G_i))$) and have their right (resp., left) end vertices in $R(G_i)$. Then $LS_i \cup RS_i$ is a subset of the internal segment set $S_I(R(G_i))$ of $R(G_i)$. Let $X(LS_i)$ (resp., $X(RS_i)$) be the set of the x-coordinates of the left (resp., right) end vertices of the segments in LS_i (resp., RS_i). Sort among the processors of G_i the segments in LS_i (resp., RS_i) based on the decreasing (resp., increasing) order of $X(LS_i)$ (resp., $X(RS_i)$).

Substep (ii) For each segment t_a in the sorted set LS_i (resp., RS_i), associate with t_a its intersection point with the left (resp., right) bounding line $L_l(R(G_i))$ (resp., $L_r(R(G_i))$) of $R(G_i)$. Then, perform a partial sum operation among the processors of G_i to find, for every t_a in LS_i (resp., RS_i), the segment t_b in LS_i

(resp., RS_i) such that among all segments $t_{a'}$ in LS_i (resp., RS_i) with $a' \geq a$, t_b has the highest intersection with $L_l(R(G_i))$ (resp., $L_r(R(G_i))$). Let $z(t_a) = t_b$.

Remark: Let $R(G_j)$ be a vertical region to the left of $R(G_i)$. Let $x_l(t_a)$ be the x-coordinate of the left end vertex of a segment $t_a \in LS_i$. For the left vertical bounding line $L_l(R(G_j))$ of $R(G_j)$, if $L_l(R(G_j))$ is between $x_l(t_a)$ and $x_l(t_{a-1})$ (with $x_l(t_0)$ being equal to the x-coordinate of $L_l(R(G_i))$), then it is clear that $z(t_a)$ is the highest crossing segment of $R(G_j)$ in $S_I(R(G_i))$.

Substep (iii) Compute the highest crossing segment h_i of $R(G_i)$ in $S_I(R(G))$:

1. Send from G_i its left (resp., right) bounding line $L_l(R(G_i))$ (resp., $L_r(R(G_i))$) to every processor of each group G_k such that $R(G_k)$ is to the right (resp., left) of $R(G_i)$. This is done by an all-to-all broadcast in G. (Hence, each processor of G_i receives $O(g)$ bounding lines from other groups G_j.)
2. The processors in G_i find, for each left bounding line $L_l(R(G_j))$ with $j < i$, the segment $t_a \in LS_i$ such that $x_l(t_a) \leq x(L_l(R(G_j))) \leq x_l(t_{a-1})$. This is done in each processor PE of G_i by a binary search for $x(L_l(R(G_j)))$ in the portion of $X(LS_i)$ stored in PE. Perform a similar computation in G_i for each $L_r(R(G_k))$ with $k > i$.
3. Send from G_i the segment $z(t_a)$ to the corresponding group G_j with $j \neq i$, such that $z(t_a)$ was found in the previous step specifically for G_j. This is done by a certain personalized all-to-all broadcast in G. (Hence, G_i receives altogether $O(g)$ crossing segments from other G_j's.)
4. The processors of G_i decide the highest crossing segment h_i of $R(G_i)$ in $S_I(R(G))$.

The correctness of the **Combining Step** follows from the remark after **Substep (ii)**, and the correctness of **Algorithm Upper-Envelope-Main**(S, n) follows from Lemma 1. It is not hard to show that **Algorithm Upper-Envelope-Main**(S, n) uses $O(1)$ communication rounds and $O(\frac{n \log n}{p})$ local computation time. To see that the scalability of the algorithm is $n/p \geq p^{\epsilon}$, observe that the algorithm only needs to use messages of length $O(n/p + g) = O(n/p)$, because $g = \min\{n/p, p\}$ and hence $g \leq n/p$.

4 Other Geometric Problems

Once the polygon visibility problem is solved, efficient coarse-grain algorithms can be obtained for several other polygon problems, including computing the convex hull and dominating maxima of a simple polygon, determination of the monotonicity of a simple polygon, computing the kernel of a simple polygon, etc. The details of these algorithms will be given in the full paper.

References

1. M. J. Atallah and D. Z. Chen. "Optimal parallel hypercube algorithms for polygon problems," *IEEE Trans. on Computers*, 44 (7) (1995), pp. 914–922.
2. M. J. Atallah, D. Z. Chen, and H. Wagener. "An optimal parallel algorithm for the visibility of a simple polygon from a point," *J. of the ACM*, 38 (3) (1991), pp. 516-533.

3. P. Bertolazzi, S. Salza, and C. Guerra. "A parallel algorithm for the visibility problem from a point," *J. Para. and Distr. Comp.*, (9) (1990), pp. 11-14.
4. E. Cohen, R. Miller, E. M. Sarraf, and Q. F. Stout. "Efficient convexity and domination algorithms for fine- and medium-grain hypercube computers," *Algorithmica*, (7) (1) (1992), pp. 51–75.
5. F. Dehne, X. Deng, P. Dymond, A. Fabri, and A. A. Khokhar. "A randomized parallel 3D convex hull algorithm for coarse grained multicomputers," *Theory of Computing Systems*, (30) (1997), pp. 547–558.
6. F. Dehne, A. Fabri, and C. Kenyon. "Scalable and architecture independent parallel geometric algorithms with high probability optimal time," *Proc. IEEE Symp. on Parallel and Distributed Processing*, Dallas, 1994, pp. 586–593.
7. F. Dehne, A. Fabri, and A. Rau-Chaplin. "Scalable parallel computational geometry for coarse grained multicomputers," *International Journal on Computational Geometry and Applications*, (6) (3) (1996), pp. 379–400.
8. X. Deng. "A convex hull algorithm on multiprocessor," *Proc. 5th International Symp. on Algorithms and Computations*, Beijing, 1994, pp. 634–642.
9. X. Deng and N. Gu. "Good programming style on multiprocessors," *Proc. IEEE Symp. on parallel and distributed processing*, 1994, pp. 538–543.
10. A. Ferreira, A. Rau-Chaplin, and S. Ubeda. "Scalable 2D convex hull and triangulation algorithms for coarse grained multicomputers," *Proc. 7th IEEE Symp. on Parallel and Distributed Processing*, 1995, pp. 561–568.
11. M. T. Goodrich. "Communication-efficient parallel sorting," *Proc. 28th ACM Symp. on Theory of Computing*, 1996, pp. 247–256.
12. M. T. Goodrich. "Randomized fully-scalable BSP techniques for multi-searching and convex hull construction," *Proc. 8th ACM-SIAM Symp. on Discrete Algorithms*, 1997, pp. 767–776.
13. *Grand Challenges: High Performance Computing and Communications.* The FY 1992 U.S. Research and Development Program. A Report by the Committee on Physical, Mathematical, and Engineering Sciences. Federal Council for Science, Engineering, and Technology. To Supplement the U.S. President's Fiscal Year 1992 Budget.
14. J. Hershberger. "Finding the upper envelope of n line segments in $O(n \log n)$ time," *Info. Proc. Letters*, (33) (1989), pp. 169–174.
15. D. T. Lee. "Visibility of a simple polygon," *Computer Vision, Graphics, and Image Processing*, 22 (1983), pp. 207–221.
16. *nCUBE 2 Programmer's Guide*, Dec. 1990, nCUBE Corporation.
17. J. O'Rourke. *Art Gallery Theorems and Algorithms*, Oxford University Press, 1987.
18. J.-J. Tsay. "Parallel algorithms for geometric problems on networks of processors," *Proc. 5th IEEE Symp. on Parallel and Distributed Processing*, 1993, pp. 200–207.
19. L. G. Valiant. "A bridging model for parallel computation," *Communications of the ACM*, (33) (1990), pp. 103–111.
20. L. G. Valiant. "General purpose parallel architectures," *Handbook of Theoretical Computer Science*, J. van Leeuwen (eds.), Elsevier/MIT Press, 1990, pp. 943–972.
21. J. Zhou, P. Dymond, and X. Deng. "A parallel convex hull algorithm with optimal communication phases," *Proc. 11th IEEE International Parallel Processing Symp.*, Geneva, 1997, pp. 596–602.

On the Bahncard Problem

Rudolf Fleischer*

Max-Planck-Institut für Informatik, Im Stadtwald, 66123 Saarbrücken, Germany.
E-mail: rudolf@mpi-sb.mpg.de.

Abstract. In this paper, we generalize the *Ski-Rental Problem* to the *Bahncard Problem* which is an online problem of practical relevance for all travelers. The Bahncard is a railway pass of the Deutsche Bundesbahn (the German railway company) which entitles its holder to a 50% price reduction on nearly all train tickets. It costs 240 DM, and it is valid for 12 months. Similar bus or railway passes can be found in many other countries.
For the common traveler, the decision at which time to buy a Bahncard is a typical online problem, because she usually does not know when and to which place she will travel next. We show that the greedy algorithm applied by most travelers and clerks at ticket offices is not better in the worst case than the trivial algorithm which never buys a Bahncard. We present two optimal deterministic online algorithms, an optimistic one and and a pessimistic one. We further give a lower bound for randomized online algorithms and present an algorithm which we conjecture to be optimal; a proof of the conjecture is given for a special case of the problem.

1 Introduction

In the *Ski-Rental Problem (SRP)* [9, p. 113], a sportsman can either rent a pair of skis for 1 DM[2] a day, or buy a pair of skis for N DM. As long as he has not bought his skis, he must decide before each trip whether to buy the skis this time or to wait until the next trip (which might never come). The SRP can be solved by algorithms for the page replication problem on two nodes A and B with distance 1 and replication cost N (initially, the file sits on node A, and all requests are to node B), or the two-server problem on a triangle with side lengths $(1, N, N)$ (nodes A and B have distance 1; initially, the two servers sit on nodes A and C, and the requests alternate between A and C). For the page replication problem, there are optimal 2-competitive deterministic [4] and $\frac{(1+\frac{1}{N})^N}{(1+\frac{1}{N})^N-1}$-competitive randomized algorithms against an oblivious adversary [1, 8]. A similar bound was obtained by Karlin et al. [7] for the problem of two servers on a $(1, N, N)$-triangle.

In this paper, we consider the *Bahncard Problem* which contains the SRP as a special case (another generalization of the SRP was given in [2]). The Bahncard

* The author was partially supported by the EU ESPRIT LTR Project No. 20244 (ALCOM-IT). He was further supported by a Habilitation Scholarship of the German Research Foundation (DFG).

[2] May 6, 1998 : 1 DM $\approx$ 0.51 Euro $\approx$ 0.56 US\$ $\approx$ 18.62 NT\$ $\approx$ 74.34 Yen\$

is a railway pass of the Deutsche Bundesbahn (the German railway company). It costs 240 DM, and it is valid for 12 months. Within this period, a traveler can buy train tickets for half of the regular price. Looking back at her travel schedule of the last few years, a traveler can easily determine when several expensive trips had been sufficiently close together to justify the additional expense of a Bahncard. Unfortunately, at any given time the traveler cannot see far into the future, so her decision when to buy a Bahncard is made with a high degree of uncertainty.

Let *BP*(C, β, T) denote the (C, β, T)*-Bahncard Problem*, where a Bahncard costs C, reduces any ticket price p to $\beta \cdot p$, and is valid for time T. For example, the *German Bahncard Problem* GBP is BP(240 DM, $\frac{1}{2}$, 1 year), and the SRP is BP$(N, 0, \infty)$ with the additional constraint that each ticket costs 1 DM.

The SRP and the Bahncard Problem are *online problems*, i.e., all decisions must be made without any knowledge of the future. The quality of an *online algorithm* is measured by the ratio of its performance and the performance of an optimal offline algorithm with full knowledge of the future. The supremum of this ratio over all possible travel request sequences is called the *competitive ratio* of the online algorithm; the smaller the competitive ratio, the better the algorithm [3, 6, 11].

We show that no deterministic online algorithm for BP(C, β, T) can be better than $(2 - \beta)$-competitive. This lower bound is achieved by SUM, a natural generalization of the optimal deterministic 2-competitive Ski-Rental algorithm. SUM is pessimistic about the future in the sense that it always buys at the latest possible time. Surprisingly, there is another optimal deterministic algorithm, OSUM, which usually buys much earlier than the pessimistic SUM (in fact, it buys at the earliest possible time). This gives the rare chance of combining competitive analysis with probabilistic analysis : A traveler with a low travel frequency should use the pessimistic algorithm, whereas a frequent traveler should use the optimistic algorithm. Then both travelers will be happy in the worst case (because both algorithms achieve an optimal competitive ratio), and on the average (because the pessimistic algorithm tries to avoid buying, in contrast to the optimistic algorithm).

Since an online algorithm must make its decisions in a state of uncertainty about future events, it seems plausible that randomization should help the algorithm (because this may help to average between good and bad unpredictable future developments). Ben-David et al. [3] defined several models for randomized competitive analysis and compared their relative strengths. In this paper, we assume an *oblivious adversary*. In this model, a request sequence is fixed in advance and the competitive ratio of a randomized algorithm is a random variable, only dependent on the random moves of the algorithm.

We show that randomized variants of SUM and OSUM are $\frac{2}{1+\beta}$-competitive against an oblivious adversary. This beats the deterministic lower bound for $\beta \in (0, 1)$, but it does not reach the lower bound of $\frac{e}{e-1+\beta}$ which we show to hold for any randomized algorithm. We give a randomized algorithm which achieves this bound in the case of $T = \infty$, i.e., a Bahncard never expires (in this case, the Bahncard Problem corresponds to a variant of the SRP where the price for renting the skis includes the daily fee for the lift, which has to be paid additionally each day after buying the skis). We conjecture that the algorithm is also optimal for the more realistic case of $T < \infty$.

We note that introducing further "real life" restrictions like limiting the number of trips or upper bounding ticket prices has no effect on the worst case behaviour of the problem as long as a single one-way ticket can already be more expensive than a Bahncard.

We apologize for not being able to give complete proofs for most of the theorems; they can be found in the full paper [5].

2 Definitions

Let $C > 0$, $T > 0$, and $\beta \in [0,1]$ be fixed constants. The *(C,β,T)-Bahncard Problem* (or shortly BP(C,β,T)) is a request-answer game between an algorithm A (the traveler) and an adversary (real life). The adversary presents a finite sequence of *travel requests* $\sigma = \sigma_1\sigma_2\cdots$. Each σ_i is a pair (t_i, p_i), where $t_i \geq 0$ is the travel time and $p_i \geq 0$ is the *regular price* of the ticket. The requests are presented in chronological order, i.e., $0 \leq t_1 < t_2 < \cdots$.

The task of A is to react to each travel request by buying a ticket (*that* cannot be avoided), but A can also decide to first buy a Bahncard. A Bahncard bought at time t is *valid* during the time interval $[t, t+T)$. A's *cost* on σ_i is

$$c_{\mathtt{A}}(\sigma_i) = \begin{cases} \beta \cdot p_i & \text{A has a valid Bahncard at time } t_i \\ p_i & \text{otherwise.} \end{cases} \quad \text{if}$$

We call $\beta \cdot p_i$ the *reduced price* of the ticket. Accordingly, σ_i is a *reduced request* for A if A already had a valid Bahncard at t_i; otherwise it is a *regular request*. Note that A might buy a Bahncard at a regular request and then pay the reduced price for the ticket.

If A buys Bahncards at times $0 \leq \tau_1 < \cdots < \tau_k$ then we call the sequence $\Gamma_{\mathtt{A}}(\sigma) = (\tau_1, \ldots, \tau_k)$ the *B-schedule* of A on σ (since σ is finite, the B-schedule is also finite). We denote the *length* k of the B-schedule by $|\Gamma_{\mathtt{A}}(\sigma)|$. The total cost of A on σ is then $c_{\mathtt{A}}(\sigma) = |\Gamma_{\mathtt{A}}(\sigma)| \cdot C + \sum_{i\geq 1} c_{\mathtt{A}}(\sigma_i)$. We do not always distinguish clearly between an algorithm A and its B-schedule $\Gamma_{\mathtt{A}}$, so $c_{\Gamma_{\mathtt{A}}}(\sigma)$ means the same as $c_{\mathtt{A}}(\sigma)$, for example.

Besides the total cost of A, we are interested in partial costs during some time interval I. Let $p^I(\sigma) = \sum_{i:t_i \in I} p_i$ be the cost of all requests in I, and let $c^I_{\mathtt{A}}(\sigma) = \sum_{i:t_i \in I} c_{\mathtt{A}}(\sigma_i)$ be the money spent by A on tickets during I. We call I *cheap* if $p^I(\sigma) < c_{crit}$, where $c_{crit} = \frac{C}{1-\beta}$ is the *critical cost*; otherwise, I is *expensive*. c_{crit} is the break-even point for any algorithm. Buying a Bahncard at the beginning of an expensive interval saves money in comparison to paying the regular price for all tickets in I. Observation 1 (b) below makes this more precise.

We are mainly interested in intervals of length T. The *T-recent-cost* (or *T-cost* for short) of σ at time t is $r^\sigma(t) = p^{(t-T,t]}(\sigma)$. The *regular T-cost* $rr^\sigma_{\mathtt{A}}(t)$ of A on σ at time t is the sum of all regular requests in $(t-T, t]$ with respect to A's B-schedule. Sometimes, we do not want the current request at time t to be included in the summation when computing $r^\sigma(t)$ or $rr^\sigma_{\mathtt{A}}(t)$. Then we speak of the T$-$*cost* at t^- instead.

For any request sequence σ there is a B-schedule $\Gamma_{\mathtt{OPT}}(\sigma)$ of minimal cost $c_{\mathtt{OPT}}(\sigma)$. In general, $\Gamma_{\mathtt{OPT}}(\sigma)$ is not unique and can only be computed by an *offline algorithm* OPT which knows the entire sequence σ in advance.

Observation 1. *Let σ be a request sequence and $\Gamma_{\mathtt{OPT}}(\sigma)$ be an optimal B-schedule for σ. Then we can assume w.l.o.g. that*

(a) OPT *never buys a Bahncard at a reduced request.*

(b) If I is an expensive time interval of length at most T then OPT *has at least one reduced request in I.* □

An *online algorithm* A must compute its B-schedule $\Gamma_{\mathtt{A}}(\sigma)$ on the fly, i.e., whenever it receives a new request σ_i it must decide immediately if it wants to add t_i to its B-schedule, without knowing future requests $\sigma_{i+1}, \sigma_{i+2}, \ldots$. Once bought, a Bahncard cannot be reimbursed, so A cannot change its B-schedule later on. If A uses randomization then the cost of A on a fixed request sequence σ is a random variable whose expected value is also denoted by $c_{\mathtt{A}}(\sigma)$. A is *d-competitive* if $c_{\mathtt{A}}(\sigma) \leq d \cdot c_{\mathtt{OPT}}(\sigma)$ for all request sequences σ. A is an *optimal online algorithm* if its *competitive ratio* d is the smallest possible among all online algorithms. If A is a randomized algorithm then this definition describes competitiveness against an *oblivious adversary* (see [3] for definitions of oblivious and adaptive adversaries and their respective strengths). Intuitively, an oblivious adversary must fix the request sequence σ before A starts serving the requests. In contrast, an *adaptive adversary* can construct the request sequence step by step, dependent on previous decisions of A. This makes it more difficult for a randomized online algorithm to be competitive. However, we do not expect real life to behave like an adaptive adversary (ignoring Murphy's Law), so we assume an oblivious adversary throughout this paper.

3 An Optimal Offline Algorithm

We note that Observation 1 (b) does not imply that an optimal algorithm will buy a Bahncard whenever it reaches the first regular request of an expensive time interval. In Fig. 1, both the intervals $[0, T)$ and $[T, 2T)$ are expensive, but if ϵ is small then the optimal algorithm would buy just one Bahncard at the second request.

Theorem 2. *Given n travel requests, we can compute an optimal B-schedule and its minimal cost in time $O(n)$.*

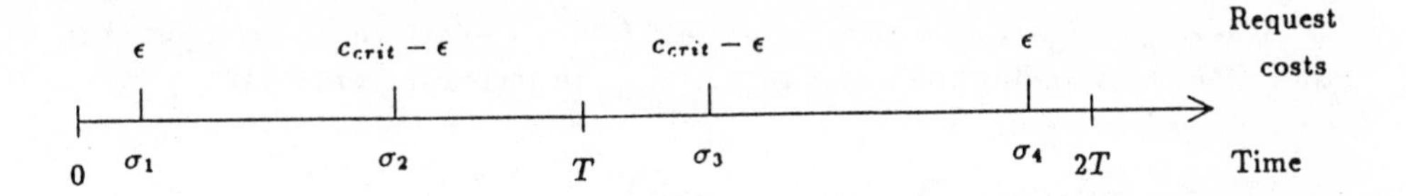

Fig. 1. Two expensive intervals, but an optimal algorithm buys at σ_2.

Proof. Let $\sigma = \sigma_1 \cdots \sigma_n$ be a sequence of n travel requests. We construct a weighted acyclic directed graph G_σ with nodes $s = \sigma_0, \sigma_1, \ldots, \sigma_n, \sigma_{n+1} = t$, where $s = (0, 0)$ and $t = (t_n + T, 0)$ are two new artificial requests. G_σ has the property that $(s \to^* t)$–paths in G_σ correspond to B-schedules, and any shortest $(s \to^* t)$–path corresponds to an optimal B-schedule.

For $i = 0, \ldots, n$, there is an edge $\sigma_i \to \sigma_{i+1}$ of weight p_i, and an edge $\sigma_i \to \sigma_i^{+T}$ of weight q_i, where σ_i^{+T} is the first request after (or at) time $t_i + T$, and q_i is the accumulated cost of buying a Bahncard at request σ_i and paying reduced ticket prices until this Bahncard expires, i.e.,

$$q_i = C + \sum_{j: t_i \leq t_j < t_i + T} \beta \cdot p_j \,.$$

Fig. 2 shows the graph G_σ corresponding to the requests of Fig. 1.

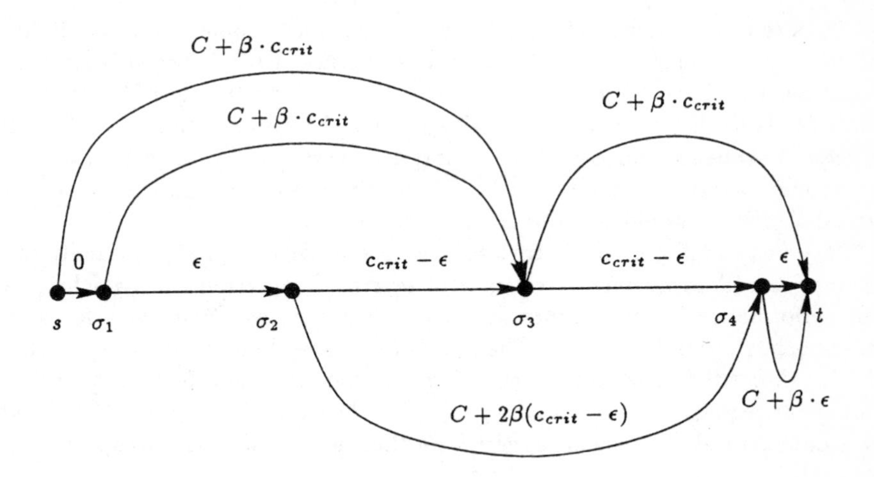

Fig. 2. The graph G_σ corresponding to the requests of Fig. 1.

The edge weights as well as a shortest $(s \rightarrow^* t)$–path can be computed in time $O(n)$ by scanning the nodes $\sigma_0, \ldots, \sigma_{n+1}$ in increasing order [12]. □

4 Deterministic Online Algorithms

The *Buy-Never-Algorithm* NEVER which never buys a Bahncard is obviously $\frac{1}{\beta}$-competitive. Before we analyze other algorithms, we show a lower bound on the deterministic competitive ratio.

Theorem 3. *No deterministic online algorithm for BP(C, β, T) can be better than $(2-\beta)$-competitive.*

Proof. Let A be an online algorithm for BP(C, β, T). Let $\epsilon > 0$ be an arbitrarily small constant. As long as A has not got a Bahncard, the adversary continues showing requests of cost ϵ (arbitrarily dense, so that all requests are in the interval $[0, T)$). If A wants to be better than $\frac{1}{\beta}$-competitive, it must eventually buy a Bahncard. Then the adversary stops showing requests. Let s be the accumulated cost of the requests so far, not including the current request. Then

$$c_{\mathtt{A}}(s) = C + s + \beta\epsilon \quad \text{and} \quad c_{\mathtt{OPT}}(s) = \begin{cases} s + \epsilon & \text{if} \quad s + \epsilon \leq c_{crit} \\ C + \beta(s+\epsilon) & \phantom{\text{if}} \quad s + \epsilon \geq c_{crit} \end{cases}.$$

Hence, $\dfrac{c_{\mathtt{A}}(s)}{c_{\mathtt{OPT}}(s)} \geq \dfrac{c_{\mathtt{A}}(c_{crit} - \epsilon)}{c_{\mathtt{OPT}}(c_{crit} - \epsilon)} = 2 - \beta - \dfrac{\epsilon(1-\beta)^2}{C} \rightarrow 2 - \beta$ for $\epsilon \rightarrow 0$.
The inequality holds because the quotient takes its minimum value at $s = c_{crit} - \epsilon$ in both cases in the definition of $c_{\mathtt{OPT}}$. □

Clerks at railway ticket offices usually advise their customers to buy a Bahncard iff they are planning to buy one or more tickets of total cost at least c_{crit}. We call this the *Ticket-Office-Algorithm* TOA. It has the advantage of being memoryless (cf. [10]), however its competitive ratio is the same as that of NEVER : If the request sequence consists of many travel requests of cost slightly less than c_{crit} within a short time interval, then TOA never buys a Bahncard, whereas the optimal algorithm would buy one at the first request.

TOA seems to fail because it tries to handle expensive requests optimally but it cannot safeguard against a sequence of several cheap requests. To achieve a good performance for both types of request sequences we must allow for non-optimal behaviour in both cases. The proof of the lower bound indicates that the following Algorithm SUM might behave better than TOA. SUM buys a Bahncard at a regular request (t, p) iff $rr^{\sigma}_{\mathtt{SUM}}(t) \geq c_{crit}$. In the example of Fig. 1, SUM would buy a Bahncard at the second request (and thus incidentally behave optimally).

Theorem 4. SUM *is $(2-\beta)$-competitive for BP(C, β, T).*

Proof. (Sketch) Let $\sigma = \sigma_1\sigma_2\cdots$ be a request sequence and let $\Gamma_{\mathtt{OPT}}(\sigma) = (\tau_1, \ldots, \tau_k)$ be an optimal B-schedule for σ. This divides time into *epochs* $[\tau_j, \tau_{j+1})$,

$0 \le j \le k$, where $\tau_0 = 0$ and $\tau_{k+1} = \infty$. Each epoch (except for, possibly, the first and last one) starts with an *expensive phase* $[\tau_j, \tau_j + T)$, followed by a *cheap phase* $[\tau_j + T, \tau_{j+1})$.

SUM will buy at most one Bahncard during any epoch. This follows from Observation 1 (b) and the fact that $(t - T, t]$ must be an expensive interval if SUM buys a Bahncard at time t. Therefore, we can upper bound SUM's total cost of buying Bahncards by assuming that SUM spends C in every expensive phase, in addition to ticket costs.

Clearly , $c^I_{\mathtt{SUM}}(\sigma) \le c^I_{\mathtt{OPT}}(\sigma)$ for a cheap phase I. So let I be any expensive phase. Let $c_{\mathtt{SUM}}$ and $c_{\mathtt{OPT}}$ denote SUM's and OPT's cost during I, respectively (including the cost of buying Bahncards). We divide I into three subphases I_1, I_2, I_3 (some of which can be empty); in I_1 and I_3, SUM has a valid Bahncard, whereas it must pay regular prices in I_2. For $i \in \{1, 2, 3\}$, let $s_i = p^{I_i}(\sigma)$ be the total cost of requests in I_i. Then

$$c_{\mathtt{SUM}} \le C + s_2 + \beta \cdot (s_1 + s_3) \qquad \text{and} \qquad c_{\mathtt{OPT}} = C + \beta(s_1 + s_2 + s_3)\,.$$

Hence

$$\frac{c_{\mathtt{SUM}}}{c_{\mathtt{OPT}}} \le \frac{C + c_{crit}}{C + \beta \cdot c_{crit}} = 2 - \beta\,,$$

because the first quotient is maximal if $s_1 = s_3 = 0$ and if s_2 is maximal, and the definition of SUM implies $s_2 \le c_{crit}$. □

So SUM is optimal for the Bahncard Problem. In particular, it is $\frac{3}{2}$-competitive for the GBP. For the SRP, it behaves like the well-known optimal 2-competitive algorithm which buys at the N-th request [4].

However, SUM tends to be pessimistic about the future : It always buys at the latest possible time, namely after it has seen enough regular requests to know for sure that an optimal algorithm would already have bought a Bahncard. In contrast to that, we consider the *Optimistic-Sum-Algorithm* OSUM which buys a Bahncard at a regular request (t, p) iff $p \ge \frac{C - s(1-\beta)}{2(1-\beta)}$, where $s = rr^{\sigma}_{\mathtt{OSUM}}(t^-)$.

Observe that OSUM will never buy its i-th Bahncard later than SUM (because OSUM buys when s reaches c_{crit}), but often will buy earlier. Consider for example the GBP. Then OSUM buys a Bahncard whenever $p \ge C - \frac{1}{2}s$. On the request sequence (Jun 22, 250 DM), (Jun 26, 100 DM), (Jul 17, 50 DM), (Jul 31, 200 DM), for example, OSUM would buy a Bahncard at the first request on Jun 22 and spend 540 DM for all four tickets (which is optimal), whereas SUM would pay the regular price for the first three tickets and buy a Bahncard only at the fourth request on Jul 31, thus spending 740 DM.

Of course, OSUM's advantage over SUM shrinks if there are many cheap requests, and if all requests are infinitesimally small then OSUM converges to SUM. Nevertheless, OSUM should be used by frequent travelers who expect to buy more tickets in the near future, whereas SUM should be preferred by sporadic travelers with a low probability of traveling. The next theorem shows that OSUM is as optimal as SUM.

Theorem 5. OSUM *is* $(2-\beta)$*-competitive for* $BP(C,\beta,T)$.

Proof. (Sketch) Augmenting the proof of Theorem 4, we define a *critical phase* as an interval $I = (t_1, t_2]$, where OSUM buys a Bahncard at time t_2, t_1 is the maximum of $t_2 - T$ and the expiration time of OSUM's previously bought Bahncard, and OPT has no valid Bahncard at any time in I. Then we can charge the cost of each of OSUM's Bahncards uniquely to either an expensive phase or a critical phase. A critical phase is induced by a request (t,p) with $p \geq \frac{C-s(1-\beta)}{2(1-\beta)} =: a$, where $s = rr^{\sigma}_{\mathtt{OSUM}}(t^-)$. Hence

$$c_{\mathtt{OSUM}} = C + s + \beta p \qquad \text{and} \qquad c_{\mathtt{OPT}} = s + p\,.$$

Therefore

$$\frac{c_{\mathtt{OSUM}}}{c_{\mathtt{OPT}}} = \frac{C + s + \beta p}{s + p} \leq \frac{C + s + \beta a}{s + a} = 2 - \beta\,,$$

because the second quotient is maximal if p is minimal. □

5 Randomized Online Algorithms

We define R-SUM (R-OSUM) as a randomized variant of SUM (OSUM) which, with probability $q = \frac{1}{1+\beta}$, buys a Bahncard at time t iff SUM (OSUM) would buy one at time t. It is easy to see from the proof of the next theorem that $\frac{1}{1+\beta}$ is the optimal choice for this probability.

Theorem 6. R-SUM *and* R-OSUM *are* $\frac{2}{1+\beta}$*-competitive for* $BP(C,\beta,T)$.

Proof. (Sketch) We only show the theorem for R-SUM. We use the same notation as in the proof of Theorem 4. Let $I = I_1 \cup I_2 \cup I_3$ be an expensive phase. Then

$$c_{\mathtt{R-SUM}} \leq qC + s_2 + (q\beta + 1 - q)\cdot(s_1 + s_3)$$

and

$$c_{\mathtt{OPT}} = C + \beta(s_1 + s_2 + s_3)\,.$$

Hence, $\frac{c_{\mathtt{R-SUM}}}{c_{\mathtt{OPT}}} \leq \frac{2}{1+\beta}$, because the first quotient is maximal if s_2 is maximal, and with $s_2 = c_{crit}$ and $q = \frac{1}{1+\beta}$ it is constant $\frac{2}{1+\beta}$. □

Note that $\frac{2}{1+\beta} < 2 - \beta$ if $\beta \in (0,1)$, so R-SUM usually beats SUM. It is $\frac{4}{3}$-competitive for the GBP, but for the SRP it is identical to the deterministic SUM algorithm.

We now consider the case that $T = \infty$, i.e., a Bahncard never expires. This makes the problem more similar to the well-understood SRP. In this case, time is no longer important, and we can w.l.o.g. assume that the behaviour of an algorithm at any moment is completely determined by the sum of all previous requests. A deterministic algorithm A can thus be described by a single positive number s_A, meaning that A buys a Bahncard if the cost has reached s_A. A randomized algorithm Q can be described by a monotone increasing function $p_Q : [0, \infty] \to [0, 1]$, where $p_Q(s)$ is the probability that Q has a Bahncard after the cost has reached s. Since small requests work in favour of the adversary, we can further assume w.lo.g. that the total ticket cost is a continous function of time (and monotone increasing, of course). Then, a request sequence is also just a positive number s, namely the sum of all requests, and the expected cost of Q on s is

$$c_Q(s) = p_Q(s) \cdot C + s - (1-\beta) \cdot \int_0^s p_Q(x)dx \, .$$

We now define the randomized algorithm RAND by

$$p_{RAND}(s) = \begin{cases} \frac{e^{\frac{s}{c_{crit}}} - 1}{e-1+\beta} & \text{if} \quad s \leq c_{crit} \\ \frac{e-1}{e-1+\beta} & \quad\quad s \geq c_{crit} \end{cases} .$$

Theorem 7. RAND *is* $\frac{e}{e-1+\beta}$*-competitive for* $BP(C, \beta, \infty)$.

Proof. (Sketch) Let s be a request sequence. If $s \leq c_{crit}$ then $c_{OPT}(s) = s$ and $c_{RAND}(s) = s \cdot \frac{e}{e-1+\beta}$. If $s \geq c_{crit}$ then $c_{OPT}(s) = C + \beta s$ and

$$\begin{aligned} c_{RAND}(s) &= c_{RAND}(c_{crit}) + (p_{RAND}(c_{crit}) \cdot \beta + 1 - p_{RAND}(c_{crit})) \cdot (s - c_{crit}) \\ &= C \cdot \frac{e}{e-1+\beta} + \beta s \cdot \frac{e}{e-1+\beta} \, . \end{aligned}$$

□

Note that RAND always beats R-SUM or R-OSUM. If $\beta = 0$ then the Bahncard Problem becomes the SRP, and RAND behaves like the optimal $\frac{e}{e-1}$-competitive randomized Ski-Rental algorithm with $N = \infty$ [1, 8].

Theorem 8. *No randomized online algorithm for* $BP(C, \beta, T)$ *can be better than* $\frac{e}{e-1+\beta}$*-competitive.*

Proof. (Sketch) We first prove the theorem for BP(C, β, ∞). If the request sequence s is cheap, i.e., $s \leq c_{crit}$, then the expected cost of a randomized algorithm Q is small if $p_Q(s')$ is small for $s' \leq s$. On the other hand, if $s \geq c_{crit}$ then Q comes out better if $p_Q(c_{crit})$ is high. In R-SUM, we chose the extreme approach : We do not buy unless the cost reaches c_{crit}. However, distributing the probability of buying at $s = c_{crit}$ over the interval $[0, c_{crit}]$ reduces the cost of expensive request sequences while raising the cost of cheap ones. One can show that the optimal probability distribution is p_{RAND} which yields the same competitive ratio of $\frac{e}{e-1+\beta}$ on *all* request sequences.

Since we can easily transform the requests sequences used in the proof above into sequences within time interval $[0, T)$, the theorem follows. □

To generalize RAND to arbitrary T, we define another randomized algorithm RAND2 : For $0 \leq \gamma \leq c_{crit}$, let $\gamma - \mathtt{SUM}$ be the deterministic algorithm which buys a Bahncard at a regular request (t, p) iff $rr^{\sigma}_{\gamma-\mathtt{SUM}}(t) \geq \gamma$ (so SUM is c_{crit}-SUM). RAND2 chooses $\gamma \in [0, c_{crit}]$ randomly such that the probability of $\gamma \in [0, s]$ is $p_{\mathtt{RAND}}(s)$, for $s \in [0, c_{crit}]$. If $T = \infty$ then RAND2 is identical to RAND and hence optimal.

Conjecture 9. RAND2 *is optimally* $\frac{e}{e-1+\beta}$*-competitive for* $BP(C, \beta, T)$. □

Acknowledgements. We want to thank Kurt Mehlhorn for his comments on a preliminary version of this paper.

References

1. S. Albers and H. Koga. New on-line algorithms for the page replication problem. In *Proceedings of the 4th Scandinavian Workshop on Algorithm Theory (SWAT'94)*. Springer Lecture Notes in Computer Science 824, pages 25–36, 1994.
2. Y. Azar, Y. Bartal, E. Feuerstein, A. Fiat, S. Leonardi, and A. Rosén. On capital investment. In *Proceedings of the 23rd International Colloquium on Automata, Languages and Programming (ICALP'96)*. Springer Lecture Notes in Computer Science 1099, pages 429–441, 1996.
3. S. Ben-David, A. Borodin, R. Karp, G. Tardos, and A. Wigderson. On the power of randomization in on-line algorithms. *Algorithmica*, 11(1):2–14, 1994.
4. D.L. Black and D.D. Sleator. Competitive algorithms for replication and migration problems. Technical Report CMU-CS-89-201, Carnegie Mellon University, 1989.
5. R. Fleischer. On the Bahncard problem. Technical Report MPI-I-97-1-018, Max-Planck-Institut für Informatik, 66123 Saarbrücken, Germany, September 1997.
6. D.S. Johnson, A. Demers, J.D. Ullman, M.R. Garey, and R.L. Graham. Worst-case performance bounds for simple one-dimensional packing algorithms. *SIAM Journal on Computing*, 3(4):299–325, 1974.
7. A.R. Karlin, M.S. Manasse, L.A. McGeoch, and S. Owicki. Competitive randomized algorithms for nonuniform problems. *Algorithmica*, 11(6):542–571, 1994.
8. C. Lund, N. Reingold, J. Westbrook, and D. Yan. On-line distributed data management. In *Proceedings of the 2nd European Symposium on Algorithms (ESA'94)*. Springer Lecture Notes in Computer Science 855, pages 202–214, 1994.
9. P. Raghavan. Lecture notes on randomized algorithms. Technical Report RC 15340 1/9/90, IBM Research Division, T.J. Watson Research Center, Yorktown Heights, New York, 1990.
10. P. Raghavan and M. Snir. Memory versus randomization in on-line algorithms. In *Proceedings of the 16th International Colloquium on Automata, Languages and Programming (ICALP'89)*. Springer Lecture Notes in Computer Science 372, pages 687–703, 1989.
11. D.D. Sleator and R.E. Tarjan. Amortized efficiency of list update and paging rules. *Communications of the ACM*, 28(2):202–208, 1985.
12. R.E. Tarjan. *Data Structures and Network Algorithms*. CBMS-NSF Regional Conference Series in Applied Mathematics. SIAM, Philadelphia, Pennsylvania, 1983.

The Ultimate Strategy to Search on m Rays?*

Alejandro López-Ortiz[1] and Sven Schuierer[2]

[1] Faculty of Computer Science, University of New Brunswick, Canada, email: alopez-o@unb.ca

[2] Institut für Informatik, Universität Freiburg, Am Flughafen 17, Geb. 051, D-79110 Freiburg, FRG, e-mail: schuiere@informatik.uni-freiburg.de

Abstract. We consider the problem of searching on m current rays for a target of unknown location. If no upper bound on the distance to the target is known in advance, then the optimal competitive ratio is $1+2m^m/(m-1)^{m-1}$. We show that if an upper bound of D on the distance to the target is known in advance, then the competitive ratio of any search strategy is at least $1+2m^m/(m-1)^{m-1} - O(1/\log^2 D)$ which is also optimal—but in a stricter sense.
We also construct a search strategy that achieves this ratio. Astonishingly, our strategy works equally well for the unbounded case, that is, if the target is found at distance D from the starting point, then the competitive ratio is $1+2m^m/(m-1)^{m-1} - O(1/\log^2 D)$ and it is not necessary for our strategy to know an upper bound on D in advance.

1 Introduction

Searching for a target is an important and well studied problem in robotics. In many realistic situations the robot does not possess complete knowledge about its environment, for instance, the robot may not have a map of its surroundings, or the location of the target may be unknown [3, 6, 10, 14, 17].

Since the robot has to make decisions about the search based only on the part of its environment that it has explored before, the search of the robot can be viewed as an *on-line* problem. One way to judge the performance of an on-line search strategy is to compare the distance traveled by the robot to the length of the shortest path from its starting point s to the target t. The ratio of the distance traveled by the robot to the optimal distance from s to t over all possible locations of the target is called the *competitive ratio* of the search strategy [18].

We are interested in obtaining upper and lower bounds on the competitive ratio of searching on m concurrent rays. Here, a point robot is imagined to stand at the origin of m rays and one of the rays contains the target t whose distance to the origin is unknown. The robot can only detect t if it stands on top of it. It can be shown that an optimal strategy visits the rays in cyclic order and increases the step length each time by a factor of $m/(m-1)$ starting with a step length of 1 [1, 4]. The competitive ratio C_m achieved by this strategy is given by $1+2m^m/(m-1)^{m-1}$. If randomization is used, the optimal competitive ratio is given by the minimum of the function $1+2a^m/((a-1)\ln a)$, for $a > 1$ [4, 9, 8].

Searching on m rays has proven to be a very useful tool for searching in a number of classes of simple polygons, such as star-shaped polygons [16], generalized streets [3, 15], HV-streets [2], and θ-streets [2, 5].

* This research is supported by the DFG-Project "Diskrete Probleme", No. Ot 64/8-1.

However, the proof of optimality for the above m-way ray searching strategy relies on the unboundedness of the rays, that is, on the fact that the target can be placed arbitrarily far away from the starting point of the rays [1,4]. But, if we consider polygons and the robot is equipped with a range finder, then it is possible to obtain an upper bound D on the distance to the target. In this case it is implicitly assumed that the strategy for searching on m-rays remains optimal though no proof of this assumption has been presented yet [2,3,15].

In this paper we provide the first lower bound proof for searching on m bounded rays; more precisely, we investigate the question if the knowledge of an upper bound on the distance to the target provides an advantage to the robot.

Let C_m^D be the optimal competitive ratio to search on m rays where the distance to the target is at most D. As mentioned above it is assumed that C_m^D approaches C_m as D goes to infinity; yet, there is only a proof for the case $m = 2$ by López-Ortiz who shows that $9 - O(1/\log D)$ is a lower bound for the competitive ratio of searching on two rays [13]. In a similar vein, Icking *et al.* investigate the maximal reach of a strategy to search on the line if the competitive ratio of the strategy is given [7]. The *reach* of a strategy X is the maximum distance D such that a target placed at a distance D to the origin is still detected by a robot using X if the competitive ratio C is given. Icking *et al.* derive a recurrence equation for the optimal reach which implies that the reach is continuous and strictly monotone in D [7]. This in turn implies that C_2^D is strictly monotone in D and assumes all values in the interval $[3,9]$.

In this paper we prove that

$$1 + 2m^m/(m-1)^{m-1} - O\left(1/\log^2 D\right) \tag{1}$$

is a lower bound on C_m^D, for general m; this also improves López-Ortiz' bound for $m = 2$. Moreover, we present a strategy that achieves a competitive ratio of the same form as Equation 1, albeit with a different constant factor in the "big-Oh" term. Here, D is the distance at which the target is discovered. Astonishingly, our strategy achieves this competitive ratio without knowing an upper bound on D in advance. These two results imply that knowing an upper bound on the distance in advance does not improve the competitive ratio significantly. Note that all previously known strategies have a competitive ratio of $1 + 2m^m/(m-1)^{m-1} - O(1/D)$ if the target is detected at distance D.

The paper is organized as follows. In the next section we present some definitions concerning searching on m rays. In Section 3 we show that an optimal strategy to search on m bounded rays is periodic and monotone. In Section 4 we first consider searching on two rays to introduce our approach to analysing the competitive ratio of an optimal strategy. In Section 5 we generalize our ideas of the case of searching on two rays to m rays. Finally, in Section 6 we present a strategy whose competitive ratio converges asymptotically as fast to $1 + 2m^m/(m-1)^{m-1}$ as the lower bound which we have shown before.

Due to the limited space all the proofs are omitted.

2 Definitions

Let X be a strategy to search on m rays. We model X as a sequence of positive real numbers, that is, $X = (x_0, x_1, x_2, \dots)$ with $x_k > 0$, for all $0 \le k < \infty$. We illustrate this for the case of a point robot searching on the real line, that is, $m = 2$.

In the beginning the position of the robot is a point s on the real line; it has to find a target t that is located somewhere to its left or right. It can only detect t if it stands on top of it. The robot starts at the origin s and travels for a distance of x_0 to one side, say to the left. The robot returns to s, travels a distance of x_1 to the right, returns, and so on. Obviously, the values x_i which denote the distance that the robot travels to the left or to the right of s suffice to characterize a search strategy completely.

The Competitive Ratio Assume that the target is discovered in Step $k+2$, say to the left of the origin. Clearly, the ray to the left of the origin was visited the last time before Step $k+2$ in Step k. Hence, the distance d to the target is greater than x_k. The distance traveled by the robot to discover t is $d+2\sum_{i=0}^{k+1} x_i$. Since obviously $d > x_k$ and the target can be placed arbitrarily close to x_k by an adversary, the highest lower bound on the competitive ratio of Step k is given by the expression

$$\sup_{d>x_k} (d+2\sum_{i=0}^{k+1} x_i)/d = \sup_{d>x_k} 1+2\sum_{i=0}^{k+1} x_i/d = 1+2\sum_{i=0}^{k+1} x_i/x_k.$$

Note that the above expression depends only on elements of X. For general m the competitive ratio is given by $1+2\sum_{i=0}^{k+m-1} x_i/x_k$ if the rays are visited cyclically.

The first step is a special case that we have not considered yet. If no information about the target is available, then one false move in the beginning may lead to an arbitrarily large competitive ratio. In order to avoid this problem we assume that a lower bound of one for the distance to the target t is known in advance.

3 Searching on m Rays

We are interested in the case that an upper bound D on the maximum distance of the target to the origin is known. We now model a strategy X as a finite sequence of positive numbers, that is, $X = (x_0, \ldots, x_n)$, for some $n \geq 0$.

3.1 Periodicity

In order to prove a lower bound on the competitive ratio, we first prove some properties of optimal strategies, that is, strategies with minimal competitive ratio. If we denote the ray that the robot visits in Step k by r_k, then a strategy is *periodic* if $r_{k+m} = r_k$, for all $0 \leq k \leq n-m$. A strategy is *monotone* if $x_{i+1} \geq x_i$, for all $0 \leq i \leq n-1$. We can show that there is an optimal strategy that is periodic and monotone.

Lemma 1. *There is an optimal strategy that is monotone and periodic up to the last step.*

By Lemma 1 it suffices to consider monotone and periodic strategies in the following. Note that if X is monotone, then the last m steps of X all have length D, that is, there is an optimal strategy with $x_{n-m+1} = \cdots x_n = D$.

3.2 A Recurrence Equation

In the following we assume that X is an optimal periodic strategy. The competitive ratio of X in Step k is given by $1+2F_k(X)$ where

$$F_k(X) = \sum_{i=0}^{k+m-1} x_i/x_k,$$

for $k = 0, \ldots, n-m$. Note that the robot visits ray r_1 in Step $n-m+1$; hence, $x_{n-m+1} = D$ and $F_{n-m+1}(X)$ is not relevant for the computation of the competitive ratio. Let $c_X = \max_{0 \leq i \leq n-m} F_i(X)$. It can be shown that if X is an optimal strategy, then the values of $F_k(X)$ are all the same, for $0 \leq k \leq n-m$ [12].

Lemma 2. *If X is an optimal strategy, then $F_k(X) = c_X$, for all $0 \leq k \leq n-m$.*

Note that if X is an optimal strategy, then $1+2c_X = C_m^D$. Lemma 2 implies that if X is an optimal strategy, then

$$x_{k+m-1} - c_m^D x_k + c_m^D x_{k-1} = 0, \tag{2}$$

for all $1 \leq k \leq n-m$, where $c_m^D = (C_m^D - 1)/2$. Equation 2 completely defines the sequence $X = (x_0, x_1, \ldots, x_n)$ if we are given the values $x_0, \ldots, x_{m-1}$ (which we do not know); however, we know the values of $x_{n-m+1}, \ldots, x_n$ since $x_i = D$ in the last m steps. Unfortunately, the value of x_n is irrelevant since x_n does not appear in Equation system (2). Instead, the m-th boundary value of Equation 2 is given by $x_{-1} = 1$ as this is the minimal distance to the target.

In order to obtain m consecutive boundary values, consider x_{n-m}. It is easy to see that $x_{n-m} \geq D/2e$. If we now require that $x_i = D/2e$, for $i = n-m, \ldots, n-1$, then it can be shown that the competitive ratio of an optimal strategy that satisfies these equations is less than the competitive ratio of an optimal strategy for the original problem. For convenience, we neglect the division by $2e$ in the following as this only influences the constant of the "big-Oh" term.

In order to make use of this information we consider the sequence Y of the values of X in reverse order, that is, $y_i = x_{n-i-1}$, for $i = 0, \ldots, n$. For simplicity we write c instead of c_m^D in the following. The values y_i satisfy the following recurrence

$$y_{k+m} - y_{k+m-1} + y_k/c = 0, \tag{3}$$

for $0 \leq k \leq n-m$.

The initial steps again have to be considered separately. The worst case competitive ratio the first time the mth ray is visited is $1 + 2\sum_{i=0}^{m-2} x_i$ which implies that $1 + 2\sum_{i=0}^{m-2} x_i \leq 1 + 2c$ and, hence, the value of $\sum_{i=n-m+1}^{n-1} y_i$ is also at most c. In addition, we note that all the values $y_0, \ldots, y_n$ have to be positive. We assume in the following that Y is given by Equation 3 which defines an infinite sequence some of whose elements may be negative.

In order to prove a lower bound on the competitive ratio $1+2c$ we show the following theorem.

Theorem 1. *If $c < m^m/(m-1)^{m-1} - O(1/\log^2 D)$, then there is no sequence Y and no $n \geq 0$ such that Y satisfies Equation 3, $\sum_{i=n-m+1}^{n-1} y_i \leq c$, $y_0 = y_1 = \cdots = y_{m-1} = D$, and $y_0, \ldots, y_n \geq 0$.*

By the construction of Y we also obtain that there is no strategy X with a competitive ratio of $1+2c$ to search on m rays in the interval $[1, D]$.

Lemma 3. *If there is no sequence Y and no $n \geq 0$ such that Y satisfies Equation 3, $\sum_{i=n-m+1}^{n-1} y_i \leq c$ and $y_0 = y_1 = \cdots = y_{m-1} = D$, and $y_m, \ldots, y_n \geq 0$, then there is no strategy X with a competitive ratio of $1+2c$ that searches on m rays for a target of distance at most D to the origin.*

3.3 The Characteristic Equation

We only consider the sequence Y in the following. Equation 3 has the characteristic equation

$$\lambda^m - \lambda^{m-1} + 1/c = 0 \quad \text{or} \quad c = 1/(\lambda^{m-1}(1-\lambda)). \tag{4}$$

We first note that since $\lambda^{m-1}(1-\lambda) < 0$, for $\lambda > 1$, there is no positive real root larger than one. On the other hand, if we set $\mu = 1/\lambda$, then $c = \mu^m/(\mu-1)$ and if there is a positive real root λ of Equation 4 with $\lambda < 1$, then $c \geq \inf_{\mu>1} \mu^m/(\mu-1) = m^m/(m-1)^{m-1}$ and we are done. This implies that we can assume in the following that there is no positive real root of Equation 4.

So we investigate the complex and negative roots of Equation 4 in more detail.

4 Solving the Recurrence Equation for $m = 2$

In order to illustrate our approach we present the case $m = 2$ in greater detail. In the following we assume that $3 < c < m^m/(m-1)^{m-1} = 4$. It is easy to see that if $c \leq 3$, then D is bounded by a constant.

4.1 An Explicit Solution

For $m = 2$ Equation 4 reduces to $\lambda^2 - \lambda + 1/c = 0$ with the solutions

$$\lambda = 1/2\left(1 + i\sqrt{(4-c)/c}\right) \quad \text{and} \quad \overline{\lambda} = 1/2\left(1 - i\sqrt{(4-c)/c}\right).$$

Here, $\overline{\lambda}$ denotes the conjugate of λ. Hence, the solution to Equation 3 in the case $m = 2$ is given by

$$y_k = a\lambda^k + \overline{a}\overline{\lambda}^k = 2Re(a\lambda^k)$$

where Re denotes the real part of a complex number. a and $\overline{a}$ are given by

$$a = D/2\left(1 - i\sqrt{c/(4-c)}\right) \quad \text{and} \quad \overline{a} = D/2\left(1 + i\sqrt{c/(4-c)}\right).$$

4.2 Polar Coordinates

If we consider the polar-coordinates of λ and $\overline{\lambda}$, that is, $\lambda = \rho e^{i\varphi}$ and $\overline{\lambda} = \rho e^{i(-\varphi)}$, then $\rho = \sqrt{1/c}$ and $\varphi = \arctan(\sqrt{(4-c)/c})$. Similarly, for $a = \sigma e^{i\theta}$ and $\overline{a} = \sigma e^{i(-\theta)}$ we obtain $\sigma = D/\sqrt{4-c}$ and $\theta = -\arctan(\sqrt{c/(4-c)})$. Hence,

$$y_k = a\lambda^k + \overline{a}\overline{\lambda}^k = \frac{2D}{\sqrt{c^k(4-c)}} \cos\left(k \arctan\left(\sqrt{\frac{4-c}{c}}\right) - \arctan\left(\sqrt{\frac{c}{4-c}}\right)\right).$$

If we visualize the above equation in the complex plane, then y_k is the projection of the vector of $2a\lambda^k$ onto the x-axis. If we multiply two complex numbers, then the radii are multiplied and the angles are added. Hence, the sequence $2a\lambda^k$ turns by an angle of φ towards the second quadrant with each iteration. Once $2a\lambda^k$ is in the second quadrant, $2Re(a\lambda^k)$ is negative. Since all elements y_i have to be positive, the first index k with $y_{k+1} < 0$ is the maximum length of Y (and X). This idea was used before to prove that there is no strategy to search on the (unbounded) line with a competitive ratio of less than nine since a strategy to search on the real line cannot be of finite length [1, 5, 7, 11]. y_k becomes negative as soon as

$$k \arctan\sqrt{\frac{4-c}{c}} - \arctan\sqrt{\frac{c}{4-c}} \in (\pi/2, 3\pi/2).$$

We show that D can be chosen large enough such that either $y_{n+1} < 0$ and $y_n > c$ or $y_{n-1}/y_n > c$. Our previous considerations imply that in both cases there is no strategy to search on the real line for a target at a distance at most D with a competitive ratio of $1+2c$.

Of course, we are interested in the smallest D for which the above inequalities holds. Let n_0 be the first index such that $y_{n_0} < 0$, that is,

$$\cos\left(n_0 \arctan\sqrt{\frac{4-c}{c}} - \arctan\sqrt{\frac{c}{4-c}}\right) < 0 \quad \text{or} \quad n_0 = \left\lceil \frac{\arctan\left(\sqrt{\frac{c}{4-c}}\right) + \frac{\pi}{2}}{\arctan\left(\sqrt{\frac{4-c}{c}}\right)} \right\rceil .$$

Some rough estimates show that $n_0 \leq 9/\sqrt{4-c}$ and $y_{n_0-2} \geq D/\sqrt{c^{9/\sqrt{4-c}}}$. With these two inequalities we can show the following result.

Lemma 4. *If* $3 < c < 4 - 81/\log^2(D/16)$, *then* $D/\sqrt{c^{9/\sqrt{4-c}}} > c^2$.

Let $3 < c < 4 - 81/\log^2(D/16)$. Lemma 4 implies that $y_{n_0-2} > c^2$ and $y_{n_0} < 0$. Hence, if $y_{n_0-1} \leq c$, then $(y_{n_0-1} + y_{n_0-2})/y_{n_0-1} > c$; otherwise $y_{n_0-1} > c$. This proves Theorem 1 for the case $m = 2$.

5 Solving the Recurrence Equation for the General Case

In the following we sketch our approach for general m. As for the case $m = 2$ we want to show that if there are only complex or negative roots of Equation 4, then the polar angle of the roots turns towards to the second quadrant. However, the details are much more involved than in the case $m = 2$ since we have many roots of Equation 4 and the roots cannot be computed explicitly. One possibility to get around this problem is to use estimates on the angles and radii of the roots.

If λ is the root with the largest radius among all roots of Equation 4, then after a sufficiently large number of steps the contribution of λ dominates the contribution of all other solutions and only the angle of λ^k determines whether the solution is positive or negative. Since the number of steps increases logarithmically with D, a large enough D yields then a negative sequence element.

Let $\lambda_0, \ldots, \lambda_{m-1}$ be the roots of Equation 4. The solution of the recurrence is given by

$$y_k = a_0\lambda_0^k + a_0\lambda_0^k + \cdots + a_{m-1}\lambda_{m-1}^k.$$

We first investigate the structure of the roots λ_i, $0 \leq i \leq m-1$.

Let λ be a complex root of Equation 4. We consider the polar coordinates of λ, that is, we set $\lambda = \rho e^{i\varphi}$. We can show the following relationship between ρ and φ.

Lemma 5. *If* $\lambda = \rho e^{i\varphi}$ *is a complex root of Equation 4, then* $\rho = \sin(m-1)\varphi/\sin m\varphi$ *and* $\lambda^{m-1}(\lambda - 1) = \rho^{m-1}(\rho\cos(m\varphi) - \cos(m-1)\varphi)$.

The Polar Angle of a Root We first concentrate on the polar angle of a root λ of Equation 4.

Lemma 6. *If* $\lambda = \rho e^{i\varphi}$ *is a complex root of Equation 4, then* $\varphi \in [2k\pi/(m-1), (2k+1)\pi/m]$, *for some* $0 \leq k \leq \lfloor m/2 \rfloor$.

We now can show that there is exactly one root λ_k for each interval $[2k\pi/(m-1),(2k+1)\pi/m]$ with $0 \leq k \leq \lfloor m/2 \rfloor$.

Lemma 7. *For each interval* $[2k\pi/(m-1),(2k+1)\pi/m]$ *with* $0 \leq k \leq \lfloor m/2 \rfloor$, *there is exactly one root* $\lambda_k = \rho_k e^{i\varphi_k}$ *of Equation 4 with* $\varphi_k \in [2k\pi/(m-1),(2k+1)\pi/m]$.

The above roots account for $\lfloor m/2 \rfloor$ roots of Equation 4. If m is odd, then there is one root $\lambda_{\lfloor m/2 \rfloor}$ with $\varphi_{\lfloor m/2 \rfloor} = 2\lfloor m/2 \rfloor \pi/(m-1) = (2\lfloor m/2 \rfloor + 1)\pi/m = \pi$, that is, $\lambda_{\lfloor m/2 \rfloor}$ is a negative real root. The remaining $\lfloor m/2 \rfloor$ roots are given by the conjugates $\overline{\lambda}_k = \rho_k e^{-i\varphi_k}$ of λ_k as in the case $m = 2$.

If $c < m^m/(m-1)^{m-1}$, then φ_0 is bounded from below as follows.

Lemma 8.

$$\varphi_0 \geq \min\left\{\frac{1}{m^{3/2}}\sqrt{\frac{m^m}{(m-1)^{m-1}} - c}, \frac{1}{\sqrt{3}m}\right\}.$$

The Radius of a Root We now consider the radius of a root of Equation 4. We can show that λ_0 is the root with the largest radius. More precisely, we have the following lemma.

Lemma 9. $\rho_0 \geq 1/3$ *and* $\rho_0/\rho_k \geq 1 + 1/(4m^3)$, *for all* $1 \leq k \leq \lceil m/2 \rceil$.

The Coefficients We finally give upper and lower bounds on the radii of the coefficients.

Lemma 10. $|a_i/a_0| \leq 4^{2m}m^m$ *and* $|a_0| > D/(2em)^{m-1}$.

Putting it all Together We now put the estimates we obtained for the radii and the angles of the roots of Equation 4 as well as the coefficients into use.

Lemma 11. *For all* $k \geq 1$,

$$2|a_0|\rho_0^k\left(\cos(\alpha_k) - \frac{4^{2m}m^{m+1}}{(1+1/(4m^3))^k}\right) \leq y_k \leq 2|a_0|\rho_0^k\left(\cos(\alpha_k) + \frac{4^{2m}m^{m+1}}{(1+1/(4m^3))^k}\right).$$

where $\alpha_k = \theta_0 + k\varphi_0$.

We claim that if

$$c < m^m/(m-1)^{m-1} - 22^2 m^8 \log^2 m/\log^2 D, \tag{5}$$

then there is a Step k such that $y_k > c^2$ and $y_{k+2} < 0$. As in the case $m = 2$ this proves Theorem 1.

In the following let $\varepsilon = \sqrt{m^m/(m-1)^{m-1} - c}$. We assume that $\varepsilon < 1$. In the case $\varepsilon \geq 1$ it is easy to see that D can be only a constant. Let k_0 be the first index greater than $4m^3(3m\log m - \log\varepsilon) + 1$ such that

$$\cos(\theta_0 + k_0\varphi_0) > 0 \text{ and } \cos(\theta_0 + (k_0+1)\varphi_0) \leq 0.$$

We can show the following bounds on y_{k_0-1} and y_{k_0+2}.

Lemma 12. $y_{k_0-1} \geq 2|a_0|\rho_0^{k_0-1}\frac{\varphi_0}{4}$ *and* $y_{k_0+2} \leq -2|a_0|\rho_0^{k_0+2}\frac{\varphi_0}{4}$.

We now bound the value of k_0. Since the distance between two consecutive transitions from positive to negative values of cosine is at most 2π and $k_0 \geq 4m^3(3m\log m - \log\varepsilon) + 1$, we have that $k_0 - 4m^3(3m\log m - \log\varepsilon) - 1 \leq 2\pi/\varphi_0$ and by Lemma 8

$$k_0 \leq 4m^3(3m\log m - \log\varepsilon) + 1 + \frac{2\pi}{\varphi_0} \leq 4m^3(3m\log m - \log\varepsilon) + 1 + \frac{2\pi m^{3/2}}{\varepsilon}. \quad (6)$$

With the above preparations our main lemma can be shown.

Lemma 13. *If c satisfies Inequality 5, then $y_{k_0-1} > c^2$ and $y_{k_0+2} < 0$.*

Since $y_{k_0+2} < 0$, the last step of the strategy is Step $k_0 + 1$. If $m \geq 4$, then the sum $\sum_{i=k_0+1-m+2}^{k_0+1} y_i$ includes y_{k_0-1} and, hence, is larger than c which proves Theorem 1. If $m = 3$ and $\sum_{i=k_0+1-3+2}^{k_0+1} y_i = y_{k_0} + y_{k_0+1} \leq c$, then $y_{k_0-1}/y_{k_0} > c$ and as in Section 4 we see that this also contradicts the existence of a strategy with a competitive ratio of $1+2c$. If we recall that we have neglected the division of D by $2e$, then we obtain the following theorem.

Theorem 2. *There is no search strategy for a target on m rays which is contained in the interval $[1,D]$ with a competitive ratio of less than*

$$1+2\left(\frac{m^m}{(m-1)^{m-1}} - \frac{22^2 m^8 \log^2 m}{\log^2(D/2e)}\right).$$

6 An Optimal Strategy

After having proven a lower bound for searching on m rays with an upper bound on the target distance, one of the questions that remains is whether there actually is an optimal strategy that achieves a competitive ratio of $1+2m^m/(m-1)^{m-1} - O(1/\log^2 D)$ and what it looks like. In this section we present a strategy to search on m rays that achieves the optimal competitive ratio even if the maximum distance D of the target to the starting point is unknown, that is, being told an upper bound on the distance to the target is not a big advantage—even if we consider the convergence rate of the competitive ratio to $1+2m^m/(m-1)^{m-1}$ as D increases.

The strategy $X = (x_1, x_2, \ldots)$[1] that achieves a competitive ratio of $1+2m^m/(m-1)^{m-1} - O(1/\log^2 D)$ is given by $x_i = \sqrt{1+i/m}\,(m/(m-1))^i$. The competitive ratio C_X of Strategy X in Step $k+m$ is bounded by $1+2c_k$ where c_k is given by

$$\frac{\sum_{j=1}^{k+m-1}\sqrt{1+\frac{j}{m}}\left(\frac{m}{m-1}\right)^j}{\sqrt{1+\frac{k}{m}}\left(\frac{m}{m-1}\right)^k} = \sum_{j=0}^{m-1}\sqrt{1+\frac{j}{k+m}}\left(\frac{m}{m-1}\right)^j + \sum_{j=1}^{k-1}\sqrt{\frac{j+m}{k+m}}\left(\frac{m-1}{m}\right)^{k-j},$$

where we assume $k \geq 1$. If we use the Taylor-expansion of $\sqrt{1+x}$, then the first sum is bounded by

$$\sum_{j=0}^{m-1}\sqrt{1+\frac{j}{k+m}}\left(\frac{m}{m-1}\right)^j \leq \frac{m^m}{(m-1)^{m-1}} - (m-1) + \frac{1}{2}\frac{(m-1)m}{k+m}.$$

[1] For convenience we start with x_1 instead of x_0.

Now we bound the second sum this time using the Taylor expansion of $\sqrt{1-x}$

$$\sum_{j=1}^{k-1}\sqrt{\frac{j+m}{k+m}}\left(\frac{m-1}{m}\right)^{k-j}$$

$$\leq m-1-m\left(\frac{m-1}{m}\right)^k-\frac{1}{2}\left(\frac{m(m-1)-(k-m-1)m\left(\frac{m-1}{m}\right)^k}{k+m}\right)$$

$$-\frac{1}{8}\left(\frac{m(m-1)(2m-1)}{(k+m)^2}-m\left(\frac{m-1}{m}\right)^k\frac{k^2+2k(m-2)+2m^2-3m+1}{(k+m)^2}\right).$$

Hence,

$$\frac{\sum_{j=1}^{k+m-1}\sqrt{1+\frac{j}{m}\left(\frac{m}{m-1}\right)^j}}{\sqrt{1+\frac{k}{m}\left(\frac{m}{m-1}\right)^k}}\leq\frac{m^m}{(m-1)^{m-1}}-\frac{1}{8}\frac{m(m-1)(2m-1)}{(k+m)^2}.$$

Finally, we relate the number of steps $k+m$ to the distance D to the target. If the target is detected in Step $k+m$, then the distance D to s is in the interval $[\sqrt{1+\frac{k}{m}(m/(m-1))^k},\sqrt{1+\frac{k+m}{m}(m/(m-1))^{k+m}}]$ and $k\leq\frac{\log D}{\log(1+1/(m-1))}\leq(m-1)\log D$. Hence,

$$\frac{\sum_{j=1}^{k+m-1}\sqrt{1+\frac{j}{m}\left(\frac{m}{m-1}\right)^j}}{\sqrt{1+\frac{k}{m}\left(\frac{m}{m-1}\right)^k}}\leq\frac{m^m}{(m-1)^{m-1}}-\frac{2m-1}{8\left(\log D+\frac{m}{m-1}\right)^2}\leq\frac{m^m}{(m-1)^{m-1}}-\frac{2m-1}{8\log^2(3D)}.$$

We have shown the following theorem.

Theorem 3. *There is a strategy X that achieves a competitive ratio of*

$$1+2\frac{m^m}{(m-1)^{m-1}}-\frac{2m-1}{4\log^2(3D)}$$

if the target is placed at distance $D>1$ to s.

By Theorem 2 the strategy we have presented above is optimal.

7 Conclusions

We present a lower bound for the problem of searching on m concurrent rays if an upper bound D on the maximal distance to the target is given. We show that in this case the competitive ratio of a search strategy is at least $1+2m^m/(m-1)^{m-1}-O(1/\log^2 D)$. Our approach is based on deriving a recursive equation for the step length in each iteration of an optimal strategy. The recursive equation gives rise to a characteristic equation whose roots determine the properties of a strategy. By computing upper and lower bounds on the radii and polar angles of the roots we can show that the competitive ratio has to be sufficiently large if the target is far away.

We also present a strategy which achieves a competitive ratio of $1+2m^m/(m-1)^{m-1}-O(1/\log^2 D)$ if the target is detected at distance D. The strategy does not need to know an upper bound on D in advance. Hence, the knowledge of an upper bound on the distance to the target only provides a marginal advantage to the robot—even the convergence rate is not improved.

An interesting open problem is to prove similar results for randomized strategies. One of the problems with randomized strategies is that there is no published proof that there is an optimal periodic strategy. It seems that this is a necessary step before the bounded distance problem can be attacked.

References

1. R. Baeza-Yates, J. Culberson, and G. Rawlins. Searching in the plane. *Information and Computation*, 106:234–252, 1993.
2. A. Datta, Ch. Hipke, and S. Schuierer. Competitive searching in polygons—beyond generalized streets. In *Proc. Sixth Annual International Symposium on Algorithms and Computation*, pages 32–41. LNCS 1004, 1995.
3. A. Datta and Ch. Icking. Competitive searching in a generalized street. In *Proc. 10th Annu. ACM Sympos. Comput. Geom.*, pages 175–182, 1994.
4. S. Gal. *Search Games*. Academic Press, 1980.
5. Ch. Hipke. Online-Algorithmen zur kompetitiven Suche in einfachen Polygonen. Master's thesis, Universität Freiburg, 1994.
6. Ch. Icking and R. Klein. Searching for the kernel of a polygon: A competitive strategy. In *Proc. 11th Annu. ACM Sympos. Comput. Geom.*, pages 258–266, 1995.
7. Ch. Icking, R. Klein, and E. Langetepe. How to find a point on a line within a fixed distance. Informatik-Bericht 220, Fernuni Hagen, November 1997.
8. M.-Y. Kao, Y. Ma, M. Sipser, and Y. Yin. Optimal constructions of hybrid algorithms. In *Proc. 5th ACM-SIAM Sympos. Discrete Algorithms*, pages 372–381, 1994.
9. M.-Y. Kao, J. H. Reif, and S. R. Tate. Searching in an unknown environment: An optimal randomized algorithm for the cow-path problem. In *Proc. 4th ACM-SIAM Sympos. Discrete Algorithms*, pages 441–447, 1993.
10. R. Klein. Walking an unknown street with bounded detour. *Comput. Geom. Theory Appl.*, 1:325–351, 1992.
11. R. Klein. *Algorithmische Geometrie*. Addison-Wesley, 1997.
12. E. Koutsoupias, Ch. Papadimitriou, and M. Yannakakis. Searching a fixed graph. In *Proc. 23rd Intern. Colloq. on Automata, Languages and Programming*, pages 280–289. LNCS 1099, 1996.
13. A. López-Ortiz. *On-line Searching on Bounded and Unbounded Domains*. PhD thesis, Department of Computer Science, University of Waterloo, 1996.
14. A. López-Ortiz and S. Schuierer. Going home through an unknown street. In S. G. Akl, F. Dehne, and J.-R. Sack, editors, *Proc. 4th Workshop on Algorithms and Data Structures*, pages 135–146. LNCS 955, 1995.
15. A. López-Ortiz and S. Schuierer. Generalized streets revisited. In M. Serna, J. Diaz, editor, *Proc. 4th European Symposium on Algorithms*, pages 546–558. LNCS 1136, 1996.
16. A. López-Ortiz and S. Schuierer. Position-independent near optimal searching and on-line recognition in star polygons. In *Proc. 4th Workshop on Algorithms and Data Structures*, pages 284–296. LNCS 1272, 1997.
17. C. H. Papadimitriou and M. Yannakakis. Shortest paths without a map. In *Proc. 16th Internat. Colloq. Automata Lang. Program.*, pages 610–620. LNCS 372, 1989.
18. D. D. Sleator and R. E. Tarjan. Amortized efficiency of list update and paging rules. *Communications of the ACM*, 28:202–208, 1985.

Better Approximation of Diagonal-Flip Transformation and Rotation Transformation (Extended Abstract)

Ming Li* and Louxin Zhang**

[1] Department of Computer Science
University of Waterloo, Canada N2L 3G1
[2] BioInformatics Center, NUS, and
Kent Ridge Digital Labs, Singapore 119613

Abstract. Approximation algorithms are developed for the diagonal-flip transformation of convex polygon triangulations and equivalently rotation transformation of binary trees. For two arbitrary triangulations in which each vertex is an end of at most d diagonals, Algorithm A has the approximation ratio $2 - \frac{2}{4(d-1)(d+6)+1}$. For triangulations containing no internal triangles, Algorithm B has the approximation ratio 1.97. Two self-interesting lower bounds on the diagonal-flip distance are also established in the analyses of these two algorithms.

1 Introduction

A *rotation* in a binary tree is a local restructuring of the tree that changes the position of a node and one of its children while the symmetric order in the tree is preserved. Such an operation has found its applications in many aspects. In data structures, rotations are the primitive used by most schemes that maintain "balance" in binary trees [13, 18]. In graphics, morphing polygons is abstracted as rotations on weighted binary trees [6, 8]. The rotation operation is also of interest from a purely mathematical point of view [11]. Further, a similar but slightly powerful operation named *nearest neighbor interchange* is used extensively for defining the dissimilarity between phylogenies and for heuristical search of optimal phylogenies in biology [2, 17].

The rotation operation on binary trees is equivalent to the diagonal-flip operation in triangulations of a convex polygon. But, the later is more intuitive(for example, see [15]). A *diagonal-flip* is an operation that converts one triangulation of a polygon into another by removing a diagonal in the triangulation and adding the diagonal that subdivides the resulting quadrilateral in the opposite way. The diagonal-flip operation was early studied by Wagner [19] in the context of arbitrary triangulated planar graphs and by Dewdney [5] in the case of

* Supported in part by the NSERC Operating Grant OGP0046506, ITRC and a CGAT grant. Address: E-mail: mli@math.uwaterloo.ca

** Part of the work was done at Waterloo and supported by a CGAT grant. E-mail: lxzhang@bic.nus.edu.sg.

graphs of genus one. They showed that any such graph can be transformed to any other by diagonal-flips. However, they did not try to accurately estimate how many flips are necessary. In [16], Sleator, Tarjan and Thurston proved that $\Omega(n \log n)$ diagonal-flips are necessary and enough for transforming a numbered triangulated planar graph into another. In another paper [15], by using hyperbolic geometry, Sleator, Tarjan and Thurston showed beautifully that $2n - 10$ diagonal flips are enough and necessary for transforming one triangulation of the n-gon into another when n is large, which improved an earlier work of Culik and Wood [3]. Since then the diagonal-flip transformation has been studied in several aspects [4, 7, 9].

Like [15], this paper works on diagonal-flips in triangulations rather than rotations on binary trees. It is open whether a shortest diagonal-flip transformation between two triangulations of a convex polygon is computable in polynomial time [14]. Our interest is to develop an approximation algorithm with ratio better than 2 for the diagonal-flip transformation, which is a hard problem raised in [14] and explicitly in [2] recently. Although there is a trivial approximation with ratio 2, any better approximation turns out to be very difficult. In this paper, we present an approximation algorithm that has better approximation ratio for triangulations of a convex polygon in which each vertex is an end of constantly many diagonals. We also study the diagonal-flip transformation for a special class of triangulations. In a triangulation of a convex polygon, a triangle is said to be *internal* if it contains three diagonals. The class of triangulation without internal triangles contains most of interesting triangulations studied in literature(for example, see [11, 7]). In fact, triangulations in this class correspond one-to-one to binary trees without degree-3 nodes. For such a class of triangulations, we presents a polynomial algorithm with approximation ratio 1.97. The ratio can further be reduced by a sophisticated argument similar to our approach. However, we are not intent on giving the best possible ratio in this short abstract. The complete proofs of these results can be found in the full version of this paper.

The rest of this extended abstract is divided into six sections. Section 2 introduces briefly the rotation on binary trees and the diagonal-flip in triangulations and shows the equivalence of these two operations. Section 3 gives two transformation primitives. Using Proposition 3 in this section, we are also able to answer negatively a problem posed by Knuth [11], who suspected that two specific triangulations without internal triangles has a large diagonal-flip distance. Section 4 presents a polynomial algorithm (Algorithm A) that has the approximation ratio $2 - \frac{2}{4(d-1)(d+6)+1}$ for triangulations in which each vertex is an end of at most d diagonals. Section 5 presents a polynomial algorithm (Algorithm B) with approximation ratio 1.97 for triangulations without any internal triangles.

2 Definitions

2.1 Binary tree rotations

A binary tree is a collection of nodes and three relations among these nodes: parent, left child and right child. A special node is called the root. Every other node has a parent and may have a left and/or right child. All nodes without any child are *leaves.* A binary tree has *size* n if it contains n nodes. (See [10] for a more complete description of binary trees and tree terminology.)

A *rotation* is an operation that changes one binary tree into another with the same size. Figure 1 shows the general rotation rule. In a tree of size n, there are $n-1$ possible rotations, each corresponding a non-root node. A rotation maintains the symmetric order of the nodes. Furthermore, a rotation is an invertible operation. We define *rotation distance* between two trees as the minimum number of rotations required to convert one tree into the other. The rotation distance between two binary trees of size n is at most $2n-6$ [15]. In [12], Pallo proposed a heuristic search algorithm for computing the rotation distance.

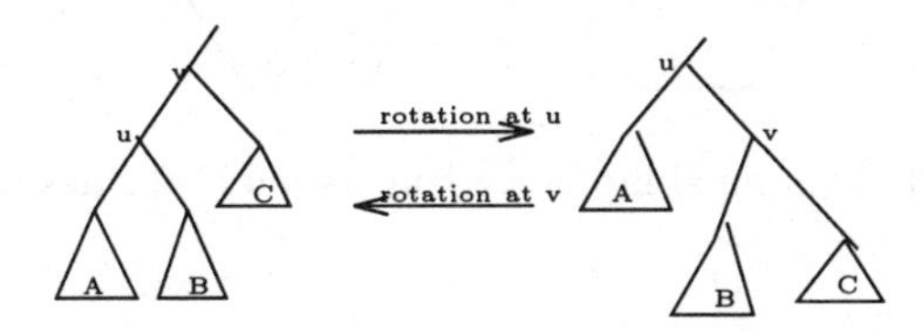

Fig. 1. The definition of rotation.

2.2 Diagonal-flips in triangulations

Binary tree rotation can be formulated with respect to different systems of combinatorial objects and their transformations. The diagonal-flip operation in triangulations is perhaps more intuitive and so supplies more insight. Consider the standard convex $(n+2)$-gon. We choose an edge of the polygon as a distinguished edge, called "root edge", and label its ends as 0 and $n+1$. We also label the other n vertices from 1 to n counterclockwise. Any triangulation of the $(n+2)$-gon has n triangles and $n-1$ diagonals. From a triangulation of the $(n+2)$-gon, we derive a binary tree of size n by assigning a node for each triangle and connecting two nodes if the corresponding triangles sharing a common diagonal. The root of the tree corresponds to the triangle containing the root edge. It is not difficult to see that the ith node of the binary tree in symmetric order corresponds to the triangle with vertices i, j and k such that $j < i < k$. In this way, we obtain a 1-1 correspondence between n-node binary trees and triangulations of the $(n+2)$-gon as illustrated in Figure 2.

A *diagonal-flip* is an operation that transforms one triangulation of a polygon into another as showed in Figure 3. The *diagonal-flip distance* between two triangulations π_1 and π_2 of a polygon is the minimum number of diagonal-flips needed to convert one triangulation into the other, which is denoted by $fd(\pi_1, \pi_2)$. Note that $fd(\pi_1, \pi_2) \leq 2n-10$ for any two triangulations π_1 and π_2 of the n-gon [15].

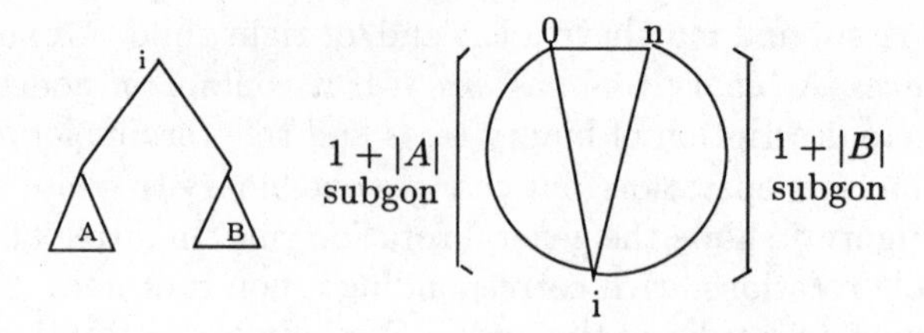

Fig. 2. A binary tree and its corresponding triangulation.

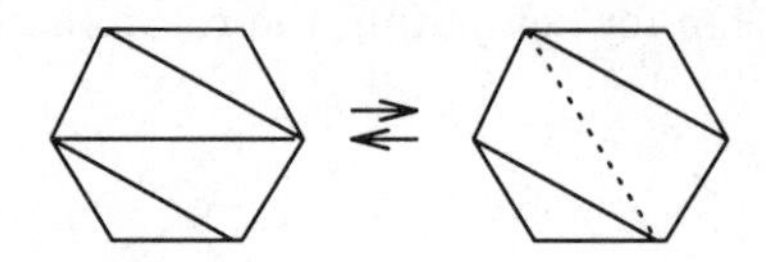

Fig. 3. A diagonal flip in a triangulation of the hexagon.

Obviously, diagonal-flips in a triangulation correspond one-to-one to rotations in the corresponding binary tree. Other interesting relationship between a triangulation of a convex polygon and its corresponding binary tree can be found in a nice survey article [1].

3 Diagonal-flip transformation primitives

3.1 Difference Graphs

Given two triangulations π_1 and π_2 of the n-gon, we define their difference graph $G = G(\pi_1, \pi_2)$ as the union of these two triangulations. Formally, the graph G has vertex set $V = \{1, 2, \cdots, n\}$, and edge set $E = \{1n, i(i+1) | i = 1, 2, \cdots, n-1\} \cup E_{\pi_1} \cup E_{\pi_2}$, where E_{π_i} denotes the set of $n-3$ internal diagonals of π_i.

If an edge shared by both triangulations π_1 and π_2, we call it a *face* edge. All the boundary edges $1n$ and $i(i+1)$, $1 \leq i \leq n-1$, are face edges. Since two triangulations may have common diagonals, there are other face edges in general. We define a subgraph inside a simple cycle consisting of face edges a *cell*. It is not difficult to see that the difference graph G can be decomposed into cells, which have disjoint diagonal edges. A difference graph $G(\pi_1, \pi_2)$ is *simple* if it has only one cell, i.e. two triangulations π_1 and π_2 don't have any common diagonals. In the rest of this section, we will focus on simple difference graphs.

In a difference graph, an edge is said to be a *diagonal edge* if it corresponds with a diagonal of one triangulation; it is a *boundary* edge otherwise. For a diagonal edge $e \in E(G)$, let $C(e) = \{e' \in E(G) \mid e' \text{ intersects } e, \}$, where two edges having only a common end are not considered to intersect each other. All edges in $C(e)$ are from the other triangulation. The cardinality $c(e)$ of $C(e)$ is called the *cross number* of e. Obviously, $c(e) \geq 1$ for any non-face edge e. We say a diagonal e *isolated* if $c(e) = 1$. Finally, a vertex $v \in G(\pi_1, \pi_2)$ is *pure* with respect to π_1 (resp. π_2) if it is only adjacent to some $\pi_1(\pi_2)$-diagonals; it is *mixed* otherwise.

3.2 Transformation primitives

Recall that $fd(\pi_1, \pi_2)$ denotes the diagonal-flip distance between π_1 and π_2.

Proposition 1. *([15]) Let π_1 and π_2 be two triangulations of a polygon and let $e \in \pi_1$ be an isolated diagonal in $G(\pi_1, \pi_2)$. Let π_2' be the triangulation created from π_2 by flipping the unique edge e' that intersects e. Then (1) π_2' has one more common diagonal with π_1 than π_2. Equivalently, the difference graph $G(\pi_1, \pi_2')$ has one more cell than $G(\pi_1, \pi_2)$; (2) $fd(\pi_1, \pi_2') = fd(\pi_1, \pi_2) - 1$.*

Next, we study properties relating with the degrees of vertices in difference graphs. Let π_1 and π_2 be two triangulations of a polygon. For a vertex $v \in G(\pi_1, \pi_2)$, the number of all diagonal edges adjacent to v is called the *degree* of v, denoted by $d(v)$. By using Proposition 1, we can prove

Proposition 2. *Let π_1 and π_2 be two triangulations of a polygon and let v be a vertex such that $d(v) = 1$ in $G(\pi_1, \pi_2)$. If the unique diagonal e adjacent to v is in π_1 (resp. π_2), then flipping e creates a triangulation π_1' (resp. π_2') which has one more diagonal in common with π_2 (resp. π_1) than π_1 (resp. π_2).*

Proposition 3. *Let π_1 and π_2 be two triangulations of a polygon and let u and v be two adjacent degree-2 vertices in $G(\pi_1, \pi_2)$. If one of u and v is pure, then it is possible to create π_1' and π_2' by flipping three of the four diagonals (which can be determined easily) adjacent to u or v such that (1) π_1' and π_2' have two more common diagonals than π_1 and π_2, and (2) $fd(\pi_1', \pi_2') \leq fd(\pi_1, \pi_2) - 2$.*

4 A better approximation for arbitrary triangulations

4.1 An approximation algorithm

In this section, we will present an approximation algorithm for the problem of transforming triangulations. Formally, the problem is defined as
Instance: Two triangulations of a polygon;
Output: A shortest diagonal-flip transformation between the given triangulations.

Our approximation algorithm is described in Table 1. Obviously, the algorithm runs in polynomial time. Now we give some basic facts which will be used in analyzing its approximation ratio. Recall that $G(\pi_1, \pi_2)$ denotes the difference graph of π_1 and π_2.

Input: Two triangulations π_1 and π_2;

Do until the following 'if' conditions fails
  if there are isolated diagonals **then**
    pick such an edge e;
    let e' be the unique diagonal that intersects e;
    if $e' \in \pi_1$ then $\pi_1 := \pi_1 + e - e'$ else $\pi_2 := \pi_2 + e - e'$;
Enddo

Let the resulting polygon triangulations have k cells $P_i (i \leq k)$, and let $\pi_j|_{P_i}$ denote the restriction of π_j on P_i for $j = 1, 2$ and $i \leq k$; assume P_i has n_i vertices.

For each cell P_i
  pick a node v;
  transform $\pi_1|_{P_i}$ into the unique triangulation π all of whose diagonals have one end at v using at most n_i steps;
  transform π into $\pi_2|_{P_i}$ reversely.
Endfor

Table 1. *Algorithm A.*

Note that Algorithm A has the following properties.

Proposition 4. *(1) Flips in Do loop does not increase the maximum degree of the difference graph $G(\pi_1, \pi_2)$. (2) Flips in Do loop will not increase the number of internal triangles in $G(\pi_1, \pi_2)$.*

4.2 Analysis of the algorithm

In this section, we shall analyze the transformation algorithm given above. Let π_1 and π_2 be two triangulations of the n-gon. Consider a sequence Π of diagonal-flips that transforms π_1 into π_2. A diagonal-flip $(ab, cd) \in \Pi$ is *auxiliary* if $cd \notin \pi_2$. We also say flip (ab, cd) *touches* the vertices a, b, c, d. Let $A(\Pi)$ denote the set of all auxiliary diagonal-flips in S. Then,

$$|\Pi| \geq |A(\Pi)| + n - 3. \tag{1}$$

Inequality 1 implies that any lower bound on the cardinality of $A(\Pi)$ induces a lower bound of $|\Pi|$. In the rest of this section, we will work on $|A(\Pi)|$ for a transformation sequence Π instead of its cardinality $|\Pi|$. Recall that a vertex v is pure with respect to π_1 if it is only an end of π_1-diagonals.

Proposition 5. *Let π_1 and π_2 be two triangulations of a polygon and let $v \in G(\pi_1, \pi_2)$ be a pure vertex with respect to π_1. If flipping any diagonal adjacent* *to v does not create a π_2-diagonal, then in any sequence Π of diagonal-flips that transforms π_1 into π_2, there are at least one auxiliary diagonal touching v.*

Proof. We consider the subpolygon consisting of v and its neighbors on boundary and adjacent nodes. Let v be adjacent to $v_1, v_2, \cdots, v_k$ and let u and w be its left and right neighbors on boundary(see Figure 4). Since flipping any diagonal vv_j does not create a π_2-diagonal, $k \geq 2$.

Since all diagonals connecting v are in π_1, $uw \in \pi_2$ and $uv_1, v_iv_{i+1}, v_kw \in \pi_1$. Consider the first flip (xy, pq) touching v in the sequence Π. If pq is adjacent to v, the flip is auxiliary because all π_2-diagonals are not adjacent to v. Otherwise, we consider the diagonal adjacent to v that is first flipped away in Π . Let such an edge be vv_j. For simplicity, we assume $1 < j < k$. By assumption, vv_{j-1} and vv_{j+1} are π_1-diagonals. If the flip $(vv_j, v_{j-1}v_{j+1})$ is not auxiliary, then $v_{j-1}v_j$ is a π_2-diagonal and has cross number 1, contradicting to the hypothesis.

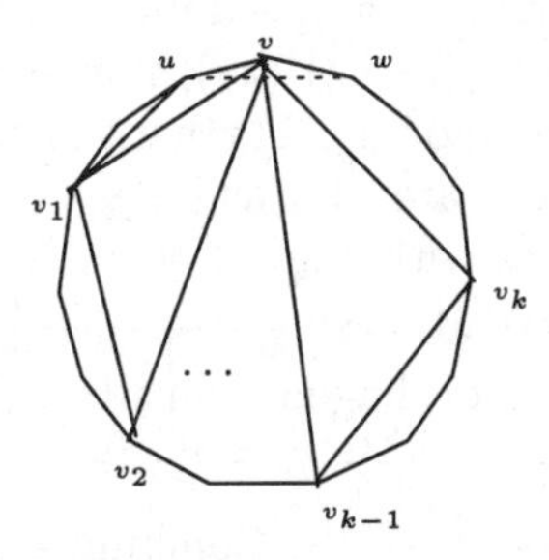

Fig. 4. *Vertex v is an end of π_1-diagonals.*

A vertex v is said to be *straddle* with respect to π_1 if for any pair of π_1-diagonals that are adjacent to v, there exists a π_2-diagonal adjacent to v between them. Otherwise, it is *non-straddle.* By definition, a degree-2 vertex is straddle only if it is mixed, i.e., an end of two diagonals from different triangulations.

Proposition 6. *Let π_1 and π_2 be two triangulations of a polygon such that $G(\pi_1, \pi_2)$ does not contain any isolated edges and let v be a non-straddle vertex with respect to π_1. If (1) v is not a vertex of any internal triangles in π_1 or π_2, (2) v is not connected with any vertices of internal triangles in π_2, and (3) flipping any π_1-diagonal adjacent to v does not create a π_2-diagonal, then in any sequence Π of diagonal-flips that transforms π_1 into π_2, there is at least one auxiliary diagonal touching v.*

Recall that a vertex $v \in G(\pi_1, \pi_2)$ is *pure* with respect to π_1 if it is only an end of π_1-diagonals. Let $V(\pi_1)$ and $V(\pi_2)$ denote the set of pure vertices with respect to π_1 and π_2 respectively.

Lemma 7. *Let π_1 and π_2 be two triangulations of the n-gon such that $G(\pi_1, \pi_2)$ does not contain any isolated edges. Then, the flip distance between π_1 and π_2 is at least $n - 3 + \frac{|V(\pi_1)|+|V(\pi_2)|}{8}$.*

Sketch of Proof. By Proposition 5, $fd(\pi_1, \pi_2) \geq n - 3 + |V(\pi_1)|/4$ and $fd(\pi_1, \pi_2) \geq n - 3 + |V(\pi_2)|/4$.

Lemma 8. *Let π_1 and π_2 be two triangulations of the n-gon in which each vertex is an end of at most d diagonals. If $G(\pi_1, \pi_2)$ does not contain any isolated edges, then $fd(\pi_1, \pi_2) \leq n - 3 + \frac{n}{4(d-1)} - \frac{(d+5)(|V(\pi_1)|+|V(\pi_2)|)}{8}$.*

Sketch of Proof. It can be proved that a triangle with no boundary edges has at least two non-straddle vertices and a triangle with one boundary edge has at least one non-straddle vertex. Since there are exactly $n - 2 - 2|V(\pi_2)|$ triangles with one boundary edge and $|V(\pi_2)|$ triangles with three diagonals and the fact any vertex can be a vertex of at most $(d-1)$ triangles with two internal diagonals connecting it, there are at least $n/(d-1)$ non-straddle vertices. Since π_1 and π_2 have $|V(\pi_1)| - 2$ and $|V(\pi_2)| - 2$ internal triangles respectively, there are at least $n/(d-1) - 3|V(\pi_1)| - 3|V(\pi_2)|$ non-straddle vertices that are not in any internal triangles of π_1 or π_2. Further, since each vertex is an end of at most d π_2-diagonals, there are at least $\frac{n}{d-1} - 3|V(\pi_1)| - 3|V(\pi_2)| - (d-1)|V(\pi_2)|$ non-traddle vertices with respect to π_1 that satisfy the conditions in Proposition 6. Thus, there are at least $\frac{n}{4(d-1)} - \frac{3|V(\pi_1)|}{4} - \frac{(d+2)|V(\pi_2)|}{4}$ auxiliary diagonal-flips in any sequence of diagonal-flips that transforms π_1 to π_2. Thus, $fd(\pi_1, \pi_2) \geq n - 3 + \frac{n}{4(d-1)} - \frac{3|V(\pi_1)|}{4} - \frac{(d+2)|V(\pi_2)|}{4}$. Similarly, $fd(\pi_1, \pi_2) \geq n - 3 + \frac{n}{4(d-1)} - \frac{3|V(\pi_2)|}{4} - \frac{(d+2)|V(\pi_1)|}{4}$. Combining these two bounds together, we have $n + \frac{n}{4(d-1)} - \frac{(d+5)(|V(\pi_1)|+|V(\pi_2)|)}{8} - 3$. This finishes the proof.

Combining Proposition 1 and 3, and Lemma 7 and 8, we have

Theorem 9. *On input of two triangulations π_1 and π_2 of the n-gon in which each vertex is an end of at most d diagonals, the transformation algorithm outputs a diagonal-flip transformation of length at most $(2 - \frac{2}{4(d-1)(d+6)+1}) fd(\pi_1, \pi_2)$.*

5 A 1.97-approximation algorithm for triangulations without internal triangles

5.1 An upper bound

First, we present an upper bound of the flip distance between π_1 and π_2 in term of the the number of mixed degree-2 vertices in $G(\pi_1, \pi_2)$.

Theorem 10. *Let π_1 and π_2 be two triangulations of the n-gon such that they do not contain any internal triangles. If $G(\pi_1, \pi_2)$ contains m mixed degree-2 vertices, then π_1 can be transformed into π_2 in at most $5(n - m) + \frac{7m}{4}$ flips.*

Input: Two triangulations π_1 and π_2 without internal diagonals;

Do until no isolated diagonals exist in $G(\pi_1, \pi_2)$
  pick an isolated edge e and flip the diagonal intersecting it;
Enddo
Let the resulting polygon triangulations have k cells $P_i (i \leq k)$, Let P_i have n_i vertices.

For each cell P_i
  if $G(\pi_1|_{P_i}, \pi_2|_{P_i})$ contains more than $0.9285|P_i|$ mixed, degree-2 vertices, **then**
    transform $\pi_1|_{P_i}$ into $\pi_j|_{P_i}$ by Thm 10;
  else
    pick a node v;
    transform $\pi_1|_{P_i}$ into the unique triangulation π all of whose diagonals
    have one end at v using at most n_i steps;
    transform π into $\pi_2|_{P_i}$ reversely.
Endfor

Table 2. *Algorithm B. A 1.97-approximation algorithm for triangulations without internal triangles.*

The upper bound in above theorem can be improved to $5(n-m) + \frac{3m}{2}$ by a more complicated technique. Here we are not intent on presenting the best possible constant.

5.2 A 1.97-approximation algorithm

Proposition 11. *Let π_1 and π_2 be two triangulations of the n-gon that do not contain any internal triangles and let the difference graph $G(\pi_1, \pi_2)$ does not contain any isolated edges. If there are m mixed, degree-2 vertices in $G(\pi_1, \pi_2)$, then the diagonal-flip distance $fd(\pi_1, \pi_2)$ is at least $n - 3 + \frac{n-m}{8}$.*

Our algorithm is described in Table 4.2. Obviously, it is a polynomial time algorithm. Using Proposition 11 and Theorem 10, we analyze its approximation ratio in the same way as Theorem 9.

Theorem 12. *On the input of two triangulations π_1 and π_2 of the n-gon that do not contain any internal triangles, Algorithm B outputs a diagonal-flip transformation of length at most $1.97 fd(\pi_1, \pi_2)$.*

Acknowledgment
The authors would like to thank B. DasGupta, X. He, T. Jiang for useful discussions at the beginning of this work, J. Tromp for his C-code for computing rotation distances and T. Yokomori for discussions and sharing his large triangulations.

References

1. D. Aldous, Triangulating the circle, at random. *Amer Math. Monthly* **89**(1994), 223-234.
2. B. DasGupta, X. He, T. Jiang, M. Li, J. Tromp and L. Zhang, On distance between phylogenetic trees. *Proc. of the SIAM-ACM 8th Annual Symposium on Discrete Algorithms*, 427-436, 1997.
3. K. Culik and D. Wood, A note on some tree similarity measures, *Information Processing Letters* **15**(1982), 39-42.
4. J.-Y. Cai and M. D. Hirsch, Rotation distance, triangulation of planar surfaces and hyperbolic geometry, *Proceeding of the 5th Inter. Symposium on Algorithm and Computation, Lecture Notes in Computer Science*, vol. 834, 172-180, 1994.
5. A. K. Dewdney, Wagner's theorem for torus graphs, *Discrete Math.* **4**(1973), 139-149.
6. L. Guibas and J. Hershberger, Morphing simple polygons, *Proceeding of the ACM 10th Annual Sym. of Comput. Geometry,* 267-276, 1994.
7. S. Hanke, T. Ottmann and S. Schuierer, The edge-flipping distance of triangulations. *Proc. of the European Workshop of Comput. Geometry*, 1996. Technical Report 76, Insttut fur Informatik, Universitat Freiburg, 1996.
8. J. Hershberger and S. Suri, Morphing binary trees. *Proceeding of the ACM-SIAM 6th Annual Symposium of Discrete Algorithms*, 396-404, 1995.
9. F. Hurtado, M. Noy, and J. Urrutia, Flipping edges in triangulations, *Proc. of the ACM 12th Annual Sym. of Comput. Geometry,* 214-223, 1996.
10. D. Knuth, *The art of computer programming, Vol. 3: Sorting and Searching*, Addison-Wesley, Reading, MA, 1973.
11. D. Knuth, Computer Musings (Videotape): The associative law, or the anatomy of rotations in binary trees, *University Video Communications*, 1993.
12. J. Pallo, On rotation distance in the lattice of binary trees, *Infor. Proc. Letters* **25**(1987), 369-373.
13. D. D. Sleator and R. Tarjan, Self-adjusting binary search trees, *J. Assoc. Comput. Mach.* **32**(1985), 652-686.
14. D. D. Sleator, R. Tarjan and W. Thurston, Rotation distance, in T.M. Cover and B. Gopinath eds, Open problems in communication and computation, Springer-Verlag, 1987.
15. D. D. Sleator, R. Tarjan and W. Thurston, Rotation distance, triangulations, and hyperbolic geometry, *Proc. of the 18th Annual ACM Symposium on Theory of Comput.*, Berkeley, CA., 122-135, 1986. Also appearing on *J. of American Math. Soc.* **1**(1988), No. 3, 647 - 681.
16. D. D. Sleator, R. Tarjan and W. Thurston, Short encoding of evolving structures, *SIAM J. Disc. Math.* **5**(1992), 428-450.
17. D. Swofford, G. J. Olsen, P. J. Waddell and D. M. Hillis, Phylogenetic Inference, In *Molecular Systematics*, 2nd Edition, edited by D. M. Hillis, C. Moritz and B. K. Mable, Sinauer Associates, Mass., 1996.
18. R. E. Tarjan, *Data structures and network algorithms*, Soc. Indus. Appl. Math., Philadelphia, PA. 1983.
19. K. Wagner, Bemerkungen zum vierfarbenproblem, *J. Deutschen Math.-Verin.* **46**(1936), 26-32.

A Cost Optimal Parallel Algorithm for Computing Force Field in N-Body Simulations

Guoliang Xue*

Department of Computer Science, The University of Vermont
Burlington, VT 05405, U.S.A. (Email: xue@cs.uvm.edu).

Abstract. We consider the following force field computation problem: given a cluster of n particles in 3-dimensional space, compute the force exerted on each particle by the other particles. Depending on different applications, the pairwise interaction could be either gravitational or Lennard-Jones. In both cases, the force between two particles vanishes as the distance between them approaches to infinity. Since there are $n(n-1)/2$ pairs, direct method requires $\Theta(n^2)$ time for force-evaluation, which is very expensive for astronomical simulations. In 1985 and 1986, two famous $\Theta(n \log n)$ time hierarchical tree algorithms were published by Appel [3] and by Barnes and Hut [4] respectively. In a recent paper, we presented a linear time algorithm which builds the oct tree bottom-up and showed that Appel's algorithm can be implemented in $\Theta(n)$ sequential time. In this paper, we present an algorithm which computes the force field in $\Theta(\log n)$ time using an $\frac{n}{\log n}$ processor CREW PRAM. A key to this optimal parallel algorithm is replacing a recursive top-down force calculation procedure of Appel by an equivalent non-recursive bottom-up procedure. Our parallel algorithm also yields a new $\Theta(n)$ time sequential algorithm for force field computation.
Keywords: Paralle algorithms, spatial tree data structures, force field evaluation, N-body simulations, PRAM, cost optimal algorithms.

1 Introduction and assumption

Fast algorithms for force field evaluation have important applications in molecular conformation, molecular dynamics, and astrophysical simulations. Given a cluster of n particles in 3-dimensional space, we need to compute the force exerted on each particle by the other particles. Since there are $n(n-1)/2$ pairs, direct method requires $\Theta(n^2)$ time for force-evaluation, which is very expensive for astronomical simulations.

In astrophysical simulations, the force exerted on one particle by another is given by the gravitational force. In molecular dynamics and molecular conformation, the Lennard-Jones potential is widely used. In both cases, the force exerted on one particle by another particle vanishes as the distance between

* The research of this author was supported in part by the National Science Foundation grants ASC-9409285 and OSR-9350540 and by the US Army Research Office grant DAAH04-9610233.

them approaches to infinity. This observation leads to several fast approximation algorithms. In 1985 and 1986, two famous $\Theta(n \log n)$ time hierarchical tree algorithms were published by Appel [3] and by Barnes and Hut [4] respectively. In 1987, Greengard and Rokhlin [8] published the fast multipole algorithm which computes the force field in $O(n)$ time. Recently, Aluru [1] showed that Greengard's algorithm is not $O(n)$. These algorithms have made great impacts on the computational study of molecular conformation/dynamics and astronomical simulations. Parallel implementations of these algorithms have been reported by many authors, including [7, 12–14, 17]. Due to the big constant in the fast multipole algorithm and the simplicity and efficiency of the tree algorithms, hierarchical tree algorithms received more attention in computational studies [2]. Therefore we concentrate on tree algorithms in this paper.

In a recent paper [16], Xue presented an algorithm which builds an oct-tree bottom-up in $\Theta(n)$ sequential time and showed that Appel's algorithm can be implemented in $\Theta(n)$ sequential time. That algorithm computes the force field top-down using a recursive procedure. It seems difficult to parallelize that linear time algorithm efficiently. In this paper, we replace the recursive top-down procedure of Appel by an equivalent non-recursive bottom-up procedure and present a parallel algorithm which computes force field in $\Theta(\log n)$ time using an $\frac{n}{\log n}$ processor CREW PRAM.

The rest of this paper is organized as follows. In section 2, we show that the oct-tree can be constructed bottom-up in $\Theta(\log n)$ time on an $\frac{n}{\log n}$ processor CREW PRAM. In section 3, we show that the force field can be computed in $\Theta(\log n)$ time on an $\frac{n}{\log n}$ processor CREW PRAM. We conclude the paper in section 4. Throughout this paper, we make the following assumption on the distribution of the particles:

Assumption 1 *There exist two positive constants c_1 and c_2 such that the minimum inter-particle distance is at least c_1 and the maximum inter-particle distance is smaller than $c_2 n^{1/3}$.*

Assumption 1 is highly believed to be true for most applications and is supported by many computer simulations. For the Lennard-Jones cluster, it is proved that the minimum inter-particle distance has positive lower bound which is independent on the number of particles in the cluster [15].

2 Building the oct-tree bottom-up in $\Theta(\log n)$ time

In section 2.1, we will describe the necessary data structure used in our algorithms. In section 2.2, we will present a $\Theta(\log n)$ time algorithm for constructing the oct-tree using an $P = \frac{n}{\log n}$ processor CREW PRAM. The time complexity of our algorithm is analyzed in section 2.3. To simplify the analysis, we assume that both n and $\log n$ are powers of 8. It is well-known that this assumption does not affect the asymptotic analysis of the algorithm.

2.1 Data structures

A *computation box* is defined by a point `base` in 3-dimensional space and a positive number `size`. Let `baseX, baseY, baseZ` be the coordinates of `base`. The computation box defined by `base` and `size` is

$$[baseX,\ baseX + size) \times [baseY,\ baseY + size) \times [baseZ,\ baseZ + size). \quad (1)$$

A computation box is illustrated in Figure 1(a). When a computation box is partitioned, we obtain 8 non-intersecting computation boxes of equal size whose union is the original computation box. An example is illustrated in Figure 1(b).

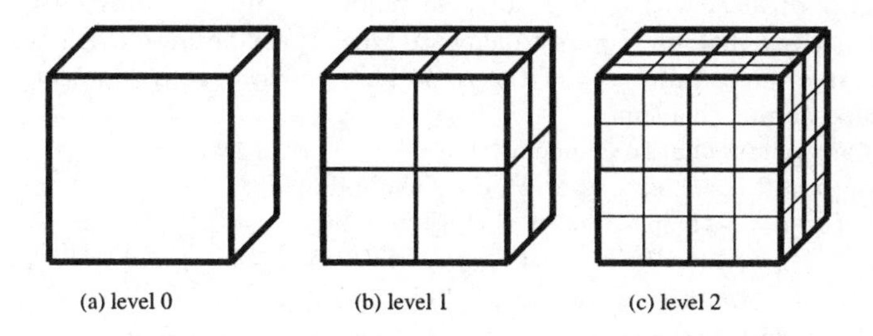

Fig. 1. Computation boxes associated with the first 3 levels of the oct-tree.

We will make reference to the following data structure during our description of the algorithm. Each node in the oct-tree is of type `NODE`.

```
typedef struct _node{
  struct _node  *parent;    struct _node  *child[8];
  int      isLeaf;  int      weight;  int      pindex;
  double  coordX;  double  coordY;  double  coordZ;
  double  forceX;  double  forceY;  double  forceZ;
  double   baseX;  double   baseY;  double   baseZ;  double   size;
}NODE;
```

Note that the partition of a computation box takes constant time. The computation box in Figure 1(a) is partitioned to 8 smaller computation boxes in Figure 1(b), which in turn are partitioned to a total of 64 even smaller computation boxes in Figure 1(c).

2.2 Building the oct-tree bottom-up in $O(\log n)$ time

We assume that the n particles are given in an array of points so that `part[i].x`, `part[i].y`, `part[i].z` represent the coordinates of particle `i` ($i = 0, 1, 2, \ldots, n-1$). Our PRAM algorithm for oct tree construction is presented as Algorithm 2.1. Since P is assumed to be a power of 8, there is an integer L such that $8^L = P$. To make the description of the algorithm easier, we assume that the P processors

Algorithm 2.1 (part 1) `{Building the oct-tree bottom-up.}`

Step_1 `{Determine the sizes of the root box and the leaf boxes}`
Using all P processors to compute the following 6 values:
$Xmin := \min_{i=0,n-1} part[i].x$; $Xmax := \max_{i=0,n-1} part[i].y$;
$Ymin := \min_{i=0,n-1} part[i].z$; $Ymax := \max_{i=0,n-1} part[i].x$;
$Zmin := \min_{i=0,n-1} part[i].y$; $Zmax := \max_{i=0,n-1} part[i].z$.
Let $\delta := \frac{\sqrt{3}}{3} c_1$. Let *maxlevel* be the smallest positive integer such that $\delta 2^{maxlevel} > \max\{Xmax - Xmin, Ymax - Ymin, Zmax - Zmin\}$. Let $\Delta := \delta \times 2^{maxlevel}$.

Step_2 `{Allocate space}`
We will use a three dimensional array of `NODE` for the nodes on each level of the oct-tree. Let $tree[l]$ be a pointer to the three dimensional array of `NODE` with $2^l \times 2^l \times 2^l$ elements ($l = 0, 1, \ldots, maxlevel$). It is clear that we require to allocate $\Theta(n)$ space because there are 8^l tree nodes on level-l of the tree. These arrays are dynamically allocated at this time.

Step_3 `{Construct the leaf nodes}`
for $p = 0$ **to** P, processor P_p does the following:
  for $t = p$ **to** $n - 1$ **step** P **do**
    Let $i = \lfloor \frac{part[t].x - Xmin}{\delta} \rfloor$; $j = \lfloor \frac{part[t].y - Ymin}{\delta} \rfloor$; $k = \lfloor \frac{part[t].z - Zmin}{\delta} \rfloor$.
    $tree[maxlevel] \rightarrow node[i][j][k].size := \delta$;
    $tree[maxlevel] \rightarrow node[i][j][k].baseX := Xmin + i\delta$;
    $tree[maxlevel] \rightarrow node[i][j][k].baseY := Ymin + j\delta$;
    $tree[maxlevel] \rightarrow node[i][j][k].baseZ := Zmin + k\delta$;
    $tree[maxlevel] \rightarrow node[i][j][k].weight := 1$;
    $tree[maxlevel] \rightarrow node[i][j][k].pindex := t$;
    $tree[maxlevel] \rightarrow node[i][j][k].coordX := part[t].x$;
    $tree[maxlevel] \rightarrow node[i][j][k].coordY := part[t].y$;
    $tree[maxlevel] \rightarrow node[i][j][k].coordZ := part[t].z$.
  endfor {t}
endfor {p}

Fig. 2. Building the oct-tree bottom-up (part 1)

are labeled as P_{IJK} where $0 \leq I < 2^L$, $0 \leq J < 2^L$, $0 \leq K < 2^L$. We have taken the liberty of treating the processors as a linear array in Step_3 of the algorithm.

Since the maximum and minimum of n numbers can be computed in $\Theta(\log n)$ time on an $\frac{n}{\log n}$ processor PRAM, the root computation box can be computed in $\Theta(\log n)$ parallel time on the PRAM. By Assumption 1, we can now decide the size of the smallest computation box as well as the size of the largest computation box, in constant time. By then, we should know an $\frac{1}{3}\log n + O(1)$ upper bound on the height of the oct-tree. Therefore we can dynamically allocate space for every possible tree node. We assume that all the fields of a tree node are initialized to zero at the time the memory is allocated. The total space allocated is $\Theta(n)$ since the number of nodes in a complete oct-tree of height $\frac{1}{3}\log n + O(1)$ is $\Theta(n)$.

Instead of inserting the particles to the tree from the root node, we insert the particles directly to the nodes corresponding to the smallest computation boxes.

Algorithm 2.1 (part 2) {`Building the oct-tree bottom-up.`}

Step_4 {`Building the tree bottom-up`}

for $l := maxlevel - 1$ **downto** L **do**

All processors P_{IJK} $(0 \leq I, J, K < 2^L)$ **do in parallel**

for $i := I2^{l-L}$ **to** $I2^{l-L} + 2^{l-L} - 1$ **do**

for $j := J2^{l-L}$ **to** $J2^{l-L} + 2^{l-L} - 1$ **do**

for $k := K2^{l-L}$ **to** $K2^{l-L} + 2^{l-L} - 1$ **do**

$tree[l] \rightarrow node[i][j][k].child[0] := tree[l+1] \rightarrow node[2i+0][2j+0][2k+0];$
$tree[l] \rightarrow node[i][j][k].child[1] := tree[l+1] \rightarrow node[2i+0][2j+0][2k+1];$
$tree[l] \rightarrow node[i][j][k].child[2] := tree[l+1] \rightarrow node[2i+0][2j+1][2k+0];$
$tree[l] \rightarrow node[i][j][k].child[3] := tree[l+1] \rightarrow node[2i+0][2j+1][2k+1];$
$tree[l] \rightarrow node[i][j][k].child[4] := tree[l+1] \rightarrow node[2i+1][2j+0][2k+0];$
$tree[l] \rightarrow node[i][j][k].child[5] := tree[l+1] \rightarrow node[2i+1][2j+0][2k+1];$
$tree[l] \rightarrow node[i][j][k].child[6] := tree[l+1] \rightarrow node[2i+1][2j+1][2k+0];$
$tree[l] \rightarrow node[i][j][k].child[7] := tree[l+1] \rightarrow node[2i+1][2j+1][2k+1];$

Also set the *parent* field for each of the 8 children of $tree[l] \rightarrow node[i][j][k]$;

$tree[l] \rightarrow node[i][j][k].baseX := Xmin + i2^{maxlevel-l}\delta;$
$tree[l] \rightarrow node[i][j][k].size := 2^{maxlevel-l}\delta;$
$tree[l] \rightarrow node[i][j][k].baseY := Ymin + j2^{maxlevel-l}\delta;$
$tree[l] \rightarrow node[i][j][k].baseZ := Zmin + k2^{maxlevel-l}\delta;$

Let $tree[l] \rightarrow node[i][j][k].weight$ be the sum of the weights of its children;
Let $tree[l] \rightarrow node[i][j][k].coordX$, $tree[l] \rightarrow node[i][j][k].coordY$, and $tree[l] \rightarrow node[i][j][k].coordZ$ be the coordinates of the weighted center of the particles contained in the computation box of the current node;

if $tree[l] \rightarrow node[i][j][k].weight == 0$ **then**

$tree[l] \rightarrow node[i][j][k].isLeaf := 1;$

elseif $tree[l] \rightarrow node[i][j][k].weight == 1$ **then**

$tree[l] \rightarrow node[i][j][k].isLeaf := 1;$

Let $tree[l] \rightarrow node[i][j][k].pindex$ be the index of the only particle contained in the current computation box;

endif

endfor {k}

endfor {j}

endfor {i}

endfor {l}

Fig. 3. Building the oct-tree bottom-up (part 2)

We then pass information from one layer of the tree to the layer above, starting from the bottom layer. Although there are $\frac{1}{3}\log n + O(1)$ layers of the tree, the amount of time required is decreased by a factor of 8 every time we move up one layer. This is the key to achieving the $\Theta(n)$ sequential time complexity. For the parallel time complexity, we have the following theorem.

Theorem 1. *Algorithm 2.1 builds the oct-tree for n particles using* $\Theta(\log n)$ *time and* $\Theta(n)$ *space, using an* $\frac{n}{\log n}$ *processor CREW PRAM, provided that the particles satisfies Assumption 1. The constant behind the asymptotic notation is proportional to* $(\frac{c_2}{c_1})^3$. □

Algorithm 2.1 (part 3) {`Building the oct-tree bottom-up.`}
Step_5 {`Building the tree bottom-up`}
for $l := L - 1$ **downto** 0 **do**
All processors P_{ijk} $(0 \le i, j, k < 2^l)$ **do in parallel**
$tree[l] \rightarrow node[i][j][k].child[0] := tree[l+1] \rightarrow node[2i+0][2j+0][2k+0];$
$tree[l] \rightarrow node[i][j][k].child[1] := tree[l+1] \rightarrow node[2i+0][2j+0][2k+1];$
$tree[l] \rightarrow node[i][j][k].child[2] := tree[l+1] \rightarrow node[2i+0][2j+1][2k+0];$
$tree[l] \rightarrow node[i][j][k].child[3] := tree[l+1] \rightarrow node[2i+0][2j+1][2k+1];$
$tree[l] \rightarrow node[i][j][k].child[4] := tree[l+1] \rightarrow node[2i+1][2j+0][2k+0];$
$tree[l] \rightarrow node[i][j][k].child[5] := tree[l+1] \rightarrow node[2i+1][2j+0][2k+1];$
$tree[l] \rightarrow node[i][j][k].child[6] := tree[l+1] \rightarrow node[2i+1][2j+1][2k+0];$
$tree[l] \rightarrow node[i][j][k].child[7] := tree[l+1] \rightarrow node[2i+1][2j+1][2k+1];$
Also set the *parent* field for each of the 8 children of $tree[l] \rightarrow node[i][j][k]$;
$tree[l] \rightarrow node[i][j][k].baseX := Xmin + i2^{maxlevel-l}\delta;$
$tree[l] \rightarrow node[i][j][k].size := 2^{maxlevel-l}\delta;$
$tree[l] \rightarrow node[i][j][k].baseY := Ymin + j2^{maxlevel-l}\delta;$
$tree[l] \rightarrow node[i][j][k].baseZ := Zmin + k2^{maxlevel-l}\delta;$
Let $tree[l] \rightarrow node[i][j][k].weight$ be the sum of the weights of its children;
Let $tree[l] \rightarrow node[i][j][k].coordX$, $tree[l] \rightarrow node[i][j][k].coordY$, and
$tree[l] \rightarrow node[i][j][k].coordZ$ be the coordinates of the weighted center of
the particles contained in the computation box of the current node;
if $tree[l] \rightarrow node[i][j][k].weight == 0$ **then**
$tree[l] \rightarrow node[i][j][k].isLeaf := 1;$
elseif $tree[l] \rightarrow node[i][j][k].weight == 1$ **then**
$tree[l] \rightarrow node[i][j][k].isLeaf := 1;$
Let $tree[l] \rightarrow node[i][j][k].pindex$ be the index of the only particle
contained in the current computation box;
endif
endfor {l}

Fig. 4. Building the oct-tree bottom-up (part 3)

3 Computing force fields in $\Theta(\log n)$ time

Given a cluster of n particles, we need to compute the potential energy function and the force exerted on each particle by the other particles. In many applications, the potential energy function of a cluster is the sum of the pair-wise potential functions. For details on the potential functions and force fields, see [16].

3.1 The two-pass algorithm

After the oct-tree is constructed, Appel's algorithm can be implemented using a bottom-up pass and a top-down pass. We assume that there is a global variable `FUNC` which is initialized to 0 and is used to accumulate the potential energy function of the cluster. We also assume that the fields `forceX`, `forceY` and `forceZ` at every tree node are all initialized to 0 before the computation. These fields are used to hold partial values of the force field during the computation.

Our algorithm is presented as Algorithm 3.1. For any two nodes A and B in the tree, a call to procedure `compGRAD(A, B)` does the following:

- Compute the force exerted on each particle in A by all particles in B and save the value in $(A.forceX, A.forceY, A.forceZ)$.
- Compute the potential between cluster A and cluster B and add this value to the global variable `FUNC`.

During each call to procedure `compGRAD(A, B)`, the potential function between cluster A and cluster B is added to the global variable `FUNC`. The force exerted on each particle in cluster A by the particles in cluster B is stored in $(A.forceX, A.forceY, A.forceZ)$. Notice that the force exerted on each particle in cluster B by the particles in cluster A is computed in the call to `compGRAD(B, A)`. Therefore, at the end of the computation, `FUNC` is 2 times the actual potential function value and $(A.forceX, A.forceY, A.forceZ)$ is the force exerted by all the other particles on the particle in node A for each leaf node A whose weight is 1. This force is exactly the force computed by Appel's algorithm [3, 16].

3.2 Time complexity

We will analyze the time complexity of Algorithm 3.1. Given any parameter $\delta > 0$ which defines well-separateness and a tree node A, the number tree nodes which are on the same level as A and which are not well-separated from A is bounded by $O(\frac{1}{\delta^3})$, which is a constant for any given δ. As a result, the inner most **for** loop in Step_1 of Algorithm 3.1 requires constant time. Therefore, the parallel run time of Step_1 of Algorithm 3.1 is $\Theta(\sum_{l=maxlevel}^{L} 8^{l-L}) = \Theta(n/P) = \Theta(\log n)$. Similarly, we can show that the parallel run time of Step_4 of Algorithm 3.1 is also $\Theta(\log n)$. In Step_2, only part of the P processors are active. The parallel runtime of Step_2 of the algorithm is $\Theta(L) = \Theta(\log P) = \Theta(\log n)$. Similarly, we can show that the parallel run time of Step_3 of Algorithm 3.1 is also $\Theta(\log n)$. To summarize, we have proved the following theorem.

Algorithm 3.1 (part 1) {`Computing force field in two passes.`}

Step_1 {`Gathering information bottom-up`}
  for $l := maxlevel$ **downto** L **do**
    All processors P_{IJK} $(0 \leq I, J, K < 2^L)$ **do** in parallel
    for $i := I2^{l-L}$ **to** $I2^{l-L} + 2^{l-L} - 1$ **do**
      for $j := J2^{l-L}$ **to** $J2^{l-L} + 2^{l-L} - 1$ **do**
        for $k := K2^{l-L}$ **to** $K2^{l-L} + 2^{l-L} - 1$ **do**
          if $tree[l] \rightarrow node[i][j][k].weight \geq 1$ **then**
            Let A be $tree[l] \rightarrow node[i][j][k]$.
            for all tree node B on level-l of the tree such that
            (1) $B.weight \geq 1$; (2) A and B are well separated;
            (3) the parents of A and B are not well separated **do**
              `compGRAD(A, B);`
            endfor
          endif
        endfor {k}
      endfor {j}
    endfor {i}
  endfor {l}

Step_2 {`Gathering information bottom-up`}
  for $l := L - 1$ **downto** 1 **do**
    All processors P_{ijk} $(0 \leq i, j, k < 2^l)$ **do** in parallel
    if $tree[l] \rightarrow node[i][j][k].weight \geq 1$ **then**
      Let A be $tree[l] \rightarrow node[i][j][k]$.
      for all tree node B on level-l of the tree such that
      (1) $B.weight \geq 1$; (2) A and B are well separated;
      (3) the parents of A and B are not well separated **do**
        `compGRAD(A, B);`
      endfor
    endif
  endfor {l}

Fig. 5. Computing force field in two passes (part 1)

Theorem 2. *Given a cluster of n particles satisfying Assumption 1, the force field can be computed using Algorithm 3.1 in* $\Theta(\log n)$ *time, on an* $\frac{n}{\log n}$ *processor CREW PRAM.* □

It is clear that Algorithms 2.1 and 3.1 yield a new linear time algorithm for computing force field for a cluster of n particles which are almost homogeneously distributed. In [5], Callahan and Kosaraju proved that a size $\Theta(n)$ sequence of *well-separated decomposition* can be computed in $\Theta(n)$ time once a *fair-split* tree is constructed. A fair-split tree for n particles can be constructed in $\Theta(n \log n)$ time using the algorithm of [5], without any restriction on the distribution of the particles. However, Algorithm 3.1 is the first $\Theta(\log n)$ time algorithm using $\frac{n}{\log n}$ processors.

Algorithm 3.1 (part 2) {Computing force field in two passes.}

```
Step_3 {Pushing information top-down}
    for l := 1 to L do
      All processors P_ijk (0 <= i, j, k < 2^l) do in parallel
      if tree[l] -> node[i][j][k].weight >= 1 then
        Let A be tree[l] -> node[i][j][k].
        for every child node B of A such that B.weight >= 1 do
          B.forceX = B.forceX + A.forceX;
          B.forceY = B.forceY + A.forceY;
          B.forceZ = B.forceZ + A.forceZ;
        endfor
      endif
    endfor {l}
Step_4 {Pushing information top-down}
    for l := L + 1 to maxlevel - 1 do
      All processors P_IJK (0 <= I, J, K < 2^L) do in parallel
      for i := I2^(l-L) to I2^(l-L) + 2^(l-L) - 1 do
        for j := J2^(l-L) to J2^(l-L) + 2^(l-L) - 1 do
          for k := K2^(l-L) to K2^(l-L) + 2^(l-L) - 1 do
            if tree[l] -> node[i][j][k].weight >= 1 then
              Let A be tree[l] -> node[i][j][k].
              for every child node B of A such that B.weight >= 1 do
                B.forceX = B.forceX + A:forceX;
                B.forceY = B.forceY + A.forceY;
                B.forceZ = B.forceZ + A.forceZ;
              endfor
            endif
          endfor {k}
        endfor {j}
      endfor {i}
    endfor {l}
```

Fig. 6. Computing force field in two passes (part 2)

4 Conclusions

In this paper, we have presented a $\Theta(\log n)$ time algorithm for computing force field in n-body simulations using an $\frac{n}{\log n}$ processor CREW PRAM, improving the previous $O(n \log n)$ time sequential algorithm of Appel. A key to this improved complexity is an $O(n)$ time bottom-up construction of the oct-tree which was constructed top-down using $O(n \log n)$ time in previous studies. We have also replaced the traditional recursive top-down force field computation with a non-recursive bottom-up computation method. We have also studied the dependency of the constant behind the asymptotic notation on the distribution parameters c_1 and c_2 and on the well-separateness parameter δ. This analysis is important because good software for these evaluations is badly needed in practice. Computational studies of the proposed algorithm on existing architectures will be reported in a forthcoming paper. Because of the space limitations, this extended abstract cannot contain all the details of the analysis of the algorithms. A full paper can be found at http://www.emba.uvm.edu/~xue.

References

1. S. Aluru, Greengard's n-body algorithm is not $O(n)$, *SIAM Journal on Scientific Computing*, Vol. 17(1996), pp. 773-776.
2. R.J. Anderson, Tree data structures for N-body simulation, *37th Annual Symposium of Foundations of Computer Science*, IEEE(1996):224-233.
3. A.W. Appel, An efficient program for many-body simulation, *SIAM Journal on Scientific and Statistical Computing*, 6 (1985):85-103.
4. J. Barnes and P. Hut, A hierarchical $O(n \log n)$ force-calculation algorithm, *Nature*, 324(1986):446-449.
5. P.B. Callahan and S.R. Kosaraju, A decomposition of multidimensional point sets with applications to k-nearest-neighbors and n-body potential fields, *Journal of the ACM*, 42(1995):67-90.
6. K. Esselink, The order of Appel's algorithm, *Information Processing Letters*, 41(1992):141-147.
7. A. Grama, V. Kumar and A. Sameh, Scalable parallel formulations of the Barnes-Hut method for n-body simulations, In *Supercomputing'94 Proceedings*, 1994.
8. L. Greengard and V. Rokhlin, A fast algorithm for particle simulations, *Journal of Computational Physics*, 73(1987):325-348.
9. L. Greengard, *The rapid evaluation of potential fields in particle systems*, The MIT Press, 1988
10. L. Greengard, Fast algorithms for classical physics, *Science*, 265(1994):909-914.
11. V. Kumar, A. Grama, A. Gupta and G. Karypis, *Introduction to Parallel Computing: Design and Analysis of Algorithms*, The Benjamin/Cummings Publishing Company, Inc. 1994.
12. J. Singh, C. Holt, T. Totsuka, A. Gupta, and J. Hennessy, Load balancing and data locality in hierarchical n-body methods, *Journal of Parallel and Distributed Computing*, 1994.
13. M. Warren and J. Salmon, Astrophysical n-body simulations using hierarchical tree data structures, In *Supercomputing'92 Proceedings*, 1992.
14. M. Warren and J. Salmon, A parallel hashed oct tree n-body algorithm, In *Supercomputing'93 Proceedings*, 1993.
15. G.L. Xue, Minimum inter-particle distance at global minimizers of Lennard-Jones clusters, *Journal of Global Optimization*, Vol. 11(1997):83-90.
16. G.L. Xue, An $O(n)$ time algorithm for computing force field in n-body simulations, *Theoretical Computer Science*, Vol. 197(1998), pp. 157-169.
17. F. Zhao and S.L. Johnson, The parallel multipole method on the connection machine, *SIAM Journal of Scientific and Statistical Computing*, Vol. 12(1991), pp. 1420-1437.

Efficient 2 and 3-Flip Neighborhood Search Algorithms for the MAX SAT

Mutsunori Yagiura and Toshihide Ibaraki

Department of Applied Mathematics and Physics, Graduate School of Informatics, Kyoto University, Kyoto 606-01, Japan
{yagiura, ibaraki}@kuamp.kyoto-u.ac.jp

Abstract. For problems SAT and MAX SAT, local search algorithms are widely acknowledged as one of the most effective approaches. Most of the local search algorithms are based on the 1-flip neighborhood, which is the set of solutions obtainable by flipping the truth assignment of one variable. In this paper, we consider r-flip neighborhoods for $r \geq 2$, and propose, for $r = 2, 3$, new implementations that reduce the number of candidates in the neighborhood without sacrificing the solution quality. For 2-flip (resp., 3-flip) neighborhood, we show that its expected size is $O(n + m)$ (resp., $O(m + t^2 n)$), which is usually much smaller than the original size $O(n^2)$ (resp., $O(n^3)$), where n is the number of variables, m is the number of clauses and t is the maximum number of appearances of one variable. Computational results tell that these estimates by the expectation well represent the real performance. These neighborhoods are then used under the framework of tabu search etc., and compared with other existing algorithms based on 1-flip neighborhood. The results exhibit good prospects of the proposed algorithms.

1 Introduction

Given n 0-1 variables x_j, $j \in N$, m clauses C_i, $i \in M$, and weights w_i (≥ 0), $i \in M$, where $N = \{1, 2, \ldots, n\}$ and $M = \{1, 2, \ldots, m\}$, the MAX SAT problem asks to determine a 0-1 assignment that maximizes the sum of the weights of satisfied clauses. Denoting $\overline{x}_j = 1 - x_j$ and $L = \cup_{j \in N}\{x_j, \overline{x}_j\}$ (the set of literals), clauses are defined by $C_i \subseteq L$ for $i \in M$ (e.g., $C_i = \{x_1, \overline{x}_3, x_8\}$). Without loss of generality, we assume that at most one of x_j and $\overline{x}_j$ is included in each clause. For a $v \in \{0,1\}^n$, let

$$P_i(v) = \{j \in N \mid x_j \in C_i \text{ and } v_j = 1, \text{ or } \overline{x}_j \in C_i \text{ and } v_j = 0\} \tag{1}$$

and

$$\theta_i(v) = \begin{cases} 1, \text{ if } |P_i(v)| \geq 1 \\ 0, \text{ otherwise.} \end{cases}$$

Then the objective function to maximize is given by

$$f(v) = \sum_{i \in M} w_i \theta_i(v). \tag{2}$$

The problem SAT is a decision problem asking whether there exists a $v \in \{0,1\}^n$ that attains $f(v) = \sum_{i \in M} w_i$; such a v is called a satisfying assignment. The problems SAT and MAX SAT are known to be NP-hard.

Many local search (abbreviated as LS) algorithms for SAT and MAX SAT problems have been proposed [4, 5, 11]. The local search starts from an initial solution v and repeats replacing v with a better solution v' in its *neighborhood* $NB(v)$ until no better solution is found in $NB(v)$, where $NB(v)$ is the set of solutions obtainable from v by slight perturbations. A solution v is called *locally optimal* (with respect to the neighborhood), if no better solution exists in $NB(v)$. We call the replacement of the current solution v by a better solution a *move*. One of the following two move strategies are commonly used: First admissible move strategy (abbreviated as FA) and best admissible move strategy (abbreviated as BA). FA scans the neighborhood $NB(v)$ according to a prespecified random order and moves to the first improved solution. BA scans the entire neighborhood and move to the best solution in $NB(v)$. The local search is often applied to a number of randomly generated initial solutions, and the best among the obtained locally optimal solutions is output. This is called the *random multi-start local search* (abbreviated as MLS).

Let

$$D(v, v') = \{j \in N \mid v_j \neq v'_j\}$$

and

$$NB_r(v) = \{v' \in \{0,1\}^n \mid |D(v, v')| \leq r\}$$

(i.e., $v' \in NB_r(v)$ is obtainable from v by flipping at most r variables). We call $NB_r(v)$ *r-flip* neighborhood. To the best of authors' knowledge, most of the existing local search algorithms for SAT and MAX SAT are based on 1-flip neighborhood [1, 3, 4, 5, 7, 9, 10, 11, 12, 13, 14, 15].

In this paper, we consider r-flip neighborhood for general r, and propose efficient implementations for these neighborhoods, which makes use of the memory structure as described in Section 3. Usually the quality of locally optimal solutions improves if larger neighborhood is used; however, the computational time to search NB_r increases exponentially with r, since $|NB_r| = O(n^r)$ holds in general. To overcome this, we propose, for $r = 2$ and 3, a method to reduce the number of candidates in the neighborhood without sacrificing the solution quality.

Let us call the computation needed for determining one move in LS as *one-round*. Let t denote the maximum number of appearances of a variable and ℓ denote the maximum number of literals in a clause. If we evaluate the objective values (2) for all the solutions in the neighborhood from scratch, taking $O(m\ell)$ time for each evaluation, one-round time is $O(m\ell|NB_r|) = O(m\ell n^r)$ for r-flip neighborhood. But this computational effort can be reduced if we use a clever memory structure. Here we assume that each memory access takes $O(1)$ time, and necessary memory cells can be stored in the linear order, whose precise definition and validity will be discussed in Section 3. With this memory structure, in the worst case, $O(2^r)$ time is needed for each evaluation and $O(rt\ell^r)$ time is required to update the memory structure after a move. Hence, one-round worst-case time becomes $O(2^r|NB_r| + rt\ell^r) = O(2^r n^r + rt\ell^r)$. The size of memory space needed for this algorithm is $O(n + m\ell^r)$.

However, the situation becomes much better if we consider the average-case complexity. Let r be a fixed constant. Then it will be shown that the expected time needed for each update of the memory structure is $O(t)$, and the expected memory requirement is $O(n + m)$. Furthermore, based on the restricted 2 and 3-flip neighborhoods (without sacrificing the solution quality), it will be shown in Section 3 that the expected size of 2-flip neighborhood is $O(n + m)$, and that of 3-flip neighborhood is $O(m + t^2 n)$. Therefore, the expected one-round time with the restricted neighborhood becomes $O(n + m)$ if $r = 2$. In the case of $r = 3$, although additional computation is needed for the restriction of the neighborhood, the expected one-round time is kept as $O(m + t^2 n)$. It is also noted that, for $r = 1$ and move strategy FA, the worst case one-round time of this algorithm is $O(t\ell)$, and its expected one-round time is $O(t)$. For 1-flip neighborhood, similar result is already reported in [3].

Computational experiments for problem instances with up to $n = 1000$ are conducted to see the effectiveness of these neighborhood restrictions. It is observed that the above probabilistic estimates represent the real performance of LS well. Computational experiments to evaluate the effectiveness of these neighborhoods are also conducted. We tested three metaheuristic frameworks: (1) Random multi-start local search (MLS), (2) Iterated local search (ILS), and (3) Tabu search (TS). It is observed, for some types of problems, that the proposed 2 and 3-flip neighborhood search algorithms with restricted neighborhoods are more effective than 1-flip neighborhood if about the same computational time is allowed. The details of these experiments, as well as comparison with other algorithms, can be found in [16].

2 Preliminaries

For convenience, let

$$CI_j = \{i \in M \mid x_j \in C_i \text{ or } \overline{x}_j \in C_i\}, \quad (3)$$

$$t = \max_{j \in N} |CI_j|,$$

$$VI_i = \{j \in N \mid x_j \in C_i \text{ or } \overline{x}_j \in C_i\}, \quad (4)$$

$$\ell = \max_{i \in M} |VI_i| = \max_{i \in M} |C_i|,$$

where t denotes the maximum number of appearances of a variable and ℓ denotes the maximum number of literals in a clause. These always satisfy $t \leq m$ and $\ell \leq n$, and in most cases $t \ll m$ and $\ell \ll n$. For a subset $S \subseteq N$ and a vector $v \in \{0,1\}^n$, let $v\updownarrow S$ denote the vector obtained from v by flipping the 0-1 assignment of variables in S (i.e., $D(v, v\updownarrow S) = S$), and call the following increase in the objective function as the flip gain,

$$\Delta f(v, S) = f(v\updownarrow S) - f(v) = \sum_{i \in M} \Delta f_i(v, S), \quad (5)$$

where

$$\Delta f_i(v, S) = w_i\{\theta_i(v\updownarrow S) - \theta_i(v)\}.$$

3 LS Implementation with Memory

In this section, we propose an efficient method to evaluate solutions in the neighborhood by making use of the memory structure. In the memory, information to evaluate each flip gain in $O(1)$ time is stored. Such information is also used to reduce the number of candidates in the neighborhood, as will be explained in 3.2 and 3.3.

3.1 Calculation of the Flip Gain

In this subsection, we propose an algorithm that calculates all flip gains (5) in one-round in $O(2^r|NB_r| + rt\ell^r)$ time in the worst case.

Intuitively, we first prepare flip gains of all candidates in 1-flip neighborhood. For the candidates in 2-flip neighborhood, instead of storing the flip gains, we evaluate the flip gain $\Delta f(v, \{j_1, j_2\})$ as the sum of $\Delta f(v, \{j_1\})$, $\Delta f(v, \{j_2\})$ and the adjustment $g(v, \{j_1, j_2\})$, whose definition will be given by (7) later. For example, consider a clause C_i where $j_1, j_2 \in VI_i$ and $P_i(v) = \emptyset$. In this case, $\Delta f_i(v, \{j_1\}) = \Delta f_i(v, \{j_2\}) = w_i$ holds, and we need the adjustment term $-w_i$ so that we obtain $\Delta f_i(v, \{j_1, j_2\}) = \Delta f_i(v, \{j_1\}) + \Delta f_i(v, \{j_2\}) - w_i = w_i$. The value $g(v, \{j_1, j_2\})$ represents the sum of such adjustments for all clauses C_i. By storing such adjustments, each solution in the 2-flip neighborhood can be evaluated in $O(1)$ time. Important point is that $g(v, \{j_1, j_2\})$ takes nonzero value only if $j_1, j_2 \in VI_i$ holds for some clause C_i. As ℓ is usually small, such cases occur rather rarely. For the same reason, the update of such $g(v, \{j_1, j_2\})$ values at each move will not be expensive. This idea can be generalized to the case of $r \geq 3$ by using the principle of inclusion and exclusion.

In general, let

$$g_i(v, S) = \begin{cases} (-1)^{|S|-|P_i(v)|+1}\, w_i, & \text{if } P_i(v) \subseteq S \subseteq VI_i \\ 0, & \text{otherwise,} \end{cases} \tag{6}$$

and

$$\begin{aligned} g(v, S) &= \sum_{i \in M} g_i(v, S) \\ &= \sum_{i \in \cap_{j \in S} CI_j \text{ s.t. } P_i(v) \subseteq S} (-1)^{|S|-|P_i(v)|+1}\, w_i. \end{aligned} \tag{7}$$

Here, it is noted that $g(v, \{j\}) = \Delta f(v, \{j\})$ holds for all $j \in N$.

Lemma 1. *The flip gain can be obtained by*

$$\Delta f(v, S) = \sum_{S' \subseteq S, S' \neq \emptyset} g(v, S'). \tag{8}$$

The important point is that we keep only nonzero $g(v,S)$. Here we assume the following property, which is usually valid (e.g., by using hash technique). More discussions are found in [16].

Assumption 1 *Each access to $g(v,S)$ is possible in $O(1)$ time and all nonzero $g(v,S)$ necessary for neighborhood $NB_r(v)$ can be stored in $O(|\{S \subseteq N|\ g(v,S) \neq 0$ and $|S| \leq r\}|)$ space.*

Lemma 2. [16] *If the current solution v is randomly chosen from $\{0,1\}^n$ with probability $1/2^n$, then, under Assumption 1, the expected number of nonzero $g(v,S)$ necessary for neighborhood $NB_r(v)$ is $O(n+m)$ for any constant r.*

Denote by $\hat{g}(S)$ the value of $g(v,S)$ stored in the memory for the current solution v. The update of $\hat{g}(S)$ for the move from v to v' with $|D(v,v')| \leq r$ is done as follows.

Algorithm UPDATE
for each $i \in \cup_{j \in D(v,v')} CI_j$ **do**
  for each S such that $P_i(v) \subseteq S \subseteq VI_i$ and $1 \leq |S| \leq r$ **do**
    Set $\hat{g}(S) := \hat{g}(S) - g_i(v,S)$
  end
  for each S such that $P_i(v') \subseteq S \subseteq VI_i$ and $1 \leq |S| \leq r$ **do**
    Set $\hat{g}(S) := \hat{g}(S) + g_i(v',S)$
  end
end.

The computation time needed for this is $O(rt\ell^r)$ in the worst case. By (8), $\Delta f(v,S)$ can be calculated in $O(2^r)$ time for each S with $|S| \leq r$. Therefore, one-round time is $O(|NB_r|2^r + rt\ell^r) = O(n^r 2^r + rt\ell^r)$ in the worst case.

Lemma 3. [16] *If the current solution v is randomly chosen from $\{0,1\}^n$ with probability $1/2^n$, then, under Assumption 1, the expected time to update $\hat{g}(S)$ is $O(t)$ for any constant r, and the expected one-round time is $O(|NB_r| + t) = O(n^r + t)$.*

Let us also consider the computational time for 1-flip neighborhood. If we store all the 1-flip candidates with $g(v,\{j\}) > 0$ (all of which give better solutions) in a linked list, it is possible in $O(1)$ time to find a better solution in $NB_1(v)$ or to conclude that v is locally optimal. The update of such list can be executed in $O(1)$ time for each change of $g(v,\{j\})$ value. Therefore, one-round time for LS with 1-flip neighborhood using FA strategy is $O(t\ell)$ in the worst case and $O(t)$ in the expected case.

3.2 Restriction of the 2-Flip Neighborhood

In this subsection, we derive a condition that reduces the number of candidates in 2-flip neighborhood without sacrificing the solution quality. The following lemma is immediate from (8).

Lemma 4. *If $\Delta f(v,\{j_1\}) \le 0$ and $\Delta f(v,\{j_2\}) \le 0$ hold for the current solution v, then $\Delta f(v,\{j_1,j_2\}) > 0$ is possible only if $g(v,\{j_1,j_2\}) > 0$ holds.*

Define

$$NB_2'(v) = \{v \updownarrow S \mid g(v,S) > 0 \text{ and } |S| \le 2\}.$$

Then, by the above lemma, we have only to check $NB_2'(v)$ to find a better solution in $NB_2(v)$ or to conclude local optimality. To do this efficiently, we store all the elements in $\{S \subseteq N \mid g(v,S) > 0 \text{ and } |S| \le 2\}$ in a linked list; thereby all candidates in $NB_2'(v)$ can be scanned in $O(|NB_2'|)$ time. The update of this list is possible in $O(1)$ time for each change of the $g(v,S)$ value. Therefore, the worst case one-round time of LS using this restricted 2-flip neighborhood is $O(|NB_2'| + t\ell^2)$ and its expected one-round time is $O(|NB_2'| + t)$. As for the size of NB_2', we have the following lemma.

Lemma 5. [16] *$|NB_2'(v)| = O(n + m\ell)$ holds for any v. If we assume that the current solution v is randomly chosen from $\{0,1\}^n$ with probability $1/2^n$, then $E(|NB_2'(v)|) = O(n+m)$.*

Based on this lemma, we have the following theorem.

Theorem 6. [16] *Under Assumption 1, the LS with the restricted 2-flip neighborhood NB_2' requires $O(n + m\ell + t\ell^2)$ one-round time and $O(n + m\ell^2)$ memory space in the worst case. If we assume that the current solution v is randomly chosen from $\{0,1\}^n$ with probability $1/2^n$, then the expected one-round time and the expected memory size both become $O(n+m)$.*

3.3 Restriction of the 3-Flip Neighborhood

In this subsection, we derive a condition that reduces the number of candidates in 3-flip neighborhood without sacrificing the solution quality.

Lemma 7. [16] *If $\Delta f(v,\{j_a\}) \le 0$ for all $a \in \{1,2,3\}$ and $\Delta f(v,\{j_a,j_b\}) \le 0$ for all $a,b \in \{1,2,3\}$ with $a \ne b$ hold, then $\Delta f(v,\{j_1,j_2,j_3\}) > 0$ is possible only if at least one of the following two conditions holds: (1) $g(v,\{j_1,j_2,j_3\}) > 0$, (2) $g(v,\{j_a,j_b\}) > 0$ and $g(v,\{j_b,j_c\}) > 0$ for some $a,b,c \in \{1,2,3\}$ (a, b and c are all distinct).*

Let

$$P2(v) = \{\{j_1,j_2,j_3\} \subseteq N \mid g(v,\{j_1,j_2\}) > 0 \text{ and } g(v,\{j_2,j_3\}) > 0\} \quad (9)$$

and

$$NB_3'(v) = \{v \updownarrow S \mid g(v,S) > 0 \text{ and } |S| \le 3\} \cup \{v \updownarrow S \mid S \in P2(v)\}.$$

Then, by Lemma 7, we have only to check $NB_3'(v)$ to find a better solution in $NB_3(v)$ or to conclude local optimality. The candidates in $NB_3'(v)$ can be efficiently scanned by using a linked list as in the case of NB_2' along with an additional data structure, whose details are omitted here (see [16]). As for the size of NB_3', we have the following lemma.

Lemma 8. [16] $|NB'_3(v)| = O(m\ell^3 + t^2\ell^2 n)$ *holds for any* v. *If we assume that the solution* v *is randomly chosen from* $\{0,1\}^n$ *with probability* $1/2^n$, *then* $E(|NB'_3(v)|) = O(m + t^2 n)$ *holds.*

Based on this lemma, we have the following theorem.

Theorem 9. [16] *Under Assumption 1, the LS with the restricted 3-flip neighborhood* NB'_3 *requires* $O(m\ell^3 + t^2\ell^2 n)$ *one-round time and memory space in the worst case. If we assume that the current solution* v *is randomly chosen from* $\{0,1\}^n$ *with probability* $1/2^n$, *then the expected one-round time and the expected memory space both become* $O(m + t^2 n)$.

4 Frameworks of ILS and TS

Based on Theorems 6 and 9, the LS with the restricted 2 and 3-flip neighborhoods can be implemented so that they become much faster and more practical than those with the ordinary 2 and 3-flip neighborhoods. We implemented random multi-start local search (MLS), iterated local search (ILS), and tabu search (TS), using the ordinary 1-flip neighborhood and the restricted 2 and 3-flip neighborhoods. Since the main target of our experiment is to compare the effect of neighborhoods, simple implementations are employed for the above three search algorithms, rather than pursuing algorithmic perfection (e.g., long-term memory is not incorporated in TS). The ILS and TS in our experiment are described as follows.

Algorithm ILS

1. Randomly generate a solution v. Set $v^* := v$.
2. Improve v by applying LS with NB'_r, and let v' be the obtained locally optimal solution.
3. If $f(v') > f(v^*)$, set $v^* := v'$.
4. If the computational time exceeds the given bound, output v^* and stop; otherwise, randomly choose a solution v'' in $NB_\kappa(v^*)$ (κ is a prespecified integer), let $v := v''$, and return to Step 2.

Algorithm TS

1. Randomly generate a solution v. Set $v^* := v$, and $TL := \emptyset$.
2. Find $v' \in NB'_r(v)\backslash TL$ with $f(v') > f(v)$ or $v' \in NB'_r(v)$ with $f(v') > f(v^*)$ using FA strategy. If no such solution exists, let v' be the best solution in $NB_1(v)\backslash TL$.
3. Set $v := v'$, and $TL := \{v\} \cup$ {solutions obtainable from v by flipping at least one variable flipped in the last τ moves} (τ is a prespecified integer). If $f(v') > f(v^*)$, set $v^* := v'$.
4. If the computational time exceeds the given bound, output v^* and stop; otherwise, return to Step 2.

In Step 2 of TS, we restrict the degrading moves to 1-flip neighborhood. Its reason is explained as follows. If $f(v') \leq f(v)$ for all $v' \in NB_r(v) \backslash TL$, the best solution in $NB_r(v) \backslash TL$ might be included in $NB_r(v) \backslash NB'_r(v)$. In other words, the restricted neighborhood is valid only for improving moves. Therefore, we avoid the use of restricted neighborhood NB'_r $(r = 2, 3)$ for degrading moves in Step 2 of TS.

5 Experimental Results

The algorithms were coded in C language and run on a workstation Sun Ultra 2 Model 2300 (300 MHz, 1 GB memory). We tested four types of problem instances: (1) randomly generated instances with unit weights (RNDU), (2) randomly generated instances with general weights (RNDW), (3) instances derived from the set covering problem (SCP) and (4) an instance derived from a time tabling problem (TTP). RNDU instances were generated according to [6]. The clauses for RNDW instances were similarly generated, and the integer weights were randomly chosen from [1, 1000] as described in [10]. SCP instances were transformed from 10 set cover instances `scp41` $\sim$ `scp410`, which were taken from the web site of OR-Library. The description of TTP instance is in [1].

Table 1 shows the average sizes of the restricted 2 and 3-flip neighborhoods for RNDW instances. The data are the average of independent 10 runs of LS starting from randomly generated solutions. The ratios between the observed values and the theoretical expectation (i.e., $E(|NB'_2(v)|) = O(n + m)$ and $E(|NB'_3(v)|) = O(m + t^2 n)$), and the ratios between the observed values and the ordinary 2 and 3-flip neighborhood sizes (i.e., $|NB_2(v)| \simeq \binom{n}{2}$ and $|NB_3(v)| \simeq \binom{n}{3}$) are also shown. It is observed that the ratio between the observed values and the theoretical expectation is almost constant (i.e., justifying the theory). It is also observed that the restriction is quite effective especially for larger n.

We then compare the performance of the restricted 2 and 3-flip neighborhoods with that of 1-flip neighborhood for MLS, ILS and TS. For comparison purposes, we also show the results of two local search algorithms called GSAT [11] and WALKSAT (abbreviated as WSAT) [14], tabu search for constraint satisfaction problem (abbreviated as TS-CSP) [8], and an exact algorithm for SAT problem called POSIT [2]. The codes of GSAT, WSAT and POSIT are taken from the web sites of the respective authors, and the code of TS-CSP is sent from the authors. All of GSAT, WSAT and TS-CSP are based on 1-flip neighborhood. ILS, TS and WSAT include parameters κ, τ and p respectively, which are carefully tuned. (The parameters κ and τ are explained in Section 4, and p is the parameter called noise in the program code, which is also written as p in [14].) Their values are also shown in the tables.

Table 2 shows the average computational time of the tested six algorithms for problem SAT (i.e., to find satisfying assignments) with RNDU instances of $n = 1000$ and $m = 7700$, where all instances have satisfying assignments. MLS with $r = 1, 2, 3$ and TS with $r = 3$ could not find any satisfying assignment for all of the 10 instances; and hence, the results are omitted. The results indicate that 1-

Table 1. Average sizes of the restricted neighborhoods for RNDW instances.

				2-flip neighborhood			3-flip neighborhood		
n	m	ℓ	t	avr. $\lvert NB_2'\rvert$	$\frac{\lvert NB_2'\rvert}{n+m}$	$\frac{\lvert NB_2'\rvert}{\binom{n}{2}}$	avr. $\lvert NB_3'\rvert$	$\frac{\lvert NB_3'\rvert}{m+t^2n}$	$\frac{\lvert NB_3'\rvert}{\binom{n}{3}}$
100	850	14	63	470.86	0.496	0.095	5488.72	0.0138	0.0339
100	1150	12	76	617.78	0.494	0.125	8714.01	0.0151	0.0539
100	2300	15	138	1088.08	0.453	0.220	22784.83	0.0119	0.1409
1000	7700	14	60	4637.18	0.533	0.009	56335.04	0.0156	0.0003
1000	11050	15	84	6493.01	0.539	0.013	102766.59	0.0145	0.0006
1000	22100	15	145	12453.54	0.539	0.025	343105.15	0.0163	0.0021

Table 2. Average computational time in seconds of six algorithms for SAT with 10 RNDU instances ($n = 1000$ and $m = 7700$). The numbers in parentheses denote how many instances were given satisfying assignments within 300 seconds.

ILS			TS		GSAT	WSAT	TS-CSP	POSIT
$r=1$	$r=2$	$r=3$	$r=1$	$r=2$				
$\kappa=16$	$\kappa=16$	$\kappa=16$	$\tau=160$	$\tau=180$		$p=0.5$		
49.5 (8)	113.6 (8)	134.4 (4)	22.0 (9)	82.5 (8)	37.7 (8)	0.13 (10)	37.5 (9)	3.7 (10)

flip neighborhood is more effective than the proposed 2 and 3-flip neighborhoods, and WSAT and POSIT are much more effective than other tested algorithms.

Tables 3 $\sim$ 5 show the average errors in % from the trivial bound $\sum_{i\in M} w_i$, i.e.,

$$\text{error of a solution } v = \left(1 - \frac{f(v)}{\sum_{i\in M} w_i}\right) \times 100\ (\%),$$

for RNDU, RNDW and SCP instances, respectively, and Table 6 shows the number of unsatisfied clauses for the TTP instance. It is known that no satisfying assignment exists for all of these instances. For such instances, the algorithm POSIT just answers 'no' and does not seek for an assignment with maximum weights; and hence, the results are omitted. Here we note that algorithms GSAT and WSAT are originally developed for unweighted instances, and weights are represented by multiplexing each clause C_i w_i times.

Table 3 shows the results for 10 RNDU instances of $n = 1000$ and $m = 11050$. We can observe that 3-flip neighborhood gives the best results for MLS, and 2-flip gives the best results for ILS; however, 1-flip is the best choice for TS. It is also observed that the performance of TS with $r = 1, 2$ is better than GSAT, WSAT and TS-CSP. Similar results are observed in Table 4 for RNDW instances of $n = 1000$ and $m = 11050$.

Table 5 shows the results for 10 SCP instances of $n = 1000$ and $m = 1200$. We can observe that 3-flip neighborhood gives the best results for all of MLS, ILS and TS, among which ILS with 3-flip neighborhood gives the best performance.

Table 3. Results for 10 RNDU instances of $n = 1000$ and $m = 11050$.

algorithm	r	error (%)	#moves	CPU secs.
MLS	1	1.2606	2079353.8	300.0
MLS	2	0.8154	297311.8	300.2
MLS	3	0.7231	15367.3	303.7
ILS ($\kappa = 8$)	1	0.5113	1736761.0	300.0
ILS ($\kappa = 16$)	2	0.5005	117468.7	300.0
ILS ($\kappa = 16$)	3	0.5213	3285.6	301.0
TS ($\tau = 80$)	1	0.4027	2022823.5	300.0
TS ($\tau = 80$)	2	0.4443	33335.2	300.0
TS ($\tau = 80$)	3	0.6271	992.2	300.7
GSAT		0.5484	5000000.0	290.9
WSAT ($p = 0$)		0.4525	25000000.0	306.5
TS-CSP		0.4606	36374.7	300.0

Table 4. Results for 10 RNDW instances of $n = 1000$ and $m = 11050$.

algorithm	r	error (%)	#moves	CPU secs.
MLS	1	0.6461	2349333.3	300.0
MLS	2	0.3525	320681.5	300.3
MLS	3	0.3056	14760.5	305.2
ILS ($\kappa = 16$)	1	0.2825	1956733.9	300.0
ILS ($\kappa = 64$)	2	0.2244	208575.2	300.1
ILS ($\kappa = 16$)	3	0.2437	4724.1	302.4
TS ($\tau = 120$)	1	0.1923	568919.4	300.0
TS ($\tau = 120$)	2	0.2205	26602.0	300.1
TS ($\tau = 140$)	3	0.2823	1229.1	301.4
GSAT		0.7348	15000.0	385.8
WSAT ($p = 20$)		0.2733	100000.0	390.6
TS-CSP		0.2812		540.0

Table 6 shows the results for the TTP instance of $n = 900$ and $m = 236539$. This instance is a real world time tabling problem [1]. The best result reported in [1] is also indicated as CIKM for comparison purpose. We can observe that 3-flip neighborhood is best for MLS, and 2-flip is best for ILS and TS. The overall best performance is given by ILS with 2-flip neighborhood, which is competitive with TS-CSP. Here we note that the size of CSP formulation of this instance has 900 0-1 variables and 774 constraints, which is much smaller than that of MAX SAT.

In summary, we have the following observation.

1. For randomly generated instances, 1-flip neighborhood under the framework of TS and WSAT gives the best performance.

Table 5. Results for 10 SCP instances of $n = 1000$ and $m = 1200$.

algorithm	r	error (%)	#moves	CPU secs.
MLS	1	6.8618	14830875.3	300.0
MLS	2	1.2164	2698703.0	300.0
MLS	3	0.7315	51890.7	301.0
ILS ($\kappa = 8$)	1	0.7606	4369174.8	300.0
ILS ($\kappa = 16$)	2	0.7261	186745.3	300.0
ILS ($\kappa = 16$)	3	0.7223	4437.3	300.8
TS ($\tau = 160$)	1	1.0586	384690.2	300.0
TS ($\tau = 100$)	2	0.8396	82043.7	300.0
TS ($\tau = 50$)	3	0.7418	2076.2	300.7
GSAT		1.5009	600000.0	461.4
WSAT ($p = 0$)		0.7797	6000000.0	368.7
TS-CSP		0.7327	210691.8	300.0
optimal		0.7214		

Table 6. Results for the TTP instance of $n = 900$ and $m = 236539$.

algorithm	r	#unsat	#moves	CPU secs.
MLS	1	124	416164	1500
MLS	2	105	207052	1500
MLS	3	99	33357	1500
ILS ($\kappa = 4$)	1	86	490890	1500
ILS ($\kappa = 1$)	2	85	126068	1500
ILS ($\kappa = 1$)	3	91	5711	1500
TS ($\tau = 20$)	1	97	788431	1500
TS ($\tau = 40$)	2	88	147049	1500
TS ($\tau = 10$)	3	94	4599	1500
CIKM		93	5000000	—
GSAT		125	3000000	2497
WSAT ($p = 0$)		94	3000000	1517
TS-CSP[1]		85	101555	1800

[1]Size of CSP formulation: 900 0-1 variables, 774 constraints.

2. For problems with structures, such as the set covering problem and the time tabling problem, the restricted 2 and 3-flip neighborhoods exhibit good prospects under various metaheuristic frameworks.

6 Conclusion

In this paper, we proposed efficient implementations of 2 and 3-flip neighborhoods for the MAX SAT. It is shown that the expected one-round time and the memory space for 2-flip (resp., 3-flip) neighborhood is $O(n+m)$ (resp., $O(m+t^2n)$). It is also shown that the expected one-round time for 1-flip neighborhood with first admissible (FA) move strategy is $O(t)$. The computational results with up to $n = 1000$ random instances indicate that these estimates by the expectation represent the real performance of LS well. The computational experiments for four types of problem instances show that the proposed 2 and 3-flip neighborhoods are effective for structured problems, while 1-flip neighborhood gives better performance for random instances.

References

1. B. Cha, K. Iwama, Y. Kambayashi and S. Miyazaki, Local search algorithms for partial MAXSAT, *Proc. AAAI* (1997) 263–268.
2. J.W. Freeman, Improvements to propositional satisfiability search algorithms, Dissertation, University of Pennsylvania, 1995.
3. A.S. Fukunaga, Variable-selection heuristics in local search for SAT, *Proc. AAAI* (1997) 275–280.
4. J. Gu, Efficient local search for very large-scale satisfiability problems, *SIGART Bulletin*, 3 (1992) 8–12.
5. P. Hansen and B. Jaumard, Algorithms for the maximum satisfiability problem, *Computing*, 44 (1990) 279–303.
6. J.N. Hooker and C. Fedjki, Branch-and-cut solution of inference problems in propositional logic, *Annals of Mathematics and Artificial Intelligence*, 1 (1990) 123–139.
7. B. Mazure, L. Saïs and É. Grégoire, Tabu search for SAT, *Proc. AAAI* (1997) 281–285.
8. K. Nonobe and T. Ibaraki, A tabu search approach to the CSP (constraint satisfaction problem) as a general problem solver, *European J. Oper. Res., Special Issue on Tabu Search* (to appear).
9. P.M. Pardalos, L.S. Pitsoulis and M.G.C. Resende, A parallel GRASP for MAX-SAT problems, *LNCS*, 1180 (1996) 575–585.
10. M.G.C. Resende, L.S. Pitsoulis and P.M. Pardalos, Approximate solution of weighted MAX-SAT problems using GRASP, *DIMACS Series on Discrete Mathematics and Theoretical Computer Science* (to appear).
11. B. Selman, H. Levesque and D. Mitchell, A new method for solving hard satisfiability problems, *Proc. AAAI* (1992) 440–446.
12. B. Selman and H.A. Kautz, An empirical study of greedy local search for satisfiability testing, *Proc. AAAI* (1993) 46–51.
13. B. Selman and H.A. Kautz, Domain-independent extensions to GSAT: solving large structured satisfiability problems, *Proc. IJCAI* (1993) 290–295.
14. B. Selman, H.A. Kautz and B. Cohen, Noise strategies for improving local search, *Proc. AAAI* (1994) 337–343.
15. B. Selman, H.A. Kautz and B. Cohen, Local search strategies for satisfiability testing, *DIMACS Series in Discrete Mathematics and Theoretical Computer Science*, 26 (1996) 521–531.
16. M. Yagiura and T. Ibaraki, Efficient 2 and 3-flip neighborhood search algorithms for the MAX SAT — Part I: Theoretical analysis, prepared for publication; Part II: Experimental results, ditto.

Algorithms for Almost-uniform Generation with an Unbiased Binary Source

Ömer Eğecioğlu[1] and Marcus Peinado[2]*

[1] Department of Computer Science, University of California, Santa Barbara, CA 93106, USA, omer@cs.ucsb.edu

[2] Institute for Algorithms and Scientific Computing, GMD National Research Center, 53754 Sankt Augustin, Germany, peinado@gmd.de

Abstract. We consider the problem of uniform generation of random integers in the range $[1, n]$ given only a binary source of randomness. Standard models of randomized algorithms (e.g. probabilistic Turing machines) assume the availability of a random binary source that can generate independent random bits in unit time with uniform probability. This makes the task trivial if n is a power of 2. However, exact uniform generation algorithms with bounded run time do not exist if n is not a power of 2.

We analyze several *almost-uniform* generation algorithms and discuss the tradeoff between the distance of the generated distribution from the uniform distribution, and the number of operations required per random number generated. In particular, we present a new algorithm which is based on a circulant, symmetric, rapidly mixing Markov chain. For a given positive integer N, the algorithm produces an integer i in the range $[1, n]$ with probability $p_i = p_i(N)$ using $O(N \log n)$ bit operations such that $| p_i - 1/n | < c\,\beta^N$, for some constant c, where

$$\beta = \frac{2^{\frac{1}{4}}}{\pi} \left(\sqrt{2\sqrt{2} - \sqrt{5 - \sqrt{5}}} \right) \approx 0.4087.$$

This rate of convergence is superior to the estimates obtainable by commonly used methods of bounding the mixing rate of Markov chains such as conductance, direct canonical paths, and couplings.

Keywords. Random number generation, uniform distribution, Markov chain, rapid mixing, eigenvalue, circulant matrix.

1 Introduction

We consider the generation of almost-uniform random integers in the range $[1, n]$, taking into account the required time, space, and number of random bits. The

* Part of the work was done at the International Computer Science Institute (ICSI), Berkeley, California

basic assumption is that independent random bits can be generated in unit time. If n is an exact power of 2, say $n = 2^m$, then the generation of a uniformly distributed random integer in the range $[1, n]$ is easily accomplished in time $O(m) = O(\log n)$ by generating m consecutive random bits. However, if n is not a power of 2, no algorithm with bounded running time can generate numbers in $[1, n]$ from the exact uniform distribution (see below).

The task of generating uniformly distributed random elements of a set whose size is not an exact power of two arises frequently in the study of randomized algorithms and is usually treated as a primitive operation. This is in part justified by the fact that simple and efficient *almost-uniform* generation algorithms are known. However, it appears that the exact costs and trade-offs between accuracy and required resources of these algorithms have not been analyzed in detail. One of our aims is to explore which options exist and to compare their costs. We present two new algorithms – one based on a rapidly mixing Markov chain and one based on a reduction from approximate counting to almost uniform generation – and compare their resource requirements with those of the well-known modular algorithms.

Sinclair [18] considers the problem on an abstract level, and shows polynomial time equivalence between almost uniform generation on probabilistic Turing machines and on a different machine model which allows biased coin flips.

One important class of applications which requires uniform generators for sets of arbitrary size is the simulation of *heat bath* Markov chains (cf. [5] for a precise definition). In practice, the size of the sets can be extremely large [15]. Heat bath Markov chains are one of the standard tools in computational physics, and are used frequently in high-precision numerical simulations. It is easy to show that a bias in the distribution of the generator translates directly into a similar bias in the output distribution of the Markov chain.

We present a new algorithm which is based on the simulation of a rapidly-mixing circulant Markov chain. Its analysis gives a direct bound on the second-largest eigenvalue of the transition matrix of the Markov chain and is of interest in its own right. In particular, we observe that the commonly used methods of bounding the mixing rate of Markov chains (conductance [18], direct canonical paths [17], couplings [2]), yield weaker bounds than the one obtained here. Direct bounds on the second-largest eigenvalue of transition matrices have been obtained previously, mostly based on algebraic properties of the underlying domain (e.g. [3]). However, the structure of our Markov chain, as well as the technique used to bound its mixing rate seem different from previous results.

The probabilistic Turing machine (PTM) is the most commonly used machine model in the study of randomized algorithms [14, 18]. It is a standard Turing machine equipped with the ability to generate (or access) random bits in unit time. A PTM is deterministic, except for special coin-tossing states in which there are exactly two possible transitions, determined by the flip of an unbiased coin.

Proposition 1. *Given $n \in \mathbb{N}$ which is not a power of 2, let A_n be a randomized algorithm (PTM) which outputs numbers in $[1, n]$ and whose running time is* *bounded by $t_n \in \mathbb{N}$. Let $r_n \leq t_n$ be an upper bound on the number of random bits used by A_n. Let p_i be the probability that A_n outputs $i \in [1, n]$. There exists $i \in [1, n]$ such that*

$$|p_i - 1/n| \geq 2^{-(r_n+1)}.$$

Proof. Omitted. See [9].

Intuitively, A_n has to place 2^{r_n} balls (elementary events) into n bins. If n is not a power of 2, some bins have to receive at least one ball more than others. The situation is slightly different for Las Vegas type algorithms whose run time is not bounded. In the simplest case, the algorithm can use r_n random bits, assign an equal number of elementary events to each number in $[1, n]$, and decide to use more random bits or, simply, not terminate with the remaining probability. We will concentrate on algorithms whose running time is bounded, and refer to Las Vegas type algorithms only where appropriate.

Since producing the exact uniform distribution on $[1, n]$ is not possible, we try to generate integers in $[1, n]$ with an *almost-uniform* distribution. We use the well-known *relative pointwise distance r.p.d.* (See e.g. [18]) to measure the closeness of the output distribution and the uniform distribution: The r.p.d. between two probability distributions p, q on a finite set X ($q_i > 0$ for all $i \in X$) is defined as

$$\Delta(p, q) = \max_{i \in X} \frac{|p_i - q_i|}{q_i}$$

In the following, q will always be the uniform distribution, and we write $\Delta(p)$ instead of $\Delta(p, q)$ to denote the r.p.d. of p from the uniform distribution. Thus $\Delta(p) = \max_{i \in X} |np_i - 1|$.

The rest of this paper is organized as follows: In Sect. 2, we describe the Markov chain algorithm. Our main result on the bound of the mixing rate of the Markov chain is stated in this section. Section 3 analyzes the resource requirements of three alternative algorithms. Remarks and conclusions are given in Sect. 4. Proofs of the results that are omitted due to space constraints can be found in the full paper [9].

2 A Rapidly Mixing Circulant Markov Chain

In this section, we describe an algorithm based on the simulation of a rapidly mixing Markov chain M. In $O(N \log n)$ time, this algorithm produces a random integer i in the range $[1, n]$ with distribution p such that

$$\Delta(p) \leq n\,\beta^N \,, \quad \text{where} \quad \beta = \frac{2^{\frac{1}{4}}}{\pi} \left(\sqrt{2\sqrt{2} - \sqrt{5 - \sqrt{5}}} \right) \approx 0.4087 \,. \qquad (1)$$

The bound $\beta \approx 0.4087$ deserves attention in two respects. Firstly, known algorithms reduce the r.p.d. only by a factor of 0.5 in each step. Similarly, standard methods for bounding the mixing rate of a Markov chain yield bounds which are worse than 0.5. These issues will be addressed below.

2.1 Construction of M

We define an $n \times n$ transition matrix $P = (\ p_{ij}\ )$ such that the corresponding Markov chain M on state space $\{1, 2, \ldots, n\}$ has the following properties: 1) M is ergodic with stationary distribution $\pi = (\frac{1}{n}, \frac{1}{n}, \ldots, \frac{1}{n})$; 2) M is rapidly mixing, i.e. the N-step transition matrix P^N converges quickly to the limiting probabilities; 3) M can be simulated efficiently. That is, the time to simulate one transition step is $O(\log n)$. The preprocessing time and space requirements for M are also $O(\log n)$. Given such P, the algorithm (referred to as Algorithm I) simulates N steps of M. The first condition guarantees that M converges to the uniform distribution, and the second condition ensures that a small number N of simulation steps is sufficient. The third condition ensures that each simulation step can be executed efficiently.

An $n \times n$ *circulant matrix* $C = C(a_1, a_2, \ldots, a_n)$ is a matrix of the form

$$\begin{bmatrix} a_1 & a_2 & \cdots & a_n \\ a_n & a_1 & \cdots & a_{n-1} \\ \vdots & & \vdots & \vdots \\ a_2 & a_3 & \cdots & a_1 \end{bmatrix}$$

where each row is a single right circular shift of the row above it [7].

Assume that n is not a power of 2, and let $m = \lfloor \log n \rfloor$. Then $\frac{n}{2} < 2^m < n$, and n can be written in the form $n = 2^m + p$ with $0 < p < 2^m$. Consider symmetric, circulant $n \times n$ 0-1 matrices $C = C(0, a_2, a_3, \ldots, a_n)$ where exactly 2^m of the entries $a_2, a_3, \ldots, a_n$ are equal to 1. Since we are forcing C to be symmetric, this imposes the condition $a_k = a_{n+2-k}$ for $k = 2, 3, \ldots, n$. For example, for $n = 7$, we have $m = 2$ and $p = 3$. In this case there are three such matrices: $C(0,1,1,0,0,1,1)$, $C(0,1,0,1,1,0,1)$, and $C(0,0,1,1,1,1,0)$. Each such matrix C defines an irreducible, aperiodic (i.e. ergodic) Markov chain M on n states $\{1, 2, \ldots, n\}$ whose transition matrix is $P = \frac{1}{2^m} C$. The symmetry of C guarantees that the stationary distribution of the corresponding Markov chain M is the uniform distribution on $\{1, 2, \ldots, n\}$. Note that the eigenvalues of P and C are related by a constant factor 2^m. Let $\bar{\lambda}_1$ denote the second largest eigenvalue of C. It is well-known that the mixing rate of M can be bounded by $\lambda_1 = 2^{-m} \bar{\lambda}_1$. The following inequality for the r.p.d. follows from [18, 6, 13]:

$$\Delta(p(N)) \leq n \lambda_1^N\ , \tag{2}$$

where $p(N)$ is the distribution on the states of M after N simulation steps. We consider the problem of picking the nonzero a_k's so that $\bar{\lambda}_1$ is minimized:

Theorem 1. *Suppose $n = 2^m + p$ with $0 < p < 2^m$. There exists a symmetric, circulant $n \times n$ 0-1 matrix $C^{\bar{s}} = C(0, a_2, a_3, \ldots, a_n)$ with 2^m nonzero entries in its first row such that*

$$2^{-m} \bar{\lambda}_1 \leq \frac{2^{\frac{1}{4}}}{\pi} \left(\sqrt{2\sqrt{2} - \sqrt{5 - \sqrt{5}}} \right) \approx 0.4087\ .$$

Furthermore, the first row of $C^{\bar{s}}$ contains at most two symmetrically placed blocks of 1's starting at column $\bar{s} = \lceil \frac{p}{10} \rceil + 1$.

The theorem is our main result. Its proof is several pages long and had to be omitted due to space restrictions. See [9].

We take $M = M^{\bar{s}}$ to be the Markov chain on $\{1, 2, \ldots, n\}$ whose transition matrix is $P = P^{\bar{s}} = \frac{1}{2^m} C^{\bar{s}}$. The structure of $C^{\bar{s}}$ is such that $p_{1j} = \frac{1}{2^m}$ if and only if

$$j \in \{\bar{s}+k \mid k = 0, 1, \ldots, 2^{m-1}-1\} \cup \{n+2-(\bar{s}+k) \mid k = 0, 1, \ldots, 2^{m-1}-1\} \ . \quad (3)$$

Since P is circulant, $p_{ij} = \frac{1}{2^m}$ if and only if j is in a translate modulo n of the set of indices in (3). Thus to move from a state i of M to state j, we only need to generate a random binary number r in the range $[0, 2^m - 1]$. We then use the high order bit to select the translate of one of the two sets of consecutive indices in (3). After this, the new state j is simply the $(r+1)$-st smallest index in the subset chosen. More formally, we describe the steps of this algorithm as Algorithm I. Let RANDOM$[0, 2^m - 1]$ denote a procedure which returns a random integer r in the range $[0, 2^m - 1]$ or, equivalently, m consecutive random bits provided by our machine model (PTM).

Algorithm I :
Input: n, N
Output: $i \in [1, n]$
begin
  $m := \lfloor \log n \rfloor$; $\bar{s} := \lceil \frac{n-2^m}{10} \rceil + 1$;
  $cur\_state := 1$;
    for $j := 1$ **to** N **do**
    begin
    $i :=$ RANDOM$[0, 2^m - 1]$;
    if $i \in [0, 2^{m-1})$ **then** $cur\_state := 1 + [\,(cur\_state - 1 + \bar{s} + i) \bmod n]$
    else $cur\_state := 1 + [(cur\_state - 1 + (n + 2 - (\bar{s} + i - 2^{m-1})) \bmod n]$
    end
  $i := cur\_state$;
  return(i);
end

The number of operations required to take one step on the Markov chain M is $O(m) = O(\log n)$. Thus, the total running time of Algorithm I is $O(N \log n)$. By (2) and Theorem 1, after starting from an arbitrary initial state and simulating N steps of M, the probability of being in some particular state j does not differ from $1/n$ by more than a constant multiple (w.r.t. N) of β^N, where β is as given in (1).

2.2 Other methods of bounding the mixing rate

We note that the bound of $\lambda_1 \leq 0.4087$ is obtained by a detailed analysis, taking special properties of circulant matrices into account. The well-known general methods for estimating mixing rates, while being useful general purpose tools, appear to be too coarse-grained to yield a similar bound. We outline this in the following paragraphs. Details can be found in [9].

The *conductance* Φ [18] which measures the expansion of the transition graph is often used to bound the second largest eigenvalue of the transition matrix via the inequality $\lambda_1 \leq 1 - \Phi^2/2$. Since, by definition, $\Phi \leq 1$, this method cannot yield a better bound than $0.5 > 0.4087$. A closer analysis for the particular case considered here shows that the conductance is significantly smaller than 1, and consequently the bound obtained in this fashion is actually much larger than 0.5.

The method of [17, 8] which bounds the second largest eigenvalue directly by a *direct canonical paths* argument (as opposed to going via the conductance) usually leads to tighter bounds than conductance-based methods. We can show by means of a counting argument that this approach does not yield a better bound than $\lambda_1 \approx 0.7$.

The *coupling* method tries to bound the mixing rate by a direct probabilistic argument and without bounding the eigenvalues. The use of the coupling method is based on [2], which bounds the mixing rate by $\mathrm{e}^{-1/(2\mathrm{e}T)}$, where T is called the coupling time. Basic but tedious steps show that $T > 2$, resulting in a mixing rate of at least $0.912 \gg 0.4087$.

3 Alternative Algorithms

In this section we analyze three alternate algorithms for the generation problem. Algorithms II and IV are straightforward modular algorithms. Algorithm III is a new algorithm and based on the reduction from almost-uniform generation to approximate counting [12].

Algorithm II: This algorithm is described in [10]: Generate a random sequence of $m = \lceil \log n \rceil$ bits. If the sequence is the binary representation of an integer i_1 in the range $[0, n-1]$, then return $i := i_1 + 1$. If not, generate another m-bit random number i_2 using the same process. If after N such trials, none of the integers $i_1, i_2, \ldots, i_N$ turns out to be in $[1, n]$, then return $i := i_N - 2^{m-1}$. A more formal description of this algorithm is given as Algorithm II.

Algorithm II :
  Input: n, N
  Output: $i \in [1, n]$
  $m := \lceil \log n \rceil$;
    for $j := 1$ **to** N **do**
      $i :=$ RANDOM$[0, 2^m - 1]$ $+1$;
      if $i \in [1, n]$ **then return**(i) and **exit**;
    return$(i - 2^{m-1})$;

Proposition 2. *Let $p_{II}(N)$ denote the output distribution of Alg. II. Then $\Delta(p_{II}(N)) \leq 2^{-N}$.*

Proof. Omitted.

Algorithm II can be run in Las Vegas mode by dropping the upper limit of N loop iterations. In this case, the expected running time $E(n)$ is

$$E(n) = \sum_{k=1}^{\infty} km \frac{n}{2^m} \left(\frac{r}{2^m}\right)^{k-1} \leq \frac{m}{(1-\frac{r}{2^m})^2} \leq \frac{m}{(1-\frac{1}{2})^2} \leq 8 \log n \ ,$$

where $r = 2^m - n$. Thus the expected running time of Algorithm II is no worse than $8 \log n$, independently of the parameter N. Using Chernoff bounds, it is easy to show that the running time is unlikely to exceed its expectation significantly.

Algorithm III: There is a close relation between almost-uniform generation problems and the corresponding approximate counting problems (computing the number of elements in the set) [12]. In our case, the solution to the counting problem is simply n, and the solutions of relevant subproblems are easily derived. This makes it possible to design a generation algorithm based on the well-known reduction from almost-uniform generation to approximate counting of [12].

Given a bitstring s, let $\text{solns}(s) = |\{x \in [0, n-1] : \exists v : sv = x\}|$ be the number of elements of $[0, n-1]$ whose binary representation begins with s. These solutions of counting subproblems are easily computed. The algorithm generates a random element of $[0, n-1]$ one bit at a time, starting with the most significant bit. At the start of the k-th round, the $k-1$ most significant bits have been determined. The invariant is that the probability of producing any given prefix is proportional to the number of elements of $[0, n-1]$ whose most significant bits coincide with this prefix. It is easy to show by induction that this relation will hold, if the next bit is set to 1 with probability solns(prefix $\circ$ 1)/solns(prefix), where $\circ$ denotes concatenation. If, at any given point, the prefix is such that $\text{solns(prefix)} = 2^i$ (for some $i > 0$), the process can be stopped. Algorithm III summarizes these steps.

Algorithm III :
    Input: $n > 1$
    Output: $\in [1, n]$
    prefix := ϵ; $k := \lceil \log n \rceil$; (* bitlength of $(n-1)$ *)
    repeat
        if solns(prefix $\circ$ 1) = 0 **then** prefix := prefix $\circ$ 0
        else
            with probability divide[solns(prefix $\circ$ 1), solns(prefix)],
                        set prefix := prefix $\circ$ 1;
            otherwise set prefix := prefix $\circ$ 0;
        $k := k - 1$;
    until $k = 0$ or solns(prefix) is a power of 2;
    prefix := prefix $\circ$ $0^{k - \log \text{solns(prefix)}}$;
    if solns(prefix) > 1 **then** prefix := prefix$\circ$RANDOM[0, solns(prefix) $-$ 1];
    return(prefix+1);

As stated, Algorithm III is an exact uniform generator. However, its running time is unbounded because the binary representation of $p1 := \frac{\text{solns}(\text{prefix} \circ 1)}{\text{solns}(\text{prefix})}$ may not be finite. One obtains an approximate version of the algorithm by truncating this fraction to some finite number m of bits. Steps similar to those of [18] show that $\Delta(p_{III}) \leq 2^{-N}$ if $m \geq 2\lceil \log n \rceil + N$. Thus, achieving an accuracy of 2^{-N} requires a total of at most $2\lceil \log n \rceil^2 + N\lceil \log n \rceil$ random bits. Note that there are only $\log n$ relevant values of $p1$. Those values can be obtained and stored in a precomputation step. Thus, the algorithm needs $O(\log^2 n + N \log n)$ time and space in the worst case.

The algorithm can be run in Las Vegas mode if the probabilistic decision ('with probability ...') is implemented appropriately. The details of this part of the algorithm can be found in [9]. For both the standard and the Las Vegas version, we have

Proposition 3. *The expected running time of Alg. III is* $O(\log n)$.

Again, one can use Chernoff bounds to show that the actual running time is concentrated around its expectation.

Algorithm IV: This algorithm appears to be widely used in practice: fix $m \gg \log n$, generate a random integer M in the range $[0, 2^m - 1]$ by generating m consecutive random bits, and output $M \bmod n$.

Algorithm IV :
  Input: n, m
  Output: $i \in [1, n]$
  begin
    $M :=$ RANDOM$[0, 2^m - 1]$;
    return$((M \bmod n) + 1)$;
  end

Proposition 4. *Let $p_{IV}(m)$ denote the output distribution of Alg. IV. Then* $\Delta(p_{IV}(m)) \leq n2^{-m}$.

We calculate the number of bit operations required by Algorithm IV. Let $b = \lfloor \log n \rfloor + 1$. Thus M and n are m-bit and b-bit integers with $m \geq b$. We can consider algorithms of varying complexity for the calculation of the remainder $R = (M \bmod n)$ depending on the sizes of the numbers involved. Straightforward division of M by n followed by a multiplication and subtraction to calculate R requires $O(mb)$ bit operations. Using the asymptotically faster Schönhage-Strassen algorithm for large integers, the integral quotient of a $2b$-bit number by a b-bit number, as well as the product of two b-bit numbers can be obtained in $O(b \log b \log \log b)$ bit operations [16, 1]. To use this algorithm for remainder calculation we prepend the binary representation of M with zeros if necessary and assume that $b - 1$ divides m. Write

$$M = \sum_{i=0}^{m/(b-1)-1} M_i 2^{(b-1)i}$$

where each M_i is $b-1$ bits. The remainders $r_i = 2^{(b-1)i} \bmod n$ for $i = 1, 2, \ldots, m/(b-1)-1$ can be computed by $m/(b-1)$ multiplications and divisions requiring $O(b \log b \log\log b)$ bit operations each, if the Schönhage-Strassen algorithm is used. After this phase, each quantity $M_i r_i \bmod n$ can be computed with an additional $O(b \log b \log\log b)$ bit operations. Finally the resulting $m/(b-1)$ remainders found are summed up modulo n in another $O(\frac{m}{b-1} b) = O(m)$ bit operations. The total number of operations required for the computation of R becomes $O(m \log b \log\log b) = O(m \log\log n \log\log\log n)$.

4 Comparisons and Conclusions

The maximum relative error between the generated distribution and the uniform distribution goes to zero geometrically for each of the four algorithms considered here: with rate 0.5 for Algorithms II, III, IV (using standard division), and with rate approximately 0.4087 for Algorithm I.

The following table compares the algorithms from a different perspective. It lists the resources required by each to produce one random number with error bound 2^{-k} (r.p.d. from uniformity), based on the bounds derived in the previous sections.

Algorithm	worst case		average case	
	time	random bits	time	random bits
I	$O((k+\log n)\log n)$	$0.775(k+\log n)\log n$	cf. worst case	cf. worst case
II	$O(k \log n)$	$k \log n$	$O(\log n)$	$O(\log n)$
III	$O((k+\log n)\log n)$	$(k+2\log n)\log n$	$O(\log n)$	$O(\log n)$
IV(simple)	$O((k+\log n)\log n)$	$k+\log n$	cf. worst case	cf. worst case
IV[16]	$O((k+\log n)\cdot\ldots$ $\cdot \log\log n \log\log\log n)$	$k+\log n$	cf. worst case	cf. worst case

It is seen that the algorithms have similar worst case running times, with Algorithm II being the fastest. The faster convergence rate of the Markov chain is hidden in the big-O notation. However, it reduces the number of random bits required by a factor of 0.775. Algorithm IV requires the smallest number of random bits and comes to within two bits of the lower bound of Prop. 1. Algorithms II and III can stop prematurely. Their average-case running times do not depend on k and are well below their worst-case times. This was reflected in a series of experiments we performed in which Algorithms II and III were significantly faster than Algorithms I and IV.

In the context of the construction of the Markov chain used in Algorithm I, one can address the problem of picking the best possible $n \times n$ 0-1 circulant matrix (in terms of the magnitude of the modulus of the second largest eigenvalue) for an arbitrary distribution of 1's in the first row. Theorem 1 only gives an upper bound for the modulus of the second largest eigenvalue and only in case the matrix is symmetric and the row sums are $2^{\lfloor \log n \rfloor}$. The advantage of the particular distribution of the 1's considered in Theorem 1 as blocks in the first row of the matrix is small storage and ease of transition selection for the associated Markov chain. Such a distribution, however, is not necessarily optimal for the solution of the more general problem in which constraints of space and constructibility are not critical issues.

References

1. A. Aho, J. Hopcroft, and J. Ullman. *Design and Analysis of Computer Algorithms.* Addison-Wesley Publishing Company, 1974.
2. D. Aldous. Random walks on finite groups and rapidly mixing Markov chains. In *Séminaire de Probabilités XVII*, Lecture Notes in Mathematics 986, pages 243–297. Springer-Verlag, 1982.
3. N. Alon and Y. Roichman. Random Cayley graphs and expanders, 1996. Manuscript.
4. N. Biggs. *Algebraic Graph Theory*, page 16. Cambridge University Press, 1974.
5. R. Bubley, M. Dyer, and C. Greenhill. Beating the 2Δ bound for approximately counting colourings: A computer-assisted proof of rapid mixing. In *Proceedings of the Ninth Annual ACM-SIAM Symposium on Discrete Algorithms*, San Francisco, California, 1998.
6. E. Çınlar. *Introduction to Stochastic Processes.* Prentice-Hall Inc., 1975.
7. P.J. Davis. *Circulant matrices.* Wiley, 1979.
8. P. Diaconis and D. Strook. Geometric bounds for eigenvalues of Markov chains. *Annals of Applied Probability*, 1:36–61, 1991.
9. Ö. Eğecioğlu and M. Peinado. Algorithms for Almost-uniform Generation with an Unbiased Binary Source. Technical Report TRCS98–04, University of California at Santa Barbara, 1998, (http://www.cs.ucsb.edu/TRs/).
10. L. Goldschlager, E.W. Mayr, and J. Ullman. Theory of parallel computation. Unpublished Notes, 1989.
11. I.S. Gradshteyn and I.M. Ryzhik. *Table of Integrals, Series, and Products*, page 30. Academic Press Inc., 1980.
12. M. Jerrum, L. Valiant, and V. Vazirani. Random generation of combinatorial structures from a uniform distribution. *Theoretical Computer Science*, 43:169–188, 1986.
13. J.G. Kemeny and J.L. Snell. *Finite Markov Chains.* Springer-Verlag, 1976.
14. R. Motwani. Lecture notes on approximation algorithms. Technical report, Stanford University, 1994.
15. M. Peinado. Random generation of embedded graphs and an extension to Dobrushin uniqueness. In *Proceedings of the 30th Annual ACM Symposium on Theory of Computing (STOC'98)*, Dallas, Texas, 1998.
16. A. Schönhage and V. Strassen. "Schnelle Multiplikation großer Zahlen", *Computing*, No. 7 (1971), 281–292.
17. A. Sinclair. Improved bounds for mixing rates of Markov chains and multicommodity flow. *Combinatorics, Probability & Computing*, 1:351–370, 1992.
18. A. Sinclair. *Algorithms For Random Generation And Counting.* Progress In Theoretical Computer Science. Birkhauser, Boston, 1993.

Improved Algorithms for Chemical Threshold Testing Problems

Annalisa De Bonis, Luisa Gargano, and Ugo Vaccaro

Dipartimento di Informatica e Applicazioni Università di Salerno 84081 Baronissi (SA), Italy {debonis,lg,uv}@dia.unisa.it

Abstract. We consider a generalization of the classical group testing problem. Let us be given a sample contaminated with a chemical substance. We want to estimate the unknown concentration c of this substance in the sample. There is a threshold indicator which can detect whether the concentration is at least a known threshold. We consider either the case when the threshold indicator does not affect the tested units and the more difficult case when the threshold indicator destroys the tested units. For both cases, we present a family of efficient algorithms each of which achieves a good approximation of c using a small number of tests and of auxiliary resources. Each member of the family provides a different tradeoff between the number of tests and the use of other resources involved by the algorithm. Previously known algorithms for this problem use more tests than most of our algorithms do. For the case when the indicator destroys the tested units, we also describe a family of efficient algorithms which estimates c using only a constant number of tubes.

1 Introduction

The well known group testing problem originated in the area of chemical analysis as a blood test technique employed to detect the infected members of a population [5, 10]. Since then, it has become clear that the group testing model occurs in a variety of situations, ranging from molecular biology applications [1, 3, 8], to multiaccess communication [11], software development [9] and many others. We refer to the monograph of Du and Hwang [6] for an excellent treatise on group testing. In the classical group testing problem, there is a set of elements each of which may be either good or defective. The problem consists of identifying all the defective elements using a minimum number of group tests. Damaschke [4] has considered a generalization of the group testing problem in which the problem consists of estimating the concentration of a chemical substance in a given sample contaminated with that substance. The search model described by Damaschke uses, as test device, a threshold indicator which gives a positive response if and only if the concentration of the tested sample is at least a fixed threshold. Tests are performed on units obtained by first diluting the original sample with water and then by either diluting or mixing previously generated units. The model described in [4] allows each unit to be tested more than once. It is the purpose of this paper to proceed further along the line of research initiated in [4], both by improving some of the results given therein and by considering a variation of the concentration problem which models the more realistic situation when units of liquid which have already been tested can not be tested again.

The model. We assume that a single unit of the sample, whose concentration we want to estimate, has been given. Tests are performed by means of a threshold indicator which gives a positive response if and only if the concentration of the tested sample is at least a fixed threshold. This threshold is adopted as measure of the concentration. For that reason, a positive response of our indicator means that the concentration in the tested sample is at least 1. Tests are performed on units of liquid obtained by means of merge operations. A merge operation may either involve units of liquid with different concentration or add units of water, that is liquid with concentration equal to 0, to units of contaminated liquid. The very first merge operation consists of diluting the original unit of sample with water. We assume that we dispose of an arbitrarily large reservoir of water. Let c denote the concentration of the original sample. A unit of liquid generated during the search process has concentration equal to rc, for some $r < 1$. The positive number r is said the *concentration ratio* of this unit of liquid. We denote k units of liquid with concentration equal to rc with the symbol $k \times r$. Therefore, a *sequence* $n_1 \times r_1, \ldots, n_m \times r_m$ denotes a situation where we dispose of n_i units with concentration ratio r_i, for $i = 1, \ldots, m$.
More precisely, the search model is defined by the following assumptions:
- Each test is performed on a single unit with unknown concentration rc. A positive answer indicates that $rc \geq 1$
- An integer number of units can be extracted from each available sample
- It is possible to merge an arbitrary number n of units. Merging n units with concentrations $c_1, c_2, \cdots, c_n$ yields n units each with concentration $\sum_{i=1}^{n} c_i/n$.

Our goal is to estimate the unknown concentration of the original sample up to some given accuracy.

Our results and outline of the paper. The main tool consists in having recognized our problem as a case of unbounded search problem [2] with particular constraints. We first consider the more interesting case when tests performed by the threshold indicator destroy the tested units and, as a consequence, units which have already been tested must be discarded. This case is treated in Sect. 2 where it is found a family of strategies which provide tradeoffs between the number of tests, the number of merge operations and the quantity of water involved in the search process. Our result implies an algorithm which finds an interval of length at most one including c, $c \geq 32$, by using $\lfloor \log c \rfloor + 2\lfloor \log \log c \rfloor + 3$ tests, at most $5\lfloor \log c \rfloor + 4$ units of water and at most $\frac{3}{2}\lfloor \log c \rfloor^2 + \frac{41}{6}\lfloor \log c \rfloor + \frac{47}{12}$ merge steps. Our algorithm compares favourably with the best algorithm given in [4], in that, even though it works under a more difficult test model, it performs a smaller number of tests while still using a logarithmic number of units of water. Both algorithms involve a logarithmic number of tubes in the search process. In an effort to reduce the number of tubes involved in the search process, in Sect.

2 we also consider the case when the algorithm may dispose only of a constant number of tubes.

The case when the threshold indicator does not affect the tested units is considered in Sect. 3. The best result given by [4] under this model, is an algorithm which allows to find an interval of length at most one including $c > 16$, using about $2\log c$ tests, $2\log c$ units of water and $\log^2 c$ merge steps. In Sect. 3, we describe a family of algorithms for approximating c with an error of at most one under this model. In particular, we present an algorithm which approximates $c \geq 8$ with this accuracy by performing $\lfloor \log c \rfloor + 2\lfloor \log\log c \rfloor + 3$ tests and by using at most $3\lfloor \log c \rfloor + 3$ units of water and at most $\frac{5}{6}\lfloor \log c \rfloor^2 + \frac{31}{6}\lfloor \log c \rfloor + \frac{10}{3}$ merge operations.

Finally, we remark that our best algorithms require a number of tests not far from the known lower bound on the number of tests for unconstrained unbounded search algorithms.

Due to space limits, most of the proofs are omitted from this extended abstract.

2 Testing in the Destructive Model

In this section we consider the problem of approximating the unknown concentration of a given sample by means of tests which affect the tested units.

In the proof of the following lemma (and of Lemma 3 of Sect. 3) we mostly use the proof technique introduced in [4], properly modified to handle the more complicated case of destructive tests not considered in [4].

Lemma 1. *Consider one unit of sample with unknown concentration $c \geq 2$, which is known to lie inside the interval $[2^{p-1}, 2^p)$. For any $t \leq 2^{p-2}$, there exists an algorithm which determines an interval of length at most $\frac{c^2}{2^{t+p}-c}$ containing c. The algorithm uses t tests, at most $\frac{(p+t+1)^2}{4} + \sum_{m=p+2}^{p+t+1} \frac{m}{3}$ merge operations and at most $p + 2t$ units of water.*

Proof. Our algorithm performs a binary search for $\frac{1}{c}$ inside the interval $(\frac{1}{2^p}, \frac{1}{2^{p-1}}]$. Each test decreases the length of the current interval containing $\frac{1}{c}$ by one half of its value before the test. After i tests, $\frac{1}{c}$ has been confined inside an interval of length $\frac{1}{2^{i+p}}$ whose left end and right end are denoted with a_i and b_i respectively. The $(i+1)$–th test is performed on a unit with concentration ratio $(a_i + b_i)/2$ obtained by merging one unit with concentration ratio a_i and one unit with concentration ratio b_i. Let a_0 and b_0 denote the values $\frac{1}{2^p}$ and $\frac{1}{2^{p-1}}$ respectively. In order to get started, the algorithm needs to generate a unit with concentration ratio a_0 and a unit with concentration ratio b_0. Starting with $r_0 = 0$ and $r_1 = 1$, the algorithm generates a sequence of concentration ratios $r_2, \cdots, r_p, r_{p+1}$ such that $r_i = \frac{1}{2^{i-1}}$. Two units with concentration ratio r_i are obtained by merging a unit with concentration ratio r_0 (water) and a unit with concentration ratio r_{i-1}, for $2 \leq i \leq p+1$. It is $a_0 = r_{p+1}$ and $b_0 = r_p$. Once $1 \times r_2, \cdots, 1 \times r_p, 2 \times r_{p+1}$ have been generated, the algorithm starts the binary search for $\frac{1}{c}$ inside $(r_{p+1}, r_p]$. The algorithm performs the i–th test on a unit with concentration ratio $r_{p+i+1} = (a_{i-1} + b_{i-1})/2$. Two units with concentration ratio r_{p+i+1} are obtained by merging one unit with concentration ratio a_{i-1} and one unit with concentration ratio b_{i-1}. If the response of the test is positive then the algorithm sets $a_i = a_{i-1}$ and $b_i = r_{p+i+1}$ otherwise it sets $a_i = r_{p+i+1}$ and $b_i = b_{i-1}$. In the former case we say that a_i is the *mate* of r_{p+i+1} while in the latter case we say that b_i is the mate of r_{p+i+1}. From this definition, one has that $\{r_{p+i+1}, \text{ mate of } r_{p+i+1}\} = \{a_i, b_i\}$. For each k, the ratio r_k coincides with one of the extremes of an interval of length $\frac{1}{2^{k-1}}$ containing $\frac{1}{c}$. Extending the definition of mate also to $r_1, \cdots, r_p$, the other extreme of the above said interval is called the mate of r_i. Observe that for any $i \leq p$ the mate of r_i is r_0. After t tests, the unknown value $\frac{1}{c}$ has been confined inside the interval $(a_t, b_t]$ of length $\frac{1}{2^{p+t}}$. Therefore, c has been confined inside the interval $[\frac{1}{b_t}, \frac{1}{a_t})$ which has length $(\frac{1}{a_t} - \frac{1}{b_t}) = \frac{(b_t - a_t)}{a_t b_t}$. Since it is $(b^t - a^t) = \frac{1}{2^{p+t}}$, $b_t \geq \frac{1}{c}$ and $a_t \geq \frac{1}{c} - \frac{1}{2^{p+t}}$ then one has $\frac{(b_t - a_t)}{a_t b_t} \leq \frac{c^2}{2^{t+p} - c}$.

Our goal is to prove that, for any $t \leq 2^{p-2}$, it is possible to generate a unit with concentration ratio r_{p+t+1} using at most $p + 2t$ units of water and at most $\frac{(p+t+1)^2}{4} + \sum_{m=p+2}^{p+t+1} \frac{m}{3}$ merge operations. Following [4], the algorithm can be described as a pebble game on an infinite acyclic directed graph. The nodes of the graph represent the concentration ratios r_i's and for each $i \geq 1$ two direct edges enter r_{i+1}, one starting from r_i, the other from the mate of r_i. We can see nodes $r_0, r_1, \cdots, r_i, \cdots$, as disposed from left to right in the graph. For each $i \geq 0$, p_i denotes the number of pebbles on r_i. A node which contains one or more pebbles is said *occupied.* Initially only r_0 and r_1 are occupied: r_0 contains an infinite number of pebbles while r_1 contains one pebble. If both r_i and its mate are occupied then it is possible to move one pebble from each of these two nodes to r_{i+1}. Let g be a node index specified by the game. The goal of the game is to deliver a pebble to r_g using a small number of steps and a small number of units of water. In our test problem a pebble on node r_i represents a unit with concentration ratio r_i. The pebbles on r_0 represent the water reservoir while the single pebble on r_1 represents a single unit of the sample whose concentration we are trying to estimate. For $i > 0$, let r_i and r_j be the *predecessors* of r_{i+1}, that is, the two nodes which the two edges entering r_{i+1} start from. The concentration ratio r_{i+1} is the arithmetic average of r_i and r_j; moreover, moving two pebbles to r_{i+1} corresponds to merging one unit with concentration ratio r_i and one with concentration ratio r_j. If a node r_i is occupied then it is possible to test a unit with concentration ratio r_i. The basic step of our game strategy consists of occupying the node with *largest* index which is empty and whose predecessors are both occupied. The game terminates when a pebble is delivered to the goal node r_g. A unit with concentration ratio r_i is tested only when r_i is occupied for the first time. As a consequence of this test which destroys the tested unit, one of the two pebbles on r_i is discarded. Notice that each node except r_0 contains at most two pebbles. The statement of the lemma is a consequence of the following claims, whose proofs are omitted.

Claim 1 Let $t \geq 0$. If $s = p+t+1$ is the index of the rightmost occupied node then $r_{p+1}, \cdots, r_s$ all together contain at most $t+2$ pebbles.
Claim 2. If s is the index of the rightmost occupied node then nodes $r_1, \cdots, r_s$ all together contain at most s pebbles. If $s > p+1$ and r_s was destination of the last move then $r_1 \cdots, r_s$ all together contain at most $s-1$ pebbles.
Claim 3. If $r_1, \cdots, r_p$ are all empty then $r_{p+1}, \cdots, r_s$ together contain $2^{p-1} - t + 1$ pebbles.
Claim 4. In order to deliver a pebble to r_s, a total of $\frac{s^2}{4} + \sum_{m=p+2}^{s} \frac{m}{3}$ merge operations are sufficient.
Let $s = p+t+1$. It is evident that if one of $p_1, \cdots, p_p$ is not zero then a further move is possible. From Claim 3 it follows that if $r_1, \cdots, r_p$ are all empty then $r_{p+1}, \cdots, r_s$ contain $2^{p-1} - t + 1$ pebbles. Claim 1 implies that $t+2 \geq 2^{p-1} - t + 1$ from which it follows that it is possible to deliver a pebble to $r_s = r_{p+t+1}$ for any $t \leq 2^{p-2}$. From Claim 4 one has that this can be done by using at most $\frac{(p+t+1)^2}{4} + \sum_{m=p+2}^{p+t+1} \frac{m}{3}$ merge operations. Since Claim 2 implies that at the end of the game $r_1, \cdots, r_s$ contain at most $s-1$ pebbles and since t pebbles have been discarded as a consequence of the t tests, then the algorithm uses at most $s - 1 + t = p + 2t$ units of water. □

Corollary 1. *Consider one unit of sample with unknown concentration c which is known to lie inside the interval* $[2^{p-1}, 2^p)$, *with* $p \geq 5$. *There exists an algorithm to find an interval containing c of length at most* $\frac{1}{3}$. *The algorithm uses* $p+2$ *tests, at most* $\frac{3}{2}p^2 + \frac{29}{6}p + \frac{47}{12}$ *merge steps and at most* $3p+4$ *units of water.*

In the following we describe a family of algorithms which determine $\lfloor \log c \rfloor$. First, we introduce some notation. Let $g(i,j)$ denote the function of two non-negative integers defined recursively as follows: $g(i,j) = i$ if $j = 0$ and $g(i,j) = 2^{g(i,j-1)}$ if $j \geq 1$. For all non-negative integers j such that $c > g(0, j-1)$, $\log^{(j)} c$ is defined as follows: $\log^{(j)} c = c$ if $j = 0$ and $\log^{(j)} c = \log \log^{(j-1)} c$ if $j > 0$.

Let j be a fixed positive integer. The following algorithm first determines, in **stage**(0), the smallest integer l ($0 \leq l \leq j$) such that $g(i-1, j-l) \leq c < g(i, j-l)$ for some $i \geq 2$. Such value of i is equal to $\lfloor \log^{(j-l)} c \rfloor + 1$. In **stage**($s$), $s = 1, \cdots, j-l-1$, the algorithm searches for $k_{j-l-s} = \lfloor \log^{(j-l-s)} c \rfloor + 1$.
Algorithm A_j ($j \geq 1$):
stage(0):
Let us consider a diluting sequence which iteratively apply the following step until a unit with the desired concentration is obtained.

step **b** Add one unit of water to a given unit of liquid while obtaining two units with half of the concentration of the given unit. Store one of these two units and use the remaining one for the successive application of step **b**.

Starting with the given unit of sample and iteratively performing step **b**, it is possible to generate the sequence of $S = \{1 \times \frac{1}{2^m}\}_{m \geq 1}$. Consider the subsequence $\{1 \times \frac{1}{g(i,j)}\}_{i \geq 1}$ of S. The i–th test of this stage involves one unit with concentration ratio $c\frac{1}{g(i,j)}$; if the result of this test is positive then the next terms of S

up to $1 \times \frac{1}{g(i+1,j)}$ are generated, otherwise no more terms are generated. Let k_j denote the index i of the last term $1 \times \frac{1}{g(i,j)}$ generated,
set $l = 0$
while $k_{j-l} = 1$ repeat the following steps:
    perform a test on a unit with concentration ratio $\frac{1}{g(1,j-l-1)}$
    if $\frac{c}{g(1,j-l-1)} < 1$ **then** set $k_{j-l-1} = 1$
        else set $k_{j-l-1} = 2$
    Increment l by one.
for $r = l+1, \cdots, j-1$
  stage$(r-l)$:
  set $L_0^r = 2^{k_{j-r+1}-1}$ and $R_0^r = 2^{k_{j-r+1}}$
  for $i = 1, \cdots, k_{j-r+1} - 1$
    set $M_i^r = \frac{L_{i-1}^r + R_{i-1}^r}{2}$
    perform a test on a unit with concentration ratio $\frac{1}{g(M_i^r, j-r)}$
    if the test is positive **then** set $L_i^r = M_i^r$ and $R_i^r = R_{i-1}^r$
        else set $L_i^r = L_{i-1}^r$ and $R_i^r = M_i^r$
  set $k_{j-r} = R_i^r$

It is easy to prove by induction that $k_m = \lfloor \log^{(m)} c \rfloor + 1$ if $c \geq g(0,m)$ and 1 otherwise. Therefore, the last stage of the algorithm yields $k_1 = \lfloor \log c \rfloor + 1$. Notice that the units tested during **stage**(0),$\cdots$,**stage**$(j-l-1)$ are among those terms of S generated during **stage**(0). This stage requires $\log(g(k_j,j)) = g(k_j, j-1)$ units of water and merge operations.

If $c \geq 32$, the algorithm A_j, for $j \geq 2$, does not need to test the unit with concentration ratio $\frac{1}{2}$. As far it concerns the algorithm A_1, the test on $1 \times \frac{1}{2}$ performed in **stage**(0) by this algorithm may be skipped. Therefore, if $c \geq 32$ then the unit with concentration ratio $\frac{1}{2}$ is preserved after the execution of A_j and the algorithm of Corollary 1 can be used to confine $\frac{c}{2}$ inside an interval of length $\frac{1}{3}$. Since it is $2^{k_1-2} \leq \frac{c}{2} < 2^{k_1-1}$ then the algorithm of Corollary 1 requires k_1+1 tests and at most $\frac{3}{2}(k_1-1)^2 + \frac{29}{6}(k_1-1) + \frac{47}{12}$ merge steps and $3k_1+1$ units of water. The above strategy finds an interval of length at most $\frac{2}{3}$ including the unknown concentration $c \geq 32$ using $k_j + l + \sum_{r=l+1}^{j-1}(k_{j-r+1}-1) + k_1 + 1$ tests, at most $g(k_j, j-1) + 3k_1 + 1$ units of water and at most $g(k_j, j-1) + \frac{3}{2}(k_1 - 1)^2 + \frac{29}{6}(k_1-1) + \frac{47}{12}$ merge steps. Hence, the following theorem holds:

Theorem 1. *For each $j \geq 1$ there exists an algorithm $\overline{A}_j$ which finds an interval of length at most 1 including an unknown concentration $c \geq 32$. Let l be the smallest integer i such that $\lfloor \log^{(j-i)} c \rfloor \geq 1$. The algorithm $\overline{A}_j$ performs $\lfloor \log^{(j-l)} c \rfloor + \lfloor \log^{(j-l-1)} c \rfloor + \cdots + \lfloor \log^{(1)} c \rfloor + \lfloor \log^{(j)} c \rfloor + l + 3$ tests and uses at most $g(k_j, j-1) + 3\lfloor \log c \rfloor + 4$ units of water and at most $g(k_j, j-1) + \frac{3}{2}\lfloor \log c \rfloor^2 + \frac{29}{6}\lfloor \log c \rfloor + \frac{47}{12}$ merge steps.*

Corollary 2. *There is an algorithm which finds an interval of length at most 1 including the unknown concentration $c \geq 32$ with $\lfloor \log c \rfloor + 2\lfloor \log \log c \rfloor + 3$ tests,* *using at most* $5\lfloor\log c\rfloor + 4$ *units of water and at most* $\frac{3}{2}\lfloor\log c\rfloor^2 + \frac{41}{6}\lfloor\log c\rfloor + \frac{47}{12}$ *merge steps.*

The above corollary provides a very good tradeoff between number of tests and number of units of water used to approximate c. Although our algorithm works under a more complicated test model than that used in [4] (i.e. the algorithm may test the same unit of liquid more then once) this corollary represents an improvement with respect to Damaschke's algorithm, in that, it uses a smaller number of tests while still using a logarithmic number of units of water.

In case our major concern is to lower the required number of tests, we must carefully choose the most appropriate algorithm A_j. If, for example, we select an algorithm A_j with a very large j when c is small, then the algorithm performs during **stage**(0) a large number of tests which give very little contribution to the search. We can use the same approach proposed by Bentley and Yao [2] in the context of unbounded search, to first decide which value of j is more appropriate and then apply the selected algorithm A_j. Following their idea, we determine $\ell^*(c) = \min h$ such that $k_h = \lfloor\log^{(h)} c\rfloor + 1$ is equal to 1.

In order to find $\ell^*(c)$ we test units with concentration ratios $\frac{1}{g(1,h)},$ $h \geq 1$, until $\frac{c}{g(1,h)} < 1$ and then set $\ell^*(c)$ equal to h. The selected algorithm is A_j with $j = \ell^*(c)-1$. The number of tests required to find j is $j+1$. Since it is known that $k_j = 2$, tests in **stage**(0) of algorithm A_j can be left out. The tests and merge steps performed to find $\ell^*(c)$ constitute **stage**(0) of this new algorithm. The successive $j-1$ stages coincides with **stage**(1),$\cdots$,**stage**($j-1$) of the algorithm A_j. The given strategy finds $k_1 = \lfloor\log c\rfloor+1$ using $j+1+\sum_{r=1}^{j-1}(k_{j-r+1}-1)$ tests. Notice that in order to find the appropriate j, we generate a sequence of units whose concentration ratios are the first $g(1,j)$ terms of the sequence S defined in **stage**(0) of A_j. The sequence of units so generated contains all the units tested in **stage**(1),$\cdots$,**stage**($j-1$) of A_j. Hence, we have the following theorem:

Theorem 2. *There exists an algorithm* A^* *which finds an interval of length at most 1 including an unknown concentration* $c \geq 32$ *with* $\ell^*(c) + \lfloor\log^{(\ell^*(c)-1)} c\rfloor + \lfloor\log^{(\ell^*(c)-2)} c\rfloor + \cdots + \lfloor\log^{(1)}(c)\rfloor + 2$ *tests, using at most* $g(1,\ell^*(c)-1) + 3\lfloor\log c\rfloor + 4$ *units of water and at most* $g(1,\ell^*(c)-1) + \frac{3}{2}\lfloor\log c\rfloor^2 + \frac{29}{6}\lfloor\log c\rfloor + \frac{47}{12}$ *merge steps.*

In the rest of this section we describe a strategy for approximating c using only a constant number of tubes.

Lemma 2. *Given two units with concentration ratio* $r_j 2^u$ *and two units with concentration ratio* $r_k 2^u$ *such that* $r_j < \frac{1}{c} \leq r_k$, *there exists an algorithm to find an interval of length at most* $\frac{c^2}{(r_j - r_k)^{-1}2^{u+4} - c}$ *including c. The algorithm uses* $u+4$ *tests,* $\frac{u^2}{2} + \frac{7u}{2} + 94$ *units of water,* $\frac{u^2}{2} + \frac{5u}{2} - 14$ *merge operations and 4 tubes.*

Proof. We describe a strategy which for each $1 \leq i \leq u+1$, preserves the following invariant:
(a) at step i it holds $a_{i-1} < \frac{1}{c} \leq b_{i-1}$ and one disposes of 4 units of liquid described by $2 \times 2^{u-i+1}a_{i-1}, 2 \times 2^{u-i+1}b_{i-1}$.

Invariant (a) is true at step 1 with $a_0 = r_j$ and $b_0 = r_k$. Suppose invariant (a) be true for i then we have $2 \times 2^{u-i+1}a_{i-1}, 2 \times 2^{u-i+1}b_{i-1}$ and $a_{i-1} < \frac{1}{c} \leq b_{i-1}$. Adding 2 units of water to each of the two concentrations we get $4 \times 2^{u-i}a_{i-1}, 4 \times 2^{u-i}b_{i-1}$. Then we merge 2 units with concentration ratio $2^{u-i}a_{i-1}$ and 2 units with concentration ratio $2^{u-i}b_{i-1}$ getting $2 \times 2^{u-i}a_{i-1}, 2 \times 2^{u-i}b_{i-1}, 4 \times 2^{u-i}\frac{a_{i-1}+b_{i-1}}{2}$. Let us consider a diluting sequence which iteratively apply the following step until a unit with the desired concentration is obtained.

step **d:** Add one unit of water to a given unit of liquid obtaining two units with half of the concentration of the given unit. Discard one of these two units and store the remaining one for the successive application of step **d**.

Starting with $1 \times 2^{u-i}\frac{a_{i-1}+b_{i-1}}{2}$ and iteratively performing step **d** $u-i$ times we obtain $1 \times \frac{a_{i-1}+b_{i-1}}{2}$. Such unit is tested and if the result of the test is positive then we set $a_i = a_{i-1}$ and $b_i = \frac{a_{i-1}+b_{i-1}}{2}$ otherwise we set $a_i = \frac{a_{i-1}+b_{i-1}}{2}$ and $b_i = b_{i-1}$. We store $2 \times a_i 2^{u-i}, 2 \times b_i 2^{u-i}$ and the invariant is restored.

After $u-5$ steps, we get $2 \times 2^5 a_{u-5}, 2 \times 2^5 b_{u-5}$ with $a_{u-5} < \frac{1}{c} \leq b_{u-5}$. Adding $2^6 - 2$ units of water to each of the two concentrations, we get $2^6 \times a_{u-5}, 2^6 \times a_{u-5}$. Let $n_{u-5} = 2^6$. For $i = u-4, \ldots, u+4$, we execute the following step:

Merge $\lceil \frac{n_{i-1}}{3} \rceil \times a_{i-1}$ with $\lceil \frac{n_{i-1}}{3} \rceil \times b_{i-1}$. Test $1 \times \frac{a_{i-1}+b_{i-1}}{2}$ and set a_i and b_i according to the result of this test. Set $n_i = \min\{n_{i-1} - \lceil \frac{n_{i-1}}{3} \rceil, 2\lceil \frac{n_{i-1}}{3} \rceil - 1\}$ and if $i < u+4$, store $n_i \times a_i, n_i \times b_i$.

The above strategy performs $u+4$ tests. After these $u+4$ tests $\frac{1}{c}$ has been confined inside an interval of size $\frac{(b_0-a_0)}{2^{u+4}}$ and by using an argument similar to the one used in the proof of Theorem 1, it is possible to see that this is equivalent to confining c inside an interval of size at most $\frac{c^2}{(b_0-a_0)^{-1}2^{u+4}-c}$. □

If we start with $2 \times r_j 2^{p-3}, 2 \times r_k 2^{p-3}$ with $r_j = \frac{1}{2^p}$ and $r_k = \frac{1}{2^{p-1}}$ then the above strategy finds an interval of length at most $\frac{c^2}{2^{2p+1}-c} < 1$.

The following algorithm determines $p - 1 = \lfloor \log c \rfloor$ for any $c \geq 8$.

Algorithm $\hat{A}$ **:**

stage(0): This stage is very similar to **stage**(0) of A^*, except we do not store all units generated during this stage. For $j \geq 2$, let $B_j(i) = \sum_{v=0}^{j-2}(g(i,v) - 1)$. $B_j(k_j)$ is an upper bound to the overall number of tests performed by **stage**(1),$\cdots$,**stage**($j-1$). Let $1 \times \frac{1}{2}, 1 \times \frac{1}{2^2}, \cdots, 2 \times \frac{1}{2^m}$, with $g(1,h-1) < 2^m \leq g(1,h)$ $(h > 2)$, denote the sequence of concentration ratios so far generated. Of this sequence we have discarded all units but $1 \times \frac{1}{2}$, $1 \times \frac{1}{2^2}$, $1 \times \frac{1}{2^{g(1,h-2)-\lfloor \log(B_{h-1}(2)-1) \rfloor}}$ (if $h > 3$), $1 \times \frac{1}{2^{g(1,h-1)-\lfloor \log(B_h(2)-1) \rfloor}}$ (if $m > g(1,h-1) - \lfloor \log(B_h(2)-1) \rfloor$) and $2 \times \frac{1}{2^m}$. If $2^m = g(1,h)$, for $h > 2$, a unit with concentration ratio 2^m is tested and consequently discarded. If $h > 3$ and the response of the test is positive, $1 \times \frac{1}{2^{g(1,h-2)-\lfloor \log(B_{h-1}(2)-1) \rfloor}}$ is discarded. At the end we are left with $1 \times \frac{1}{2}, 1 \times \frac{1}{2^2}, 1 \times \frac{1}{2^{g(1,\ell^*(c)-2)-\lfloor \log(B_{(\ell^*(c)-1)}(2)-1) \rfloor}}, 1 \times \frac{1}{2^{g(1,\ell^*(c)-1)-\lfloor \log B_{(\ell^*(c))}(2)-1) \rfloor}}, 1 \times \frac{1}{g(1,\ell^*(c))}$. We dilute $1 \times \frac{1}{2^{g(1,\ell^*(c)-2)-\lfloor \log(B_{(\ell^*(c)-1)}(2)-1) \rfloor}}$ with $2^{\lceil \log B_{(\ell^*(c)-1)}(2) \rceil} - 1$ units of water thus obtaining $2^{\lceil \log B_{(\ell^*(c)-1)}(2) \rceil} \times \frac{1}{2^{g(1,\ell^*(c)-2)+1}}$.

stage(r) (for $r = 1, \cdots, \ell^*(c) - 2$) : this stage is similar to the corresponding stage of A^*. The only difference is that the tested units are not generated during

stage(0). When a unit is needed, it is generated from a unit with concentration ratio $\frac{1}{2^{g(1,\ell^*(c)-2)+1}}$ by successive iterations of step **d**.

Since $k_{\ell^*(c)-1} = 2$, then $B_{\ell^*(c)-1}(2)$ is an upper bound on the overall number of tests performed by **stage**(1),$\cdots$,**stage**($\ell^*(c)-2$), and, as a consequence, we can generate as many test units as needed by these stages. Algorithm $\hat{A}$ uses at most $g(1,\ell^*(c)-1)+2B_{(\ell^*(c)-1)}(2)-2+(g(1,\ell^*(c)-1)-g(1,\ell^*(c)-2)-2)\sum_{r=1}^{\ell^*(c)-2}(k_{\ell^*(c)-r}-1)$ units of water and at most $g(1,\ell^*(c)-1)+1+(g(1,\ell^*(c)-1)-g(1,\ell^*(c)-2)-2)\sum_{r=1}^{\ell^*(c)-2}(k_{(\ell^*(c)-r)}-1)$ merge steps.

Notice that **stage**(0) of algorithm $\hat{A}$ does not test units with concentration ratio $\frac{1}{2}$ and $\frac{1}{2^2}$ which are therefore still available. Consider a search strategy which first applies algorithm $\hat{A}$ to the initial sample to find $p = \lfloor \log c \rfloor + 1$ and then executes the algorithm of Lemma 2 starting with the units $2 \times \frac{1}{2^3}$ and $2 \times \frac{1}{2^2}$ which can be obtained from units $1 \times \frac{1}{2^2}, 1 \times \frac{1}{2}$. Then the following theorem holds:

Theorem 3. *There exists an algorithm which finds an interval of length at most 1 including an unknown concentration $c \geq 8$ using 6 tubes and*

- $\ell^*(c)+\lfloor\log^{(\ell^*(c)-1)}c\rfloor+\lfloor\log^{(\ell^*(c)-2)}c\rfloor+\cdots+\lfloor\log^{(1)}(c)\rfloor+2$ *number of tests,*
- *at most* $\frac{(\lfloor\log c\rfloor-2)^2}{2}+\frac{7(\lfloor\log c\rfloor-2)}{2}+94+g(1,\ell^*(c)-1)+2B_{(\ell^*(c)-1)}(2)-2+(g(1,\ell^*(c)-1)-g(1,\ell^*(c)-2)-2)\sum_{r=1}^{\ell^*(c)-2}\lfloor\log^{(\ell^*(c)-r)}c\rfloor+2$ *units of water,*
- *at most* $\frac{(\lfloor\log c\rfloor-2)^2}{2}+\frac{5(\lfloor\log c\rfloor-2)}{2}-14+g(1,\ell^*(c)-1)+1+(g(1,\ell^*(c)-1)-g(1,\ell^*(c)-2)-2)\sum_{r=1}^{\ell^*-2}\lfloor\log^{(\ell^*(c)-r)}c\rfloor+2$ *merge steps.*

3 The Conservative Model

In this section we consider the case when tests do not destroy the tested units. This is the same model considered by Damaschke [4]. As in Sect. 2, we describe a family of algorithms $\tilde{A}_j$ which provide tradeoffs between the number of tests, the number of units of water and the number of merge steps. The first phase of the algorithm $\tilde{A}_j$ performs the same steps executed by the algorithm A_j of Sect. 2. The algorithm of Lemma 3 is then used to perform the binary search for $\frac{1}{c}$ inside the interval $(\frac{1}{2^{\lfloor\log c\rfloor+1}}, \frac{1}{2^{\lfloor\log c\rfloor}}]$. Along the same lines as Lemma 1 we can prove the following result:

Lemma 3. *Consider one unit of sample with unknown concentration $c > 2$, which after an application of A_j is known to lie inside the interval $[2^{p-1}, 2^p)$. For any $t \leq 2^{p-1}-3$ there exists an algorithm which determines an interval of length at most $\frac{c^2}{2^{t+p-c}}$ containing c. The algorithm uses t tests and at most $\frac{(p+t+1)^2}{4}-\frac{p^2}{6}-\frac{p}{2}$ merge operations and at most $t+1$ units of water.*

Corollary 3. *Consider one unit of sample with unknown concentration c which after an application of A_j is known to lie inside the interval $[2^{p-1}, 2^p)$, with $p \geq 4$. There exists an algorithm to find an interval containing c of length at most 1. The algorithm uses $p+1$ tests, $\frac{(2p+2)^2}{4}-\frac{p^2}{6}-\frac{p}{2}$ merge steps and at most $p+2$ units of water.*

Theorem 4. *For each $j \geq 1$ there exists an algorithm $\tilde{A}_j$ which finds an interval of length at most 1 including an unknown concentration $c \geq 8$. Let l be the smallest integer i such that $\lfloor \log^{(j-i)} c \rfloor \geq 1$. The algorithm performs $\lfloor \log^{(j-l)} c \rfloor + \lfloor \log^{(j-l-1)} c \rfloor + \cdots + \lfloor \log^{(1)} c \rfloor + \lfloor \log^{(j)} c \rfloor + l + 3$ tests and uses at most $g(k_j, j-1) + \lfloor \log c \rfloor + 3$ units of water and at most $g(k_j, j-1) + \frac{5}{6}\lfloor \log c \rfloor^2 + \frac{19}{6}\lfloor \log c \rfloor + \frac{10}{3}$ merge steps.*

Proof. Algorithm $\tilde{A}_j$ is obtained by applying algorithm A_j followed by an application of the algorithm of Corollary 3. □

In the special case $j = 1$, the algorithm $\tilde{A}_1$ attains the same performances of Damaschke's algorithm [4]. Moreover, setting $j = 2$ in the statement of Theorem 4 yields the following improvement:

Corollary 4. *There is an algorithm which finds an interval of length at most 1 including an unknown concentration $c \geq 8$ with $\lfloor \log c \rfloor + 2\lfloor \log \log c \rfloor + 3$ tests, using at most $3\lfloor \log c \rfloor + 3$ units of water and at most $\frac{5}{6}\lfloor \log c \rfloor^2 + \frac{31}{6}\lfloor \log c \rfloor + \frac{10}{3}$ merge steps.*

As in Sect. 2 we can optimize the choice of j to get the following theorem:

Theorem 5. *There exists an algorithm $\tilde{A}^*$ which finds an interval of length at most 1 including an unknown concentration $c \geq 8$ with $\ell^*(c) + \lfloor \log^{(\ell^*(c)-1)} c \rfloor + \lfloor \log^{(\ell^*(c)-2)} c \rfloor + \cdots + \lfloor \log^{(1)}(c) \rfloor + 2$ tests, using at most $g(1, \ell^*(c)-1) + \lfloor \log c \rfloor + 3$ units of water and at most $g(1, \ell^*(c)-1) + \frac{5}{6}\lfloor \log c \rfloor^2 + \frac{19}{6}\lfloor \log c \rfloor + \frac{10}{3}$ merge steps.*

References

1. E. Barillot *et al.*, "Theoretical analysis of library screening using a n-dimensional pooling strategy", *Nucleic Acids Research*, (1991), 6241-6247
2. J.L. Bentley and A. Yao, "An almost optimal algorithm for unbounded search", *IPL* 5, (1976), 82–87
3. W. J. Bruno *et al.*, "Design of efficient pooling experiments", *Genomics*, vol., 26, (1995), 21–30
4. P. Damaschke, "The algorithmic complexity of chemical threshold testing", in: *CIAC '97*, LNCS, 1203, Springer–Verlag, (1997), 215–216
5. R. Dorfman, "The detection of defective members of large populations", *Ann. Math. Stat.*, 14 (1943), 436–440
6. D.Z. Du and F.K. Hwang, " Competitive group testing", *Discrete Applied Math.*, 45 (1993), 221–232
7. D.Z. Du and F.K. Hwang, *Combinatorial Group Testing and its Applications*, World Scientific Publishing, (1993)
8. M. Farach *et al.*, "Group testing problems with sequences in experimental molecular biology", in: *Proc. Compr. and Compl. of Sequences '97*, B. Carpentieri, A. De Santis, J. Storer, and U. Vaccaro, (Eds.), IEEE Computer Society, pp. 357–367.
9. F.K. Hwang and P.J. Wan, "Comparing file copies with at most three disagreeing pages" *IEEE Transactions on Computers*, 46, No. 6, (June 1997), 716–718
10. M. Sobel and P.A. Groll, "Group testing to eliminate efficiently all defectives in a binomial sample", *Bell System Tech. J.* 38, (1959), 1179–1252
11. J. K. Wolf, "Born again group testing: Multiacces communications", *IEEE Trans. Inf. Th.* IT-31, (1985), 185–191

Min-Max-Boundary Domain Decomposition*

Marcos Kiwi[1]** and Daniel A. Spielman[2]*** and Shang-Hua Teng[3]†

[1] Dept. de Ingeniería Matemática, Fac. de Ciencias Físicas y Matemáticas, U. de Chile, Casilla 170/3, Correo 3, Santiago, Chile. (mkiwi@dim.uchile.cl, http://www.dim.uchile.cl/~mkiwi).

[2] Department of Mathematics, M.I.T., Cambridge, MA 02139, USA. (spielman@math.mit.edu, http://www-math.mit.edu/~spielman).

[3] Department of Computer Science, University of Illinois, Urbana IL 61801, USA. (steng@cs.umn.edu, http://www.cs.umn.edu/~steng).

Abstract. Domain decomposition is one of the most effective and popular parallel computing techniques for solving large scale numerical systems. In the special case when the amount of computation in a subdomain is proportional to the volume of the subdomain, domain decomposition amounts to minimizing the surface area of each subdomain while dividing the volume evenly. Motivated by this fact, we study the following *min-max boundary* multi-way partitioning problem: Given a graph G and an integer $k > 1$, we would like to divide G into k subgraphs $G_1, \ldots, G_k$ (by removing edges) such that (i) $|G_i| = \Theta(|G|/k)$ for all $i \in \{1, \ldots, k\}$; and (ii) the maximum boundary size of any subgraph (the set of edges connecting it with other subgraphs) is minimized.

We provide an algorithm that given G, a well-shaped mesh in d dimensions, finds a partition of G into k subgraphs $G_1, \ldots, G_k$, such that for all i, G_i has $\Theta(|G|/k)$ vertices and the number of edges connecting G_i with the other subgraphs is $O((|G|/k)^{1-1/d})$. Our algorithm can find such a partition in $O(|G| \log k)$ time. Finally, we extend our results to vertex-weighted and vertex-based graph decomposition. Our results can be used to simultaneously balance the computational and memory requirement on a distributed-memory parallel computer without sacrificing the communication overhead.

1 Introduction

Domain decomposition is one of the most effective and popular technique for solving large scale numerical systems on parallel computers [6, 8]. This technique is used for finding solutions to partial differential equations by iteratively solving subproblems defined on smaller subdomains. Thus, it is a divide-and-conquer technique. When applying this technique, it is desirable to decompose the domain into subdomains with approximately the same computational work associated to them (for balancing

* Part of this work was done while Daniel Spielman and Shang-Hua Teng were visiting Universidad de Chile.

** Supported in part by Fondecyt No. 1981182, and Fondap in Applied Mathematics 1998.

*** Supported in part by an Alfred P. Sloan Research Fellowship

† Supported in part by an NSF CAREER award (CCR-9502540), an Alfred P. Sloan Research Fellowship, and an Intel research grant.

the load) and to minimize communication among subdomains (for reducing total communication and communicational bottleneck) [6].

We first focus in the special case where the amount of computational work associated to a subdomain is proportional to the volume of the subdomain. Here, domain decomposition amounts to minimizing the surface area of each subdomain while dividing the volume evenly.

The ratio of the measure of the boundary to the measure of the computational work of a subdomain is sometimes referred to as the *surface-to-volume ratio* or the *communication-to-computation ratio* of the subdomain. Minimizing this ratio plays a key role in efficient parallel iterative methods [8].

To solve partial differential equations numerically, one discretizes the domain into a mesh of well-shaped elements such as simplices or hexahedral elements. The density of mesh points, and hence the size of mesh elements, may vary within the domain giving rise to unstructured meshes [4, 13, 17]. Obtaining good partitions of unstructured meshes is, in general, significantly more challenging than partitioning their uniform/regular counterparts.

The main result established in this work is that every d-dimensional well-shaped unstructured mesh has a k-way partition where the surface-to-volume ratio of every sub-mesh is as small as that of a regular d-dimensional grid that has the same number of nodes.

In Section 2, we introduce the problem of min–max–boundary multi–way partitioning. In Section 3, we describe a multi–way partitioning algorithm and present our main result. In Section 4, we extend the results of Section 3 to graphs with non–negative weights at each vertex. More precisely, we propose an efficient algorithm that partitions vertex–weighted graphs into subgraphs of similar total weight and vertex size and at the same time achieves low surface–to–volume ratio in all subgraphs. Such multi–way partitioning algorithm can be used to simultaneously balance the computational work and the memory requirements on a distributed–memory parallel computer without sacrificing communication overhead. In Section 5, we address the vertex–based partitioning problem in order to handle graphs with large vertex degree.

2 Multi–way Partitioning

A *bisection* of a graph G is a division of its vertices into two disjoint subsets whose sizes differ by at most one. In general, for every integer $k > 1$, a *k-way partition* of G is a division of its vertex set into k disjoint subsets of size $\lceil |G|/k \rceil$ or $\lfloor |G|/k \rfloor$, where $|G|$ denotes the number of vertices in G.

Partitions that evenly divide the vertices are not necessary in most applications [15]. In most cases, *balanced partitions* suffice. Given a graph $G = (V, E)$ and an integer $k > 1$ and a real number $\beta \geq 1$, a partition $P = \{G_1, \ldots, G_k\}$ is a (β, k)-partition of G if $|V_i| \leq \beta \lceil |G|/k \rceil$, for all $i \in \{1, \ldots, k\}$, where V_i is the vertex set of G_i. We denote by $\partial_V(G_i)$ the set of *boundary-vertices* of G_i, i.e. the set of vertices in V_i that are connected by an edge of G to a vertex not in V_i; we denote by $\partial_E(G_i)$ the *boundary-edges* of G_i, i.e. the set of edges in G exactly one of whose endpoints is in V_i.

We consider the following two costs associated with a (β, k)-partition:

$$\text{total-boundary}_E(P) = \left(\sum_{i=1}^{k} |\partial_E(G_i)|\right)/2$$

$$\text{max-boundary}_V(P) = \max_{i=1,\dots,k} |\partial_E(G_i)|.$$

The problem of *min-total-boundary (multi-way) partitioning* is to construct a (β, k)-partition that minimizes total-boundary, while *min-max-boundary (multi-way) partitioning* is to construct a (β, k)-partition that minimizes max-boundary.

3 Bounds for Min–Max–Boundary Partitioning

We first introduce some terminology. Let $\mathcal{G}$ be a family of graphs that is closed under the subgraph operation, i.e. every subgraph of a graph $G \in \mathcal{G}$ belongs to $\mathcal{G}$. For $0 < \alpha < 1$, we say *$\mathcal{G}$ has an n^α-separator theorem* or *$\mathcal{G}$ is n^α-separable* if there is a constant c such that every n-node graph in $\mathcal{G}$ has a bisection of cut-size at most cn^α. Moreover, we refer to the latter type of bisections as n^α-separators. (More information concerning small separators can be found in [11, 16].)

We denote by $\mathcal{G}(\alpha)$ a family of graphs that is n^α-separable and closed under the subgraph operation. Examples include bounded-degree planar graph ($O(n^{1/2})$-separable) [11], graphs with bounded genus ($n^{1/2}$-separable) [10], graphs with no h-clique minor for a constant h ($n^{1/2}$-separable) [1], well-shaped meshes in $\mathbb{R}^d$ and nearest neighbor graphs in $\mathbb{R}^d$ (($n^{1-1/d}$)–separable) [13, 14, 16].

The min–total–boundary partitioning problem has been addressed in the literature. The following lemma has been shown in [15].

Lemma 1. *Let k be an integer such that $k > 1$. Then, for every graph G in $\mathcal{G}(\alpha)$ a k-way partition P such that* total-boundary$_E(P) = O(k^{1-\alpha}|G|^\alpha)$ *can be constructed.*

A closed related problem is bifurcators [5]. A graph G has an $(F_0, F_1, \dots, F_r)$-*decomposition tree* if G can be decomposed into two subgraphs G_0 and G_1 by removing no more than F_0 edges from G, and in turn, both G_0 and G_1 can be decomposed into smaller subgraphs by removing no more than F_1 edges from each, and so on. An n-node graph has an *α-bifurcator* of size F if it has an $(F, F/\beta, F/\beta^2, \dots .1)$-decomposition tree. Bhatt and Leighton [5] showed that every graph in $\mathcal{G}(\alpha)$ has a $\sqrt{2}$-bifurcator of size $O(\sqrt{n})$ if $\alpha \leq 1/2$, and has a $\sqrt{2}$-bifurcator of size $O(n^\alpha)$ if $\alpha > 1/2$.

The following is the main result of this section which establishes a separator theorem for min–max–boundary partitioning.

Theorem 2 main. *Let k be an integer such that $k > 1$. Then, every bounded-degree graph G in $\mathcal{G}(\alpha)$ has a $(2, k)$-partition P such that* max-boundary$_E(P) = O((|G|/k)^\alpha)$.

Notice that Theorem 2 implies Lemma 1. Thus, our main result can be seen as an extension of the result of [15] cited above.

3.1 Simultaneous Partition of Vertices and Boundary

We first examine a simple example. Consider a $\sqrt{n} \times \sqrt{n}$ grid in two dimensions where we assume both k and $\sqrt{n}$ are powers of two. One way of partitioning the grid is to divide it into two $\sqrt{n} \times \sqrt{n}/2$ grids by removing the edge in the middle of every row (a $\sqrt{n}$-separator), and then divide each of the two sub-grids into two $\sqrt{n}/2 \times \sqrt{n}/2$ sub-grids by removing the middle edge of every column. This process can continue by recursively dividing the sub-grids until k disconnected sub-grids are found. Clearly, each final sub-grid has n/k vertices and at most $4\sqrt{n/k}$ boundary-edges. However, the naive recursive application of the separator Theorem of Lipton and Tarjan does not, in general, guarantee the generation of a k-way partition P with max-boundary$_E(P) = O(\sqrt{n/k})$ for all bounded degree n-node planar graphs. The following somewhat stronger version of the small-separator Theorem was used in partitioning the 2D grid: at every stage of the divide-and-conquer, (1) Each subgraph was divided into two subgraphs of the same size by removing a set of edges whose size is on the order of the square-root of the size of the subgraph (*a la* standard Lipton-Tarjan Theorem). (2) The boundary-vertices of the subgraphs were divided evenly.

Our method is motivated by the latter observation, more formally given below.

Lemma 3. *Let $k > 1$ be a power of two. Let G be a bounded-degree graph in $\mathcal{G}(\alpha)$ such that $|G|$ is a power of two. If in every stage of a divide-and-conquer partitioning procedure the vertices and boundary-vertices of each subgraph are evenly divided by a separator, whose size is on the order of the α-th power of the size of the subgraph, then the divide-and-conquer procedure, on input G, will generate a k-way partition whose* max-boundary *is $O((|G|/k)^\alpha)$.*

Proof: Let $s(i)$ be the maximum possible number of boundary-vertices for graphs at level i of the divide-and-conquer partitioning procedure. It follows from the assumption of the lemma that there exists a constant c such that $s(1) \le c(|G|/2)^\alpha$, and if $i \ge 1$,

$$\begin{aligned} s(i) &\le s(i-1)/2 + c \cdot (|G|/2^i)^\alpha \\ &\le c \cdot (|G|/2^i)^\alpha \left(\sum_{j=0}^{i-1} 2^{j(\alpha-1)} \right). \end{aligned}$$

Since $\alpha < 1$, we get that $s(i) = O((|G|/2^i)^\alpha)$. Fixing $i = \log k$, we have $s(i) = O((|G|/k)^\alpha)$. The lemma follows from the assumption that G is a bounded-degree graph. □

Unfortunately, we may not always be able to find a small separator that evenly divides both vertices and boundary-vertices. We show that this simultaneous partition can be achieved approximately.

A variation of the following lemma were given in Lipton and Tarjan [11].

Lemma 4. *Let $G = (V, E)$ be a graph in $\mathcal{G}(\alpha)$ such that $|G|$ is a power of two. Let S be a subset of V. Then, one can find an $O(|G|^\alpha)$-separator that divides G into two subgraphs $G_1 = (V_1, E_1)$ and $G_2 = (V_2, E_2)$ such that $|S \cap V_1| = \lfloor |S|/2 \rfloor$ and $|S \cap V_2| = \lceil |S|/2 \rceil$.*

Lemma 5. *Let $0 < \epsilon < 1/2$. Let $G = (V, E)$ be a graph in $\mathcal{G}(\alpha)$ such that $|G|$ is a power of two. Let $S \subseteq V$ be a subset of V. Then, one can find an $O(|G|^{\alpha})$-separator that divides G into two subgraphs $G_1 = (V_1, E_1)$ and $G_2 = (V_2, E_2)$ such that $|S \cap V_1| = \lfloor |S|/2 \rfloor$, $|S \cap V_2| = \lceil |S|/2 \rceil$, and $|V_1|, |V_2| \leq (1+\epsilon)|G|/2$.*

Proof: Let t be the smallest integer such that $1/2^t \leq \epsilon$. Divide G into $T = 2^t$ subgraphs $G'_1, \ldots, G'_T$ of equal vertex size by recursively using n^{α}-separators. By Lemma 1 this can be done so that the total number of edges removed is $O(T^{1-\alpha}|G|^{\alpha}) = O(|G|^{\alpha})$. Now, divide each $G'_i = (V'_i, E'_i)$ into two subgraphs $G'_{i,1}$ and $G'_{i,2}$ by Lemma 4, so that $S \cap V'_i$ is evenly partitioned. Without loss of generality assume $|G'_{i,1}| \leq |G'_{i,2}|$. Consider the following procedure for dividing G into two subgraphs G_1 and G_2 satisfying the conditions stated in the lemma:

1. Let G_1, G_2 be empty graphs.
2. For $i = 1$ to T,
   If $|G_1| \geq |G_2|$, then let $G_1 = G_1 \cup G'_{i,1}$ and $G_2 = G_2 \cup G'_{i,2}$; otherwise let $G_1 = G_1 \cup G'_{i,2}$ and $G_2 = G_2 \cup G'_{i,1}$.

First S is evenly divided between G_1 and G_2. Moreover, there are at most $O(|G|^{\alpha})$ edges of G connecting G_1 and G_2. We now show that $|V_1|, |V_2| \leq (1+1/2^t)|G|/2$. We will prove that at the end of every iteration of the for–loop in the above procedure, $||G_1| - |G_2|| \leq |G|/2^t$. The proof is by induction on the for–loop counter i. Let u_i and v_i be the size of G_1 and G_2 after the i-th iteration, respectively. The claim is true when $i = 1$ because $|G'_{1,1}|, |G'_{1,2}| \leq |G|/2^t$ and hence $|u_1 - v_1| = ||G'_{1,1}| - |G'_{1,2}|| \leq |G|/2^t$. By the induction hypothesis, we get $|u_{i-1} - v_{i-1}| \leq |G|/2^t$. WLOG, assume $u_{i-1} \geq v_{i-1}$. Thus $u_i = u_{i-1} + |G'_{i,1}|$ and $v_i = v_{i-1} + |G'_{i,2}|$. If $u_i \geq v_i$, then since $|G'_{i,1}| \leq |G'_{i,2}|$, we get that

$$\begin{aligned} u_i - v_i &= (u_{i-1} + |G'_{i,1}|) - (v_{i-1} + |G'_{i,2}|) \\ &= (|G'_{i,1}| - |G'_{i,2}|) + (u_{i-1} - v_{i-1}) \leq u_{i-1} - v_{i-1} \leq |G|/2^t . \end{aligned}$$

If $u_i < v_i$, then since $|G'_{i,1}| + |G'_{i,2}| = |G|/2^t$, we get that

$$\begin{aligned} v_i - u_i &= (v_{i-1} + |G'_{i,2}|) - (u_{i-1} + |G'_{i,1}|) \\ &= (|G'_{i,2}| - |G'_{i,1}|) + (v_{i-1} - u_{i-1}) \leq |G'_{i,2}| - |G'_{i,1}| \leq |G|/2^t . \end{aligned}$$

□

3.2 An Algorithm for and the Proof of the Main Theorem

Let $G = (V, E)$ be a graph. Let $\Theta = \lceil |G|/k \rceil$ and ϵ be a constant satisfying the conditions of Lemma 5. Assume $|G|$ is a power of two and that we know a bisection of G of cut size $O(|G|^{\alpha})$. Consider the following recursive procedure:

Algorithm: `min-max-boundary-partition`(G, Θ, ϵ)

1. If $|G| \leq \Theta$ then return G.
2. Apply the procedure of Lemma 5 to divide G into $G_1 = (V_1, E_1)$ and $G_2 = (V_2, E_2)$ where S is chosen to be the set of all boundary–vertices in G (at the first level of the recursion let S be the boundary–vertices of the known bisection of G).

3. Let the set of boundary–vertices of G_1 and G_2 be those boundary–vertices inherited from G and those produced by the partition of the previous step.
4. Recursively call `min-max-boundary-partition`(G_1, Θ, ϵ).
5. Recursively call `min-max-boundary-partition`(G_2, Θ, ϵ).
6. If more than k subgraphs were generated, repeatly merge the two smallest subgraphs until only k subgraphs remain.

Note that the partitioning procedure of Step 2 evenly divides the boundary–vertices and approximately divides the vertex set. We now prove our main separator theorem.
Proof: The recursive procedure above defines a separator tree T. The size of the subgraph at a leaf is at least $(1-\epsilon)|G|/2k$ but at most $\lceil |G|/k \rceil$. The graph associated to the root of the separator tree is G itself. Let the *level* of a node in the tree be its distance to the root. Let c' be a constant such that every graph H in $\mathcal{G}(\alpha)$ has a separator of cut size at most $c'|H|^\alpha$. We now prove, by induction on the levels of the separator tree, that there is a constant c such that for every node v of T, $\partial_V(G_v) \leq c|G_v|^\alpha$. The claim is true for the two children of the root, provided $c \geq c'$, since we can find a bisection of G of size at most $c'|G|^\alpha$. Assume that the claim is true for every internal node v at level $i-1$. Let u and w be the two children of v. The algorithm divides G_v into G_u and G_w. Let c_1 be the constant hidden in the O-notation of Lemma 5. Hence, if G' denotes either G_u or G_w, we have that

$$\begin{aligned} \partial_V(G') &\leq \partial_V(G_v)/2 + c_1|G_v|^\alpha \\ &\leq (c/2)|G_v|^\alpha + c_1|G_v|^\alpha \\ &= (c/2 + c_1)|G_v|^\alpha \\ &\leq (2^\alpha(c/2 + c_1)/(1-\epsilon)^\alpha)\,|G'|^\alpha . \end{aligned}$$

The last inequality follows since Lemma 5 insures that $|G'| \geq (1-\epsilon)|G_v|/2$. To conclude, recall that G is a bounded-degree graph and choose c such that $c \geq 2^\alpha(c/2 + c_1)/(1-\epsilon)^\alpha$, i.e.,

$$c \cdot ((1-\epsilon)^\alpha/2^\alpha - 1/2) \geq c_1 .$$

This can be done as long as $\epsilon < 1 - 2^{1-1/\alpha}$. □

Corollary 6. *Let k be an integer such that $k > 1$. Then, every n-node well–shaped mesh or nearest neighbor graph has a $(2,k)$-partition P with* max-boundary$_E(P) = O((n/k)^{1-1/d})$*; every n-node bounded–degree planar graph, graph with bounded genus, and graph with bounded forbidden minor has a $(2,k)$-partition P with* max-boundary$_E(P) = O(\sqrt{n/k})$.

4 Partitioning Weighted Graphs

The following are two examples where partitioning of weighted graphs are needed. In adaptive numerical formulation, in order to efficiently achieve a desired solution accuracy, sophisticated *adaptive* strategies that vary the solution or discretization technique within each finite element are used. For example, the p-refinement technique applies a higher order basis function in those elements having a rapidly changing solution or a large error. The h-refinement technique involves subdivision of the mesh elements themselves. (The $p-$ and hybrid hp-refinement [3] techniques can be used to efficiently find accurate solutions to problems in areas such as computational plasticity.) Strategies such as $p-$ and hp-refinement may cause the work to vary at different elements in the domain. This variation may be as high as one or two orders of magnitude [3].

In N-body simulations for non–uniformally distributed particles [2, 7, 18], particles will be grouped into clusters based on their geometric location. The interaction between particles in a pair of well–separated clusters will be approximated by the interaction between their clusters. The amount of calculations associated with some cluster/particle may be much higher than the amount of calculations needed in some other cluster/particle.

Consider a graph where every vertex is assigned a weight that is proportional to the amount of computation needed at the vertex. Let the *total weight* of a graph be the sum of the weight of its vertices. Rather than partitioning the graph into subgraphs of equal vertex size we would now like to partition it into subgraphs with "equal" total weight. However, partitioning according to weights alone may cause an imbalance in the size of the resulting subgraphs. In some applications, this may cause an imbalance on local memory requirements since, in general, all vertices need a similar amount of storage even though the computational work associated to them may vary. We consider the problem of partitioning vertex–weighted graphs into subgraphs and simultaneously balancing the total weight and the size of the vertex set of the resulting subgraphs wile minimizing the max–boundary.

4.1 Simultaneous Partition of Vertices and Weights

Let $G = (V, E, w)$ be a vertex-weighted graph, where $w : V \to \mathbb{R}_+$ is a positive weight vector. For any subgraph $G' = (V', E')$ of G, we denote by $w(G')$ or $w(V')$ the total weight of G', i.e. $w(G') = w(V') = \sum_{v \in V'} w(v)$.

A variant of the following lemma was given in Lipton and Tarjan [11].

Lemma 7. *Let $0 < \lambda < 1/2$. Let $G = (V, E)$ be a bounded-degree graph in $\mathcal{G}(\alpha)$ and $w : V \to \mathbb{R}_+$ be a weight-vector such that $w(v) < \lambda w(G)$ for all $v \in V$. Then, one can find an $O(|G|^\alpha)$-separator that divides G into two subgraphs $G_1 = (V_1, E_1)$ and $G_2 = (V_2, E_2)$ such that $w(G_1), w(G_2) \leq (1 + \lambda)w(G)/2$.*

Lemma 8. *Let $0 < \epsilon < 1/2$ and $0 < \lambda < 1/2$. Let $G = (V, E)$ be a bounded-degree graph in $\mathcal{G}(\alpha)$ such that $|G|$ is a power of two. Let $w : V \to \mathbb{R}_+$ be a weight-vector such that $w(v) < \lambda w(G)$ for all $v \in V$. Then, one can find an $O(|G|^\alpha)$-separator that divides G into two subgraphs $G_1 = (V_1, E_1)$ and $G_2 = (V_2, E_2)$ such that $|V_1|, |V_2| \leq (1 + \epsilon)|V|/2$ and $w(G_1), w(G_2) \leq (1 + \lambda)w(G)/2$.*

The proof is similar to that of Lemma 5.

Let k be an integer such that $k > 1$. Let $G = (V, E, w)$ be a vertex–weighted graph. Let $P = \{G_1, \ldots, G_k\}$ be a collection of subgraphs $G_i = (V_i, E_i)$ of G that have disjoint vertex sets. We say that P is a (β, δ, k)-partition of G if the V_i's cover all of V, and for all $i \in \{1, \ldots, k\}$ it holds that $|V_i| \leq \beta\lceil |G|/k \rceil$ and $w(G_i) \leq \delta w(G)/k$.

The following corollary follows by recursively applying Lemma 8.

Corollary 9. *Let $0 < \epsilon < 1/2$ and $0 < \lambda < 1/2$. Let k be an integer such that $k > 1$. Let $G = (V, E)$ be a graph in $\mathcal{G}(\alpha)$ such that $|G|$ is a power of two. Let $w : V \to \mathbb{R}_+$ be a weight-vector such that $w(v) < \lambda w(G)$ for all $v \in V$. Then, a $(1+\epsilon, 1+\lambda, k)$-partition P of G such that* total-boundary$_E(P) = O(k^{1-\alpha}|G|^\alpha)$ *can be constructed.*

4.2 Min–Max–Boundary Partition of Weighted Graphs

Theorem 10. *Let k be an integer such that $k > 1$ and λ be a constant such that $0 < \lambda < 1/2$. Let $G = (V, E)$ be a bounded-degree graph in $\mathcal{G}(\alpha)$ such that $|G|$ is a power of two. Let $w : V \to \mathbb{R}_+$ be a weight-vector such that $w(v) < \lambda w(G)$ for all $v \in V$. Then, a $(2, 1+\lambda, k)$-partition P of G such that* max-boundary$_E(P) = O((|G|/k)^\alpha)$ *can be constructed.*

To prove the latter theorem we will follow the same argument used in Section 3.2 to prove Theorem 2. The algorithm recursively applies the following lemma to simultaneously partition weights, vertices, and boundary. Details will be given in the full version.

Lemma 11. *Let $0 < \epsilon < 1/2$ and $0 < \lambda < 1/2$. Let $G = (V, E)$ be a bounded-degree graph in $\mathcal{G}(\alpha)$ such that $|G|$ is a power of two and $w : V \to \mathbb{R}_+$ be a weight-vector such that $w(v) < \lambda w(G)$ for all $v \in V$. Let $S \subseteq V$ be a subset of V. Then, one can find an $O(|G|^\alpha)$-separator that divides G into two subgraphs $G_1 = (V_1, E_1)$ and $G_2 = (V_2, E_2)$ such that $|S \cap V_1| = \lfloor |S|/2 \rfloor$, $|S \cap V_2| = \lceil |S|/2 \rceil$, $|V_1|, |V_2| \leq (1+\epsilon)|G|/2$, and $w(G_1), w(G_2) \leq (1+\lambda)w(G)/2$.*

The proof follows the basic idea developed in the proof of Lemma 5.

5 Vertex-Based Decomposition

An alternative way to partition a graph is by removing vertices rather than by removing edges. Vertex–based decomposition has been used in nested dissection for solving sparse linear systems [12] and overlapping domain decomposition [8]. Lipton, Rose, and Tarjan [12] gave the following scheme to recursively divide a graph using vertex separators:

Algorithm: LRT(G, Θ)

1. If $|G| \leq \Theta$ then return G.
2. Find a small vertex separator $C \subseteq V$ of $G = (V, E)$ that partitions V into two disjoint subsets A and B such that $|A| \leq |V|/2$ and $|B| \leq |V|/2$.
3. Let G_1 and G_2 be the subgraphs of G induced by the vertex sets $A \cup C$ and $B \cup C$ respectively.
4. Recursively call LRT(G_1, Θ).
5. Recursively call LRT(G_2, Θ).

The procedure above decomposes the input graph G into k subgraphs $G_1, \ldots, G_k$, for some $k \geq 1$. Vertices used in separators may occur in two or more subgraphs.

Motivated by this procedure, we define the following vertex-based decomposition problem. Given a graph $G = (V, E)$ and an integer $k > 1$, we say that $D = \{V_1, \ldots, V_k\}$ is a *(β, k)-decomposition* of V if the subgraphs $G_i = (V_i, E_i)$ of G induced by the V_i's are such that $\cup_{i=1}^{k} V_i = V$, $\cup_{i=1}^{k} E_i = E$, and $|V_i| \leq \beta\lceil |V|/k \rceil$, for all $i \in \{1, \ldots, k\}$. Note that in such a decomposition $G_1, \ldots, G_k$ may be pair-wise overlapping.

In this section, we denote by $\partial(G_i)$ the set of vertices in V_i that are also nodes of some other subgraph G_j, $j \neq i$. As in multi-way graph partitioning, we consider the following two costs associated with a (β, k)-decomposition:

$$\text{total-boundary}_V(D) = \sum_{i=1}^{k} |\partial(G_i)|$$

$$\sum_{i=1}^{k} |V_k| = |V| + \text{total-boundary}_V(D) - |\bigcup_{i=1}^{k} \partial(G_i)|.$$

Given a graph $G = (V, E)$, a subset C of V is called a *vertex-bisector* of G if it is a vertex separator of G that partitions V into two disjoint subsets of size at most $|V|/2$. Again, let $\mathcal{G}$ denote a family of graphs (not necessarily of bounded-degree) that is closed under the subgraph operation. Let $0 < \alpha < 1$. We say that $\mathcal{G}$ has an *n^α-vertex-separator theorem* or $\mathcal{G}$ is *n^α-vertex-separable* if there is a constant c such that every n-node graph in $\mathcal{G}$ has a vertex-bisector of size cn^α. Moreover, we refer to the latter type of bisectors as n^α-vertex-separators. There are several families of graphs that have n^α-vertex-separators and are closed under the subgraph operation. Examples includeplanar graphs [11], graphs with bounded genus [10], graphs with no h-clique minor for a constant h [1], well-shaped meshes in $\mathbb{R}^d$ and nearest neighbor graphs in $\mathbb{R}^d$ [13, 14, 16].

Below we state two vertex-separator results similar in spirit to those presented in Section 3. Their proofs follow from the same type of arguments as those developed in Section 3. Thus, we omit the proofs.

Lemma 12. *Let $0 < \epsilon < 1/2$. Let $\mathcal{G}$ be a family of n^α-vertex-separable graphs closed under the subgraph operation. Let $G = (V, E)$ be a graph in $\mathcal{G}$ such that $|G|$ is a power of two. Let $S \subseteq V$ be a subset of V. Then, one can find an $O(n^\alpha)$-vertex-separator that divides G into two subgraphs $G_1 = (V_1, E_1)$ and $G_2 = (V_2, E_2)$ such that $|S \cap V_1| = \lfloor |S|/2 \rfloor$, $|S \cap V_2| = \lceil |S|/2 \rceil$, and $|V_1|, |V_2| \leq (1 + \epsilon)|V|/2$.*

Theorem 13. *Let k be an integer such that $k > 1$ and ϵ be a constant such that $0 < \epsilon < \min\{1/2, 1 - 2^{1-1/\alpha}\}$. Let $\mathcal{G}$ be a family of n^α-vertex-separable graphs closed under the subgraph operation. Then, for every graph $G = (V, E)$ in $\mathcal{G}$ a $(2, k)$-decomposition D such that* max-boundary$_V(D) = O((|V|/k)^\alpha)$ *can be constructed.*

6 Conclusions

We have conducted experiments on several variations of our algorithm presented in this paper. On various finite element meshes in both two and three dimensions, the experiments show that the constant of the Big-O on the boundary size in all the separator theorems presented in the paper is less than 1.5.

References

1. N. Alon, P. Seymour, and R. Thomas. A separator theorem for graphs with an excluded minor and its applications. In *STOC90*, pages 293–299. ACM, 1990.
2. J. Barnes and P. Hut. A hierarchical $O(n \log n)$ force calculation algorithm. *Nature*, (324):446–449, 1986.
3. M. Benantar, R. Biswas, J. E. Flaherty, and M. S. Shephard. Parallel computation with adaptive methods for elliptic and hyperbolic systems. *Comp. Methods Applied Mech. and Eng.*, pages 73–93, 1990.
4. M. Bern, D. Eppstein, and J. R. Gilbert. Provably good mesh generation. In *FOCS90*, pages 231–241. IEEE, 1990.
5. S. N. Bhatt and F. T. Leighton. A framework for solving VLSI graph layout Problems. *JCSS*, 28, pp 300-343, 1984.
6. J. H. Bramble, J. E. Pasciak, and A. H. Schatz. An iterative method for elliptic problems on regions partitioned into substructures. *Math. Comp.*, 46:361–9, 1986.
7. J. Carrier, L. Greengard, and V. Rokhlin. A fast adaptive multipole algorithm for particle simulations. *SIAM J. Sci. Statist. Comput.* 9:669–686, 1988.
8. T. F. Chan and T. P. Mathew. Domain decomposition algorithms. *Acta Numerica*, pages 61-144, 1994.
9. J. R. Gilbert, G. L. Miller, and S.-H. Teng. Geometric mesh partitioning: Implementation and experiments. In *SIAM J. Sci. Comput.*, to appear, 1999.
10. J.R. Gilbert, J.P. Hutchinson, and R.E. Tarjan. A separation theorem for graphs of bounded genus. *Journal of Algorithms*, 5:391–407, 1984.
11. R. J. Lipton and R. E. Tarjan. A separator theorem for planar graphs. *SIAM J. of Appl. Math.*, 36:177–189, April 1979.
12. R. J. Lipton, D. J. Rose, and R. E. Tarjan. Generalized nested dissection. *SIAM J. on Numerical Analysis*, 16:346–358, 1979.
13. G. L. Miller, S.-H. Teng, W. Thurston, and S. A. Vavasis. Finite element meshes and geometric separators. *SIAM J. Sci. Comput.*, to appear, 1999.
14. G. L. Miller, S.-H. Teng, W. Thurston, and S. A. Vavasis. Separators for sphere-packings and nearest neighborhood graphs. *J. ACM*, Jan. 1997.
15. H. D. Simon and S.-H. Teng. How good is recursive bisection? *SIAM J. Sci. Comput.*, to appear, 1996.
16. D. A. Spielman and S.-H. Teng. Spectral partitioning works: planar graphs and finite element meshes. In *FOCS96*, pages 96–107, IEEE, 1996,
17. G. Strang and G. J. Fix. *An Analysis of the Finite Element Method.* Prentice-Hall, Englewood Cliffs, New Jersey, 1973.
18. S.-H. Teng. Provably good partitioning and load balancing algorithms for parallel adaptive n-body simulation. *SIAM J. Scientific Computing*, to appear, 1999.

On Boolean Lowness and Boolean Highness

Steffen Reith and Klaus W. Wagner

Lehrstuhl für Theoretische Informatik
Universität Würzburg
Am Exerzierplatz 3
97072 Würzburg, Germany
[streit,wagner]@informatik.uni-wuerzburg.de

Abstract. The concepts of lowness and highness originate from recursion theory and were introduced into the complexity theory by Schöning [Sch85]. Informally, a set is low (high, resp.) for a relativizable class $\mathcal{K}$ of languages if it does not add (adds maximal, resp.) power to $\mathcal{K}$ when used as an oracle. In this paper we introduce the notions of boolean lowness and boolean highness. Informally, a set is boolean low (boolean high, resp.) for a class $\mathcal{K}$ of languages if it does not add (adds maximal, resp.) power to $\mathcal{K}$ when combined with $\mathcal{K}$ by boolean operations. We prove properties of boolean lowness and boolean highness which show a lot of similarities with the notions of lowness and highness. Using Kadin's technique of hard strings (see [Kad88,Wag87,CK96,BCO93]) we show that the sets which are boolean low for the classes of the boolean hierarchy are low for the boolean closure of Σ_2^p. Furthermore, we prove a result on boolean lowness which has as a corollary the best known result (see [BCO93]; in fact even a bit better) on the connection of the collapses of the boolean hierarchy and the polynomial-time hierarchy: If $BH = NP(k)$ then $PH = \Sigma_2^p(k-1) \oplus NP(k)$.

Keywords: Computational complexity, lowness, highness, boolean lowness, boolean highness, boolean hierarchy, polynomial-time hierarchy, hard/easy, advice, collapse.

1 Introduction

The concept of lowness and highness was originally studied in a recursion theoretic context ([Coo74] and [Soa74]). At that time the question arose how to measure the content of information of an oracle, used by a Turing machine. An oracle was called *low* for a given class $\mathcal{K}$ if it does not add power to the machines accepting sets from $\mathcal{K}$. It was called *high* for $\mathcal{K}$ if it adds (in a sense) maximal power to $\mathcal{K}$. These notions have been studied particularly in the context of the arithmetic hierarchy.

Many ideas and concepts were translated from recursion theory into the terms of complexity theory. So Schöning ([Sch83],[Sch85]) introduced the notions of lowness and highness into complexity theory. In the context of the polynomial-time hierarchy he defined $Low_k^p =_{\text{df}} \{A \in NP | (\Sigma_k^p)^A = \Sigma_k^p\}$ and $High_k^p =_{\text{df}} \{A \in NP | (\Sigma_k^p)^A = \Sigma_{k+1}^p\}$ for $k \geq 0$. It was shown in [Sch85] that the Low_k^p as well as the $High_k^p$ classes build a (possibly non-proper) hierarchy, that the collapse of the polynomial-time hierarchy is related to some properties of Low_k^p and $High_k^p$ sets, and that the lower classes of these hierarchies can be characterized in well-known terms.

In Section 3 we introduce the notions of boolean lowness and boolean highness. Informally, a set is boolean low (boolean high, resp.) for a class $\mathcal{K}$ of languages if it does not add (adds maximal, resp.) power to $\mathcal{K}$ when combined with $\mathcal{K}$ by boolean operations. We make this precise in the context of the classes $NP(k)$ of the boolean hierarchy. Let $\mathcal{R}_m^p(A)$ be the class of all languages which are $\leq_m^p$-reducible to A and let $\hat{\mathcal{R}}_m^p(A) =_{\text{df}} \mathcal{R}_m^p(A) \cup P$. For classes $\mathcal{K}$ and $\mathcal{K}'$ let further $co\text{-}\mathcal{K} =_{\text{df}} \{\overline{A} | A \in \mathcal{K}\}$, $\mathcal{K} \wedge \mathcal{K}' =_{\text{df}} \{A \cap B | A \in \mathcal{K}, B \in \mathcal{K}'\}$, and $\mathcal{K} \vee \mathcal{K}' =_{\text{df}} \{A \cup B | A \in \mathcal{K}, B \in \mathcal{K}'\}$. For $k \geq 0$, a set $A \in NP$ is in low_k^p iff $NP(k) \wedge \mathcal{R}_m^p(A) = NP(k) \vee \mathcal{R}_m^p(A) = NP(k) \wedge co\text{-}\mathcal{R}_m^p(A) = NP(k) \vee co\text{-}\mathcal{R}_m^p(A) = NP(k)$; a set $A \in NP$ is in $high_k^p$ iff $NP(k) \wedge \mathcal{R}_m^p(A) = NP(k) \wedge NP$, $NP(k) \vee \mathcal{R}_m^p(A) = NP(k) \vee NP$, $NP(k) \wedge co\text{-}\mathcal{R}_m^p(A) = NP(k) \wedge co\text{-}NP$, and $NP(k) \vee co\text{-}\mathcal{R}_m^p(A) = NP(k) \vee co\text{-}NP$. For the classes low_k^p and $high_k^p$ we prove results which are very similar to those for the classes Low_k^p and $High_k^p$ (in the context of the boolean hierarchy rather than the polynomial-time hierarchy).

In Section 4 we relate boolean lowness to lowness. Using Kadin's technique of hard strings (see [Kad88,Wag87,CK96,BCO93]) we prove that $A \in low_k^p$ implies $(\Sigma_2^p)^A \subseteq \Sigma_2^p(2k-1)$ where $\Sigma_2^p(k)$ denotes the k-th level of the boolean hierarchy over Σ_2^p. Hence every low_k^p set is low for the boolean closure of Σ_2^p, and consequently $low_k^p \subseteq Low_3^p$ for all $k \geq 0$.

These results have interesting consequences to the connection between the collapses of the boolean hierarchy and the polynomial-time hierarchy. Kadin [Kad88] showed that $BH = NP(k)$ implies $PH = \Delta_3^p$. This was improved in [Wag87] where Δ_3^p could be replaced by the boolean closure of Σ_2^p and independently in [CK96] where Δ_3^p was replaced by $\Sigma_2^p(k)$. Eventually, Beigel, Chang, and Ogihara [BCO93] proved that $BH = NP(k)$ implies the collapse of the polynomial-time hierarchy to the class of all languages that are computable in polynomial time with $k-1$ parallel queries to a Σ_2^p set and an unbounded number of queries in NP. Adapting their method we prove in Section 5 that $low_k^p \cap High_1^p \neq \emptyset$ implies $PH = \Sigma_2^p(k-1) \oplus \Delta_2^p$ for $k \geq 1$, where $\mathcal{K} \oplus \mathcal{K}' =_{\text{df}} \{A \triangle B | A \in \mathcal{K}, B \in \mathcal{K}'\}$. From this and a recent result by Chang [Cha97] one can conclude: If $BH = NP(k)$ then $PH = \Sigma_2^p(k-1) \oplus NP(k)$, which even slightly improves the [BCO93] result.

2 Preliminaries

With P^B (NP^B) we denote the class of all languages accepted by a deterministic polynomial time (nondeterministic polynomial time, resp.) oracle machine using the oracle B. If the queries to an oracle depend on the answers to previous queries, we will call them *adaptive* queries. The queries are said to be made in *parallel*, if a list of all queries is calculated before the machine asks the oracle. The class of all languages accepted by deterministic polynomial time (nondeterministic polynomial time) oracle machines making only parallel queries to an oracle B will be denoted by $P_{\|}^{B}$ ($NP_{\|}^{B}$). For a class $\mathcal{K}$ of languages we define $P^{\mathcal{K}} =_{\mathrm{df}} \bigcup_{B\in\mathcal{K}} P^{B}$, $NP^{\mathcal{K}} =_{\mathrm{df}} \bigcup_{B\in\mathcal{K}} NP^{B}$ and $P_{\|}^{\mathcal{K}} =_{\mathrm{df}} \bigcup_{B\in\mathcal{K}} P_{\|}^{B}$.

For a function $r : \mathbb{N} \mapsto \mathbb{N}$ we denote by $P^{B}[r]$ the class of languages accepted by some deterministic polynomial time oracle machine asking the oracle only $r(n)$ times for an input of length n. Similarly let $P_{\|}^{B}[r]$ be the class of languages accepted by deterministic polynomial oracle machines making only $r(n)$ parallel queries to B.

For classes $\mathcal{K}$ and $\mathcal{K}'$ of languages we define *co-*$\mathcal{K} =_{\mathrm{df}} \{\overline{A} | A \in \mathcal{K}\}$, $\mathcal{K} \wedge \mathcal{K}' =_{\mathrm{df}} \{A \cap B | A \in \mathcal{K}, B \in \mathcal{K}'\}$, $\mathcal{K} \vee \mathcal{K}' =_{\mathrm{df}} \{A \cup B | A \in \mathcal{K}, B \in \mathcal{K}'\}$, and $\mathcal{K} \oplus \mathcal{K}' =_{\mathrm{df}} \{A \triangle B | A \in \mathcal{K}, B \in \mathcal{K}'\}$. The *boolean hierarchy* over a complexity class $\mathcal{K} \supseteq P$ consists of the classes $\mathcal{K}(k)$ and *co-*$\mathcal{K}(k)$ for $k = 0, 1, \ldots$ which are defined inductively by $\mathcal{K}(0) =_{\mathrm{df}} P$ and $\mathcal{K}(k+1) =_{\mathrm{df}} co\text{-}\mathcal{K}(k) \wedge \mathcal{K}$. Furthermore let $BH(\mathcal{K}) =_{\mathrm{df}} \bigcup_{k\geq 0} \mathcal{K}(k)$.

With $\mathcal{K} = NP$ we obtain the well known boolean hierarchy over NP (see [CH86], [WW85], [Köb85], [KSW87] and [CGH+88]). For convenience we set $BH =_{\mathrm{df}} BH(NP)$.

We will need the following lemma which is proved for $\mathcal{K} = NP$ in [KSW87]. The proofs remain valid in the general case. The equality in statement 5 can be found in [Wag97].

Lemma 1 *Let $\mathcal{K} \supseteq P$ be closed under union and intersection, and let $k \geq 0$.*

1. $\mathcal{K}(2k+1) = \mathcal{K}(2k) \vee \mathcal{K}$
2. $\mathcal{K}(2k+2) = \mathcal{K}(2k+1) \wedge co\text{-}\mathcal{K}$
3. $\mathcal{K}(k+2) = \mathcal{K}(k) \vee (\mathcal{K} \wedge co\text{-}\mathcal{K}) = \mathcal{K}(k) \wedge (\mathcal{K} \vee co\text{-}\mathcal{K})$
4. $\mathcal{K}(k) = \underbrace{\mathcal{K} \oplus \mathcal{K} \oplus \ldots \oplus \mathcal{K}}_{k\ times}$
5. $\mathcal{K}(k) \subseteq P_{\|}^{\mathcal{K}}[k] = P \oplus \mathcal{K}(k) \subseteq \mathcal{K}(k+1)$

The *polynomial-time hierarchy* consists of the classes Σ_k^p, Π_k^p, Δ_k^p and Θ_k^p by defining inductively $\Sigma_0^p =_{\mathrm{df}} \Pi_0^p =_{\mathrm{df}} \Delta_0^p =_{\mathrm{df}} \Theta_0^p =_{\mathrm{df}} P$ and $\Sigma_{k+1}^p =_{\mathrm{df}} NP^{\Sigma_k^p}$, $\Pi_{k+1}^p =_{\mathrm{df}} co\text{-}\Sigma_{k+1}^p$, $\Delta_{k+1}^p =_{\mathrm{df}} P^{\Sigma_k^p}$, $\Theta_{k+1}^p =_{\mathrm{df}} P^{\Sigma_k^p}[O(\log n)]$. Obviously $\Sigma_k^p \cup \Pi_k^p \subseteq BH(\Sigma_k^p) \subseteq \Theta_{k+1}^p \subseteq \Delta_{k+1}^p \subseteq \Sigma_{k+1}^p \cap \Pi_{k+1}^p$ for all $k \geq 0$. Finally let $PH =_{\mathrm{df}} \bigcup_{k\geq 0} \Sigma_k^p$.

3 Definitions and properties

In a general way, one can define the notions of lowness and highness as follows:

Definition 2

- *For any relativizable class $\mathcal{K}$, the set A is* low for $\mathcal{K}$ *if and only if $\mathcal{K}^A = \mathcal{K}$. Let $Low(\mathcal{K})$ be the class of all sets which are low for $\mathcal{K}$.*
- *For any relativizable class $\mathcal{K}$ and any class $\mathcal{M}$, the set $A \in \mathcal{M}$ is* high for $\mathcal{K}$ with respect to $\mathcal{M}$ *if and only if $\mathcal{K}^A = \mathcal{K}^{\mathcal{M}}$.*
  Let $High(\mathcal{K}, \mathcal{M})$ be the class of all sets which are high for $\mathcal{K}$ with respect to $\mathcal{M}$.

In [Sch85] Schöning introduced the classes $Low_k^p =_{df} Low(\Sigma_k^p) \cap NP$ and $High_k^p =_{df} High(\Sigma_k^p, NP)$ for $k \geq 0$, where $(\Sigma_0^p)^A =_{df} P^A$ and $(\Sigma_{k+1}^p)^A =_{df} NP^{(\Sigma_k^p)^A}$. The following facts are known about the classes Low_k^p and $High_k^p$.

Theorem 3 ([Sch85]) *For $k \geq 0$, the following are equivalent:*
- $PH = \Sigma_k^p$
- $Low_k^p = NP$
- $High_k^p = NP$
- $Low_k^p \cap High_k^p \neq \emptyset$

Theorem 4 ([Sch85]) *For all $k \geq 0$,*
1. $Low_k^p \subseteq Low_{k+1}^p$
2. $High_k^p \subseteq High_{k+1}^p$

Theorem 5 ([Sch85])
1. $Low_0^p = P$
2. $Low_1^p = NP \cap co\text{-}NP$
3. $High_0^p = \{A | A \leq_T^p \text{-complete for } NP\}$

Now we want to have notions of boolean lowness and boolean highness, i.e. notions which are not based on oracle constructions (which build the polynomial-time hierarchy) but on the boolean construction which build the boolean hierarchy. Let $\mathcal{R}_m^p(A)$ be the class of all languages which are $\leq_m^p$-reducible to A and let $\hat{\mathcal{R}}_m^p(A) =_{df} \mathcal{R}_m^p(A) \cup P$. Note that $\hat{\mathcal{R}}_m^p(A) = \mathcal{R}_m^p(A)$ if $A \neq \emptyset$ and $\overline{A} \neq \emptyset$, and that $\hat{\mathcal{R}}_m^p(A) = P$ otherwise.

Definition 6
- *For any class $\mathcal{K} \supseteq P$, the set A is* boolean low for $\mathcal{K}$ *if and only if $\mathcal{K} \wedge \hat{\mathcal{R}}_m^p(A) = \mathcal{K} \vee \hat{\mathcal{R}}_m^p(A) = \mathcal{K} \wedge co\text{-}\hat{\mathcal{R}}_m^p(A) = \mathcal{K} \vee co\text{-}\hat{\mathcal{R}}_m^p(A) = \mathcal{K}$.*
  Let $low(\mathcal{K})$ be the class of all sets which are boolean low for $\mathcal{K}$.
- *For any class $\mathcal{K} \supseteq P$ and any $\mathcal{M} \supseteq P$ which is closed under $\leq_m^p$, the set $A \in \mathcal{M}$ is* boolean high for $\mathcal{K}$ with respect to $\mathcal{M}$ *if and only if $\mathcal{K} \wedge \hat{\mathcal{R}}_m^p(A) = \mathcal{K} \wedge \mathcal{M}$, $\mathcal{K} \vee \hat{\mathcal{R}}_m^p(A) = \mathcal{K} \vee \mathcal{M}$, $\mathcal{K} \wedge co\text{-}\hat{\mathcal{R}}_m^p(A) = \mathcal{K} \wedge co\text{-}\mathcal{M}$, and $\mathcal{K} \vee co\text{-}\hat{\mathcal{R}}_m^p(A) = \mathcal{K} \vee co\text{-}\mathcal{M}$.*
  Let $high(\mathcal{K}, \mathcal{M})$ be the class of all sets which are boolean high for $\mathcal{K}$ with respect to $\mathcal{M}$.

The following properties are easy to prove.

Proposition 7 *Let $\mathcal{K} \supseteq P$ and $\mathcal{M} \supseteq P$. For sets A and B such that $A \leq_m^p B$,*
1. *If $B \in low(\mathcal{K})$ then $A \in low(\mathcal{K})$.*
2. *If $A \in high(\mathcal{K}, \mathcal{M})$ and $B \in \mathcal{M}$ then $B \in high(\mathcal{K}, \mathcal{M})$.*

Proposition 8 *Let $\mathcal{K} \supseteq P$ be closed under union, intersection, and $\leq_m^p$.*
1. *$low(\mathcal{K}) = \mathcal{K} \cap co\text{-}\mathcal{K}$.*
2. *If $\mathcal{M} \cup co\text{-}\mathcal{M} \subseteq \mathcal{K}$ then $high(\mathcal{K}, \mathcal{M}) = \mathcal{M}$.*

Consequently, boolean lowness and boolean highness are interesting mainly for classes $\mathcal{K}$ which are not closed under union and intersection. The classes of the boolean hierarchy (besides the levels 0 and 1) most likely have this properties.

For them we define special classes low_k^p and $high_k^p$ as analogues for the classes Low_k^p and $High_k^p$ for the polynomial-time hierarchy.

Definition 9 *For* $k \geq 0$,

1. $low_k^p =_{\mathrm{df}} low(NP(k)) \cap NP$
2. $high_k^p =_{\mathrm{df}} high(NP(k), NP)$

Boolean lowness and boolean highness for the class $NP(k)$ are strongly connected with collapse properties for the boolean hierarchy.

Theorem 10 *For* $k \geq 0$, *the following are equivalent:*

- $BH = NP(k)$
- $low_k^p = NP$
- $high_k^p = NP$
- $low_k^p \cap high_k^p \neq \emptyset$

The next result exhibits the nature of the classes low_0^p, low_1^p and $high_0^p$.

Theorem 11

1. $low_0^p = P$
2. $low_1^p = NP \cap co\text{-}NP$
3. $high_0^p = \begin{cases} \{A | A \text{ is } \leq_{\mathrm{m}}^{\mathrm{p}}\text{-complete for } NP\} & \text{if } P \neq NP \\ NP & \text{if } P = NP \end{cases}$

Thus these classes are similar to the corresponding Low^p and $High^p$ classes. More precisely: $low_0^p = Low_0^p$, $low_1^p = Low_1^p$ and $high_0^p \subseteq High_0^p$. The next lemma demonstrates, that the inclusion structure of the low^p classes ($high^p$ classes, resp.) is similar to that of Low^p classes ($High^p$ classes, resp.).

Theorem 12 *For all* $k \geq 0$,

1. $low_k^p \subseteq low_{k+1}^p$
2. $high_k^p \subseteq high_{k+2}^p$
3. $high_{2k}^p \subseteq high_{2k+1}^p$

Proof: Use Lemma 1 to decompose $NP(k+1)$, $NP(k+2)$, and $NP(2k+1)$ in such a way that the assumption can be used. □

Unfortunately, we are not able to prove $high_{2k+1}^p \subseteq high_{2k+2}^p$; even the easiest case $high_1^p \subseteq high_2^p$. This is equivalent with proving that $NP \wedge co\text{-}\hat{\mathcal{R}}_m^p(A) = NP \wedge co\text{-}NP$ and $NP \vee co\text{-}\hat{\mathcal{R}}_m^p(A) = NP \vee co\text{-}NP$ implies $(NP \wedge co\text{-}NP) \vee \hat{\mathcal{R}}_m^p(A) = (NP \wedge co\text{-}NP) \vee NP$ and $(NP \wedge co\text{-}NP) \vee co\text{-}\hat{\mathcal{R}}_m^p(A) = (NP \wedge co\text{-}NP) \vee co\text{-}NP$. By Theorem 11 it is obvious that the classes low_0^p and low_1^p are closed under complement. The next theorem shows the consequences of other low_k^p and $high_k^p$ classes to be closed under complement. The behavior of low_k^p and $high_k^p$ classes seems to differ from the one of Low_k^p and $High_k^p$ classes, which are obviously closed under complement.

Theorem 13

1. *For* $k \geq 2$, $low_k^p = co\text{-}low_k^p \Leftrightarrow low_k^p = NP \cap co\text{-}NP$
2. *For* $k \geq 0$, $high_k^p = co\text{-}high_k^p \Leftrightarrow NP = co\text{-}NP$.

4 low_k^p sets are low for $BH(\Sigma_2^p)$

In this section we show, that our low_k^p sets are low for $BH(\Sigma_2^p)$ and hence in Low_3^p. To prove this, we use similar *hard* and *easy* arguments as introduced in [Kad88] (see also: [BCO93], [CK96], [BF96], [Cha97], [HHH97a] and [HHH97b]).
The projection function of the ith to the jth component of a k tuple will be denoted by $\langle x_1, \ldots, x_k\rangle_{(i,j)} = \langle x_i, \ldots, x_j\rangle$. The projection function of the ith component will be denoted by the shortcut $\langle x_1, \ldots, x_k\rangle_{(i)} =_{\text{df}} \langle x_1, \ldots, x_k\rangle_{(i,i)}$.
For every set A and $k \geq 0$ we define the set A_k by

$$A_0 =_{\text{df}} \overline{A}$$
$$A_{k+1} =_{\text{df}} \{\langle x_1, \ldots, x_{k+2}\rangle | \langle x_1, \ldots, x_{k+1}\rangle \in \overline{SAT_k} \wedge x_{k+2} \in \overline{A}\} \ (k \geq 0)$$

Lemma 14 *For $k \geq 0$ and $A \in NP \backslash \{\emptyset, \Sigma^*\}$, if $A \in low_{k+1}^p$ then $A_k \leq_{\text{m}}^{\text{p}} \overline{\mathbf{SAT}_k}$.*

Proof: Define $NP'(2k) =_{\text{df}} NP(2k)$ and $NP'(2k+1) =_{\text{df}} co\text{-}NP(2k+1)$ for $k \geq 0$. We observe that $NP'(k+1) = co\text{-}NP'(k) \wedge co\text{-}NP$. Hence $\mathbf{SAT}_k$ is $\leq_{\text{m}}^{\text{p}}$-complete for $NP'(k+1)$ for $k \geq 0$, and $A_k \in co\text{-}NP'(k) \wedge co\text{-}\hat{\mathcal{R}}_m^p(A)$ for $k \geq 1$. Using these facts and $A \in low_{k+1}^p$ we conclude $A_k \in co\text{-}NP'(k+1) \wedge co\text{-}\hat{\mathcal{R}}_m^p(A) = co\text{-}NP'(k+1) = co\text{-}\mathcal{R}_m^p(\mathbf{SAT}_k) = \mathcal{R}_m^p(\overline{\mathbf{SAT}_k})$. □

Similar to [CK96] we define the notion of hard sequences. Note that a hard sequence wrt. h for length m there corresponds to a hard sequence wrt. $(\mathbf{SAT}, h, m)$ here.

Definition 15

- *Let $k \geq 1$, $m \geq 1$, $j = 1, \ldots, k-1$, $A \subseteq \Sigma^*$, and $h : (\Sigma^*)^k \rightarrow (\Sigma^*)^k$. We call $\langle x_1, \ldots, x_j\rangle$ a* hard sequence *wrt. (A, h, m) iff $j = 0$ or*
  1. *$1 \leq j \leq k-1$,*
  2. *$|x_j| \leq m$,*
  3. *$x_j \in \overline{A}$ if $j = 1$ and $x_j \in \overline{\mathbf{SAT}}$ if $j > 1$,*
  4. *$\forall y_1, \ldots, y_{k-j} \in \Sigma^{\leq m}(h(y_1, \ldots, y_{k-j}, x_j, \ldots, x_1)_{(k-j+1)} \in \overline{\mathbf{SAT}})$,*
  5. *$\langle x_1, \ldots, x_{j-1}\rangle$ is a hard sequence wrt. (A, h, m).*
- *We call j* order *of the hard sequence $\langle x_1, \ldots, x_j\rangle$.*
- *If there exists no hard sequence wrt. (A, h, m) of order greater than j then every hard sequence wrt. (A, h, m) of order j is called a* maximal hard sequence *wrt. (A, h, m).*

For out main results (Theorem 20 and Corollary 21) we need some technical lemmas whose proofs follows the ideas developed in [Kad88], [CK96] and [BCO93]. Because of the page restriction we omit these proofs here.
Let $k \geq 1$ and $A \in low_k^p$. By Lemma 14 we get $A_{k-1} \leq_{\text{m}}^{\text{p}} \overline{\mathbf{SAT}_{k-1}}$. The next lemma shows, that under this assumption one can reduce also A_{k-j-1} to $\overline{\mathbf{SAT}_{k-j-1}}$ by using a hard sequence of order j.

Lemma 16 *Let $A \in NP\backslash\{\emptyset, \Sigma^*\}$, $k \geq 2$, $m \geq 1$, $j = 1, \dots, k-1$, and $A_{k-1} \leq^{\mathrm{p}}_{\mathrm{m}} \overline{\mathbf{SAT}_{k-1}}$ via $h \in FP$. If $\langle x_1, \dots, x_j\rangle$ is a hard sequence wrt. (A, h, m) then for all $y_1, \dots, y_{k-j} \in \Sigma^{\leq m}$:*

$$\langle y_1, \dots, y_{k-j}\rangle \in \mathbf{SAT}_{k-j-1} \Leftrightarrow h(y_1, \dots, y_{k-j}, x_j, \dots, x_1)_{(1,k-j)} \in \overline{\mathbf{SAT}_{k-j-1}}.$$

The next lemma shows, that we can use maximal hard sequences to reduce $\overline{A}$ to **SAT**.

Lemma 17 *Let $A \in NP\backslash\{\emptyset, \Sigma^*\}$, $k \geq 1$, and $A_{k-1} \leq^{\mathrm{p}}_{\mathrm{m}} \overline{\mathbf{SAT}_{k-1}}$ via $h \in FP$. There exist a set $B \in NP$ and a polynomial r such that for every n: If $\langle x_1, \dots, x_j\rangle$ is a maximal hard sequence wrt. $(A, h, r(n))$ then for all $y \in \Sigma^{\leq n}$:*

$$y \in \overline{A} \Leftrightarrow \langle y, 1^n, x_1, \dots, x_j\rangle \in B$$

Lemma 18 *Let $A \in NP$, $k \geq 1$, and $A_{k-1} \leq^{\mathrm{p}}_{\mathrm{m}} \overline{\mathbf{SAT}_{k-1}}$ via $h \in FP$. For every set $L \in NP^A$ there exists a set $C \in NP$ and a polynomial s such that for every n: If $\langle x_1, \dots, x_j\rangle$ is a maximal hard sequence wrt. $(A, h, s(n))$ then for all $w \in \Sigma^{\leq n}$:*

$$w \in L \Leftrightarrow \langle w, 1^n, x_1, \dots, x_j\rangle \in C$$

Lemma 19 *Let $A \in NP$, $k \geq 1$, and $A_{k-1} \leq^{\mathrm{p}}_{\mathrm{m}} \overline{\mathbf{SAT}_{k-1}}$ via $h \in FP$. For every set $L \in (\Sigma^p_2)^A$ there exists a set $D \in \Sigma^p_2$ and a polynomial t such that for every n: If $j \in \{0, 1, \dots, k-1\}$ is the order of a maximal hard sequence wrt. $(A, h, t(n))$ then for all $z \in \Sigma^{\leq n}$:*

$$z \in L \Leftrightarrow \langle z, 1^n, j\rangle \in D$$

If j is greater than the order of a maximal hard sequence wrt. $(A, h, t(n))$ then $\langle z, 1^n, j\rangle \notin D$.

Theorem 20 *Let $k \geq 1$ and let $A \in low^p_k$.*
1. *$(\Sigma^p_2)^A \subseteq \Sigma^p_2(2k-1)$*
2. *$(\Delta^p_2)^A \subseteq \Sigma^p_2(k-1) \oplus \Delta^p_2$*

Proof: Let $A \in low^p_k$. By Lemma 14 we get $A_{k-1} \leq^{\mathrm{p}}_{\mathrm{m}} \overline{\mathbf{SAT}_{k-1}}$ via a suitable function $h \in FP$.

For $L \in (\Sigma^p_2)^A$ we obtain by Lemma 19 a set $D \in \Sigma^p_2$ and a polynomial t such that for every n: If $j \in \{0, 1, \dots, k-1\}$ is the order of a maximal hard sequence wrt. $(A, h, t(n))$ then $z \in L \Leftrightarrow \langle z, 1^n, j\rangle \in D$ for all $z \in \Sigma^{\leq n}$ and $(z, 1^n, j) \notin D$ if j is greater than the order of a maximal hard sequence wrt. $(A, h, t(n))$. We define the Σ^p_2 set

$$T =_{\mathrm{df}} \{(1^m, j) | \exists x_1, \dots, x_j \in \Sigma^{\leq m} \text{ and } \langle 1^m, x_1, \dots, x_j\rangle \text{ is a hard sequence wrt. } (A, h, t(n))\},$$

and we obtain:

$$
\begin{aligned}
z \in L \Leftrightarrow & (((z, 1^{|z|}, 0) \in D) \vee ((1^{t(|z|)}, 1) \in T))) \oplus \\
& ((1^{t(|z|)}, 1) \in T) \oplus (((z, 1^{|z|}, 1) \in D) \vee ((1^{t(|z|)}, 2) \in T))) \oplus \\
& \vdots \\
& ((1^{t(|z|)}, k-2) \in T) \oplus (((z, 1^{|z|}, k-2) \in D) \vee ((1^{t(|z|)}, k-1) \in T))) \oplus \\
& ((1^{t(|z|)}, k-1) \in T) \oplus ((z, 1^{|z|}, k-1) \in D)
\end{aligned}
$$

Consequently, $L \in \underbrace{\Sigma_2^p \oplus \Sigma_2^p \oplus \ldots \oplus \Sigma_2^p}_{2k-1} = \Sigma_2^p(2k-1)$ (Lemma 1.4).

For $L \in P^{NP^A}$ let M be a deterministic polynomial time machine accepting L with oracle $L' \in NP^A$ and running time bounded by a polynomial t. By Lemma 14 we have $A_{k-1} \leq_m^p \overline{\mathbf{SAT}_{k-1}}$ via a suitable function $h \in FP$. By Lemma 18 there exists a set $C \in NP$ and a polynomial s such that for every n: If $\langle x_1, \ldots, x_j \rangle$ is a maximal hard sequence wrt. $(A, h, s(n))$ then $w \in L' \Leftrightarrow \langle w, 1^n, x_1, \ldots, x_j \rangle \in C$ for all $w \in \Sigma^{\leq n}$. Hence there a set $E \in P^{NP}$ such that for every n: If $\langle x_1, \ldots, x_j \rangle$ is a maximal hard sequence wrt. $(A, h, s(t(n)))$ then $z \in L \Leftrightarrow \langle z, 1^{t(n)}, x_1, \ldots, x_j \rangle \in E$ for all $z \in \Sigma^{\leq n}$. To use the *mind-change technique* we consider on the set $(\Sigma^*)^{<k}$ a partial order $\sqsubseteq$ defined by $\langle x_1, \ldots, x_i \rangle \sqsubseteq \langle y_1, \ldots, y_j \rangle$ iff ($i \leq j$ and $x_l = y_l$ for $l = 1, \ldots, i$). Define

$$
\begin{aligned}
F =_{\mathrm{df}} \{(z, 1^n, j) | z \in \Sigma^{\leq n} \wedge & \exists \overline{x_1} \ldots \exists \overline{x_j} (\overline{x_1} \sqsubseteq \ldots \sqsubseteq \overline{x_j} \wedge \overline{x_j} \\
& \text{is hard wrt. } (A, h, s(t(n))) \wedge \\
& c_E(z, 1^{t(n)}) \neq c_E(z, 1^{t(n)}, \overline{x_1}) \neq \ldots \neq c_E(z, 1^{t(n)}, \overline{x_j}))\}.
\end{aligned}
$$

Obviously $F \in \Sigma_2^p$, and we obtain

$$
z \in L \Leftrightarrow (z, 1^{t(n)}) \in E \oplus (z, 1^{t(n)}, 1) \in F \oplus \cdots \oplus (z, 1^{t(n)}, k-1) \in F
$$

for all $z \in \Sigma^{\leq n}$. Hence $L \in \Delta_2^p \oplus \underbrace{\Sigma_2^p \oplus \Sigma_2^p \oplus \ldots \oplus \Sigma_2^p}_{k-1} = \Sigma_2^p(k-1) \oplus \Delta_2^p$ (Lemma 1.4). □

Now we are able to relate boolean lowness to classical lowness.

Corollary 21 *For $k \geq 0$, if $A \in low_k^p$ then $BH(\Sigma_2^p)^A = BH(\Sigma_2^p)$, i.e. $low_k^p \subseteq Low(BH(\Sigma_2^p))$.*

Proof: By Theorem 20 and Lemma 1.5 we conclude $(\Sigma_2^p(m))^A = (\Sigma_2^p)^A(m) \subseteq (\Sigma_2^p(2k-1))(m) \subseteq P_{||}^{\Sigma_2^p}[(2k-1) \cdot m] \subseteq \Sigma_2^p(2km - m + 1)$. □

Corollary 22 *$low_0^p = Low_0^p \subseteq low_1^p = Low_1^p \subseteq low_2^p \subseteq \ldots \subseteq Low(BH(\Sigma_2^p)) \subseteq Low_3^p$.*

5 Consequences to collapse results

From Theorem 20.1 we get immediately a statement on the connection between the collapses of the boolean hierarchy and the polynomial-time hierarchy.

Corollary 23 *For $k \geq 1$, if $BH = NP(k)$ then $PH = \Sigma_2^p(2k-1)$.*

Proof: From $BH = NP(k)$ we get $\mathbf{SAT} \in low_k^p$ by Theorem 10 and $\Sigma_3^p = (\Sigma_2^p)^{\mathbf{SAT}} \subseteq \Sigma_2^p(2k-1)$ by Theorem 20.1. Hence $PH = \Sigma_2^p(2k-1)$. □

This improves Kadin's original result from [Kad88] but it is not as good as the result $BH = NP(k) \Rightarrow PH = \Sigma_2^p(k)$ from [CK96] or the further improvement in [BCO93]. It is obvious that an improvement of Theorem 20.1 by replacing $\Sigma_2^p(2k-1)$ by a smaller class $\mathcal{K}$ yields the improvement $BH = NP(k) \Rightarrow PH = \mathcal{K}$ of Corollary 23. However, to follow the idea from [BCO93] to improve Theorem 20.1 we have to combine Theorem 20.1 and Theorem 20.2. But this cannot be done without the additional assumption that A is not "too easy".

Lemma 24 *For $k \geq 1$, if $A \in low_k^p \cap High_1^p$ then $(\Sigma_2^p)^A \subseteq \Sigma_2^p(k-1) \oplus \Delta_2^p$.*

Proof: By Theorem 20 we obtain $(\Sigma_2^p)^A \subseteq \Sigma_2^p(2k-1)$ and $(\Delta_2^p)^A \subseteq \Sigma_2^p(k-1) \oplus \Delta_2^p$. Since $A \in High_1^p$ we can conclude $(\Sigma_2^p)^A \subseteq P^{\Sigma_2^p} = P^{NP^A} \subseteq \Sigma_2^p(k-1) \oplus \Delta_2^p$. □

Theorem 25 *For $k \geq 1$, if $low_k^p \cap High_1^p \neq \emptyset$ then $PH = \Sigma_2^p(k-1) \oplus \Delta_2^p$.*

Proof: For $A \in low_k^p \cap High_1^p$ we obtain $(\Sigma_2^p)^A \subseteq \Sigma_2^p(k-1) \oplus \Delta_2^p$ by Lemma 24 and $(\Sigma_2^p)^A = \Sigma_3^p$ because of $High_1^p \subseteq High_2^p$. Hence $\Sigma_3^p = \Sigma_2^p(k-1) \oplus \Delta_2^p$ and consequently $PH = \Sigma_2^p(k-1) \oplus \Delta_2^p$. □

Now we get as a corollary the best known result on the connection between the collapses of the boolean hierarchy and the polynomial-time hierarchy. This can be found implicitly in [BCO93].

Corollary 26 *For $k \geq 1$, if $BH = NP(k)$ then $PH = \Sigma_2^p(k-1) \oplus NP(k)$.*

Proof: From $BH = NP(k)$ we get $\mathbf{SAT} \in low_k^p$ by Theorem 10. Since $\mathbf{SAT} \in High_1^p$ we obtain $PH = \Sigma_2^p(k-1) \oplus \Delta_2^p$ by Theorem 25. However, it is known from [Cha97] that $BH = NP(k)$ implies $\Delta_2^p = NP(k)$. □

Since all the above results are relativizable we have also[1]

Corollary 27 *For $m, k \geq 1$, if $BH(\Sigma_m^p) = \Sigma_m^p(k)$ then $PH = \Sigma_{m+1}^p(k-1) \oplus \Sigma_m^p(k)$.*

[1] After completion of this work the authors were informed by Lane A. Hemaspaandra about the fact that this result was independently obtained by Hemaspaandra, Hemaspaandra, and Hempel [HHH98].

References

[BCO93] R. Beigel, R. Chang, and M. Ogihara. A relationship between difference hierarchies and relativized polynomial hierarchies. *Mathematical Systems Theory*, 26:293 - 310, 1993.

[BF96] H. Buhrmann and L. Fortnow. Two queries. TR 96-20, University of Chicago, Department of Computer Science, 1996.

[CGH+88] J.-Y. Cai, T. Gundermann, J. Hartmanis, L. A. Hemachandra, V. Sewelson, K. Wagner, and G. Wechsung. The boolean hierarchy I: Structural properties. *SIAM Journal on Computing*, 17(6):1232 - 1252, 1988.

[CH86] J. Cai and L. Hamachandra. The boolean hierarchy: Hardware over NP. *Proceedings of the Structure in Complexity Theory Conference*, 223:105 - 124, 1986.

[Cha97] R. Chang. Bounded queries, approximation and the boolean hierarchy. TR CS 97-04, University of Maryland, Department of Computer Science and Electrical Engineering, 1997.

[CK96] R. Chang and J. Kadin. The boolean hierarchy and the polynomial hierarchy: A closer connection. *SIAM Journal on Computing*, 25(2):340, 1996.

[Coo74] S. B. Cooper. Minimal pairs and high recursively enumerable degree. *Journal of Symbolic Logic*, 39:655 - 660, 1974.

[HHH97a] L. A. Hemaspaandra, E. Hemaspaandra, and H. Hempel. A downward translation in the polynomial hierarchy. In *Proc. of the 14th STACS*, volume 1200 of *LNCS*. Springer Verlag, 1997.

[HHH97b] L. A. Hemaspaandra, E. Hemaspaandra, and H. Hempel. Translating equality downwards. TR 657, University of Rochester, Department of Computer Science, 1997.

[HHH98] E. Hemaspaandra, L. Hemaspaandra, and H. Hempel. What's up with downward collapse: Using the easy-hard technique to link boolean and polynomial hierarchy collapses. TR 682, University of Rochester, Department of Computer Science, February 1998.

[Kad88] J. Kadin. The polynomial time hierarchy collapses if the boolean hierarchy collapses. *SIAM Journal on Computing*, 17(6):1263 - 1282, 1988.

[Köb85] J. Köbler. Untersuchungen verschiedener polynomieller Reduktionsklassen von NP. Diploma Thesis, Universität Stuttgart, 1985.

[KSW87] J. Köbler, U. Schöning, and K. W. Wagner. The difference and truth-table hierarchies for NP. *R.A.I.R.O. Theoretical Informatics and Applications*, 21(4):419 - 435, 1987.

[Sch83] U. Schöning. A low and a high hierarchy within *NP*. *Journal of Computer and System Sciences*, 27:14 - 28, 1983.

[Sch85] U. Schöning. *Complexity and Structure*, volume 211 of *LNCS*. Springer-Verlag, 1985.

[Soa74] R. I. Soare. Automorphisms of the lattice of recursively enumerable sets. *Bulletin of the American Mathematical Society*, 80:53 - 58, 1974.

[Wag87] K. W. Wagner. Number-of-query hierachies. TR 158, University of Augsburg, 1987.

[Wag97] K. W. Wagner. A note on bounded queries and the difference hierarchy. TR 137, Institut für Informatik, Universität Würzburg, 1997.

[WW85] G. Wechsung and K. Wagner. On the boolean closure of NP, manuscript. Extended abstract as: Wechsung, G., On the boolean closure of NP. In *Proc. of the Conference of Foundations on Computation Theory*, volume 199 of *LNCS*, pages 485–493. Springer Verlag, 1985.

The Inherent Dimension of Bounded Counting Classes

Ulrich Hertrampf

Institut für Informatik, Universität Stuttgart
Breitwiesenstr. 20-22, D-70565 Stuttgart
Germany

Abstract. Cronauer et al. [2] introduced the chain method to separate counting classes by oracles. Among the classes, for which this method is applicable, are NP, coNP, MOD-classes, ... As these counting classes are defined via subsets of $\mathbb{N}^k$, it is natural to ask for the minimum value of k, such that a given class can be defined via such a set. We call this value the inherent dimension of the respective class.
The inherent dimension is a very natural concept, but it is quite hard to check the value for a given bounded counting class. Thus, we complement this notion by the notion of type-3 dimension, which is less natural than inherent dimension, but very easy to check. We compare type-3 dimension and inherent dimension, with the result that for classes of inherent dimension less than 3, both notions coincide, and generally the inherent dimension is never greater than the type-3 dimension.
For $k \leq 2$ we can completely solve the questions, whether a given class has inherent dimension k, and which are the minimal classes with that dimension. For $k \geq 3$ we give a sufficient condition for a class being of dimension at least k. We disprove the conjecture that this is also a necessary condition by a counterexample.

1 Introduction

The term "Counting Classes" has been used for a lot of complexity classes which have some counting process inherent in their definition; for an overview see [4]. These classes can be defined by putting certain restrictions on the outcomes of #P-functions, where #P denotes Valiant's basic counting class [12]. The following general framework has been introduced in [2]: Let V be a (finite or infinite) set of integers. Then the class (V)P is the class of all languages L for which there exists a function $f \in$ #P such that $x \in L \iff f(x) \in V$. In other words: L is a language from (V)P if there is a nondeterministic Turing machine M which has a number of accepting paths from the set V exactly for inputs which are words in L. (See also [3].) In general, we will consider sets V of vectors of natural numbers and vectors of #P functions; this corresponds to Turing machines as follows: Given some input and the according computation of our machine, for some alphabet Σ, we count the number of outputs of all symbols from Σ in M's accepting computations. An input is accepted iff the vector of numbers formed in this way belongs to V.

Investigating complexity classes in the area between P and PSPACE always leads to the difficulty that one cannot hope to find (unconditional) separations, due to the fact that all this area might still collapse. Thus, one of the main techniques to develop relations among such complexity classes is the oracle separation technique. We make this precise by calling two classes equal, if and only if they are equal under all relativizations, and the same for inclusion. In this terminology, it has been shown in [2] for instance that all bounded counting classes are not closed under complement.

The starting point for these results was given in 1992 by Bovet, Crescenzi, and Silvestri [1], who presented a uniform way to define complexity classes, which nowadays is referred to as the *leaf language approach,* see also [8–10] and the recent textbook [11], and a general and sufficient criterion for two classes defined in such a way to be separable by an oracle was given. A similar result has been obtained independently by Vereshchagin [13].

Though these results considerably simplified questions whether oracles with certain properties exist by reducing them to combinatorial questions, the argumentations still were sometimes a bit clumsy. In [5], however, a technique was developed, by which the criterion from [1] is much easier applicable in the case of bounded counting classes. In [6], things were pushed a bit further and an algorithm was presented, that given two explicit bounded counting classes decides whether they can be separated by an oracle, or not. However, this algorithm has the drawback that it does not give explicit answers when the classes are given in a parameterized form.

Finally in [2], using a refinement of the main result from [6], two very easily applicable methods, called the First and Second Chain Method, were developed, which allow for a lot of bounded counting classes, to decide whether they can be separated by an oracle or not. By these methods, for example the question which relativizable inclusions between classes of the Boolean Hierarchy over NP and other bounded counting classes exist, was completely resolved.

The current paper builds on these techniques and investigates the question, given a subset U of $\mathbb{N}^k$, what is the minimum value of k', such that a subset V of $\mathbb{N}^{k'}$ exists, for which $(U)\mathrm{P} \subseteq (V)\mathrm{P}$ (relative to all oracles). We call this value the inherent dimension of the complexity class $(U)\mathrm{P}$. We introduce another characteristic of $(U)\mathrm{P}$, which we call the *type-3 dimension.* We show that the type-3 dimension never can be greater than the inherent dimension, and that the type-3 dimension is equal to the inherent dimension, if the inherent dimension is less than 3. Using this result, we show that there are exactly two minimal elements in the set of all classes of (inherent) dimension at least 1, namely NP and coNP (and these two classes are incomparable). Similarly, we exhibit four explicit classes (NP(2), coNP(2), NP $\wedge$ co-1-NP, coNP $\vee$ 1-NP), which are all minimal elements in the set of all classes of (inherent) dimension at least 2.

The technique generalizes to give 2^d minimal elements in the set of all classes of type-3 dimension at least d for $d > 2$. However, finally we show that this does not say much about inherent dimensions, by giving an example of a class, which is of inherent dimension 3, but of type-3 dimension 2.

2 Preliminaries

2.1 Notations

We assume familiarity with basic concepts of complexity theory. We will use the following notations for complexity classes:

- The classes P, NP, coNP, 1-NP, #P, and MOD-classes should be well known.
- If $\mathcal{C}$ is a complexity class, then the class co-$\mathcal{C}$ is defined as the class of all languages L, such that $\overline{L} \in \mathcal{C}$.
- If $\mathcal{C}_1$ and $\mathcal{C}_2$ are two complexity classes, then the class $\mathcal{C}_1 \wedge \mathcal{C}_2$ is defined as the class of all languages L, such that there are languages $L_1 \in \mathcal{C}_1$ and $L_2 \in \mathcal{C}_2$ satisfying $L = L_1 \cap L_2$. $\mathcal{C}_1 \vee \mathcal{C}_2$ is analogously defined via $L = L_1 \cup L_2$.
- The classes of the Boolean Hierarchy over a class $\mathcal{C}$ are the classes, which can be obtained by iterated application of the operators $\wedge$, $\vee$, and co- to the class $\mathcal{C}$, e.g. $\mathcal{C} \wedge$ co-$\mathcal{C}$ or $(\mathcal{C} \wedge$ co-$\mathcal{C}) \vee \mathcal{C}$.
- The classes NP(k) and coNP(k) are the classes of the Boolean Hierarchy over NP.

2.2 Complexity Classes Defined by Counting

Many well-known complexity classes can be defined by counting as done in the following definition.

Definition 1. *For $V \subseteq \mathbb{N}^k$ the complexity class $(V)\mathrm{P}$ is defined as the class of all languages L such that there exist functions $f_1, \ldots, f_k \in \#\mathrm{P}$ satisfying*

$$x \in L \iff (f_1(x), \ldots, f_k(x)) \in V.$$

Using the class $\#\mathrm{P}^{\mathrm{A}}$ ($\#\mathrm{P}$ relative to oracle A) instead of $\#\mathrm{P}$, we obtain the relativized complexity class $(V)\mathrm{P}^A$.

If there is a bound $m \in \mathbb{N}$, such that $v \in V$ if and only if $\min(v, m) \in V$ (the minimum taken componentwise), then we call $(V)\mathrm{P}$ a bounded counting class.

2.3 Relativized Equality and Inclusion

It should be emphasized once again that, whenever we speak of inclusion (or equality) of classes, it means inclusion (or equality) under all relativizations.

Accordingly, whenever we speak of inequality (or noninclusion) of classes, we mean that there is an oracle, which makes the classes different.

3 The First and Second Chain Theorem

We first recall the chain theorems and the definitions needed for them from [2]:

Definition 2. *A sequence* $v_1, v_2, \ldots, v_l \in \mathbb{N}^k$ *is an* alternating chain of length l with respect to V, *if* $v_i \leq v_{i+1}$ *and* $v_i \in V \iff v_{i+1} \notin V$ $(1 \leq i < l)$. *We say that this chain has* positive signature *if* $v_1 \in V$, *otherwise, we say that it has* negative signature.

Theorem 1 (First Chain Theorem). *Let* $V \subseteq \mathbb{N}^k$ *and* $U \subseteq \mathbb{N}^{k'}$ *be of bounded significance, and suppose that* $(V)\mathrm{P} \subseteq (U)\mathrm{P}$ *(in all relativizations). If there is an alternating chain with respect to* V*, then there is an alternating chain of the same length and the same signature with respect to* U.

Definition 3. *Let* $V \subseteq \mathbb{N}^k$ *be of bounded significance with bound* m. *We say, a sequence* $v_1, \ldots, v_s$ *is a* type-2 alternating chain with respect to V of length s, *if*

- $v_1 \leq \cdots \leq v_s \leq (\underbrace{m, m, \ldots, m}_{k \text{ times}})$,
- $v_i \in V \iff v_{i+1} \notin V$ *for* $i = 1, \ldots, s-1$, *and*
- *if* v_i *and* v_{i+1} *differ in the* j*-th component, then the* j*-th component of* v_{i+1} *is equal to* m $(i = 1, \ldots, s-1,\ j = 1, \ldots, k)$.

If v_1 *belongs to* V*, then we say that this chain has* positive signature, *otherwise it has* negative signature.

Theorem 2 (Second Chain Theorem). *Let* $V \subseteq \mathbb{N}^k$ *and* $U \subseteq \mathbb{N}^{k'}$ *be of bounded significance, and suppose that* $(V)\mathrm{P} \subseteq (U)\mathrm{P}$ *(in all relativizations). If there is a type-2 alternating chain w.r.t.* V*, then there exists a type-2 alternating chain of the same length and the same signature w.r.t.* U.

Proposition 1. *If* $V \subseteq \mathbb{N}^k$ *is of bounded significance with bound* m*, then the length of every alternating chain w.r.t.* V *is bounded by* $mk + 1$*, and the length of every type-2 alternating chain w.r.t.* V *is bounded by* $k + 1$.

4 Type-3 Dimension

In [2] the chain theorems served to obtain a lot of separation results between classes from different hierarchies. Looking at these theorems with the aim to define "the dimension" of bounded counting classes it seems evident that the ordinary alternating chains from the first chain theorem do not help, as there can be arbitrarily long alternating chains in one-dimensional sets V (i.e. $V \subseteq \mathbb{N}$). We leave the obvious proof of this fact as an exercise.

How well are type-2 chains suited to serve as basis for a definition of dimension? In [2] a series of rather complex classes was shown to have only type-2 chains of length at most 2, namely the classes $\bigvee_k$ 1-NP and co-$\bigvee_k$ 1-NP. It is easy to see that these classes can not be represented in the form $(V)\mathrm{P}$ for any $V \subseteq \mathbb{N}^{k-1}$. So again we conclude that the existence of type-2 chains does not reflect the intuitive idea of "dimension" very well.

So neither ordinary alternating chains nor type-2 alternating chains should be used to define the dimension of a bounded counting class. The following definition looks more reasonable:

Definition 4. *A* type-3 alternating chain of length l with respect to $V \subseteq \mathbb{N}^k$ *(where V is of bounded significance with bound m) is a sequence of pairs*

$$(v_1, v_1'), (v_2, v_2'), \ldots, (v_l, v_l'),$$

where all v_i and v_i' are elements from $\mathbb{N}^k$ with components at most m, such that for all $i \in \{1, \ldots, l-1\}$ we have $v_i < v_i' \leq v_{i+1} < v_{i+1}'$, further for all $i \in \{1, \ldots, l\}$ we have $v_i \in V \iff v_i' \notin V$, and if v_i and v_i' differ in the j-th component, then v_i' has value m in the j-th component. The signature *of a type-3 alternating chain is a sequence $t \in \{+,-\}^l$, $t = t_1 \ldots t_l$, where $t_i = +$, if $v_i' \in V$ and $t_i = -$, if $v_i' \notin V$.*

Note that unlike the situation for ordinary or type-2 chains the signature of a longest type-3 alternating chain is not uniquely determined.

Proposition 2. *Let $V \subseteq \mathbb{N}^k$ be of bounded significance. Then any type-3 alternating chain w.r.t. V has length at most k.*

Proof. Just observe that in any pair (v_i, v_i') of the chain at least one component has to change from less than m to m, where m is the bound of V. Together with the monotonicity of the entire chain, we obtain that after l pairs the vector v_l' would necessarily equal $(m, \ldots, m)$, so a vector v_{l+1}' cannot exist.

4.1 The Third Chain Theorem

Theorem 3 (Third Chain Theorem). *Let $U \subseteq \mathbb{N}^k$, $V \subseteq \mathbb{N}^{k'}$ be of bounded significance and let $(U)\mathrm{P} \subseteq (V)\mathrm{P}$ (relativizably). If there is a type-3 alternating chain of length l with signature t w.r.t. U, then there is a type-3 alternating chain of length l with signature t w.r.t. V.*

Proof. We build on a theorem from [6], where it was shown that under our assumptions a monotone mapping f from $\mathbb{N}^k$ to $\mathbb{N}^{k'}$ must exist, which can in each component be written as a linear combination of multinomial coefficients, such that $u \in U \iff f(u) \in V$. Especially, this means that the mapping f is "superlinear" in the sense that for all $u, v \in \mathbb{N}^k$ with $u \leq v$ and for every $r \in \mathbb{N}$

$$f(u + r \cdot (v - u)) \geq f(u) + r \cdot (f(v) - f(u)).$$

Now, in $\mathbb{N}^k$ let any type-3 alternating chain w.r.t. U be given. Let m be the bound of U, and let m' be the bound of V. We may increase the bound of U to $m \cdot (m'+1)$. In our type-3 chain, we now have to change all components of value m to value $m \cdot (m'+1)$. This does not change the property of being a type-3 alternating chain, but now the difference between v_i and v_i' in all pairs from this chain has increased at least by a factor of m'. Thus, because of the superlinearity of f, under f all these vectors are mapped in such a way that whenever $f(v_i)$ and $f(v_i')$ differ in the j-th component, then the j-th component of $f(v_i')$ is at least m'. Replacing all vectors $f(v_i)$ and $f(v_i')$ by $\max(f(v_i), m')$ and $\max(f(v_i'), m')$, respectively, results in a type-3 alternating chain with respect to V of the same length and signature as our original chain with respect to U.

4.2 Definition of Type-3 Dimension

The third chain theorem motivates our definition of a type-3 dimension:

Definition 5. *Let $\mathcal{C}$ be a bounded counting class, $\mathcal{C} = (V)\mathrm{P}$ for $V \subseteq \mathbb{N}^k$ with bounded significance. Let l be the length of a longest type-3 alternating chain w.r.t. V. Then l is called the type-3 dimension of $\mathcal{C}$.*

Lemma 1. *The type-3 dimension of a bounded counting class $\mathcal{C}$ is well defined, i.e. it does not depend on the choice of the set V with $\mathcal{C} = (V)\mathrm{P}$.*

Proof. We have to show that if $\mathcal{C} = (U)\mathrm{P} = (V)\mathrm{P}$ (in all relativizations), then the longest type-3 alternating chains in U and in V are equally long. Because of symmetry, it suffices to show that if $(U)\mathrm{P} \subseteq (V)\mathrm{P}$, then the longest type-3 alternating chain w.r.t. U is not longer than the longest type-3 alternating chain w.r.t. V. But the latter is a direct consequence of Theorem 3.

We want to compute the type-3 dimension for all classes from the Boolean Hierarchies over NP and 1-NP. Let for $k \geq 1$,

$$A_k = \{(n_1, \ldots, n_k) \mid \#\{i \mid n_i > 0\} \text{ is odd}\} \quad D_k = \{(n_1, \ldots, n_k) \mid \#\{i \mid n_i = 1\} \text{ is even}\}$$
$$B_k = \{(n_1, \ldots, n_k) \mid \#\{i \mid n_i > 0\} \text{ is even}\} \quad E_k = \{(n_1, \ldots, n_k) \mid \#\{i \mid n_i = 1\} > 0\}$$
$$C_k = \{(n_1, \ldots, n_k) \mid \#\{i \mid n_i = 1\} \text{ is odd}\} \quad F_k = \{(n_1, \ldots, n_k) \mid \#\{i \mid n_i = 1\} = 0\}.$$

Then $(A_k)\mathrm{P} = \mathrm{NP}(k)$, $(B_k)\mathrm{P} = \text{co-NP}(k)$, $(C_k)\mathrm{P} = \text{1-NP}(k)$, $(D_k)\mathrm{P} = \text{co-1-NP}(k)$, $(E_k)\mathrm{P} = \bigvee_k \text{1-NP}$, $(F_k)\mathrm{P} = \text{co-}\bigvee_k \text{1-NP}$. (Readers, who do not know some of the classes named here, may take the above vector sets as their definitions.) Clearly, in all cases the longest type-3 alternating chains have length k. Thus we obtain

Theorem 4. *The classes* $\mathrm{NP}(k)$, $\mathrm{coNP}(k)$, $\text{1-NP}(k)$, $\text{co-1-NP}(k)$, $\bigvee_k \text{1-NP}$, *and* $\text{co-}\bigvee_k \text{1-NP}$ *have type-3 dimension k.*

4.3 Minimal Elements

In this subsection we want to exhibit minimal classes of a certain type-3 dimension. Note that there are no non-trivial classes of type-3 dimension 0. Thus we first investigate the case of type-3 dimension 1.

Theorem 5. *The classes* NP *and* coNP *have type-3 dimension* 1. *They are incomparable, and for every class $\mathcal{C}$ of type-3 dimension greater or equal to* 1, *we have either* $\mathrm{NP} \subseteq \mathcal{C}$ *or* $\mathrm{coNP} \subseteq \mathcal{C}$.

See the full version [7] for a proof of Theorem 5.

Theorem 6. *The classes* $\mathrm{NP}(2)$, $\mathrm{coNP}(2)$, $\mathrm{NP} \wedge \text{co-1-NP}$, *and* $\mathrm{coNP} \vee \text{1-NP}$ *have type-3 dimension* 2. *They are pairwise incomparable, and for every class $\mathcal{C}$ of type-3 dimension greater or equal to* 2, *we have that at least one of the given four classes is contained in $\mathcal{C}$.*

Proof. We describe these four classes by subsets V_i of $\mathbb{N}^2$ with bound 2, respectively, s.t. $(V_1)\mathrm{P} = \mathrm{NP}(2)$, $(V_2)\mathrm{P} = \mathrm{coNP}(2)$, $(V_3)\mathrm{P} = \mathrm{NP} \wedge \text{co-1-NP}$, and $(V_4)\mathrm{P} = \mathrm{coNP} \vee \text{1-NP}$.

$$V_1: \begin{array}{c|ccc} 2 & + & - & - \\ 1 & - & - & - \\ 0 & - & - & - \\ \hline & 0 & 1 & 2 \end{array} \qquad V_2: \begin{array}{c|ccc} 2 & - & + & + \\ 1 & + & + & + \\ 0 & + & + & + \\ \hline & 0 & 1 & 2 \end{array} \qquad V_3: \begin{array}{c|ccc} 2 & + & - & + \\ 1 & - & - & - \\ 0 & - & - & - \\ \hline & 0 & 1 & 2 \end{array} \qquad V_4: \begin{array}{c|ccc} 2 & - & + & - \\ 1 & + & + & + \\ 0 & + & + & + \\ \hline & 0 & 1 & 2 \end{array}$$

It is an easy exercise to show that these sets describe the four classes as desired, and that by the third chain theorem, they are incomparable.

Now let $\mathcal{C}$ be any bounded counting class of type-3 dimension 2. Let (v_1, v_1'), (v_2, v_2') be a type-3 chain w.r.t. $U \subseteq \mathbb{N}^k$, where $\mathcal{C} = (U)\mathrm{P}$. We may w.l.o.g. assume that $v_2' = (m, \ldots, m)$, where m is the bound of U.

Let this chain be of signature t. We claim that

1) $t = +- \implies (V_1)\mathrm{P} \subseteq \mathcal{C}$ 3) $t = ++ \implies (V_3)\mathrm{P} \subseteq \mathcal{C}$

2) $t = -+ \implies (V_2)\mathrm{P} \subseteq \mathcal{C}$ 4) $t = -- \implies (V_4)\mathrm{P} \subseteq \mathcal{C}$

Only cases 1) and 3) have to be proven, because the other two cases follow by complementation. We give only the proof for case 1) now. Case 3) is similar, though a little more complicated.

In case 1) we have $v_1 \notin U$, $v_1' \in U$, $v_2 \in U$, $v_2' \notin U$. We show that $(V_1)\mathrm{P} \subseteq (U)\mathrm{P}$ by providing a map f of admissible form, which maps $\mathbb{N}^2$ to $\mathbb{N}^k$ in such a way that $(x, y) \in V_1$ if and only if $f(x, y) \in U$. The map is

$$f(x, y) = v_1 + \binom{y}{2} \cdot (v_1' - v_1) + x \cdot v_2'.$$

We have to prove that $(x, y) \in V_1$ if and only if $f(x, y) \in U$. Let $x = 0$ and $y \geq 2$. Then $(x, y) \in V_1$. We obtain $f(x, y) = v_1 + \binom{y}{2} \cdot (v_1' - v_1)$, which is in U, if and only if $v_1 + 1 \cdot (v_1' - v_1)$ is in U. But the latter equals v_1', so it definitely is in U.

Now, let $x = 0$ and $y < 2$. Then we obtain $f(x, y) = v_1$, which is not in U.

Finally, let $x > 0$. Then $f(x, y) \geq v_2' = (m, \ldots, m)$, and thus $f(x, y) \in U \iff v_2' \in U$, but by assumption we have $v_2' \notin U$, and so $f(x, y) \notin U$.

In all cases we obtained $f(x, y) \in U \iff (x, y) \in V_1$, so f in fact shows that $\mathrm{NP}(2) = (V_1)\mathrm{P} \subseteq (U)\mathrm{P} = \mathcal{C}$.

Theorems 5 and 6 can be generalized as follows:

Theorem 7. *There are 2^d classes of type-3 dimension d, which are pairwise incomparable, and for every class $\mathcal{C}$ of type-3 dimension greater or equal to d, we have that at least one of these 2^d classes is contained in $\mathcal{C}$.*

These classes may be constructed in such a way that for every string $t \in \{+, -\}^d$ exactly one of these classes has a type-3 alternating chain of length d and signature t.

5 Inherent Dimension

In Section 4 we introduced a notion of dimension which has the advantage to be quite easily applicable, because for every subset $U \subseteq \mathbb{N}^k$ all longest type-3 alternating chains can be constructed by trying all finitely many possibilities. In the current section we will introduce another notion, which is the obvious notion intuitively associated with the term "dimension", but with the major disadvantage that it is not at all clear, how to compute this dimension for a given bounded counting class.

Definition 6. *Let C be a bounded counting class, and let k be minimal such that there is a set $U \subseteq \mathbb{N}^k$ of bounded significance with $C = (U)\mathrm{P}$. Then k is called the* inherent dimension *of the class C.*

This definiton together with Proposition 2 yields

Lemma 2. *Let C be a bounded counting class with inherent dimension k and type-3 dimension d. Then $k \geq d$.*

Like the type-3 dimension, also the inherent dimension gives us the natural value of k for the classes from the Boolean Hierarchies over NP and over 1-NP, i.e. for NP(k), coNP(k), 1-NP(k), co-1-NP(k), $\bigvee_k$ 1-NP, and co-$\bigvee_k$ 1-NP:

Lemma 3. *All of the classes* NP(k), coNP(k), 1-NP(k), co-1-NP(k), $\bigvee_k$ 1-NP, *and* co-$\bigvee_k$ 1-NP *have inherent dimension k.*

Proof. The description of these classes by the sets A_k, B_k, C_k, D_k, E_k, and F_k in Subsection 4.2 shows that their inherent dimension is at most k, since all these sets were subsets of $\mathbb{N}^k$.

But, on the other hand Theorem 4, together with Lemma 2 shows that their inherent dimension is at least k.

This lemma might give some hope that type-3 dimension and inherent dimension coincide in all cases. This would be very convenient, since the inherent dimension is the more natural notion, while the type-3 dimension is the one that can be checked more easily. In fact, we will show that the two notions coincide for all classes of inherent dimension less than 3, but they do not coincide for greater dimensions.

5.1 Classes of Inherent Dimension 1

Let C be a bounded counting class of inherent dimension 1. We want to show that in this case also the type-3 dimension is 1. From Lemma 2 we already know that the type-3 dimension is at most 1. But it cannot be 0, since only the trivial classes (the class containing only the empty set, and the class containing only Σ^*) have type-3 dimension 0.

In fact the converse is also true, but this is not obvious. See the full version [7] for a proof.

Theorem 8. *Let C be a bounded counting class. C has inherent dimension* 1, *if and only if C has type-3 dimension* 1.

Thus, using Theorem 5, we trivially obtain:

Theorem 9. *The classes* NP *and* coNP *have inherent dimension* 1. *They are incomparable, and for every class* $\mathcal{C}$ *of inherent dimension greater or equal to* 1, *we have either* $\mathrm{NP} \subseteq \mathcal{C}$ *or* $\mathrm{coNP} \subseteq \mathcal{C}$.

5.2 Classes of Inherent Dimension 2

Similar as Theorem 9, we transfer Theorem 6 to the case of inherent dimension:

Theorem 10. *The classes* NP(2), coNP(2), $\mathrm{NP} \wedge \text{co-1-NP}$, *and* $\mathrm{coNP} \vee \text{1-NP}$ *have inherent dimension* 2. *They are pairwise incomparable, and for every class* $\mathcal{C}$ *of inherent dimension greater or equal to* 2, *we have that at least one of the given four classes is contained in* $\mathcal{C}$.

5.3 Classes of Inherent Dimension greater than 2

The conjecture that the minimal classes are the same for inherent dimension and for type-3 dimension in every stage is destroyed by the following counterexample:

We define a class $\mathcal{C}$ by a subset $U \subseteq \mathbb{N}^3$ with bound 3. The definition is given by tables showing the (x, y)-projection of U for the four cases $z = 0, \ldots, z = 3$:

$z = 0:$				
3	−	+	−	+
2	−	+	−	−
1	+	+	+	+
0	−	+	−	−
	0	1	2	3

$z = 1:$				
3	+	+	+	+
2	+	+	+	+
1	+	+	+	+
0	+	+	+	+
	0	1	2	3

$z = 2:$				
3	−	+	−	+
2	−	+	−	−
1	+	+	+	+
0	−	+	−	−
	0	1	2	3

$z = 3:$				
3	+	+	+	+
2	−	+	−	+
1	+	+	+	+
0	−	+	−	+
	0	1	2	3

Theorem 11. *The bounded counting class* $\mathcal{C} = (U)\mathrm{P}$ *with the set* U *as given above has type-3 dimension* 2, *but inherent dimension* 3.

The proof of Theorem 11 can be found in the full paper [7].

6 Conclusion

In this paper, we introduced the notion of dimension for bounded counting classes. As these classes can generally be described by a nondeterministic polynomial time machine with accepting paths of k different kinds (or k different *outputs*), and possibly also rejecting paths, one might naturally ask for the minimum k, such that the given class can be defined using this model. We call this number the inherent dimension of the counting class.

We proved that for small dimensions this dimension can be checked by looking for the longest alternating monotone chain with respect to the defining set U.

In dimensions 1 and 2 we explicitly found the minimal elements in the set of bounded counting classes of that dimension, namely NP and coNP for dimension 1, and NP(2), coNP(2), $\mathrm{NP} \wedge \text{co-1-NP}$, and $\mathrm{coNP} \vee \text{1-NP}$ for dimension 2.

For dimension at least 3, there are 2^d classes which can be defined in analogy to the lower dimension cases by taking a type-3 chain of any given signature (thus we obtain 2^d cases) and leaving everything else trivial. However, our last result showed that these classes for dimension 3 do not form a complete set of minimal elements; there are other classes, which cannot contain any of these 8 classes.

Acknowledgement. I am very grateful to Katja Cronauer for several helpful ideas, and to Heribert Vollmer and Klaus Wagner for interesting comments on this paper's subject.

References

1. Bovet, D. P., Crescenzi, P., Silvestri, R.: A Uniform Approach to Define Complexity Classes. Theoretical Computer Science **104** (1992) 263–283
2. Cronauer, K., Hertrampf, U., Vollmer, H., Wagner, K.W.: The Chain Method to Separate Counting Classes. Theory of Computing Systems **31** (1998) 93–108
3. Fenner, S., Fortnow, L. Kurtz, S.: Gap-definable Counting Classes. Journal of Computer and System Sciences **48** (1994) 116–148
4. Gundermann, T., Nasser, N.A., Wechsung, G.: A Survey on Counting Classes. Proceedings of the 5th Structure in Complexity Theory Conference. IEEE (1990) 140–153
5. Hertrampf, U.: Complexity Classes with Finite Acceptance Types. Proceedings of the 11th Symp. on Theoretical Aspects of Computer Science. LNCS 775 (1994) 543–553
6. Hertrampf, U.: Classes of Bounded Counting Type and Their Inclusion Relations. Proceedings of the 12th Symp. on Theoretical Aspects of Computer Science. LNCS 900 (1995) 60–70.
7. Hertrampf, U.: The Inherent Dimension of Bounded Counting Classes. Technical Report. University of Stuttgart (1998)
8. Hertrampf, U., Lautemann, C., Schwentick, T., Vollmer, H., Wagner, K.W.: On the Power of Polynomial Time Bit-Reductions. Proceedings of the 8th Structure in Complexity Theory Conference. IEEE (1993) 200–207.
9. Hertrampf, U., Vollmer, H., Wagner, K.W.: On Balanced vs. Unbalanced Computation Trees. Mathematical Systems Theory **29** (1996) 411–421
10. Jenner, B., McKenzie, P., Thérien, D.: Logspace and Logtime Leaf Languages. Proceedings of the 9th Structure in Complexity Theory Conference. IEEE (1994) 242–254
11. Papadimitriou, C.H.: Computational Complexity. Addison-Wesley, Reading, Mass. (1994)
12. Valiant, L.G.: The Complexity of Computing the Permanent. Theoretical Computer Science **8** (1979) 189–201
13. Vereshchagin, N.K.: Relativizable and Non-relativizable Theorems in the Polynomial Theory of Algorithms. (In Russian.) Izvestija Rossijskoj Akademii Nauk **57** (1993) 51–90

An Exact Characterization of Symmetric Functions in $qAC^0[2]$

Chi-Jen Lu

Computer Science Department
University of Massachusetts at Amherst
cjlu@cs.umass.edu

Abstract. $qAC^0[2]$ is the class of languages computable by circuits of constant depth and quasi-polynomial ($2^{\log^{O(1)} n}$) size with unbounded fan-in AND, OR, and PARITY gates. Symmetric functions are those functions that are invariant under permutations of the input variables. Thus a symmetric function $f_n : \{0,1\}^n \to \{0,1\}$ can also be seen as a function $f_n : \{0,1,\cdots,n\} \to \{0,1\}$. We give the following characterization of symmetric functions in $qAC^0[2]$, according to how $f_n(x)$ changes as x grows from 0 to n. A symmetric function $f = (f_n)$ is in $qAC^0[2]$ if and only if f_n has period $2^{t(n)} = \log^{O(1)} n$ except within both ends of length $\log^{O(1)} n$.

1 Introduction

Proving lower bounds is one of the most fundamental tasks in complexity theory. However, it appears to be a rather difficult one, and so far people can only show lower bounds for very restricted classes, mainly for variants of constant depth circuits.

Let AC^0 denote the class of languages computable by constant depth polynomial size circuits with AND and OR gates. For any constant $p \in \mathbf{N}$, by allowing AND, OR, and MOD^p gates, we have the class $AC^0[p]$. Allowing a MAJORITY gate on the top but only AND and OR gates for the remaining, we get the class PERCEPTRON. If a letter q is added before any of the class name above, the circuit size is now allowed to be quasi-polynomial, or $2^{\log^{O(1)}} n$.

The first significant lower bound on the size of such circuits came from Furst, Saxe, and Sipser [7] and Ajtai [1], showing that the PARITY function is not in AC^0. This was later improved by Yao [12] and Håstad [9], showing that PARITY is even outside of qAC^0. Razborov [10] considered $qAC^0[2]$ and showed that the MAJORITY function is not in it. Smolensky [11] showed that for any prime p and any constant c not divisible by p, MOD^c is not in $qAC^0[p]$. Barrington and Straubing [3], generalizing the result of Green [8] and Aspnes *et al.* [2], proved that for any constant c, MOD^c is not in qPERCEPTRON.

Note that all these lower bounds are for symmteric functions. A boolean function $f : \{0,1\}^* \to \{0,1\}$, can be seen as a sequence of functions $f_n : \{0,1\}^n \to \{0,1\}$ for $n \in \mathbf{N}$, and vice versa. A function $f_n : \{0,1\}^n \to \{0,1\}$ is called symmetric if its value depends merely on the number of 1's in its input, and we say that $f : \{0,1\}^* \to \{0,1\}$ is symmetric if for all $n \in \mathbf{N}$, f_n is symmetric. A symmetric function f_n can also be seen as a function from $[n] = \{0, 1, \ldots, n\}$ to $\{0,1\}$. We will abuse the notation and also use f_n to denote this function from $[n]$ to $\{0,1\}$. That is, for $k \in [n]$, $f_n(k) = f_n(1^k 0^{n-k})$. It turns out to be very useful to look at how $f_n(x)$ changes as x grows from 0 to n. The sequence $f_n(0) f_n(1) \cdots f_n(n)$ is called the *weight spectrum* of f_n.

Symmetric functions in some circuit classes appear to be special, as there are some neat characterizations for them. Fagin, Klawe, Pippenger, and Stockmeyer [6] pioneered the study of symmetric functions in AC^0 in terms of their weight spectra. Brustmann and Wegener [4] followed this approach, equiped with Håstad's lower bound, and found an exact characterization for symmetric functions in AC^0. A careful analysis of their proof shows that all the symmetric functions in qAC^0 are also in AC^0. So we have the following:

- Suppose that f is a symmetric function. Then f is in AC^0 iff f is in qAC^0 iff f_n is constant except within both ends of length $\log^{O(1)} n$.

For qPERCEPTRON, Zhang, Barrington, and Tarui [13] gave the following characterization:

- A symmetric function f is in qPERCEPTRON iff f_n has $\log^{O(1)} n$ many value changes.

Damm and Lenz [5] attempted a characterization of $AC^0[2]$ but did not quite succeed. Their characterization was based on some unproven assumption. We proceed along their line and succeed in characterizing symmetric functions in $qAC^0[p]$ for any fixed prime p:

- A symmetric function f is in $qAC^0[p]$ iff f_n has period $p^{t(n)} = \log^{O(1)} n$ except within both ends of length $\log^{O(1)} n$.

It's not clear if for symmetric functions qPERCEPTRON would collapse to PERCEPTRON, or $qAC^0[2]$ would collapse to $AC^0[2]$. However, our result implies the following:

- The set of symmetric functions in $qAC^0[2]$ is equal to the set of symmetric functions in $AC^0[2]$ iff $\sigma_k \in AC^0[2]$ for all $k = \log^{O(1)} n$.

2 Preliminaries

Let sB_n denote the class of symmetric functions from $\{0,1\}^n$ to $\{0,1\}$, and let sB denote the class of symmetric functions from $\{0,1\}^*$ to $\{0,1\}$. sB_n can be seen as a vector space of dimension $n+1$ over $\mathbf{Z_2}$, and there are two natural bases for it. The first one is $\{\epsilon_k \mid 0 \leq k \leq n\}$, where $\epsilon_k \in \mathrm{sB}_n$ is defined as

$$\epsilon_k(x) = \begin{cases} 1 & \text{if } x = k, \\ 0 & \text{otherwise.} \end{cases}$$

The second one is $\{\sigma_k \mid 0 \le k \le n\}$, where $\sigma_k \in \mathrm{sB}_n$ is defined as

$$\sigma_k(x) = \binom{x}{k} \bmod 2.$$

One can check that σ_k is the kth symmetric polynomial over $\mathbf{Z_2}$, that is,

$$\sigma_k(x_1, \ldots, x_n) = \bigoplus_{I \in [n], |I| = k} \prod_{i \in I} x_i.$$

So for $k = \log^{O(1)} n$, both ϵ_k and σ_k are in $qAC^0[2]$.

Let $f_n \in \mathrm{sB}_n$. Clearly $f_n = \sum_{k \in [n]} f_n(k)\epsilon_k$. Let $v(f_n)$ denote the weight spectrum $f_n(0)f_n(1)\cdots f_n(n) \in \mathbf{Z_2}^{n+1}$ of f_n. Define $C(f_n)$ to be the smallest integer k such that $v(f_n)$ is constant except within both ends of length k, that is, the smallest k such that $f_n(x_1 \ldots x_{n-2k}1^k0^k)$ is a constant.

For $f_n = \sum_{k \in [n]} \hat{f}_n(k)\sigma_k$, define the degree of f_n, denoted as $D(f_n)$, to be the largest k with $\hat{f}_n(k) \neq 0$. The period of f_n, denoted as $P(f_n)$, is defined as the smallest k such that $f_n(x) = f_n(x+k)$ for $0 \le x \le n-k$. The following proposition shows the period of σ_k. It can be proved using Lucas' theorem: $\binom{x}{k} \equiv \binom{\lfloor x/2 \rfloor}{\lfloor k/2 \rfloor}\binom{x \bmod 2}{k \bmod 2} \pmod 2$.

Proposition 1 *The period of σ_k is 2^t where t is the smallest integer such that $k \le 2^t - 1$. That is, $P(\sigma_k) = 2^{\lceil \log_2(k+1) \rceil}$.*

Corollary 1 *Every function in sB_n of degree k has period $2^{\lceil \log_2(k+1) \rceil}$.*

The following can easily be proved using dimension arguments.

Proposition 2 *Every function in sB_n with period 2^t has degree less than 2^t.*

We will use a measure for functions in sB_n, which is slightly different from that used by Damm and Lenz [5]. For $f_n \in \mathrm{sB}_n$, define $b(f_n)$ as the smallest integer k such that $f_n \in$ span $\{\epsilon_i, \epsilon_{n-i}, \sigma_i \mid 0 \le i \le k\}$, over $\mathbf{Z_2}$.

For $f_n \in \mathrm{sB}_n$, it's useful to divide $v(f_n)$ into three parts with a periodic middle part. Let $v(f_n) = \alpha\beta\gamma$, where $|\alpha| \le |\beta| \le |\gamma| \le |\alpha| + 1$. Let $g_n \in \mathrm{sB}_n$ be the function with the smallest period such that $v(g_n) = \alpha'\beta\gamma'$ for some α' and γ' with $|\alpha'| = |\alpha|$ and $|\gamma'| = |\gamma|$. Define $h_n = f_n \oplus g_n$. The decomposition $f_n = g_n \oplus h_n$ is called a *standard decomposition.* So for a standard decomposition $f_n = g_n \oplus h_n$, $P(g_n), C(h_n) \le \lceil (n+1)/3 \rceil$.

Let MAJ_n denote the MAJORITY function on n boolean variables, which outputs 1 iff at least $n/2$ input variables are 1. We will need the following lower bound of Smolensky [11].

Lemma 1. *For any fixed prime p, any depth d circuit with AND, OR, and MOD^p gates for MAJ_n must have size $2^{\Omega(n^{1/2d})}$.*

Let MOD_n^q denote the function on n boolean variables that outputs 1 iff the number of 1's in the input is a multiple of q. For any fixed prime $q \neq p$, the above size lower bound also holds for the function MOD_n^q, as proved in [11]. However, we need a slightly stronger lemma instead, which can be proved by slightly modifying Smolensky's argument. Notice the point in the statement where q is quantified.

Lemma 2. *Let p be a fixed prime. There exist contants n_0 and c, such that for any $n > n_0$ and for any prime $q \neq p$ with $q \leq n/2$, any depth d circuit with AND, OR, and MOD^p gates computing MOD_n^q must have size at least $2^{cn^{1/2d}}$.*

3 Main Results

Lemma 3. *Suppose that $f \in sB$ and for each $n \in \mathbf{N}$, $f_n = g_n \oplus h_n$ is a standard decompostion. If f is in $qAC^0[2]$, then $P(g_n) = \log^{O(1)} n$.*

Proof: Let $b = P(g_n)$, which is a function of n. We will construct the function MAJ_b from f_n, and a circuit for MAJ_b from the circuit for f_n without blowing up the size too much. So if f is in $qAC^0[2]$, then b must be small because of the lower bound for MAJ_b.

Consider the interval of length b centered at $\lfloor n/2 \rfloor$. Clearly f_n and g_n agree at inputs from this interval. Let $A = \{i_1, i_2, \ldots, i_l\}$, for some integer l, be the set of indices i in that interval where $g_n(i) = 1$. The only index x in that interval such that $g_n(x + i_1 - i_j) = 1$ for each $i_j \in A$ is $x = i_1$, for otherwise g_n has a smaller period than $b = P(g_n)$.

Let $k = \lceil 3b/2 \rceil$. Consider those l functions on k variables derived from f_n by fixing $i_j - \lceil b/2 \rceil$ variables to 1 and $n - b - i_j$ variables to 0, for $1 \leq j \leq l$. The AND of these l functions has weight spectrum $\alpha 1 0^{b-1}$, for some α of length $\lceil b/2 \rceil$. By negating all its variables we get a function with weight spectrum $0^{b-1} 1 \beta$, where β is the reverse of α. More precisely, define

$$f'_k(x_1 \cdots x_k) = \bigwedge_{1 \leq i \leq l} f_n(\overline{x}_1 \cdots \overline{x}_k 1^{i_j - \lceil b/2 \rceil} 0^{n-b-i_j}).$$

Then $v(f'_k) = 0^{b-1} 1 \beta$, for some β of length $\lceil b/2 \rceil$. Next define

$$f''_b(x_1 \cdots x_b) = \bigvee_{1 \leq i \leq \lceil b/2 \rceil} f'_k(x_1 \cdots x_b 1^{i-1} 0^{k-b-i+1}).$$

Then $f''_b = \text{MAJ}_b$. If f_n has a $qAC^0[2]$ circuit of depth d and size $2^{\log^c n}$, MAJ_b has a depth $d+2$ size $2^{\log^{c+2} n}$ circuit. From Lemma 1, $P(g_n) = b = \log^{O(1)} n$. □

Lemma 4. *Suppose that $f \in sB$ and for each $n \in \mathbf{N}$, $f_n = g_n \oplus h_n$ is a standard decompostion. If f is in $qAC^0[2]$, than $P(g_n)$ is a power of 2 for all but finitely many n.*

Proof: Suppose $f \in qAC^0[2]$. We know that $P(g_n) = \log^{O(1)} n$ from the previous lemma. Now suppose that $P(g_n)$ is not a power of 2 for an infinite number of n. We will show that this leads to a contradiction.

Consider the function derived from f_n by discarding both ends of length $C(h_n)$. That is, for $l = n - 2C(h_n)$, define

$$f'_l(x_1 \cdots x_l) = f_n(x_1 \cdots x_l 1^{C(h_n)} 0^{C(h_n)}).$$

Then f'_l is a periodic function with $P(f'_l) = P(g_n)$. As $C(h_n) \leq \lceil (n+1)/3 \rceil$, $l = n - 2C(h_n) = \Omega(n)$. So $P(f'_n) = \log^{O(1)} n$, and it is not a power of 2 for an infinite number of n. Observe that there exist constants n_0, d, c_1, c_2, c_3, such that when $n > n_0$, all the following hold:

- f'_n can be computed by a $qAC^0[2]$ circuit of depth d and size $2^{\log^{c_1} n}$, as f is assumed to be in $qAC^0[2]$.
- $P(f'_n) \leq \log^{c_2} n$.
- $1 + \log^{c_2} n 2^{\log^{c_1} n} \leq 2^{\log^{c_3} \lfloor \frac{n - 2P(f'_n)}{P(f'_n)} \rfloor}$.
- For any prime $q \neq p$ with $q \leq n/2$, the function MOD_n^q is not computable by any $qAC^0[2]$ circuit of depth $d+1$ and size $2^{\log^{c_3} n}$, from Lemma 2.

As $P(f'_n)$ is not a power of 2 infinitely often, there exist an $m > n_0$ and a prime $q \neq 2$ such that $q \mid P(f'_m)$ and $\lfloor \frac{m - 2P(f'_m)}{P(f'_m)/q} \rfloor > n_0$. Let $b = P(f'_m)$, $r = m - 2b$, and $k = \lfloor \frac{r}{b/q} \rfloor$. As f'_m is a function of period b, we can construct the function MOD_r^b and then the function MOD_k^q as the following:

$$\mathrm{MOD}_r^b(x_1 \cdots x_r) = \bigwedge_{i \in [b-1], f'_m(i)=1} f'_m(x_1 \cdots x_r 1^i 0^{2b-i}, \text{and}$$

$$\mathrm{MOD}_k^q(x_1 \cdots x_k) = \mathrm{MOD}_b^r(x_1^{b/q} \cdots x_k^{b/q} 0^{r-kb/q}).$$

Then MOD_k^q can be computed by a $qAC^0[2]$ circuit of depth $d+1$ and size $1 + b2^{\log^{c_1} m} \leq 1 + \log^{c_2} m 2^{\log^{c_1} m} \leq 2^{\log^{c_3} \lfloor \frac{m-2b}{b/q} \rfloor} \leq 2^{\log^{c_3} k}$, a contradiction. □

Lemma 5. *Suppose $f \in sB$. If f is in $qAC^0[2]$, than $b(f_n) = \log^{O(1)} n$.*

Proof: Suppose that f is in $qAC^0[2]$. For each $n \in \mathbf{N}$, let $f_n = g_n \oplus h_n$ be a standard decompostion. From Lemmas 3 and 4 we know that $P(g_n) = \log^{O(1)} n$ and is a power of 2 for almost every n. So from Proposition 2, $D(g_n) = \log^{O(1)} n$. Then $g = (g_n)$ is in $qAC^0[2]$, and so is $h = (h_n) = (f_n \oplus g_n)$. Let $b = C(h_n)$, which is a function of n. We will construct the function MAJ_b from h_n, and use the lower bound for MAJ_b to upper bound b.

Let $v(h_n) = v_0 \cdots v_b 0^{n-2b-1} v_{n-b} \cdots v_n$. Assume without loss of generality that $v_{n-b} = 1$ (otherwise consider the function $h_n(\overline{x}_1 \cdots \overline{x}_n)$). For $1 \leq i \leq \lceil b/2 \rceil$, by fixing $n - 2b + i$ variables to 1 and $b - i$ variables to 0, we get a function with weight spectrum $0^{b-i} 1 v_{n-b} \cdots v_{n-b+i-2}$. The OR of these functions has weight spectrum $0^{\lfloor b/2 \rfloor} 1^{\lceil b/2 \rceil}$, that of MAJ_b. More precisely, define

$$h'_b(x_1 \cdots x_b) = \bigvee_{1 \leq i \leq \lceil b/2 \rceil} h_n(x_1 \cdots x_b 1^{n-2b+i} 0^{b-i}).$$

Then $h'_b = \text{MAJ}_b$. If h_n has a depth d size $2^{\log^c n}$ circuit, MAJ_b has a depth $d+1$ size $2^{\log^{c+1} n}$ circuit. From Lemma 1, $b = \log^{O(1)} n$. So $b(f_n) = \max\{D(g_n), C(h_n) - 1\} = \log^{O(1)} n$. □

Fagin, Klawe, Pippenger, and Stockmeyer [6] showed that for $i = \log^{O(1)} n$, ϵ_i and ϵ_{n-i} are in AC^0. Also for $i = \log^{O(1)} n$, σ_i can be computed by a PARITY of $2^{\log^{O(1)} n}$ ANDs. So for $f \in \text{sB}$, $b(f_n) = \log^{O(1)} n$ implies that f_n can be computed by the PARITY of some subset of $\{\epsilon_i, \epsilon_{n-i}, \sigma_i \mid i = \log^{O(1)} n\}$. So we have our main theorem and a normal form theorem for $\text{sB} \cap qAC^0[2]$.

Theorem 1 *For $f \in sB$, $f \in qAC^0[2]$ iff $b(f_n) = \log^{O(1)} n$ iff f_n has period $2^{t(n)} = \log^{O(1)} n$ except at both ends of length $\log^{O(1)} n$.*

Theorem 2 *Any function in $sB \cap qAC^0[2]$ can be computed by circuits that are PARITY of quasi-polynomial number of AC^0 circuits.*

There is nothing special about MOD^2. In fact for any prime p, we have the following similar theorem. The proof is almost identical.

Theorem 3 *For $f \in sB$, $f \in qAC^0[p]$ iff f_n has period $p^{t(n)} = \log^{O(1)} n$ except at both ends of length $\log^{O(1)} n$.*

4 Acknowledgements

We would like to thank Dave Barrington for some helpful comments.

References

1. M. Ajtai, $\sum_1^1$-formula on finite structures, *Annals of Pure and Applied Logic*, 24, pages 1-48, 1983.
2. J. Aspnes, R. Beigel, M. Furst, and S. Rudich, The expressive power of voting polynomials, in *Proceedings of the 23rd Annual ACM Symposium on Theory of Computing*, pages 402-409, 1991.
3. D. A. Mix Barrington and H. Straubing, Complex polynomials and circuit lower bounds for modular counting, In *Proceedings of the 1st Latin Amercan Symposium on Theoretical Informatics*, pages 24-31, 1992.
4. B. Brustmann and I. Wegener, The complexity of symmetric functions in bounded-depth circuits, *Information Processing Letters*, 25, pages 217-219, 1987.
5. C. Damm and K. Lenz, Symmetric functions in $AC^0[2]$, *manuscript.*
6. R. Fagin, M. M. Klawe, N. J. Pippenger, and L. Stockmeyer, Bounded depth, polynomial size circuits for symmetric functions, *Theoretical Computer Science*, 36, pages 239-250, 1985.
7. M. Furst, J. Saxe, and M. Sipser, Parity, circuits, and the polynomial time hierarchy, *Mathematical System Theory*, 17, pages 13-27, 1984.
8. F. Green, An oracle separating $\oplus P$ from PP^{PH}, In *Proceedings of Fifth Annual Conference on Structure in Complexity Theory*, pages 295-298, 1990
9. J. Håstad, Computational limitations of small-depth circuits, *MIT Press*, 1986.

10. A. A. Razborov, Lower bounds for the size of bounded depth with basis $\{\wedge, \oplus\}$, *Mathematical Notes of the Academy of Sciences of the USSR*, 41, pages 598-607, 1987.
11. R. Smolensky, Algebraic methods in the theory of lower bounds for boolean circuit complexity, In *Proceedings of the 19th Annual ACM Symposium on Theory of Computing*, pages 77-82, 1987.
12. A. C.-C. Yao, Separating the polynomial-time hierachy by oracles, In *Proceedings of the 26th Annual IEEE Symposium on Foundations of Computer Science*, pages 1-10, 1985.
13. Z.-L. Zhang, D. A. Mix Barrington, and J. Tarui, Computing symmetric functions with AND/OR circuits and a single MAJORITY gate, In *Proceedings of the 10th Annual Symposium on Theoretical Aspects of Computing*, pages 535-544, 1993.

Robust Reductions*

Jin-Yi Cai[1], Lane A. Hemaspaandra[2], and Gerd Wechsung[3]

[1] Department of Computer Science, State University of New York at Buffalo, Buffalo, NY 14260, USA cai@cs.buffalo.edu

[2] Department of Computer Science, University of Rochester, Rochester, NY 14627, USA lane@cs.rochester.edu

[3] Institut für Informatik, Friedrich-Schiller-Universität Jena, 07740 Jena, Germany wechsung@informatik.uni-jena.de

Abstract. We continue the study of robust reductions initiated by Gavaldà and Balcázar. In particular, a 1991 paper of Gavaldà and Balcázar [6] claimed an optimal separation between the power of robust and nondeterministic strong reductions. Unfortunately, their proof is invalid. We re-establish their theorem.

Generalizing robust reductions, we note that robustly strong reductions are built from two restrictions, robust underproductivity and robust overproductivity, both of which have been separately studied before in other contexts. By systematically analyzing the power of these reductions, we explore the extent to which each restriction weakens the power of reductions. We show that one of these reductions yields a new, strong form of the Karp-Lipton Theorem.

1 Introduction

Reductions are the key tools used in complexity theory to compare the difficulty of problems. Beyond that, reductions play a central role in countless theorems of complexity theory, and to understand the power of such theorems we must understand the relationships between reductions. For example, Karp and Lipton [11] proved that if SAT Turing-reduces to some sparse set then the polynomial hierarchy collapses. A more careful analysis reveals that the same result applies under the weaker hypothesis that SAT robustly-strong-reduces to some sparse set. In fact, the latter result is simply a relativized version of the former result [8], though the first proofs of the latter result were direct and quite complex [1,10]. As another example, in the present paper—but not by simply asserting relativization—we will note that various theorems, among them the Karp-Lipton Theorem, indeed hold for certain reductions that are even more flexible than robustly strong reductions.

In this paper, we continue the investigation of robust reductions started by Gavaldà and Balcázar [6]. We now briefly mention one way of defining strong reduction [16,14] and robustly strong reduction [6]. Definition 1 provides a formal definition of the same notions in terms of concepts that are central to this paper. We say that a nondeterministic Turing machine is a nondeterministic polynomial-time Turing machine (NPTM) if

* A complete version of this paper, including full proofs, is available as [4]. Research supported in part by grants DAAD-315-PRO-fo-ab/NSF-INT-9513368, NSF-CCR-9057486, NSF-CCR-9319093, and NSF-CCR-9322513, and an Alfred P. Sloan Fellowship.

there is a polynomial p such that, for each oracle A and for each integer n, the nondeterministic runtime of N^A on inputs of size n is bounded by $p(n)$. (Requiring that the polynomial upper-bounds the runtime *independent of the oracle* is superfluous in the definition of $\leq_{\mathrm{T}}^{\mathrm{SN}}$, but may be a nontrivial restriction in the definition of $\leq_{\mathrm{T}}^{\mathrm{RS}}$; see the discussion of this point in Section 6. The definitions used here agree with those in the previous literature.) Consider NPTMs with three possible outcomes on each path: **acc**, **rej**, and **?**. We say *A strong-reduces to B*, $A \leq_{\mathrm{T}}^{\mathrm{SN}} B$, if there is an NPTM N such that, for every input x, it holds that (a) if $x \in A$ then $N^B(x)$ has at least one **acc** path and no **rej** paths, and (b) if $x \notin A$ then $N^B(x)$ has at least one **rej** path and no **acc** paths. (Note that in either case the machine may also have some **?** paths.) Furthermore, we say *A robustly strong-reduces to B*, $A \leq_{\mathrm{T}}^{\mathrm{RS}} B$, if there is an NPTM N such that $A \leq_{\mathrm{T}}^{\mathrm{SN}} B$ via N (in the sense of the above definition) and, moreover, for every oracle O and every input x, $N^O(x)$ is *strong*, i.e., it either has at least one **acc** path and no **rej** paths, or has at least one **rej** path and no **acc** paths. This paper is concerned with the relative power of these two reductions, and with reductions whose power is intermediate between theirs.

In particular, it is claimed in [6] that the following strong separation holds with respect to the two reductions: *For every recursive set* $A \notin \mathrm{NP} \cap \mathrm{coNP}$, *there is a recursive set B such that A strong-reduces to B but A does not robustly strong-reduce to B [6, Theorem 11].* Unfortunately, there is a subtle but apparently fatal error in their proof. One of the main contributions of this paper is that we re-establish their sweeping theorem. Note that the zero degrees of these reducibilities are identical, namely the class $\mathrm{NP} \cap \mathrm{coNP}$ [6]. Thus, in a certain sense, the above claim of Gavaldà and Balcázar is optimal (if it is true, as we prove it is), as if $A \in \mathrm{NP} \cap \mathrm{coNP}$ then A strong-reduces to every B and A also robustly strong-reduces to every B.

Section 3 re-establishes the above claim of Gavaldà and Balcázar. The proof (available in the full version of this paper [4]) is delicate, and is carried out in three stages: First, we establish the result for all $A \in \mathrm{EXP} - (\mathrm{NP} \cap \mathrm{coNP})$, where $\mathrm{EXP} = \cup_{k>0} \mathrm{DTIME}[2^{n^k}]$. Here, the set B produced from the proof is not necessarily recursive. Second, we remove the restriction of $A \in \mathrm{EXP}$, by showing that if the result fails for $A \notin \mathrm{EXP}$ then indeed $A \in \mathrm{EXP}$, yielding a contradiction. The proof so far only establishes the existence of some B, which is not necessarily recursive. Finally, with the certainty that *some* B exists, we can recast the proof and show that for every recursive A a recursive B can be constructed.

The notion of "robustly strong" is made up of two components—one stating that for all sets and all inputs the reducing machine has at least one non-**?** path, and the other stating that for all sets and all inputs the reducing machine does not simultaneously have **acc** and **rej** paths. Each component has been separately studied before in the literature, in different contexts. By considering each of these two requirements in conjunction with strong reductions, we obtain two natural new reductions whose power falls between that of strong reductions and that of robustly strong reductions. Section 4 studies the relative power of Turing reductions, of strong reductions, of robustly strong reductions, and of our two new reductions. In some cases we establish absolute separations. In other cases, we see that the relative computation power is tied to the $\mathrm{P} = \mathrm{NP}$ question. Curiously, the two new reductions are deeply asymmetric in terms of what is currently provable about their properties. For one of the new reductions, we show that if it differs from

Turing reductions then $\mathrm{P} \neq \mathrm{NP}$. For the other, we show that the reduction does differ from Turing reductions. In Section 5, we discuss some issues regarding what collapses of the polynomial time hierarchy occur if sparse sets exist that are hard or complete for NP with respect to the new reductions. One of the new reductions extends the reach of hardness results.

2 Two New Reducibilities

For each NPTM N and each set $D \subseteq \Sigma^*$, define $out_{N^D}(x) = \{y \mid y \in \{\mathbf{acc}, \mathbf{rej}, \mathbf{?}\} \wedge$ some computation path of $N^D(x)$ has outcome $y\}$. As is standard, for each nondeterministic machine N and each set $D \subseteq \Sigma^*$, let $L(N^D)$ denote the set of all x for which $\mathbf{acc} \in out_{N^D}(x)$. For each nondeterministic machine N and each set $D \subseteq \Sigma^*$, let $L_{rej}(N^D)$ denote the set of all x for which $\mathbf{rej} \in out_{N^D}(x)$. A computation $N^D(x)$ is called *underproductive* if $\{\mathbf{acc}, \mathbf{rej}\} \not\subseteq out_{N^D}(x)$. That is, $N^D(x)$ does not have as outcomes both **acc** and **rej**. N^D is said to be *underproductive* if, for each string x, $N^D(x)$ is underproductive. That is, $L(N^D) \cap L_{rej}(N^D) = \emptyset$. Underproductive machines were introduced by Buntrock his 1989 Ph.D. thesis. Allender et al. [2] have shown underproductivity to be very useful in the study of almost-everywhere complexity hierarchies for nondeterministic time classes. A computation $N^D(x)$ is called *overproductive* if $out_{N^D}(x) \neq \{\mathbf{?}\}$. A machine N^D is said to be *overproductive* if, for each string x, $N^D(x)$ is overproductive. Equivalently, $L(N^D) \cup L_{rej}(N^D) = \Sigma^*$. We say that N is *robustly overproductive* if for each $D \subseteq \Sigma^*$ it holds that N^D is overproductive. We say that N is *robustly underproductive* if for each $D \subseteq \Sigma^*$ it holds that N^D is underproductive. $\leq_{\mathrm{T}}^{\mathrm{p}}$ as always has its routine definition. Using underproductivity, overproductivity, and robustness, we may now define strong and robustly strong reductions, which have been previously studied. We also introduce two intermediate reductions, obtained by limiting the robustness to just the overproductivity or the underproductivity.[1] The trivial containment relationships are shown in Proposition 1. In this paper we will ask whether some of the containments of Proposition 1 might in fact be equalities, and in particular we seek necessary conditions and sufficient conditions for such.

Definition 1. *1. [14], see also [16] ("strong reductions") $A \leq_{\mathrm{T}}^{\mathrm{SN}} B$ if there is an NPTM N such that N^B is overproductive, N^B is underproductive, and $A = L(N^B)$. 2. [6] ("robustly strong reductions") $A \leq_{\mathrm{T}}^{\mathrm{RS}} B$ if $A \leq_{\mathrm{T}}^{\mathrm{SN}} B$ via an NPTM N, and N is both robustly overproductive and robustly underproductive. 3. ("strong and robustly underproductive reductions" or, for short, "U-reductions") $A \leq_{\mathrm{T}}^{\mathrm{U}} B$ if $A \leq_{\mathrm{T}}^{\mathrm{SN}} B$ via an NPTM N that is robustly underproductive. 4. ("strong and robustly overproductive reductions" or, for short, "O-reductions") $A \leq_{\mathrm{T}}^{\mathrm{O}} B$ if $A \leq_{\mathrm{T}}^{\mathrm{SN}} B$ via an NPTM N that is robustly overproductive.*

---

[1] The literature contains various notations for strong reductions (also known as strong nondeterministic reductions). We adopt the notation of Long's paper [14], i.e., $\leq_{\mathrm{T}}^{\mathrm{SN}}$. However, we note that some papers use other notations, such as $\leq^{\mathrm{SN}}$, $\leq_{\mathrm{T}}^{\mathrm{sn}}$, and $\leq_{\mathrm{T}}^{\mathrm{p,NP\cap coNP}}$. For the three other reductions we discuss, we replace the SN with a mnemonic abbreviation. For robustly strong we follow Gavaldà and Balcázar [6] and use RS. For brevity, we use O as our abbreviation for our "strong and robustly overproductive" reductions, and we use U as our abbreviation for our "strong and robustly underproductive" reductions.

Notation 1 *For each well-defined reduction $\leq_a^b$, let $\leq_a^b$ denote $\{(A,B) \mid A \leq_a^b B\}$.*

Proposition 1. $\leq_T^P \subseteq \leq_T^{RS} \begin{matrix} \subseteq \leq_T^U \subseteq \\ \subseteq \leq_T^O \subseteq \end{matrix} \leq_T^{SN}$.

Using different terminology, robust underproductivity (though not $\leq_T^U$) has been introduced into the literature by Beigel ([3], see also [7]), and the following theorem will be of use in the present paper.

Theorem 2. *([3], see also [7]) If NPTM N is robustly underproductive, then $(\forall A)(\exists L \in P^{SAT \oplus A})[L_{rej}(N^A) \subseteq L \subseteq L(N^A)]$.*

Theorem 2 says that if a machine is robustly underproductive, then for every oracle there is a relatively simple set that separates its acceptance set from its L_{rej} set. In particular, if P $=$ NP and N is a robustly underproductive machine, then for every oracle A it holds that $L(N^A)$ and $L_{rej}(N^A)$ are P^A-separable.

As is standard, we say that a set S is *sparse* if there is a polynomial r such that, for each n, $||S^{\leq n}|| \leq r(n)$. Using different terminology, "robust with respect to sparse sets"-overproductivity (though not $\leq_T^O$) has been introduced into the literature by Hartmanis and Hemachandra [7], and the following theorem will be of use in the present paper.

Theorem 3. *[7] If NPTM N is such that for each sparse set S it holds that N^S is overproductive, then for every sparse set S there exists a binary predicate b computable in $FP^{SAT \oplus S}$ such that, for all x, $\{x \mid b(x)\} \subseteq L(N^S)$ and $\{x \mid \neg b(x)\} \subseteq L_{rej}(N^S)$, where FP denotes the polynomial-time computable functions.*

Theorem 3 says that if a machine is "robustly with respect to sparse oracles"-overproductive, then for every sparse oracle there is a relatively simple function that for each input correctly declares either that the machine has accepting paths or that the machine has rejecting paths. Crescenzi and Silvestri [5] show via Sperner's Lemma that Theorem 3 fails when the sparseness condition is removed, and their proof approach will be of use in this paper.

It is known that SN reductions and RS reductions have nonuniform characterizations. In particular, for every reducibility $\leq_a^b$ and every class $\mathcal{C}$, let $R_a^b(\mathcal{C}) = \{A \mid (\exists B \in \mathcal{C})[A \leq_a^b B]\}$. Gavaldà and Balcázar proved the following result.

Theorem 4. *[6] 1.* $R_T^{SN}(\text{SPARSE}) = \text{NP/poly} \cap \text{coNP/poly}$. *2.* $R_T^{RS}(\text{SPARSE}) = (\text{NP} \cap \text{coNP})/\text{poly}$.

We note in passing that the downward closures of the sparse sets under our two new reductions have analogous characterizations, albeit somewhat stilted ones. We say $A \in \text{NP/poly} \cap \text{coNP/poly}$ via the pair (M,N) of NPTMs if there is a sparse set S such that $A = L(M^S)$ and $A = \overline{L(N^S)}$. Hartmanis and Hemachandra [7] defined *robustly Σ^*-spanning pairs of machines* (M,N) to be pairs having the property $L(M^X) \cup L(N^X) = \Sigma^*$ for every oracle X, and *robustly disjoint pairs* to be pairs having the property $L(M^X) \cap L(N^X) = \emptyset$ for every oracle X. Using these notions we note the following characterizations. $A \in R_T^O(\text{SPARSE})$ if and only if $A \in \text{NP/poly} \cap \text{coNP/poly}$ via some robustly Σ^*-spanning pair (M,N) of NPTMs. $A \in R_T^U(\text{SPARSE})$ if and only if $A \in \text{NP/poly} \cap \text{coNP/poly}$ via some robustly disjoint pair (M,N) of NPTMs.

3 A Strong Separation of $\leq_{\rm T}^{\rm SN}$ and $\leq_{\rm T}^{\rm RS}$

It follows from each of Section 4's Theorems 8 and 13, both of which have relatively simple proofs, that the reducibilities $\leq_{\rm T}^{\rm SN}$ and $\leq_{\rm T}^{\rm RS}$ are distinct. However, more can be said. The separation of these two reductions turns out to be extremely strong, namely, for every recursive set $A \notin \rm NP \cap coNP$, there exists a recursive set B such that A is strongly reducible to B but A is not robustly strong reducible to B. This is Theorem 6. As noted in Section 1, this claim cannot be generalized to include NP ∩ coNP since NP ∩ coNP is the zero degree of $\leq_{\rm T}^{\rm RS}$, as has been pointed out by Gavaldà and Balcázar [6]. Theorem 6 was first stated in Gavaldà and Balcázar's 1991 paper [6]. The diagonalization proof given there correctly establishes $A \not\leq_{\rm T}^{\rm RS} B$, but it fails to establish $A \leq_{\rm T}^{\rm SN} B$. The main error is the following: In the proof there is a passage where the minimum word x is searched for that witnesses that the machine under consideration does not strongly reduce A to B. If such an x is found, then B is augmented by some suitably chosen word (triple). Now it is true that such an x must always exist. However, it might be huge, and then between this x and the previous one, say x', no coding has been done, i.e., for all z between x' and x, no triple $\langle z, y, 0\rangle$ or $\langle z, y, 1\rangle$ with $|z| = |y|$ has been added to B. Thus, the condition "(i)" of [6, p. 6], which is intended to guarantee $A \leq_{\rm T}^{\rm SN} B$, is violated.

We state as Theorem 5 a key claim. In our full version of this paper [4] we prove that and then build on that to achieve our main result, Theorem 6.

Theorem 5. $(\forall \text{ recursive } A \notin \rm NP \cap coNP)(\exists B)[A \leq_{\rm T}^{\rm SN} B \wedge A \not\leq_{\rm T}^{\rm RS} B]$.

Theorem 6. $(\forall \text{ recursive } A \notin \rm NP \cap coNP)(\exists \text{ recursive } B)[A \leq_{\rm T}^{\rm SN} B \wedge A \not\leq_{\rm T}^{\rm RS} B]$.

More generally, our proof [4] actually shows that a B recursive in A can be found to satisfy the theorem.

One can ask whether the difference of $\leq_{\rm T}^{\rm SN}$ and $\leq_{\rm T}^{\rm RS}$ is so strong that the following statement holds: $(\forall \text{ recursive } B \notin \rm NP \cap coNP)(\exists \text{ recursive } A)[A \leq_{\rm T}^{\rm SN} B \wedge A \not\leq_{\rm T}^{\rm RS} B]$. This can be reformulated in terms of reducibility downward closures: $(\forall \text{ recursive } B \notin \rm NP \cap coNP)[\rm R_{T}^{RS}(B) \subsetneq R_{T}^{SN}(B)]$. However, this claim is false. Intuitively, if B is chosen to be sufficiently complex, the differences between the two reductions may be too fine to still be distinguishable in the presence of B. Indeed, if for instance B is an EXPSPACE-complete set, and thus is certainly not contained in NP ∩ coNP, then for every $A \in \rm R_{T}^{SN}(B) = NP^{B} \cap coNP^{B} = EXPSPACE$ we have $A \leq_{\rm m}^{\rm p} B$ and hence $A \leq_{\rm T}^{\rm RS} B$, i.e., $\rm R_{T}^{SN}(B) = R_{T}^{RS}(B)$.

4 Comparing the Power of the Reductions

Long [14] proved that strong and Turing polynomial-time reductions differ. More precisely, he proved the following result.

Theorem 7. *[14]* $(\forall \text{ recursive } A \notin \rm P)(\exists \text{ recursive } B)[A \leq_{\rm T}^{\rm SN} B \wedge A \not\leq_{\rm T}^{\rm p} B]$.

Consequently, at least one of the edges in Figure 1 must represent a strict inclusion. Indeed, we can show that strong reductions differ from both overproductive and underproductive reductions.

Theorem 8. *1.* $(\exists$ recursive $A)(\exists$ recursive $B)[A \leq_{\mathrm{T}}^{\mathrm{SN}} B \wedge A \not\leq_{\mathrm{T}}^{\mathrm{O}} B]$. *Indeed, we may even achieve this via a recursive sparse set B and a recursive tally set A.*

2. $(\exists$ recursive $A)(\exists$ recursive $B)[A \leq_{\mathrm{T}}^{\mathrm{SN}} B \wedge A \not\leq_{\mathrm{T}}^{\mathrm{U}} B]$. *Indeed, we may even achieve this via a recursive sparse set B and a recursive tally set A.*

Next we consider the relationship between $\leq_{\mathrm{T}}^{\mathrm{O}}$ and $\leq_{\mathrm{T}}^{\mathrm{RS}}$. Let M be an NPTM. By interchanging the accept and the reject states of M we get a new NPTM machine N such that $L_{rej}(M) = L(N)$. If M is robustly strong, we have $L(M^A) = \overline{L(N^A)}$ for every oracle A. The pair (M, N) is what Hartmanis and Hemachandra [7] call a robustly complementary pair of machines. For such a pair, the following is known.

Theorem 9. *[7] If (M,N) is a robustly complementary pair of machines, then* $(\forall A)[L(M^A) \in \mathrm{P}^{\mathrm{SAT} \oplus A}]$.

Gavaldà and Balcázar [6] noted that, in view of the preceding discussion, one gets as an immediate corollary the following.

Corollary 1. *[6]* $(\forall A, B)[A \leq_{\mathrm{T}}^{\mathrm{RS}} B \longrightarrow A \in \mathrm{P}^{\mathrm{SAT} \oplus B}]$.

In fact, the proof of Theorem 9 still works if M is an underproductive machine reducing A to B. Thus, we have the following.

Theorem 10. $(\forall A, B)[A \leq_{\mathrm{T}}^{\mathrm{U}} B \longrightarrow A \in \mathrm{P}^{\mathrm{SAT} \oplus B}]$.

Not only is the proof of Theorem 9 not valid for $\leq_{\mathrm{T}}^{\mathrm{O}}$, but indeed the statement of Theorem 10 with $\leq_{\mathrm{T}}^{\mathrm{U}}$ replaced by $\leq_{\mathrm{T}}^{\mathrm{O}}$ is outright false. This follows as a corollary to a proof of Crescenzi and Silvestri [5, Theorem 3.1] in which they give a very nice application of Sperner's Lemma.

Theorem 11. $(\exists A, E)[A \leq_{\mathrm{T}}^{\mathrm{O}} E \wedge A \notin \mathrm{P}^{\mathrm{SAT} \oplus E}]$.

As mentioned earlier, Theorem 11 follows from the *proof* of [5, Theorem 3.1], but not from the theorem itself.

The preceding two theorems have the consequence of showing a deep asymmetry between $\leq_{\mathrm{T}}^{\mathrm{O}}$ and $\leq_{\mathrm{T}}^{\mathrm{U}}$. This asymmetry—that $\leq_{\mathrm{T}}^{\mathrm{O}} \neq \leq_{\mathrm{T}}^{\mathrm{P}}$, yet to prove the analog for $\leq_{\mathrm{T}}^{\mathrm{U}}$ would resolve the P $\neq$ NP question—contrasts with the seemingly symmetric definitions of these two notions. We now turn to some results that will lead to the establishment of this asymmetry.

Theorem 12. *Overproductive and underproductive reductions differ in such a way that* $\leq_{\mathrm{T}}^{\mathrm{O}} \not\subseteq \leq_{\mathrm{T}}^{\mathrm{U}}$.

An immediate consequence of Theorem 12 is the following.

Theorem 13. $\leq_{\mathrm{T}}^{\mathrm{RS}} \neq \leq_{\mathrm{T}}^{\mathrm{O}}$.

We conjecture that Theorem 13 can be stated in the much stronger form of Theorem 6, where $\leq_{\mathrm{T}}^{\mathrm{SN}}$ is replaced with $\leq_{\mathrm{T}}^{\mathrm{O}}$. From Theorem 13, it follows that $\leq_{\mathrm{T}}^{\mathrm{O}} \neq \leq_{\mathrm{T}}^{\mathrm{P}}$. It is interesting to note that, although we know that $\leq_{\mathrm{T}}^{\mathrm{O}}$ and $\leq_{\mathrm{T}}^{\mathrm{P}}$ differ, it may be extremely hard to prove them to differ with a sparse set on the right-hand side. More precisely, we have the following.

Theorem 14. $(\exists B \in \mathrm{SPARSE})[\mathrm{R}^{\mathrm{O}}_{\mathrm{T}}(B) \neq \mathrm{R}^{\mathrm{p}}_{\mathrm{T}}(B)] \longrightarrow \mathrm{P} \neq \mathrm{NP}$.

The fact $\leq^{\mathrm{O}}_{\mathrm{T}} \neq \leq^{\mathrm{p}}_{\mathrm{T}}$, stated above, sharply contrasts with the following.

Theorem 15. $\leq^{\mathrm{U}}_{\mathrm{T}} \neq \leq^{\mathrm{p}}_{\mathrm{T}} \longrightarrow \mathrm{P} \neq \mathrm{NP}$.

Theorem 15 strengthens in two ways the statement, noted by Gavaldà and Balcázar [6], that if $\leq^{\mathrm{RS}}_{\mathrm{T}}$ differs from $\leq^{\mathrm{p}}_{\mathrm{T}}$ anywhere on the recursive sets then $\mathrm{P} \neq \mathrm{NP}$. In particular, we have these two improvements of that statement of Gavaldà and Balcázar: (a) we improve from $\leq^{\mathrm{RS}}_{\mathrm{T}}$ to $\leq^{\mathrm{U}}_{\mathrm{T}}$, and (b) we remove the "on the recursive sets" scope restriction.

Below, we use $X \not\subset Y$ to denote that it is not the case that $X \subsetneq Y$.

Corollary 2. *1.* $\leq^{\mathrm{U}}_{\mathrm{T}} \not\subset \leq^{\mathrm{O}}_{\mathrm{T}} \longrightarrow \mathrm{P} \neq \mathrm{NP}$. *2.* $\leq^{\mathrm{U}}_{\mathrm{T}} \neq \leq^{\mathrm{RS}}_{\mathrm{T}} \longrightarrow \mathrm{P} \neq \mathrm{NP}$.

So proving $\leq^{\mathrm{U}}_{\mathrm{T}} \neq \leq^{\mathrm{p}}_{\mathrm{T}}$, $\leq^{\mathrm{U}}_{\mathrm{T}} \neq \leq^{\mathrm{RS}}_{\mathrm{T}}$, or $\leq^{\mathrm{U}}_{\mathrm{T}} \not\subset \leq^{\mathrm{O}}_{\mathrm{T}}$ amounts to proving $\mathrm{P} \neq \mathrm{NP}$. In particular, we cannot hope to strengthen Theorem 6 so that it is valid for $\leq^{\mathrm{U}}_{\mathrm{T}}$ rather than $\leq^{\mathrm{SN}}_{\mathrm{T}}$. Although we know that $\leq^{\mathrm{O}}_{\mathrm{T}} \not\subseteq \leq^{\mathrm{U}}_{\mathrm{T}}$, it is also difficult to show that they differ with respect to a sparse set on the right hand side, because we have $(\exists B \in \mathrm{SPARSE})[\mathrm{R}^{\mathrm{O}}_{\mathrm{T}}(B) \not\subseteq \mathrm{R}^{\mathrm{U}}_{\mathrm{T}}(B)] \longrightarrow \mathrm{P} \neq \mathrm{NP}$, which is a consequence of Theorem 14.

Theorem 16. *1.* $\leq^{\mathrm{RS}}_{\mathrm{T}} \neq \leq^{\mathrm{p}}_{\mathrm{T}} \longrightarrow \mathrm{P} \neq \mathrm{NP}$. *2.* $\leq^{\mathrm{RS}}_{\mathrm{T}} = \leq^{\mathrm{p}}_{\mathrm{T}} \longrightarrow \mathrm{P} = \mathrm{NP} \cap \mathrm{coNP}$.

5 Overproductive Reductions and the Classic Hardness Theorems

As is standard, the polynomial hierarchy is defined as follows: (a) $\Sigma^{\mathrm{p}}_0 = \mathrm{P}$; (b) for each $i \geq 0$, $\Sigma^{\mathrm{p}}_{i+1} = \mathrm{NP}^{\Sigma^{\mathrm{p}}_i}$; (c) for each $i \geq 0$, $\Pi^{\mathrm{p}}_i = \{L \mid \overline{L} \in \Sigma^{\mathrm{p}}_i\}$; and (d) $\mathrm{PH} = \cup_{i \geq 0} \Sigma^{\mathrm{p}}_i$. $\Theta^{\mathrm{p}}_2 = \{L \mid L \leq^{\mathrm{p}}_{\mathrm{tt}} \mathrm{SAT}\}$, where $\leq^{\mathrm{p}}_{\mathrm{tt}}$ denotes polynomial-time truth-table reduction. ZPP denotes expected polynomial time. It is well-known that $\mathrm{NP} \subseteq \Theta^{\mathrm{p}}_2 \subseteq \mathrm{P}^{\mathrm{NP}} \subseteq \mathrm{ZPP}^{\mathrm{NP}} \subseteq \Sigma^{\mathrm{p}}_2$.

It is very natural to ask whether the existence of sparse hard or complete sets with respect to our new reductions would imply collapses of the polynomial hierarchy similar to those that are known to hold for $\leq^{\mathrm{p}}_{\mathrm{T}}$. That is, are our reductions useful in extending the key standard results? To study this question, we must first briefly review what is known regarding the consequences of the existence of sparse NP-hard sets. The classic result in this direction was obtained by Karp and Lipton, and more recent research has yielded three increasingly strong extensions of their result.

Theorem 17. *1. [11]* $\mathrm{NP} \subseteq \mathrm{R}^{\mathrm{p}}_{\mathrm{T}}(\mathrm{SPARSE}) \longrightarrow \mathrm{PH} \subseteq \Sigma^{\mathrm{p}}_2$. *2. (implicit in [11], see [15] and the discussion in [8]; explicitly achieved in [1,10])* $\mathrm{NP} \subseteq \mathrm{R}^{\mathrm{RS}}_{\mathrm{T}}(\mathrm{SPARSE}) \longrightarrow \mathrm{PH} \subseteq \Sigma^{\mathrm{p}}_2$. *3. [13]* $\mathrm{NP} \subseteq \mathrm{R}^{\mathrm{RS}}_{\mathrm{T}}(\mathrm{SPARSE}) \longrightarrow \mathrm{PH} \subseteq \mathrm{ZPP}^{\mathrm{NP}}$. *4. [12] If A has self-computable witnesses and* $A \in (\mathrm{NP}^B \cap \mathrm{coNP}^B)/\mathrm{poly}$, *then* $\mathrm{ZPP}^{\mathrm{NP}^A} \subseteq \mathrm{ZPP}^{\mathrm{NP}^B}$.

To explain why part 4 of this theorem is stronger than part 3, we mention that Köbler and Watanabe [13] state part 3 in the form $\mathrm{NP} \subseteq (\mathrm{NP} \cap \mathrm{coNP})/\mathrm{poly} \longrightarrow \mathrm{PH} \subseteq \mathrm{ZPP}^{\mathrm{NP}}$, which is equivalent to the statement of part 3 in light of Theorem 4.

It remains open whether parts 3 or 4 of Theorem 17 can be extended from robustly strong reductions to overproductive reductions. However, as Theorem 18 we extend part 2 of Theorem 17 to overproductive reductions. As a consequence, there is at the present time no single strongest theorem on this topic; Theorem 18 and the final two parts of Theorem 17 seem to be incomparable in strength.

Theorem 18. $\mathrm{NP} \subseteq \mathrm{R}_{\mathrm{T}}^{\mathrm{O}}(\mathrm{SPARSE}) \longrightarrow \mathrm{PH} \subseteq \Sigma_2^{\mathrm{p}}$.

It remains open whether Theorem 18 can in some way be extended to underproductive reductions (our proof does not extend to that case). An analog for strong nondeterministic reductions is implicitly known, but has a far weaker conclusion.

Theorem 19. *(implicit in [13])* $\mathrm{NP} \subseteq \mathrm{R}_{\mathrm{T}}^{\mathrm{SN}}(\mathrm{SPARSE}) \longrightarrow \mathrm{PH} \subseteq \mathrm{ZPP}^{\Sigma_2^{\mathrm{p}}}$.

In contrast with the above results regarding sparse hard sets for NP, in the case of sparse *complete* sets for NP we have just as strong a collapse for $\leq_{\mathrm{T}}^{\mathrm{SN}}$-reductions as we have for $\leq_{\mathrm{T}}^{\mathrm{p}}$-reductions.

Theorem 20. *([9])* $\mathrm{NP} \subseteq \mathrm{R}_{\mathrm{T}}^{\mathrm{SN}}(\mathrm{SPARSE} \cap \mathrm{NP}) \longrightarrow \mathrm{PH} = \Theta_2^{\mathrm{p}}$.

As mentioned earlier, we leave as an open problem whether one can establish the collapse $\mathrm{PH} \subseteq \Sigma_2^{\mathrm{p}}$ (or, better still, $\mathrm{PH} \subseteq \mathrm{ZPP}^{\mathrm{NP}}$) under the assumption $\mathrm{NP} \subseteq \mathrm{R}_{\mathrm{T}}^{\mathrm{SN}}(\mathrm{SPARSE})$, or even under the stronger assumption that $\mathrm{NP} \subseteq \mathrm{R}_{\mathrm{T}}^{\mathrm{U}}(\mathrm{SPARSE})$. We conjecture that no such extension is possible.

6 Conclusions and Open Problems

Define the runtime of a nondeterministic machine on a given input to be the length of its longest computation path. (Though in most settings this is just one of a few equivalent definitions, we state it explicitly here as for the about-to-be-defined notion of local-polynomial machines, it is not at all clear that this equivalence remains valid.) Recall that we required that NPTMs be such that for each NPTM, N, it holds that there exists a polynomial p such that, for each oracle D, the runtime of N^D is bounded by p. Call such a machine "global-polynomial" as there is a polynomial that globally bounds its runtime. Does this differ from a requirement that for a machine N it holds that, for each oracle D, there is a polynomial p (which may depend on D) such that the runtime of N^D is bounded by p? Call such a machine "local-polynomial" as, though for every oracle it runs in polynomial time, the polynomial may depend on the oracle.

In general, these notions do differ, notwithstanding the common wisdom in complexity theory that one may "without loss of generality" assume enumerations of machines come with attached clocks independent of the oracle. (The subtle issue here is that the notions in fact usually do not differ on enumerations of machines that will be used with only one oracle.) The fact that they in general differ is made clear by the following theorems. These theorems show that there is a language transformation that can be computed by a local-polynomial machine, yet each global-polynomial machine will, for some target set, fail almost everywhere to compute the set's image under the language transformation. We write $A =^* B$ if A and B are equal almost everywhere, i.e., if $(A - B) \cup (B - A)$ is a finite set.

Theorem 21. *There is a function $f_N : 2^{\Sigma^*} \to 2^{\Sigma^*}$ (respectively, $f_D : 2^{\Sigma^*} \to 2^{\Sigma^*}$) such that (1) there is a nondeterministic (respectively, deterministic) local-polynomial Turing machine $\widehat{M}$ such that for each oracle A it holds that $L(\widehat{M}^A) = f_N(A)$ (respectively, $L(\widehat{M}^A) = f_D(A)$), and (2) for each NPTM, i.e., each nondeterministic global-polynomial Turing machine M (respectively, DPTM, i.e., each deterministic global-polynomial Turing machine M) it holds that there is a set $A \subseteq \Sigma^*$ such that $L(M^A) =^* \overline{f_N(A)}$ (respectively, $L(M^A) =^* \overline{f_D(A)}$).*

Though this claim may at first seem counterintuitive, its proof is almost immediate if one is given f_N and f_D, and so we simply give functions f_N and f_D satisfying the theorem. In particular, we can use $f_N(A) = \{x \mid (\exists y)[(|y| \leq \log|x|) \wedge (y \text{ is the lexicographically first string in } A) \wedge (\exists z)[|z| = |x|^{|y|} \wedge xz \in A]]\}$ and $f_D(A) = \{x \mid (\exists y)[(|y| \leq \log|x|) \wedge (y \text{ is the lexicographically first string in } A) \wedge (\exists z)[(z \text{ is one of the } |x|^{|y|} \text{ lexicographically smallest length } |x|^{|y|} \text{ strings in } \Sigma^*) \wedge xz \in A]]\}$.

The difference between global-polynomial machines and local-polynomial machines in general mappings, as just proven, may make one wonder whether the fact that robust strong reduction is defined in terms of global-polynomial (as opposed to local-polynomial) machines makes a difference and, if so, which definition is more natural. Regarding the former issue, we leave it as an open question. (The above theorems do not resolve this issue, as they deal with language-to-language transformations defined specifically over all of 2^{Σ^*}, but in contrast a robustly strong reduction must accept a specific language only for one oracle, and for all others merely has to be underproductive and overproductive, plus it must have the global-polynomial property.) That is, the open question is: Does there exist a pair of sets A and B such that $A \leq_{\mathrm{T}}^{\mathrm{RS}} B$ (which by definition involves a global-polynomial machine) and no nondeterministic local-polynomial Turing machine N has the properties that $L(N^B) = A$ and $(\forall D \subseteq \Sigma^*)[N^D$ is both underproductive and overproductive]? Regarding the question of naturalness, this is a matter of taste. However, we point out that the global-polynomial definition is exactly that of Gavaldà and Balcázar [6], and that part 2 of Theorem 4, Gavaldà and Balcázar's [6] natural characterization of robustly strong reductions to sparse sets in terms of the complexity class $(\mathrm{NP} \cap \mathrm{coNP})/\mathrm{poly}$, seems to depend crucially on the fact that one's machines are global-polynomial.

On the other hand Theorem 18, though its proof seems on its surface to be dependent on the fact that $\leq_{\mathrm{T}}^{\mathrm{O}}$ is defined via global-polynomial machines, in fact remains true even if $\leq_{\mathrm{T}}^{\mathrm{O}}$ is redefined via local-polynomial machines.

Acknowledgments: We are grateful to Yenjo Han, who first noticed the error in the proof of Gavaldà and Balcázar. We thank Edith Hemaspaandra and Jörg Rothe for many important suggestions, and we are grateful to Osamu Watanabe for extremely helpful comments and corrections, including pointing out that a step of the proof of Theorem 5 uses the self-reducibility of EXP.

References

[1] M. Abadi, J. Feigenbaum, and J. Kilian. On hiding information from an oracle. *Journal of Computer and System Sciences*, 39:21–50, 1989.

[2] E. Allender, R. Beigel, U. Hertrampf, and S. Homer. Almost-everywhere complexity hierarchies for nondeterministic time. *Theoretical Computer Science*, 115:225–241, 1993.

[3] R. Beigel. On the relativized power of additional accepting paths. In *Proceedings of the 4th Structure in Complexity Theory Conference*, pages 216–224. IEEE Computer Society Press, June 1989.

[4] J. Cai, L. Hemaspaandra, and G. Wechsung. Robust reductions. Technical Report TR-666, Department of Computer Science, University of Rochester, Rochester, NY, December 1997. Available on-line at http://www.cs.rochester.edu/trs/theory-trs.html.

[5] P. Crescenzi and R. Silvestri. Sperner's lemma and robust machines. In *Proceedings of the 8th Structure in Complexity Theory Conference*, pages 194–199. IEEE Computer Society Press, May 1993.

[6] R. Gavaldà and J. Balcázar. Strong and robustly strong polynomial time reducibilities to sparse sets. *Theoretical Computer Science*, 88(1):1–14, 1991.

[7] J. Hartmanis and L. Hemachandra. Robust machines accept easy sets. *Theoretical Computer Science*, 74(2):217–226, 1990.

[8] L. Hemaspaandra, A. Hoene, A. Naik, M. Ogiwara, A. Selman, T. Thierauf, and J. Wang. Nondeterministically selective sets. *International Journal of Foundations of Computer Science*, 6(4):403–416, 1995.

[9] J. Kadin. $P^{NP[\log n]}$ and sparse Turing-complete sets for NP. *Journal of Computer and System Sciences*, 39(3):282–298, 1989.

[10] J. Kämper. Non-uniform proof systems: a new framework to describe non-uniform and probabilistic complexity classes. *Theoretical Computer Science*, 85(2):305–331, 1991.

[11] R. Karp and R. Lipton. Some connections between nonuniform and uniform complexity classes. In *Proceedings of the 12th ACM Symposium on Theory of Computing*, pages 302–309, April 1980.

[12] J. Köbler and U. Schöning. High sets for NP. In D. Zu and K. Ko, editors, *Advances in Algorithms, Languages, and Complexity*, pages 139–156. Kluwer Academic Publishers, 1997.

[13] J. Köbler and O. Watanabe. New collapse consequences of NP having small circuits. In *Proceedings of the 22nd International Colloquium on Automata, Languages, and Programming*, pages 196–207. Springer-Verlag *Lecture Notes in Computer Science #944*, 1995.

[14] T. Long. Strong nondeterministic polynomial-time reducibilities. *Theoretical Computer Science*, 21:1–25, 1982.

[15] T. Long and A. Selman. Relativizing complexity classes with sparse oracles. *Journal of the ACM*, 33(3):618–627, 1986.

[16] A. Selman. Polynomial time enumeration reducibility. *SIAM Journal on Computing*, 7(4):440–457, 1978.

Approaches to Effective Semi-continuity of Real Functions

(Extended Abstract)

Vasco Brattka, Klaus Weihrauch, Xizhong Zheng*

Theoretische Informatik I,
FernUniversität Hagen,
58084 Hagen, Germany

Abstract. By means of different effectivities of the epigraphs and hypographs of real functions we introduce several effectivizations of the semi-continuous real functions. We call a real function f lower semi-computable of type one if its hypograph $\text{hypo}(f) := \{(x,y) : f(x) > y \ \& \ x \in \text{dom}(f)\}$ is recursively enumerably open in $\text{dom}(f) \times \mathbb{R}$; f is lower semi-computable of type two if its closed epigraph $\text{Epi}(f) := \{(x,y) : f(x) \leq y \ \& \ x \in \text{dom}(f)\}$ is recursively enumerably closed in $\text{dom}(f) \times \mathbb{R}$ and f is lower semi-computable of type three if $\text{Epi}(f)$ is recursively closed in $\text{dom}(f) \times \mathbb{R}$. These semi-computabilities and computability of real functions are compared. We show that, type one and type two semi-computability are independent and that type three semi-computability plus effectively uniform continuity implies computability which is false for type one and type two instead of type three. We show also that the integral of a type three semi-computable real function on a computable interval is not necessarily computable.

1 Introduction

In recursive analysis, real numbers x are usually represented by fast convergent Cauchy sequences of rational numbers which converge to x. This representation is denoded by ρ and the sequence corresponding to x is called a ρ-name of x. Then, a real number x is computable (more precisely ρ-computable), iff it has a computable ρ-name. A (partial) real function $f :\subseteq \mathbb{R} \to \mathbb{R}$ is computable, iff there is an algorithm M which produces a ρ-name of $f(x)$ from any ρ-name of x, for $x \in \text{dom}(f)$ (cf. [4, 5, 8]). Such kind of algorithms can be described by Type-2 Turing machines (TT-machines) which generalize classic Turing machines in such a way that their inputs and outputs can be infinite sequences as well as finite strings (see [8, 9]). Thus, similar to the case of number-theoretical functions, a real function f is computable iff there is a TT-machine M to compute it by means of the representation ρ (so called (ρ, ρ)-computability). Because any finite initial segment of the output of a TT-machine depends only on a finite

* Contact author. Email address: xizhong.zheng@fernuni-hagen.de

portion of the input, any computable real function is continuous. In some sense, computability of real functions is a kind of effectivization of continuity of real functions.

For the effectivization of semi-continuity, Ge and Nerode [2] introduced a notion of recursively semi-continuity. A function $f : [a; b] \longrightarrow \mathbb{R}$ is *recursively lower semi-continuous* if its closed epigraph $\mathrm{Epi}(f) := \{(x, y) \in \mathbb{R}^2 : f(x) \leq y \ \&\ x \in [a; b]\}$ is recursively closed in the sense of [14, 10]. Another effectivization of semi-continuity is given by Weihrauch and Zheng [11], where a real function $f :\subseteq \mathbb{R} \longrightarrow \mathbb{R}$ is called *lower semi-computable* if there is a TT-machine M such that M outputs a rational left cut of $f(x)$ from the input of any ρ-name of $x \in \mathrm{dom}(f)$, i.e., f is $(\rho, \rho_<)$-computable, where $\rho_<$ is a representation of real numbers by means of the left Dedekind cut. Equivalently, f is lower semi-computable iff its hypograph $\mathrm{hypo}(f) := \{(x, y) \in \mathbb{R}^2 : f(x) > y \ \&\ x \in \mathrm{dom}(f)\}$ is recursively enumerably (r.e.) open (in $\mathrm{dom}(f) \times \mathbb{R}$), i.e. there is a computable sequence $(B_n : n \in \mathbb{N})$ of rational open balls of $\mathbb{R}^2$ such that $\forall n \in \mathbb{N}(B_n \subseteq \mathrm{hypo}(f))$ and $\bigcup_{n\in\mathbb{N}} B_n = \mathrm{hypo}(f)$. Because a recursively closed set is always r.e. closed, a recursively lower semi-continuous function is also lower semi-computable. Besides, it is also very natural to introduce another effective version of semi-continuous real function by requesting recursive enumerability of the closed epigraph $\mathrm{Epi}(f)$. Then we have altogether three kinds of effectivizations of lower semi-continuity of real function f: 1. by r.e. openess of $\mathrm{hypo}(f)$, 2. by r.e. closedness of $\mathrm{Epi}(f)$ and 3. by recursive closedness of $\mathrm{Epi}(f)$. In this paper we will call them *type 1, type 2* and *type 3 lower semi-computability* (*1-, 2-* and *3-l.s.comp.* in short), respectively. Accordingly, we can introduce *type 1, type 2* and *type 3 upper semi-computabilities* (*1-, 2-* and *3-u.s.comp.* in short.)

The basic properties of above lower semi-computabilities and their relationships will be discussed in this paper. We will show that, type 1 and type 2 lower semi-computabilities are independent. In many respects, type 1 semi-computability looks more "natural" than type 2. For example, it is shown in [11], that if the function f is both 1-l.s.comp. and 1-u.s.comp., then f is computable; a 1-l.s.comp. function f maps every computable sequence of real numbers to a computable sequence of $\rho_<$-computable real numbers (i.e., f is sequentially $\rho_<$-computable) and the integral $\int_a^b f(x)dx$ of a 1-l.s.comp. function f on a computable interval $[a; b]$ is $\rho_<$-computable. But all of these fail for 2-l.s.comp. function. It is well known that a computable real function $f : [0; 1] \longrightarrow \mathbb{R}$ is effectively uniformly continuous (see e.g., [5]), i.e., there is a recursive function $e : \mathbb{N} \longrightarrow \mathbb{N}$ such that $|f(x) - f(y)| < 2^{-n}$ holds whenever $|x - y| < 1/e(n)$ for any $x, y \in [0; 1]$ and $n \in \mathbb{N}$. This is not the case for the continuous semi-computable function. In fact, we show that if a 3-l.s.comp. function f is effectively uniformly continuous, then f must be computable.

For the integral, Ge [3] asked whether a 3-l.s.comp. function f has always a computable integral $\int_a^b f(x)dx$ on a computable interval $[a; b]$. By a finite injury priority construction, we construct a 3-l.s.comp. function $f : [0; 1] \longrightarrow \mathbb{R}$ such that the integral $\int_0^1 f(x)dx$ is $\rho_<$-computable but not ρ-computable.

In the next section, we will recall at first some definitions and basic facts about computable real subsets of $\mathbb{R}^2$ and semi-continuous real functions. The precise definition of semi-computable real function and some of their basic properties are given in Section 3. Section 4 discusses effectively uniformly continuity of semi-computable functions. In the last Section 5 we discuss the integral of semi-computable functions.

2 Preliminaries

We define some notions which will be used in this paper at first. For any $n \neq 0$ and $a_1, \ldots, a_n, b_1, \ldots, b_n \in \mathbb{R}$, the set $I^{(n)} := \{(x_1, \ldots, x_n) \in \mathbb{R}^{(n)} : a_i < x_i < b_i \text{ for } i = 1, \ldots, n\}$ is called an *open n-cuboid.* Its left and right boundaries are denoted by $l_i(I^{(n)}) := a_i$ and $r_i(I^{(n)}) := b_i$, respectively. If all boundaries of $I^{(n)}$ are rational numbers, then $I^{(n)}$ is an *rational open n-cuboid. Rational closed n-cuboids* can be defined similarly and denoted by $\bar{I}^{(n)}$ accordingly. The sets of all rational open and closed n-cuboids are denoted by *Int*(n) and $\overline{Int}(n)$, respectively. A sequence $(r_n : n \in \mathbb{N})$ of rational numbers is *computable* if there are recursive functions $f, g, h : \mathbb{N} \to \mathbb{N}$ such that $r_n = (f(n) - g(n))/(h(n) + 1)$ for all $n \in \mathbb{N}$. A double sequence $(r_{nm} : n, m \in \mathbb{N})$ of rational numbers is *computable* if $(r'_n : n \in \mathbb{N})$, with $r'_{\langle n,m\rangle} := r_{nm}$, is computable, where $\langle \cdot, \cdot \rangle : \mathbb{N}^2 \to \mathbb{N}$ is a computable pairing function with computable inverse functions $\pi_1, \pi_2 : \mathbb{N} \to \mathbb{N}$. A sequence $(x_n : n \in \mathbb{N})$ of real numbers is called *computable* if there is a double computable sequence $(r_{nm} : n, m \in \mathbb{N})$ of rational numbers such that $|x_n - r_{nm}| < 2^{-m}$ holds for all $n, m \in \mathbb{N}$. A sequence $(I_m^{(n)} : m \in \mathbb{N})$ of rational open (closed) n-cuboids is called *computable* if the sequences $(l_i(I_m^{(n)}) : m \in \mathbb{N})$ and $(r_i(I_m^{(n)}) : m \in \mathbb{N})$, $1 \leq i \leq n$, of their boundaries are computable sequences of rational numbers.

Suppose that $(I_m^{(n)} : m \in \mathbb{N})$ is an effective enumeration of all rational open n-cuboids of $\mathbb{R}^n$. An open set $A \subseteq \mathbb{R}^n$ is called *recursively enumerably open* (*r.e. open*) if the set $\{m \in \mathbb{N} : \bar{I}_m^{(n)} \subseteq A\}$ is recursively enumerable. A closed set $B \subseteq \mathbb{R}^n$ is called *recursively enumerably closed* (*r.e. closed*) if the set $\{m \in \mathbb{N} : I_m^{(n)} \cap B \neq \emptyset\}$ is recursively enumerable. If $A \subseteq \mathbb{R}^n$ is r.e. closed (open) and its complement A^c is r.e. open (closed), then A is called *recursively closed* (*open*). By recursive invariance of r.e. sets (cf. [7]), these definitions are independent of the recursive enumeration of $(I_m^{(n)} : m \in \mathbb{N})$.

A sequence $(A_n : n \in \mathbb{N})$ of r.e. closed sets is called *computable* if there is a computable double sequence $(J_{ij} : i, j \in \mathbb{N})$ of rational open n-cuboids of $\mathbb{R}^n$ such that $(J_{ij} : j \in \mathbb{N})$ enumerates all rational open n-cuboids of $\mathbb{R}^n$ which intersect A_i for all $i \in \mathbb{N}$. *Computable sequences of r.e. open sets* and *computable sequences of recusive open (closed) sets* can be definied accordingly.

R.e. open and r.e. closed sets have following useful characterizations.

Theorem 1 (Brattka and Weihrauch [10]). *(1) An open set $A \subseteq \mathbb{R}^n$ is r.e. open iff there is a computable sequence $(J_n : n \in \mathbb{N})$ of closed rational n-cuboids of $\mathbb{R}^n$ such that $\bigcup_{n\in\mathbb{N}} J_n = A$;*

(2) A closed set $A \subseteq \mathbb{R}^n$ is r.e. closed iff there is a computable sequence $(x_n : n \in \mathbb{N})$ of real points of $\mathbb{R}^n$ such that $\{x_n : n \in \mathbb{N}\}$ forms a dense subset of A.

The next theorem gives a convenient way to construct an r.e. closed real set.

Theorem 2. *Let $(A_n : n \in \mathbb{N})$ be a computable sequence of r.e. closed sets. Then the set $A := \mathrm{cl}(\bigcup_{i\in\mathbb{N}} A_i)$ is r.e. closed, where $\mathrm{cl}(B)$ means the closure of set B.*

Now we recall the definition of semi-continuous real functions by means of epigraphs and hypographs. For any real function $f :\subseteq \mathbb{R} \to \mathbb{R}$, its *closed epigraph* Epi(f), *epigraph* epi(f), *closed hypograph* Hypo(f) and *hypograph* hypo(f) are defined, respectively, by:

$$\begin{aligned}
\mathrm{Epi}(f) &:= \{(x,y) \in \mathbb{R}^2 : f(x) \leq y \;\&\; x \in \mathrm{dom}(f)\};\\
\mathrm{epi}(f) &:= \{(x,y) \in \mathbb{R}^2 : f(x) < y \;\&\; x \in \mathrm{dom}(f)\};\\
\mathrm{Hypo}(f) &:= \{(x,y) \in \mathbb{R}^2 : f(x) \geq y \;\&\; x \in \mathrm{dom}(f)\};\\
\mathrm{hypo}(f) &:= \{(x,y) \in \mathbb{R}^2 : f(x) > y \;\&\; x \in \mathrm{dom}(f)\}.
\end{aligned}$$

A real function $f : X \to \mathbb{R}$ is called *lower semi-continuous* (*l.s.c.* in short) if its hypograph hypo(f) is open in $X \times \mathbb{R}$, or equivalently, its closed epigraph Epi(f) is closed in $X \times \mathbb{R}$. f is called *upper semi-continuous* (*u.s.c.* in short) if its closed hypograph Hypo(f) is closed in $X \times \mathbb{R}$, or equivalently, its epigraph epi(f) is open in $X \times \mathbb{R}$. The sets of all l.s.c. and u.s.c. functions defined on X are denoted by LSC(X) and USC(X), respectively.

Fix Σ to be an alphabet which contains $0, 1$ and all the other symbols we need later. Σ^* and Σ^ω are the sets of all finite strings and all infinite sequences of elements from Σ, respectively. An infinite sequence $p \in \Sigma^\omega$ is called *computable* if there is a computable function $f : \mathbb{N} \to \Sigma$ such that $p = f(0)f(1)f(2)\cdots$. Let $\nu_Q :\subseteq \Sigma^* \to \mathbb{Q}$ be some standard notation of set $\mathbb{Q}$ of rational numbers. $\nu_Q(u)$ is denoted usually by $\bar{u}$. For the real number set $\mathbb{R}$, besides the representation ρ by fast convergent Cauchy sequences mentioned in Section 1 which means that $x = \rho(p)$ iff $p = \sharp u_0 \sharp u_1 \sharp u_2 \sharp \ldots$ such that $\lim_{n\to\infty} \bar{u}_n = x$ and $\forall m \forall n \geq m(|\bar{u}_m - \bar{u}_n| < 2^{-n})$, we will use another representation ρ_1 defined by $x = \rho_1(p)$ iff $p = \sharp u_0 ¢ v_0 \sharp u_1 ¢ v_1 \sharp \cdots$ and $\{(u_n, v_n) : n \in \mathbb{N}\} = \{(u,v) \in \mathrm{dom}(\nu_Q)^2 : \bar{u} < x < \bar{v}\}$, i.e., an ρ_1-name of x *enumerates* all paires of $(u,v) \in \mathrm{dom}(\nu_Q)^2$ such that $\bar{u} < x < \bar{v}$. Because ρ and ρ_1 are recursively equivalent (see e.g., [11]), we often do not distinguish them explicitly.

Other useful representations of real numbers are $\rho_<$ and $\rho_>$, which are defined by $\rho_<(p) = x$ ($\rho_>(p) = x$) iff p enumerates all $u \in \mathrm{dom}(\nu_Q)$ such that $\bar{u} < x$ ($\bar{u} > x$). That is, an $\rho_<$ ($\rho_>$)-name enumerates the left (right) Dedekind cut of x. Let γ and γ' be two representations of $\mathbb{R}$. A real number x is called γ-*computable* if there is a computable sequence $p \in \Sigma^\omega$ such that $x = \gamma(p)$; and a function $f :\subseteq \mathbb{R} \to \mathbb{R}$ is (γ,γ')-*computable* iff there is a TT-machine M such that $f(\gamma(p)) = \gamma'(f_M(p))$ for any $p \in \mathrm{dom}(f \circ \gamma)$, where $f_M :\subseteq \Sigma^\omega \to \Sigma^\omega$ is the function computed by M.

The next lemma is useful to construct counterexamples.

Lemma 3 (Weihrauch & Zheng [13]). *There are a $\rho_<$-computable real number a_1 and a $\rho_>$-computable real number a_2 such that the sum $a := a_1 + a_2$ is neither $\rho_<$-computable nor $\rho_>$-computable.*

For other unexplained notations please refer to [5, 8, 10, 11].

3 Semi-Computabilities of Real Functions

In this section we define several semi-computabilities of real functions by means of different effectivities of their epigraphs and hypographs. Their basic properties and mutual relations are dicsussed.

Definition 4. Let $X \subseteq \mathbb{R}$ and $f : X \to \mathbb{R}$.

(1) f is called *lower semi-computable of type one (1-l.s.comp.)* if its hypograph hypo(f) is r.e. open in $X \times \mathbb{R}$ and f is called *upper semi-computable of type one (1-u.s.comp.)* if its epigraph epi(f) is r.e. open in $X \times \mathbb{R}$.

(2) f is called *lower semi-computable of type two (2-l.s.comp.)* if its closed epigraph Epi(f) is r.e. closed in $X \times \mathbb{R}$ and f is called *upper semi-computable of type two (2-u.s.comp.)* if its closed hypograph Hypo(f) is r.e. closed in $X \times \mathbb{R}$.

(3) f is called *lower semi-computable of type three (3-l.s.comp.)* if it is both 1-l.s.comp. and 2-l.s.comp. or equivalently, if Epi(f) is recursively closed and f is called *upper semi-computable of type three (3-u.s.comp.)* if Hypo(f) is recursively closed.

Remark: In the following, we will often say simply "open" ("closed") instead of "open in dom(f) $\times \mathbb{R}$" ("closed in dom(f) $\times \mathbb{R}$").

Semi-computability of the third type for an upper-bounded real function defined on a computable interval $[a; b]$ was introduced by Ge and Nerode [2] and they use the name *recursive semi-continuity*. And first type lower semi-computability was introduced by Weihrauch and Zheng [11], where they call them *lower semi-computable*.

It is obvious that for any $f : X \to \mathbb{R}$, $X \subseteq \mathbb{R}$, if f is 1-, 2- or 3-l.s.comp., then f is lower semi-continuous, and f is 1-, 2- or 3-l.s.comp. iff $-f$ is 1-, 2- or 3-u.s.comp., respectively.

Theorem 5 (Weihrauch & Zheng [11]). *Let $X \subseteq \mathbb{R}$ and $f : X \to \mathbb{R}$.*

(1) f is 1-l.s.comp. (1-u.s.comp) iff f is $(\rho, \rho_<)$- ($(\rho, \rho_>)$-) computable;

(2) If X is a computable interval, then f is 1-l.s.comp (1-u.s.comp) iff there is a computable increasing (decreasing) sequence $(pg_n : n \in \mathbb{N})$ of rational polygon functions on X such that $f(x) = \lim_{n\to\infty} pg_n(x)$ for any $x \in X$.

Theorem 6 (Weihrauch & Zheng [11]). *Let $X \subseteq \mathbb{R}$ and $f : X \to \mathbb{R}$ be 1-l.s.comp..*

(1) f is computable iff f is also 1-u.s.comp.;

(2) f is sequentially $\rho_<$-computable, i.e., $(f(x_n) : n \in \mathbb{N})$ is a computable sequence of $\rho_<$-computable real numbers whenever $(x_n : n \in \mathbb{N})$ is a computable sequence of real numbers;

(3) For any computable real number $a \in \mathbb{R}$, the set $\text{hypo}(f, a) := \{x \in X : f(x) > a\}$ *is r.e. open in X;*

(4) If $\liminf_{x \to +a} f(x) > f(a)$ $(\liminf_{x \to -a} f(x) > f(a))$, then a is an $\rho_>$ $(\rho_<)$-computable real number for any $a \in X$.

(5) If X is a computable interval, then $\min\{f(x) : x \in X\}$ *is $\rho_<$-computable.*

The assertions (1) – (3) of Theorem 6 do not hold accordingly for 2-l.s.comp. functions. By Theorem 2, a constant function f which takes a noncomputable but $\rho_>$-computable real value is a 2-l.s.comp. function, hence 2-l.s.comp. functions are not necessarily sequentially $\rho_<$-computable. Furthermore, let $(a_n : n \in \mathbb{N})$ be a computable sequence of rational numbers such that $a := \lim_{n \to \infty} a_n$ is neither $\rho_<$-computable nor $\rho_>$-computable (see Lemma 3), and define a function $f : [0; 1] \to \mathbb{R}$ by $f(1/2) = a$; $f(n/(2n + 1)) := f((n + 1)/(2n + 1)) := a_n$ for any $n \in \mathbb{N}$ and f is linear on all intervals $[n/(2n + 1); (n + 1)/(2n + 3)]$ and $[(n+2)/(2n+3); (n+1)/(2n+1)]$ for $n \in \mathbb{N}$. Then f is both 2-l.s.comp. and 2-u.s.comp. but neither 1-l.s.comp. nor 1-u.s.comp., hence not computable, because $f(1/2) = a$ is neither $\rho_<$-computable nor $\rho_>$-computable. From this example, we can obtain immediately the following corollary.

Corollary 7. *(1) There is a 2-l.s.comp. and 2-u.s.comp. real function which is neither sequentially $\rho_<$-computable nor sequentially $\rho_>$-computable.*

(2) There is a real function which is both 2-l.s.comp. and 2-u.s.comp. but not computable.

For (3) of Theorem 6 we have the following negative result.

Theorem 8. *There is a continuous 2-l.s.comp. function $f : [0; 1] \to \mathbb{R}$ and a computable real number $a \in \mathbb{R}$ such that* $\text{Epi}(f, a) := \{x \in [0; 1] : f(x) \leq a\}$ *is not r.e. closed.*

Part (4) and (5) of Theorem 6 hold accordingly for 2-l.s.comp. functions by exchanging $\rho_>$ and $\rho_<$.

Theorem 9. *Let $X \subseteq \mathbb{R}$ and $f : X \to \mathbb{R}$ be a 2-l.s.comp. real function.*

(1) For any $a \in X$, if $\liminf_{x \to +a} f(x) > f(a)$ $(\liminf_{x \to -a} f(x) > f(a))$, then a is $\rho_<$-computable ($\rho_>$-computable);

(2) If f is lower bounded, then $\inf\{f(x) : x \in X\}$ *is $\rho_>$-computable. Especially, if X is a computable interval, then* $\min\{f(x) : x \in X\}$ *is $\rho_>$-computable.*

Corollary 10. *(1) If $f : X \to \mathbb{R}$ is 3-l.s.comp. and X is a computable interval, then* $\min\{f(x) : x \in X\}$ *is computable;*

(2) There are real functions $f, g : X \to \mathbb{R}$ such that f is 1-l.s.comp. but not 2-l.s.comp. and g is 2-l.s.comp, but not 1-l.s.comp.;

(3) Let $f : X \to \mathbb{R}$ be 3-l.s.comp. and $a \in X$. If $\liminf_{x \to +a} f(x) > f(a)$ or $\liminf_{x \to -a} f(x) > f(a)$, then a is computable.

Next theorem is about the integral of 1-l.s.comp. function.

Theorem 11. *Let $f : [a;b] \to \mathbb{R}$ be 1-l.s.comp. real function, where $a, b \in \mathbb{R}$ are computable. Then the integral $\int_a^b f(x)dx$ is $\rho_<$-computable.*

Theorem 11 is not true for the 2-l.s.comp. function. In Section 5 we will see that there is a 2-l.s.comp. function $f : [0;2] \to \mathbb{R}$ such that $\int_0^2 f(x)dx$ is neither $\rho_<$- nor $\rho_>$-computable.

4 Effectively Uniform Continuity

This section discusses uniform continuity of semi-computable real functions. A function $f : X \to \mathbb{R}$ is called *uniformly continuous* if

$$\forall \varepsilon \in \mathbb{R}^+ \exists \delta \in \mathbb{R}^+ \forall x, y \in X(|x - y| < \delta \Longrightarrow |f(x) - f(y)| < \varepsilon). \tag{1}$$

f is *effectively uniformly continuous* if there is a recursive function $e : \mathbb{N} \to \mathbb{N}$ such that

$$\forall n \in \mathbb{N} \forall x, y \in X \left(|x - y| < 2^{-e(n)} \Longrightarrow |f(x) - f(y)| < 2^{-n}\right). \tag{2}$$

Function e is called a *modulus of uniform continuity of f*. We will see that although a continuous 3-l.s.comp. function defined on a closed interval $[a;b]$ is uniformly continuous, it is not necessarily effectively uniformly continuous. But, if a 3-l.s.comp. function $f : [a;b] \to \mathbb{R}$ is effectively uniformly continuous, then it must be computable.

Theorem 12. *There is a continuous 3-l.s.comp. function $f : [0;1] \to \mathbb{R}$ such that f is not effectively uniformly continuous.*

Proof. (schedule) Let $(a_n : n \in \mathbb{N})$ be a computable increasing sequence of rational numbers such that $\lim_{n\to\infty} a_n := a$ is a $\rho_<$-computable but not a computable real number. Define a function $f : [0;1] \to \mathbb{R}$ by

$$f(x) := \begin{cases} a & \text{if } x = 1/2; \\ a_n & \text{if } x = n/(2n+1) \text{ or } x = (n+1)/(2n+1)) \text{ for } n \in \mathbb{N}, \end{cases}$$

and f is linear on intervals $[n/(2n+1); (n+1)/(2n+3)]$ and $[(n+2)/(2n+3); (n+1)/(2n+1)]$ for any $n \in \mathbb{N}$. We can show that f is 3-l.s.comp. but f is not computable because $f(1/2) = a$ is not computable and any computable function maps a computable real number to a computable one (cf. [5]).

Now assume by contradiction that f is effectively uniformly continuous, i.e., there is a recursive function $e : \mathbb{N} \to \mathbb{N}$ which statisfies condition (2). Define a recursive function $i : \mathbb{N} \to \mathbb{N}$ by $i(n) := \mu m \left(\left|\frac{1}{2} - \frac{m}{2m+1}\right| < 2^{-e(n)}\right)$, for any $n \in \mathbb{N}$. Let $r_n := \frac{i(n)}{2i(n)+1}$. Then we have, for any $n \in \mathbb{N}$, $|a - a_{i(n)}| = |f(1/2) - f(r_n)| < 2^{-n}$ because (2) and $\forall n \in \mathbb{N}(|1/2 - r_n| < 1/e(n))$, i.e., $(a_{i(n)} : n \in \mathbb{N})$ is a computable Cauchy sequence of rational numbers which converges effectively to a. This contradicts to noncomputability of a. So f is not effectively uniformly continuous.

Next theorem shows that effectively uniform continuity guarantees computability of a 3-l.s.comp. real function.

Theorem 13. *Let $f : [0;1] \to \mathbb{R}$ be a 3-l.s.comp. function. If f is effectively uniformly continuous, then f is computable.*

5 Non-computability of the Integrals

It is well known that any computable real function has a computable integral on a computable interval. In Section 3 we have shown, that $\int_a^b f(x)dx$ is $\rho_<$-computable if f is a 1-l.s.comp. function on computable interval $[a;b]$. Ge [3] asked that, whether the integral $I = \int_0^1 f(x)dx$ is always computable, if $f : [0;1] \to \mathbb{R}$ is recursively lower semi-continuous in their sense [2], i.e., 3-l.s.computable. The following results give a negative answer to the question.

Theorem 14. *There is an 3-l.s.comp. function $f : [0;1] \to \mathbb{R}$ such that the integral $\int_0^1 f(x)dx$ is $\rho_<$-computable but not computable.*

Proof. We will define a l.s.c. function $f : [0;1] \to \mathbb{R}$ such that hypo(f) is r.e. open, Epi(f) is r.e. closed and $\int_0^1 f(x)dx$ is $\rho_<$-computable but not computable.

Let $a \in (0;1)$ be an $\rho_<$-computable but not computable real number and $(a_n : n \in \mathbb{N})$ an increasing sequence of rational numbers which converges to a. Suppose that $((u_n, v_n) : n \in \mathbb{N})$ is a computable sequence of rational points which enumerates all rational points of $[0;1] \times \mathbb{R}$. Assume w.l.o.g. that $a_0 = 0$, $(u_0, v_0) = (0,0)$.

We define a function $f : [0;1] \to \mathbb{R}$ by

$$f(x) = \begin{cases} 0 & x \geq a \vee \exists n(a_n \leq x \ \& \ \exists i \leq n(x = u_i)) \\ 1 & \text{otherwise.} \end{cases}$$

Note that there are only finitely many rational numbers $x < a_n$ with $f(x) = 0$ for any $n \in \mathbb{N}$. This "finite injury priority"-like trick makes sure that Epi(f) is r.e. closed and hypo(f) is r.e. open.

We prove now that the function f defined above satisfies the properties of the theorem. At first we show that f is l.s.c., i.e., $\liminf_{y \to x} f(y) \geq f(x)$ for any $x \in [0;1]$.

If $x \in [a;1]$, we have obviously that $\liminf_{y \to x} f(y) = 0 = f(x)$. For $x \in [0;a)$, there is an $n \in \mathbb{N}$ such that $x < a_n$. Because there are at most finitely many $y \in [0;a_n)$ such that $f(y) = 0$ and for almost all other $y \in [0;a_n)$, we have always $f(y) = 1$, hence $\liminf_{y \to x} f(y) = 1 \geq f(x)$. So f is a l.s.c. function.

Now we show that f is 1-l.s.comp., i.e., hypo(f) is r.e. open. Define, for any $b \in [0;1]$, a step function $f_b : [0;1] \to \mathbb{R}$ by $f_b(x) := 1$ if $x \in [0;b)$ and $f_b(x) := 0$ is $x \in [b;1]$. Obviously, f_b is l.s.c. for any $b \in [0;1]$. Especially, $f_n : [0;1] \to \mathbb{R}$ defined by $f_n(x) := \min\{f(x), f_{a_n}(x)\}$ is 3-l.s.comp. and (hypo(f_n) $: n \in \mathbb{N}$) is a computable sequence of r.e. open sets such that hypo(f) $= \bigcup_{n \in \mathbb{N}}$ hypo(f_n)

since $(a_n : n \in \mathbb{N})$ is a computable sequence of rational numbers. It follows immediately that hypo(f) is also r.e. open and hence f is 1-l.s.comp.

Next, we show that f is 2-l.s.comp., i.e., Epi(f) is r.e. closed. Since Epi(f) is closed by lower semi-continuity, it suffices to show that there is a computable sequence of rational points of $[0;1] \times \mathbb{R}$ which forms a dense subset of Epi(f) by Theorem 1. We define such a computable sequence $((x_n, y_n) : n \in \mathbb{N})$ inductively: $(x_0, y_0) := (u_0, v_0) = (0, 0)$ and, for $n+1$, if (u_{n+1}, v_{n+1}) satisfies one of the following conditions:

$$0 \leq u_{n+1} \leq 1 \ \& \ v_{n+1} \geq 1; \tag{3}$$

$$a_{n+1} \leq u_{n+1} \leq 1 \ \& \ v_{n+1} \geq 0; \text{ or} \tag{4}$$

$$\exists m \leq n(u_{n+1} = x_m \ \& \ y_m < 1) \ \& \ v_{n+1} \geq 0, \tag{5}$$

then define $(x_{n+1}, y_{n+1}) := (u_{n+1}, v_{n+1})$. Otherwise, let $(x_{n+1}, y_{n+1}) := (u_0, v_0)$. Obviously, $((x_n, y_n) : n \in \mathbb{N})$ is a computable sequence of rational points of $[0;1] \times \mathbb{R}$. We prove now by induction on n that the following hold for any $n \in \mathbb{N}$:

$$(x_n, y_n) \in \text{Epi}(f); \text{ and} \tag{6}$$

$$(u_n, v_n) \in \text{Epi}(f) \Longrightarrow (x_n, y_n) = (u_n, v_n). \tag{7}$$

For $n = 0$: It is true because $f(0) = 0$.

For $n \to n+1$: Suppose that (6) and (7) hold for all $m \leq n$. If (u_{n+1}, v_{n+1}) satisfies one of the conditions (3)–(5), then $(u_{n+1}, v_{n+1}) \in \text{Epi}(f)$ by the definition of f. Thus $(x_{n+1}, y_{n+1}) \in \text{Epi}(f)$ since $(x_{n+1}, y_{n+1}) = (u_{n+1}, v_{n+1})$. Otherwise, $(u_{n+1}, v_{n+1}) \notin \text{Epi}(f)$ and $(x_{n+1}, y_{n+1}) := (u_0, v_0)$. In both cases, (6) and (7) hold for $n+1$.

It follows from (6) and (7) that $((x_n, y_n) : n \in \mathbb{N})$ is a computable sequence of rational points which consists of exactly all rational points of Epi(f). To see that $\{(x_n, y_n) : n \in \mathbb{N}\}$ is dense in Epi(f), consider any point $(x, y) \in \text{Epi}(f)$ and any open 2-cuboid $(c_1; c_2) \times (d_1; d_2)$ which contains point (x, y), i.e., $c_1 < x < c_2 \ \& \ d_1 < y < d_2$. If $x \geq a$ or $y \geq 1$, then there are rational numbers u, v such that $x < u < c_2$ and $y < v < d_2$. Thus $(u, v) \in \text{Epi}(f) \cap (c_1; c_2) \times (d_1; d_2)$. If $x < a$ and $y < 1$, then $f(x) = 0$ because $f(x) \leq y < 1$ and rang$(f) = \{0, 1\}$. By the definition of f, there is $m \in \mathbb{N}$ such that $x = u_m$. Choose a rational number v such that $y < v < d_2$, then we have also $(u_m, v) \in \text{Epi}(f) \cap (c_1; c_2) \times (d_1; d_2)$. In both cases we have shown that there is a rational point which is contained in Epi$(f) \cap (c_1; c_2) \times (d_1; d_2)$. Because the sequence $((x_n, y_n) : n \in \mathbb{N})$ consists of all rational points of Epi(f), it follows that $\{(x_n, y_n) : n \in \mathbb{N}\} \cap (c_1; c_2) \times (d_1; d_2) \neq \emptyset$. This means that $\{(x_n, y_n) : n \in \mathbb{N}\}$ is a dense subset of Epi(f). By Theorem 1 and Definition 4, f is a 2-l.s.comp., hence 3-l.s.comp. function.

At last, we show that the integral $\int_0^1 f(x)dx = a$. Since $\forall x \in [0;1](f(x) \leq f_a(x)$ holds for the step function f_a, we have that $\int_0^1 f(x)dx \leq \int_0^1 f_a(x)dx = a$ at first. On the other hand, let $f_n : [0;1] \to \mathbb{R}$ be defined by $f_n(x) := \min\{f(x), f_{a_n}(x)\}$ for any $x \in [0;1]$ and $n \in \mathbb{N}$. Then $\forall x \in [0;1](f(x) \geq f_n(x))$ holds for all $n \in \mathbb{N}$ and there are only finitely many $x \in [0;1]$ which are among the rational numbers $u_0, u_1, \ldots u_{n-1}$ such that $f_n(x) = 0 \neq 1 = f_{a_n}(x)$. It follows that $\int_0^1 f(x)dx \geq \int_0^1 f_n(x)dx = \int_0^1 f_{a_n}(x)dx = a_n$. Hence $\int_0^1 f(x)dx \geq a$ because $\lim_{n\to\infty} a_n = a$. Therefore $\int_0^1 f(x)dx = a$, i.e., the integral $\int_0^1 f(x)dx$ is $\rho_<$-computable but not computable. This completes the proof of our theorem.

Corollary 15. *There is a 2-l.s.comp. function* $f : [0; 2] \to \mathbb{R}$ *such that the integral* $\int_0^2 f(x)dx$ *is neither* $\rho_<$*-computable nor* $\rho_>$*-computable.*

References

1. V. Brattka Computable Invariance. *COCOON'97*, Shanghai, China, August 1997.
2. X. Ge & A. Nerode Effective content of the calculus of variations I: semicontinuity and the chattering lemma. *Annals of Pure and Applied Logic*, 79(1996), no. 1–3, 127–146.
3. X. Ge Private Communication, 1997.
4. A. Grzegorczy On the definitions of recursive real continuous functions. *Fund. Math.* 44(1957), 61–71.
5. M. Pour-El & J. Richards *Computability in Analysis and Physics.* Springer-Verlag, Berlin, Heidelberg, 1989.
6. J. Ian Richards & Q. Zhou Computability of closed and open sets in Euclidean space, Preprint, 1992.
7. H. Jr. Rogers *Theory of Recursive Functions and Effective Computability* McGraw-Hill, Inc. New York, 1967.
8. K. Weihrauch *Computability.* EATCS Monographs on Theoretical Computer Secience Vol. 9, Springer-Verlag, Berlin, Heidelberg, 1987.
9. K. Weihrauch A foundation for computable analysis. in Douglas S. Bridges, Cristian S. Calude, Jeremy Gibbons, Steves Reeves and Ian H. Witten (editors), *Combinatorics, Complexity, Logic, Proceedings of DMTCS'96*, pp 66-89, Springer-Verlag, Sigapore, 1997.
10. K. Weihrauch & V. Brattka Computable Subsets of Euclidean Space. to appear in *Theoret. Comput. Sci.*
11. K. Weihrauch & X. Zheng Computability on Continuous, Lower Semi-Continuous and Upper Semi-Continuous Real Functions, to appear in *Theoret. Comput. Sci.* 211(1998). (Extended abstract in *Proc. of COCOON'97* Shanghai, China, August 1997, 166-175.)
12. K. Weihrauch & X. Zheng Effectivity of the Global Modulus of Continuity on Metric Spaces, *CTCS'97*, Santa Margherita Ligure, Italy, September 1997, 210–219.
13. K. Weihrauch & X. Zheng A Finite Hierarchy of Recursively enumerable real numbers, submitted.
14. Q. Zhou Computability of closed and open sets in Euclidean space, *Math. Log. Quart.* 42(1996), 379 - 409.

On the Power of Additive Combinatorial Search Model

Vladimir Grebinski

INRIA-Lorrainne, 615, rue du Jardin Botanique,
BP 101, 54602 Villers-lès-Nancy, France
E-mail: Vladimir.Grebinski@inria.fr

Abstract. We consider two generic problems of combinatorial search under the additive model. The first one is the problem of reconstructing bounded–weight vectors. We establish an optimal upper bound and observe that it unifies many known results for coin–weighing problems. The developed technique provides a basis for the graph reconstruction problem. Optimal upper bound is proved for the class of k–degenerate graphs.

1 Introduction

In many practical situations, one needs to obtain some information indirectly available through some physical device. Sometimes this implies costly or lengthy experiments so that the viability of the method crucially depends on the total number of them. Such problems are studied in the field of combinatorics called *combinatorial search.* We refer to monographs [2, 6] for a detailed account of modern methods and results in this area.

Informally, a general combinatorial search problem is described by three parameters: a universe of objects, a set of queries to the oracle and a set of possible answers. Objects are accessible only by the oracle. As every query to the oracle yields some information about the object, we repeat the process until we have enough information in order to uniquely identify the object. Our goal is to minimize the number of queries to the oracle.

One can distinguish two major classes of combinatorial search problems, namely the adaptive and non-adaptive ones. The latter class contains all algorithms which make all queries in advance, before any answer is known. In contrast, an adaptive algorithm takes into account outcomes of previous queries in order to form a next one. The non-adaptive algorithms form a subclass of adaptive ones and they are generally weaker. Surprisingly, in many cases non-adaptive algorithms achieve the power of adaptive ones. This will be the case for our problems.

In this paper we concentrate on two sets of objects. The first one is the set of d-bounded weight vectors $\Omega(n,d)$, which consists of all n–dimensional, non-negative integer–valued vectors of the total weight (sum of components) at most d. The second class is the set $\mathcal{G}_{n,k}$ of k-degenerate graphs on n vertices $v_1,\ldots,v_n$.

The definition of k-degenerate graphs is given below. Terms "d-bounded weight vector reconstruction problem" and "k-degenerate graph reconstruction problem" will refer to these two sets respectively.

The set of allowed queries and the set of oracle's answers are crucial for the complexity of the combinatorial search problem. For the set $\Omega(n, d)$, an allowed query is a subset $S \subset \{1, \ldots, n\}$ of vector positions. The answer to such a query S is the sum of entries corresponding to indices in S and will be denoted $\mu_{\boldsymbol{v}}(S)$. That is, if the unknown vector is $\boldsymbol{v} = (a_1, \ldots, a_n)$, then $\mu_{\boldsymbol{v}}(S) = \sum_{i \in S} a_i$. For $\mathcal{G}_{n,k}$, an allowed query is a subset of vertices $Q \subset \{v_1, \ldots, v_n\}$. For a graph $G = (V, E) \in \mathcal{G}_{n,k}$, the answer to the query $Q \subset V$ is the number of edges with both endpoints in Q, we denote $\mu_G(Q) = |(Q \times Q) \cap E|$. Such a choice of queries and answers corresponds to the *additive* or *quantitative model* of combinatorial search.

Historically, the additive model takes roots in a coin-weighing problem, posed by Södenberg and Shapiro in 1963 (see [2]). In this problem there is a finite number of coins, defective and authentic ones. The goal is to find the set of defective coins by possibly minimal number of weighings (or experiments). Each experiment consists in weighing an arbitrary subset of coins which reveals the number of defective ones. The problem was solved by B. Lindström [11], who gave an explicit optimal construction for the set of $\frac{2n}{\log_2 n}$ queries. A probabilistic proof can be found in [7]. This result was extended in several ways. In [10] Lindström obtained an explicit construction of a d-detecting matrix, which provides an optimal reconstruction algorithm for vectors with each entry bounded by d. This construction can be shown to be optimal for the class of non-adaptive algorithms (see [9]). Paper [9] studies the coin–weighing problem where the number of defective coins is bounded by a constant d_0. The upper bound of $4\frac{d_0}{\log d_0} \log n$ was established for the non-adaptive version of this problem. The naive information–theoretic lower bound for non-adaptive algorithms was improved in [3] to $2\frac{d_0}{\log d_0 - c} \log n$ for all $d_0 < n$ and some constant c. Again, this class of objects is a proper subclass of d_0–bounded weight vectors.

To introduce main results of this paper, we point out a connection between coin–weighing and vector reconstruction problems. Namely, assuming all counterfeit coins are heavier, we can associate with every coin its "degree of falsity", that is the difference between the coin weight and the weight of an authentic one. Our goal is to reconstruct the degree of falsity of every coin, i.e. the vector of coin overweights. A weighing of a subset of coins reveals the total overweight which is equal to the sum of corresponding entries of the coin overweights vector. This establishes correspondence between coin–weighing and vector reconstruction problems.

In the first part of this paper we extend previous results in the following direction: we show that an optimal algorithm exists for the problem when only the total overweight is known and the overweight of each individual coin is not bounded. Furthermore, the optimal upper bound can be achieved by a non-adaptive algorithm. This bound is of the same order as for the classical coin-weighing problem where degrees of falsity are restricted to $(0, 1)$ only. Thus, we gain a uniform viewpoint to all previously mentioned results.

In the second part of the paper, we apply the results for bounded–weight vectors to reconstruction of graphs. Reconstruction of graphs covers a broad class of combinatorial search problems. Note that the problem of graph reconstruction is different from that of *verifying* a graph property [2].

In [8, 9] optimal algorithms were proposed for some classes of graphs. For example, it was shown that d-bounded degree graphs have reconstruction complexity $O(dn)$ which can be reached by a non-adaptive algorithm. Another example is provided by general graphs, where the universe of objects is the set of all labeled graphs on n vertices. This class has complexity $O(\frac{n^2}{\log n})$ matched by a non-adaptive algorithm. The same problem was already considered in [1] in a slightly different setting.

While these results already cover many classes of graphs, they all assume some local restriction (except for the extremal case of the class of all graphs). In particular, the maximum degree of a vertex turns out to be the main parameter in complexity bounds. We get rid of this restriction, but require a graph to be k-degenerate (see Definition 2). We prove that for this graph reconstruction problem, the lower and upper bounds asymptotically coincide up to a multiplicative factor. Furthermore, this can be achieved by a non-adaptive algorithm.

Definitions and Conventions

The following notation will be used throughout the paper. We assume implicitly that all graphs are labeled and simple, i.e. without loops or multiple edges. The *weight* of a vector is the sum of its entries, $\mathbf{wt}(v) = \sum_{i=1}^{n} v_i$ if $v = \{v_1, \ldots, v_n\}$. The non-zero positions of a vector represent its *support*, $\mathbf{sp}(v) = \{i | v_i \neq 0\}$. All logarithms are natural unless the base is indicated. Finally, all considered matrices are $(0, 1)$-matrices over the ring of integers.

Throughout the paper we make several assumptions about the range of parameters. In the first part of the paper, we consider only n–dimensional vectors, whose weight is bounded by a $n^{1+\epsilon}$, for an $\epsilon > 0$. This choice excludes the range of values where a trivial construction can be applied. In the second part of this paper we consider only k–degenerate graphs with $k \leq n^{\alpha}$, with $\alpha < 1$, the choice is motivated by similar considerations.

2 Non-adaptive Vector Reconstruction Problem

In this section we give a lower and upper bounds for the complexity of reconstruction of bounded-weight vectors by a non-adaptive algorithm. Recall that a d-bounded weight vector is a vector $v = (v_1, \ldots, v_n)$, with non-negative integer components $v_i \in \{0\} \cup \mathbb{N}$ and $\sum v_i \leq d$. The set of all such vectors will be denoted by $\Omega(n, d)$ or Ω. An algorithm tries to reconstruct a vector from Ω by asking for a sum of entries with indices in a set $S \subset \{1, \ldots, n\}$ which it is free to choose. The complexity measure of the algorithm is the number of queries and will be denoted by $k(n, d)$.

2.1 Separating Matrices and Bounded Weight Vectors

The notion of separating matrix plays a central role in the study of non-adaptive algorithms for coin-weighing problems.

Definition 1. *A matrix $M \in (0,1)^{k\times n}$ with n columns and k rows is called* separating *for a set of vectors V iff the function $\boldsymbol{v} \to M \cdot \boldsymbol{v}$ is injective on V.*

The importance of this notion is due to the following simple observation:

Proposition 1. *Constructing a non-adaptive algorithm for a coin-weighing problem with n coins under the additive model is equivalent to constructing a separating matrix with n columns.*

Indeed, let V be the set of all possible input vectors. Each query can be represented as an incidence $(0,1)$–vector of the objects that are put in the query. Consider the matrix M, whose rows correspond to queries and columns to objects. A crucial observation is that the vector of answers for configuration $\boldsymbol{v}$ coincides with the vector $M \cdot \boldsymbol{v}$ (in the additive model). Since the algorithm must distinguish between different vectors $\boldsymbol{v}_1 \neq \boldsymbol{v}_2$ we have $M \cdot \boldsymbol{v}_1 \neq M \cdot \boldsymbol{v}_2$. Thus, M is a separating matrix for V. On the other hand, given a separating matrix M for a set of vectors V we obtain a non-adaptive algorithm, by treating rows of M as incidence vectors of queries. □

2.2 Lower Bound

In this section we obtain a lower bound using the second moment method [4]. This lower bound is the factor of two away from the upper bound which will be obtained later. The idea of the proof is to consider the set of all vectors of the weight d as a uniform probabilistic space. Then, an estimation of a certain variance will show that the image $M \cdot \boldsymbol{w}$ of at least a half of vectors $\boldsymbol{w} \in \Omega$ belong to a sphere of small radius if $M \in (0,1)^{k\times n}$ is a separating matrix for Ω. We then obtain an estimation of the dimension of the matrix.

Let $\Omega = \Omega(n,d) = \{(d_1,\ldots,d_n) | \sum_{i=1}^n d_i = d\}$ be a probabilistic space with uniform distribution (here we consider only vectors of weight exactly d.) The $\mathbb{P}[d_1 = i] = \binom{n+d-2-i}{n-2} / \binom{n+d-1}{n-1}$, and a simple calculation shows that $E[d_i] = \frac{d}{n}$ and $Var[d_i] = \frac{n-1}{n+1} \cdot \frac{(n+d)d}{n^2}$. Consider a random vector $\boldsymbol{w} = (d_1,\ldots,d_n) \in \Omega$, and let $\boldsymbol{v} = M \cdot \boldsymbol{w}$, where $\boldsymbol{v} = (v_1,\ldots,v_k)$. The first step is to estimate $Var[v_i]$. Suppose there are exactly m non-zero entries in i-th line of the matrix M. The symmetric structure of Ω imposes that $Var[v_i] = Var[d_{i_1} + \cdots + d_{i_m}] = Var[d_1 + \cdots + d_m]$. A direct calculation shows that

$$Var[d_1 + \cdots + d_m] = \frac{d(n+d)}{n^2(n+1)} \cdot m \cdot (n-m) \leq \frac{d(n+d)}{n^2(n+1)} \cdot \frac{n^2}{4} \tag{1}$$

Together with the linearity of expectation this gives:

$$E_{\boldsymbol{w}\in\Omega}\left[\sum_{i=1}^{k}(v_i - E[v_i])^2\right] \leq k\frac{d(n+d)}{4(n+1)} \tag{2}$$

From Markov inequality it follows that:

$$\mathbb{P}\left[\sum_{i=1}^{k}(v_i - E[v_i])^2 \le k\frac{d(n+d)}{2(n+1)}\right] \ge \frac{1}{2} \tag{3}$$

Hence, at least $\frac{1}{2}\binom{n+d-1}{n-1}$ vectors $\boldsymbol{v}$ belong to a k-dimensional sphere of radius $\sqrt{\frac{k \cdot d(n+d)}{2(n+1)}}$. The volume of k-dimensional sphere is known to be $\left(\frac{2c_1 \cdot R^2}{k}\right)^{k/2}$, for a constant c_1. Therefore, by volume argument,

$$\left(\frac{c_1 d(n+d)}{(n+1)}\right)^{k/2} \ge \frac{1}{2}\binom{n+d-1}{n-1} \tag{4}$$

From this we obtain:

$$k \ge 2\frac{\min(n-1,d)\log\left(1+\frac{\max(n-1,d)}{\min(n-1,d)}\right)}{\log d + \log(1+\frac{d}{n+1}) + \log c_1} \tag{5}$$

Considering two cases of $d < n-1$ and $d \ge n-1$ and taking into account that $d \le n^{1+\epsilon}$, we can further simplify the last expression and formulate the result in the following theorem:

Theorem 1. *There exists an absolute constant c, such that for all $n \to \infty$ and $d \le n^{1+\epsilon}$:*

$$k(n+1,d) \ge 2\frac{\min(n,d)\log\left(1+\frac{\max(n,d)}{\min(n,d)}\right)}{(1+2\epsilon)\log\min(n,d) + c} \tag{6}$$

2.3 Upper Bound for the Vector Reconstruction Problem

In this section we apply the probabilistic method [7, 4] to obtain an upper bound on the dimension of a separating matrix M for the set $\Omega(n,d)$ of d-bounded weight vectors. The general idea is to consider a set of "bad" events, defined by *critical pairs*, and estimate the probability for a uniformly drawn matrix that any of them takes place. When this probability is strictly below 1 there is a matrix where no "bad" events occurs. Thus, we will estimate the dimension of the matrix M.

For two different vectors $\boldsymbol{v}_1, \boldsymbol{v}_2 \in V$ and a matrix M, we define a characteristic function $\chi(\boldsymbol{v}_1, \boldsymbol{v}_2, M)$:

$$\chi(\boldsymbol{v}_1, \boldsymbol{v}_2, M) = \begin{cases} 1 \text{ if } M\boldsymbol{v}_1 = M\boldsymbol{v}_2, \\ 0 \text{ otherwise.} \end{cases}$$

For a matrix M which is *not* a separating matrix for Ω we can find two witness vectors $\boldsymbol{a} = (a_1, \ldots, a_n)$, $\boldsymbol{b} = (b_1, \ldots, b_n)$ that enjoy two additional properties:

1. $\mathbf{sp}(\boldsymbol{a}) \cap \mathbf{sp}(\boldsymbol{b}) = \emptyset$. Otherwise, consider $(\boldsymbol{a}', \boldsymbol{b}')$, where $\boldsymbol{a}' = (a'_1, \ldots, a'_n)$, $\boldsymbol{b}' = (b'_1, \ldots, b'_n)$, $a'_i = a_i - \min(a_i, b_i)$, $b'_i = b_i - \min(a_i, b_i)$. Obviously, $\mathbf{wt}(\boldsymbol{a}') \le \mathbf{wt}(\boldsymbol{a})$, $\mathbf{wt}(\boldsymbol{b}') \le \mathbf{wt}(\boldsymbol{b})$ and $M\boldsymbol{a}' = M\boldsymbol{b}'$ when $M\boldsymbol{a} = M\boldsymbol{b}$.

2. $\mathbf{wt}(\boldsymbol{a}) = \mathbf{wt}(\boldsymbol{b})$. This can be insured by adding to M an additional row with all entries equal to 1. We implicitly assume this row is always present in the matrix M.

An ordered pair of vectors $\boldsymbol{v}_1, \boldsymbol{v}_2 \in \Omega(n,d)$ satisfying the two properties above is said to be a *critical* pair. Let $\mathcal{C} = \mathcal{C}(\Omega)$ be the set of all critical pairs. We have

$$\mathbb{P}\left[M \text{ is not separating for } \Omega\right] = \mathbb{P}\left[\bigvee_{(\boldsymbol{v}_1,\boldsymbol{v}_2)\in\mathcal{C}(\Omega)} (\chi(\boldsymbol{v}_1,\boldsymbol{v}_2,M)=1)\right] \tag{7}$$

We estimate this probability from above:

$$\mathbb{P}\left[\bigvee_{(\boldsymbol{v}_1,\boldsymbol{v}_2)\in\mathcal{C}} (\chi(\boldsymbol{v}_1,\boldsymbol{v}_2,M)=1)\right] \leq \frac{1}{2}\sum_{(\boldsymbol{v}_1,\boldsymbol{v}_2)\in\mathcal{C}} \mathbb{P}\left[\chi(\boldsymbol{v}_1,\boldsymbol{v}_2,M)=1\right] \tag{8}$$

From now on we assume the uniform distribution over $k \times n$ matrices M, except for the implicit row of all 1's mentioned above. The idea of obtaining an upper bound is to find the smallest k which makes the above sum smaller than 1. The first step is to obtain an upper bound for $\mathbb{P}\left[\chi(\boldsymbol{v}_1,\boldsymbol{v}_2,M)=1\right]$.

Lemma 1. *Given a critical pair* $(\boldsymbol{v}_1,\boldsymbol{v}_2)$ *and* M *uniformly distributed over* $(0,1)^{k\times n}$

$$\mathbb{P}_M\left[\chi(\boldsymbol{v}_1,\boldsymbol{v}_2,M)=1\right] \leq \left(\frac{8}{9\cdot|\mathbf{sp}(\boldsymbol{v}_1)|}\right)^{k/4} \cdot \left(\frac{8}{9\cdot|\mathbf{sp}(\boldsymbol{v}_2)|}\right)^{k/4} \tag{9}$$

Proof. Let $\xi_1,\ldots,\xi_n$ be a set of independent random variables with $\mathbb{P}[\xi_i=0] = \mathbb{P}[\xi_i=1] = 1/2$. The event $M\boldsymbol{v}_1 = M\boldsymbol{v}_2$ is equivalent to k independent events corresponding to the equality in each row. Therefore,

$$\mathbb{P}\left[M\boldsymbol{v}_1 = M\boldsymbol{v}_2\right] = \mathbb{P}\left[\langle \boldsymbol{s},\boldsymbol{v}_1\rangle = \langle \boldsymbol{s},\boldsymbol{v}_2\rangle\right]^k, \tag{10}$$

where $\boldsymbol{s} = (\xi_1,\ldots,\xi_n)$, and $\langle \boldsymbol{s},\boldsymbol{v}_i\rangle$ is the inner product of $\boldsymbol{s}$ and $\boldsymbol{v}_i$. Since $\mathbf{sp}(\boldsymbol{v}_1)\cap\mathbf{sp}(\boldsymbol{v}_2)=\emptyset$, then $\langle \boldsymbol{s},\boldsymbol{v}_1\rangle$ and $\langle \boldsymbol{s},\boldsymbol{v}_2\rangle$ are independent and

$$\mathbb{P}\left[\langle \boldsymbol{s},\boldsymbol{v}_1\rangle = \langle \boldsymbol{s},\boldsymbol{v}_2\rangle\right] = \sum_i \mathbb{P}\left[\langle \boldsymbol{s},\boldsymbol{v}_1\rangle = i\right]\cdot\mathbb{P}\left[\langle \boldsymbol{s},\boldsymbol{v}_2\rangle = i\right] \leq \tag{11}$$

$$\sqrt{\sum_i \mathbb{P}\left[\langle \boldsymbol{s},\boldsymbol{v}_1\rangle = i\right]^2}\cdot\sqrt{\sum_i \mathbb{P}\left[\langle \boldsymbol{s},\boldsymbol{v}_2\rangle = i\right]^2} \tag{12}$$

The sum $\sum_i \mathbb{P}\left[\langle \boldsymbol{s},\boldsymbol{v}_j\rangle = i\right]^2$, $j=1,2$, can be bounded from above by $\max_i \mathbb{P}\left[\langle \boldsymbol{s},\boldsymbol{v}_j\rangle = i\right]$. Indeed, consider an arbitrary integer-valued random variable ξ and let $p_{max}(\xi) = \max_{i\in\mathbb{Z}} \mathbb{P}[\xi=i]$. Then $\sum_i \mathbb{P}[\xi=i]^2 \leq p_{max}(\xi)\sum_i \mathbb{P}[\xi=i] = p_{max}(\xi)$. Therefore, we can weaken (12) to

$$\mathbb{P}\left[\langle \boldsymbol{s},\boldsymbol{v}_1\rangle = \langle \boldsymbol{s},\boldsymbol{v}_2\rangle\right] \leq \sqrt{p_{max}(\langle \boldsymbol{s},\boldsymbol{v}_1\rangle)}\cdot\sqrt{p_{max}(\langle \boldsymbol{s},\boldsymbol{v}_2\rangle)} \tag{13}$$

To estimate $p_{max}(\langle \boldsymbol{s},\boldsymbol{v}\rangle)$ we need the following technical proposition.

Proposition 2. *Let t be a natural number, $a_1,\ldots,a_t > 0$, and $\xi_1,\ldots,\xi_t$ be independent random variables with* $\mathbb{P}[\xi_i=0] = \mathbb{P}[\xi_i=1] = 1/2$. *Then*

1. $2^{-t}\binom{t}{\lfloor t/2\rfloor} \le \sqrt{\frac{8}{9t}}$, *for all* $t \ge 1$,
2. $p_{max}(\xi_1 + \cdots + \xi_t) = 2^{-t}\binom{t}{\lfloor t/2\rfloor}$,
3. $p_{max}(a_1\xi_1 + \cdots + a_t\xi_t) \le p_{max}(\xi_1 + \cdots + \xi_t)$,

Proof. 1. For big t the inequality easily follows from Stirling formula. The constant was chosen to satisfy the inequality for *all* $t \ge 1$.
2. This is obvious since $\mathbb{P}[\xi_1 + \cdots + \xi_t = i] = 2^{-t}\binom{t}{i} \le 2^{-t}\binom{t}{\lfloor t/2\rfloor}$.
3. Let $P^{\max} = p_{max}(a_1\xi_1 + \cdots + a_t\xi_t)$. By definition, there is a value s such that $\mathbb{P}[a_1\xi_1 + \cdots + a_t\xi_t = s] = P^{\max}$. Consider the family $\mathcal{F} = \{A | \sum_{i\in A} a_i = s\}$. Clearly, $card(\mathcal{F}) \cdot 2^{-t} = P^{\max}$. Since $a_i > 0$, $\mathcal{F}$ is a Sperner family of sets, that is there is no two sets $A, B \in \mathcal{F}$ such that $A \subset B$. By Sperner's theorem [4], $card(\mathcal{F}) \le \binom{t}{\lfloor t/2\rfloor}$.

We return to the proof of Lemma 1. To bound $p_{max}(\langle \boldsymbol{s}, \boldsymbol{v}_j\rangle)$, $j = 1, 2$, we apply Proposition 2 with $t = \mathbf{sp}(\boldsymbol{v}_j)$. We have

$$p_{max}(\langle \boldsymbol{s}, \boldsymbol{v}_j\rangle) \le 2^{-t}\binom{t}{\lfloor t/2\rfloor} \le \left(\frac{8}{9\cdot|\mathbf{sp}(\boldsymbol{v}_j)|}\right)^{1/2}$$

By (10), (13), the Lemma follows. □

Let $C_w = \{(\boldsymbol{v}_1, \boldsymbol{v}_2) | \mathbf{wt}(\boldsymbol{v}_1) = \mathbf{wt}(\boldsymbol{v}_2) = w \text{ and } \mathbf{sp}(\boldsymbol{v}_1) \cap \mathbf{sp}(\boldsymbol{v}_2) = \emptyset\}$ and rewrite the right-hand side of (8) as

$$\sum_{(\boldsymbol{v}_1,\boldsymbol{v}_2)\in C} \mathbb{P}[\chi(\boldsymbol{v}_1, \boldsymbol{v}_2, M) = 1] = \sum_{w=1}^{d} \sum_{C_w} \mathbb{P}[\chi(\boldsymbol{v}_1, \boldsymbol{v}_2, M) = 1] \tag{14}$$

Using Lemma 1, we bound the inner sum for some fixed w.

$$\sum_{C_w} \mathbb{P}[\chi(\boldsymbol{v}_1, \boldsymbol{v}_2, M) = 1] \le \sum_{C_w} \left(\frac{8}{9\cdot|\mathbf{sp}(\boldsymbol{v}_1)|}\right)^{\frac{k}{4}} \cdot \left(\frac{8}{9\cdot|\mathbf{sp}(\boldsymbol{v}_2)|}\right)^{\frac{k}{4}} \le \tag{15}$$

$$\left(\sum_{\substack{\boldsymbol{v}_1,\\ \mathbf{wt}(\boldsymbol{v}_1)=w}} \left(\frac{8}{9\cdot|\mathbf{sp}(\boldsymbol{v}_1)|}\right)^{\frac{k}{4}}\right)^2 = \left(\sum_{s=1}^{w} \sum_{\substack{\boldsymbol{v}_1,\\ \mathbf{wt}(\boldsymbol{v}_1)=w,\\ |\mathbf{sp}(\boldsymbol{v}_1)|=s}} \left(\frac{8}{9s}\right)^{\frac{k}{4}}\right)^2 = \left(\sum_{s=1}^{w} \binom{n}{s}\binom{w-1}{s-1}\left(\frac{8}{9s}\right)^{\frac{k}{4}}\right)^2 \tag{16}$$

The last inequality is obtained by dropping the condition $\mathbf{sp}(\boldsymbol{v}_1) \cap \mathbf{sp}(\boldsymbol{v}_2) = \emptyset$. Next we used the fact that there are $\binom{n}{s}\binom{w-1}{s-1}$ vectors $\boldsymbol{v}_1$ of weight w with $|\mathbf{sp}(\boldsymbol{v}_1)| = s$, which follows from simple combinatorial considerations.

Now we are left with the technical problem of finding a possibly minimal k which makes (16) smaller than $\frac{2}{d}$. This will make (14) smaller than 1 and achieve our goal. Finding such k requires some routine calculations that we omit. The following proposition gives the final result.

Theorem 2. *There exist absolute constants* C_1, C_2, C_3 *such that for all* n, d *there exists a* $k \times n$ *separating matrix for the set of* d*-bounded weight vectors with* $k(n, d)$ *bounded as*

$$k(n, d) \le \frac{4\min(n, d)\log(C_1 \cdot \max(n, d)/\min(n, d))}{\log\min(n, d) + C_2} + C_3 \log d. \tag{17}$$

Comparing (17) with lower bound (6), we conclude that upper bound (17) is within the factor of $2(1 + 2\epsilon)$ from the lower bound provided that $d < n^{1+\epsilon}$ for our fixed parameter $\epsilon > 0$.

3 Non-adaptive Reconstruction of k-Degenerate Graphs

In this section we study the complexity of non-adaptive algorithms which reconstruct the class of k-degenerate graphs. This class of graphs is large enough to contain k-bounded degree graphs, sums of $k/2$ trees and other interesting structures.

Definition 2. *A graph $G = (V, E)$ is called* k-degenerate *if there exists an ordering of vertices $V = \{v_1, v_2, \ldots, v_n\}$ such that for every i we have $deg(v_i) \leq k$ in the subgraph induced by the vertices $\{v_i, v_{i+1}, \ldots, v_n\}$.*

The class of k-degenerate graphs on n vertices will be denoted $\mathcal{G}_{n,k}$. For example, every tree is 1-degenerate, planar graphs are 5-degenerate (see [5]). Note that our definition is equivalent to the one in [5]. We mention that k-degenerate graphs are $k+1$-colorable and have at most $n \cdot k - \binom{k+1}{2}$ edges. For other properties of k-degenerate graphs see [5].

Let $\mu_G(X)$ be the query function, i.e. the number of edges of the graph G with endpoints in X. The complexity $c(\mathcal{G})$ of graph reconstruction for a class of graphs $\mathcal{G}$ is the number of queries sufficient to uniquely identify every graph in $\mathcal{G}$.

Theorem 3. *For any constant $\alpha < 1$ there are two constants b_α and c_α such that for all $k \leq n^\alpha$*

$$b_\alpha \leq \frac{c(\mathcal{G}_{n,k})}{nk} \leq c_\alpha \tag{18}$$

We start the proof by establishing the lower bound. Next we reformulate our problem in terms of bipartite graphs and finally apply the techniques developed for bounded weight vectors.

Proof of the lower bound: To establish the information–theoretic lower bound we need to estimate from below the number $N(n, k)$ of k-degenerate graphs with n vertices. To obtain a k-degenerate graph with $m + 1$ vertices one can take a k-degenerate graph with m vertices and choose any k vertices to be adjacent to the new vertex v_{m+1}. Since this can be done in $\binom{m}{k}$ ways, we obtain the following estimation

$$N(n+1, k) \geq \prod_{i=k+1}^{n} \binom{i}{k} \geq \prod_{i=1}^{n} \left(\frac{i}{k}\right)^k = \frac{(n!)^k}{k^{nk}} \tag{19}$$

As it was mentioned above, the number of edges in a k-degenerate graph is at most $kn - k(k+1)/2$. From (19), our assumption $k \leq n^\alpha$ and asymptotic $n! \approx (n/e)^n$ we obtain the following information–theoretic lower bound:

$$\log_{k(n+1-\frac{k+1}{2})} N(n+1, k) \geq \log_{nk} \left(\frac{n}{ke}\right)^{nk} = \frac{nk(\log n - \log k - 1)}{\log n + \log k} \geq \frac{1-\alpha}{1+\alpha} nk + o(nk)$$

Therefore, we can set $b_\alpha = \frac{1-\alpha}{1+\alpha}$. □

Proof of the upper bound: In order to prove the upper bound, we reduce our problem to a problem of reconstructing a bipartite graph of special form. Specifically, we reduce the graph $G = (V, E)$ and query function $\mu(X)$ to a bipartite graph $G' = (V', V'', E')$ and a new query function μ'. Here G' is the bipartite representation of G, i.e. V' and V'' are copies of V, and there is an edge between $v' \in V'$ and $v'' \in V''$ iff $(v', v'') \in E$. The query function $\mu'(X, Y)$ for $X \subset V'$ and $Y \subset V''$ is *defined* to be $\mu'(X, Y) = |E' \cap (X \times Y)|$, the number of edges between X and Y.

Lemma 2. *One query $\mu'(\cdot, \cdot)$ can be evaluated by five queries $\mu(\cdot)$.*

Proof. In [9] it was shown that for arbitrary $X \subset V'$, $Y \subset V''$ one query μ' can be simulated by five queries μ:

$$\mu'(X,Y) = \mu((X \setminus Y) \cup (Y \setminus X)) - 2(\mu(X \setminus Y) + \mu(Y \setminus X)) + \mu(X) + \mu(Y). \quad \square$$

We are going to explicitly describe a family of queries $\mu'_{G'}(X_i, Y_i)$ that reconstruct G' uniquely provided that G' corresponds to a k-degenerate graph G as above. Let $\{Q_j\}_{j=1}^{m}$ be a family of sets corresponding to rows of a matrix that is separating for the set of k-bounded weight vectors. Theorem 2 states that $m = O(k \frac{\log n}{\log k})$ as $n \to \infty$. Recall that for a given k–bounded weight vector $v = (v_1, \ldots, v_n)$, values $s_j = \sum_{i \in Q_j} v_i$ uniquely define v. Let $\{P_i\}_{i=1}^{l}$ be a family of sets corresponding to rows of a matrix which is separating for the set of $(2nk)$–bounded weight vectors. Theorem 2 implies that $l = O(n \frac{\log k}{\log n})$.

Lemma 3. *Values $\{\mu'_{G'}(P_i, Q_j)\}_{i=1..l}^{j=1..m}$ uniquely identify graph G'.*

Proof. The proof relies on the following essential properties of reconstruction of bounded-weight vectors:

1. For fixed j, the value of $\mu'(\{v_r\}, Q_j)$ can be uniquely reconstructed for all $r = 1 \ldots n$. Indeed, $\sum_{r=1}^{n} \mu'(\{v_r\}, Q_j) \leq \sum_{r=1}^{n} \mu'(\{v_r\}, V'') = \mu'(V', V'') = 2\mu_G(V) \leq 2nk$. Consider a vector $w = (w_1, \ldots, w_n)$, where $w_r = \mu'(\{v_r\}, Q_j)$. By the choice of $\{P_i\}$, vector w is uniquely defined by values of the sum $\sum_{r \in P_i} w_r$ for $i = 1 \ldots l$, which are known, since by definition of μ', $\sum_{r \in P_i} w_r = \mu'(P_i, Q_j)$.
2. Fix an order on vertices of $V' = \{v_1, v_2, \ldots, v_n\}$, which is compatible with the definition of k-degenerate graph. Thus $\mu'(\{v_i\}, \{v_{i+1}, \ldots, v_n\}) \leq k$.
3. Consider a vertex $v_1 \in V'$ and vector $e = (e_1, \ldots, e_n)$, where $e_i = \mu'(\{v_1\}, \{v''_i\})$, the incidence vector of v_1 in G'. If one reconstructs e one will find all vertices adjacent to v_1. By Step 3, v_1 has at most k adjacent vertices in V'', so the values $\sum_{k \in Q_j} e_k = \mu'(\{v_1\}, Q_j)$ $(j = 1 \ldots m)$ uniquely define e by the property of $\{Q_j\}$. According to Step 3, the values $s_j = \mu'(\{v_1\}, Q_j)$ can be reconstructed for all $j = 1 \ldots m$, which proves that vector e can be reconstructed and all vertices adjacent to v_1 can be found.
4. To proceed to vertex v_2, we "exclude" vertex v_1 from graph G and update $\mu'(P_i, Q_j)$. This can be done without additional queries due to the additive nature of μ'. Namely, given an edge (v_1, w), we subtract 2 from $\mu'(P_i, Q_j)$

if both v_1 and w belong to P_i and Q_j, we subtract 1 if exactly one of v_1 or w belongs to P_i and the other to Q_j, and we do not change the value if $\{(v_1, w) \cup (w, v_1)\} \cap (P_i \times Q_j) = \emptyset$.
5. We repeat the process for $v_2, v_3, \dots v_{n-1}$.
6. It is possible that there are several orders on vertices compatible with the definition of k-degenerate graphs. The uniqueness of reconstruction follows from the fact that at the i-th step we reconstruct *exactly* those edges which are adjacent to v_i in the graph. This implies that different graphs have different values $\{\mu'(P_i, Q_j)\}$. □

The total number of queries μ' is $m \cdot l = O(nk)$. The reduction between μ' and μ gives a factor of 5, according to Lemma 2. Thus, Theorem 3 follows. □

4 Open Problems

A plausible conjecture is that the result of Theorem 3 holds for the graphs with a specified number of edges (i.e. $|E| = nk$), but we are unable to prove it with our technique.

5 Acknowledgments

I am grateful to Gregory Kucherov for numerous valuable discussions and remarks about these topics.

References

[1] Alok Aggarwal, Don Coppersmith, and Dan Kleitman. A generalized model for understanding evasiveness. *Information Processing Letters*, 30:205–208, 1989.
[2] Martin Aigner. *Combinatorial Search.* John Wiley & Sons, 1988.
[3] Noga Alon. Separating matrices. private communication, May 1997.
[4] Noga Alon and Joel Spencer. *The Probabilistic Method.* John Wiley & Sons, 1992.
[5] Béla Bollobás. *Extremal Graph Theory.* Academic Press, 1978.
[6] Ding-Zhu Du and Frank K. Hwang. *Combinatorial Group Testing and its applications*, volume 3 of *Series on applied mathematics.* World Scientific, 1993.
[7] Paul Erdős and Joel Spencer. *Probabilistic Methods in Combinatorics.* 1974.
[8] Vladimir Grebinski and Gregory Kucherov. Optimal query bounds for reconstructing a hamiltonian cycle in complete graphs. In *Proceedings of the 5th Israeli Symposium on Theory of Computing and Systems*, pages 166–173. IEEE Press, 1997.
[9] Vladimir Grebinski and Gregory Kucherov. Optimal reconstruction of graphs under the additive model. In Rainer Burkard and Gerhard Woeginger, editors, *Algorithms - ESA'97*, volume 1284 of *LNCS*, pages 246–258. Springer, 1997.
[10] Bernt Lindström. On Möbius functions and a problem in combinatorial number theory. *Canad. Math. Bull.*, 14(4):513–516, 1971.
[11] Bernt Lindström. Determining subsets by unramified experiments. In J.N. Srivastava, editor, *A Survey of Statistical Designs and Linear Models*, pages 407–418. North Holland, Amsterdam, 1975.

The Number of Rearrangements in a 3-stage Clos Network Using an Auxiliary Switch

Frank K. Hwang and Wen-Dar Lin

Department of Applied Mathematics,
National Chiao Tung University, HsinChu 30050,Taiwan, Republic of China
fhwang@math.nctu.edu.tw
wdlin@csie.nctu.edu.tw

Abstract. We consider the problem raised by Bassalygo: "What is the maximum number of rearrangements required by a rearrangeable 3-stage Clos network when there is an auxiliary middle switch carrying a light load?" For a 3-stage Clos network with an auxiliary middle switch carrying s connections, he claimed that the maximum number of rearrangements $\varphi_1(n,n,r;s)$ is less than $s+\sqrt{2s}+1$. In this paper, we give a lower bound $3\times\lfloor s/2\rfloor$ and an upper bound $2s+1$, where the lower bound shows that the upper bound given by Bassalygo does not hold in general.

1 Introduction

The 3-stage Clos network $v(n,m,r)$ is the most basic multistage interconnection network and has been widely studied. The first stage of $v(n,m,r)$ consists of r $m\times m$ crossbars, the second stage m $r\times r$ crossbars, and the third stage r $m\times m$ crossbars, and the connections between adjacent stages are complete bipartite graphs (see Fig. 1).

A *request* is a pair of input and output requesting a connection. The requests come in sequentially; once connected, they turn into *connections* which can be released any time. A network is *rearrangeable* if under the condition that existing connections are allowed to be rearranged (rerouted), then a request can always be routed, meaning there exist link-disjoint paths one for each connection (including the request). It is well known [2] that $v(n,n,r)$ is rearrangeable, while $v(n,n-1,r)$ is not.

Let $\varphi(n,n,r)$ denote the maximum number of rearrangements required to guarantee the routing of any request for any network states. So far, all rearrangement algorithms use Paull's matrix [3] (see Sec 2.) with all connections rearranged lying on one path in Paull's matrix. We call this a one-path algorithm and let $\varphi_1(n,n,r)$ denote the minimum number of rearrangements required under a *one-path algorithm.* Paull proved

$$\varphi_1(n,n,r)\le 2(r-1), \tag{1}$$

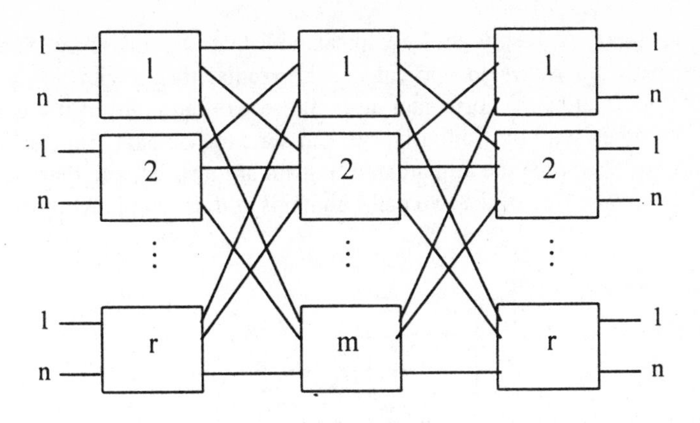

Fig. 1. $\nu(n,,m,r)$

and Benes [2] improved to $r-1$.

Bassalygo [1] considered the case that there exists a third middle switch c with a light load s. By taking advantage of the existence of c, he claimed

$$\varphi_1(n,n,r;s) \leq s+\sqrt{2s}+1. \tag{2}$$

He gave an outline of proof but no detail. He also commented that

$$\varphi(n,n,r;s) = s+1 \tag{3}$$

is probable. In this paper we give a lower bound

$$\varphi_1(n,n,r;s) \geq 3\times\lfloor s/2 \rfloor. \tag{4}$$

Note that

$$3\times\lfloor s/2 \rfloor \geq s+\sqrt{2s}+1 \text{ for } s \geq 16 \tag{5}$$

, thus invalidating the upper bound of Bassalygo. We also give an example that

$$\varphi(n,n,r;s) \geq s+2. \tag{6}$$

2 One-path algorithm

The Paull's matrix P is an $r\times r$ matrix, where rows are indexed by input switches I_i, columns by output switches O_j, and cell p_{ij} by the set of middle switches each carrying a connection (I_i,O_j). By convention, the request to be routed is (I_i,O_j) and we may assume that there exist two middle switches a and b, one appearing in the first row and the other the first column, but not both. Paull's method is to switch the connection carried by a in the first row to be carried by b; then the request can be routed through a. But switching that a to b means that if there is already a b in the same column as that a, then we must switch that b to a. Again, we need to check whether there is already an a colinear with that b, and so on. Such a path, called an ab-path, alternates in vertical and horizontal turns, with an a at every vertical turn and b at every horizontal turn, and stops if and only if it hits a line not containing the other symbol. Similarly, we can start the ba-path from the b in the first column (Fig. 2). Since the ab-path and ba-path are disjoint, and there are at most $2r-2$ a's and b's. One of the two paths contains at most $r-1$ symbols.

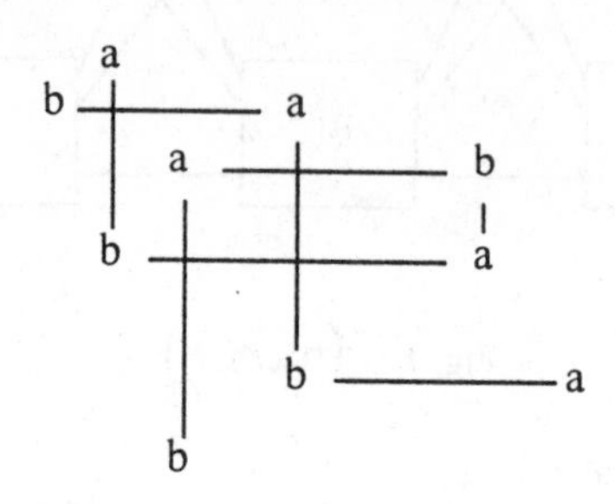

Fig. 2. The ab-path and the ba-path

Bassalygo's idea is to alter an ab-path or a ba-path at some point to an ac-path or a bc-path, and use the scarcity of c to show the existence of a shorter path. He gave the following rules for an one-path algorithm:

1. A path must start with an ab-path or a ba-path.
2. Once it turns into an ac-path or a bc-path, it cannot turn back to ab-path or ba-path.
3. When the first c appears in a path, the line containing two letters preceding it does not contain letter c.
4. Every matrix letter is visited at most once.
5. A path stops if and only if either it hits a line not containing the other symbol, or an ab-path (or ba-path) hits a symbol whose row and column contain no c.

Definition: We call a path a *rerouting path* if the path follows the rules given by Bassalygo. And a rerouting path can be divided into two parts, where the first part is an ab-path(or ba-path) and the second part is an ac-path(or bc-path). Note that a rerouting path can have null second part.

For example, there are three rerouting paths in the Paull's matrix of Fig. 3. They are $ababab$, $abcb$, and $babababababab$. Note that the path $acaca$ starting from cell $(1,12)$ is not a rerouting path, because by definition a rerouting path must start with an ab-path or a ba-path. And the a's in cell $(9,4)$ and $(10,5)$ can't turn to an ac-path because it will revisit an a already on the path (violating rule 4). After we choose a rerouting path, we can reroute the existing connections lying on the rerouting path, the scheme is to exchange a's and b's in the first part, which then starts an exchange of c's and the other letter of the second part. If a rerouting path has null second part, which means the first part (an ab-path or ba-path) hits a symbol whose row and column contain no c, we change the last letter to a c.

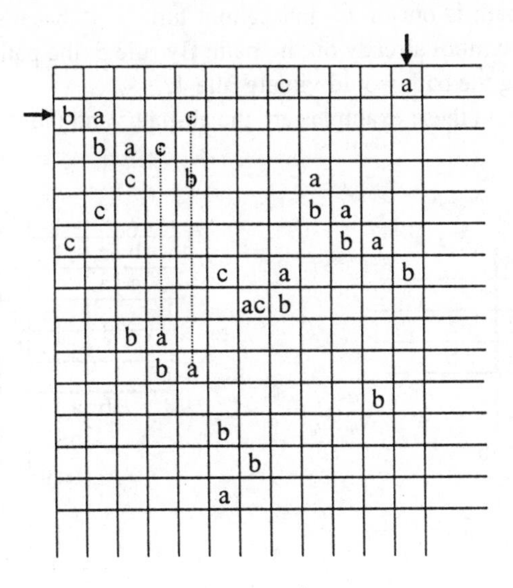

Fig. 3. A Paull's matrix

3 The lower bound and upper bound

Before giving the construction for $\varphi_1(n,n,r;s) \geq 3 \times \lfloor s/2 \rfloor$, we define the function φ_1 more clearly.

Definition: A *network state* is a legitimate arrangement of a Paull's matrix (each row and each column has at most n entries, all disjoint).

Definition: $\phi(x) = \min(\{\text{lengths of rerouting paths of network state } x\})$

Definition: $\varphi_1(n,n,r;s) = \max_{x \in N(n,n,r;s)}(\phi(x))$, $N(n,n,r;s)$ denotes the set of all states of $v(n,n,r)$ where the minimum-load middle switch carries s connections.

Now we give the serial construction that $\varphi_1(n,n,r;2k) \geq 3k$. The general pattern of construction should be clear from the examples given in Fig. 4~8.

Three points should be noted for understanding these examples:

1. We have permuted the rows and columns such that the first s rows and the first s columns all contain c's. The heavy lines denote this $s \times s$ matrix C. By rule 3, a rerouting path can turn to c only when it is out of C.
2. Sometimes, the path is out of C but cannot turn to c because the c lies in the same line with a symbol already on the path. By rule 5, the path cannot stop at this c. But continuing the path would violate rule 4.
3. The shortest paths in these examples are the ab-paths and the ba-paths.

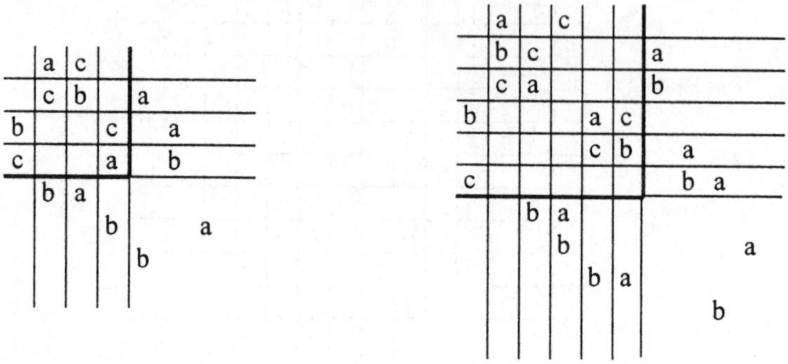

Fig. 4. $k = 2$ and $k = 3$

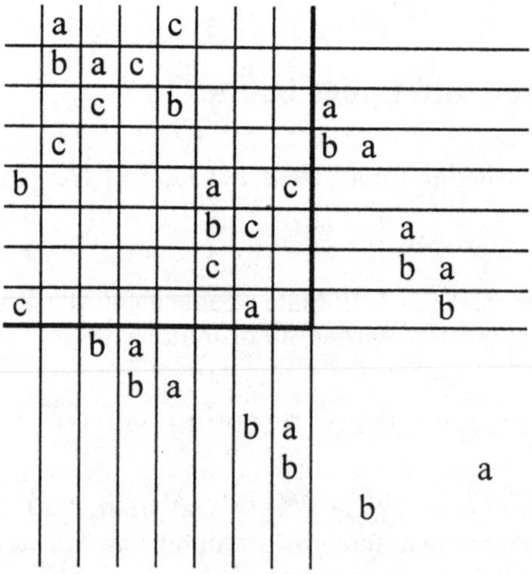

Fig. 5. $k = 4$

Fig. 6. $k = 5$

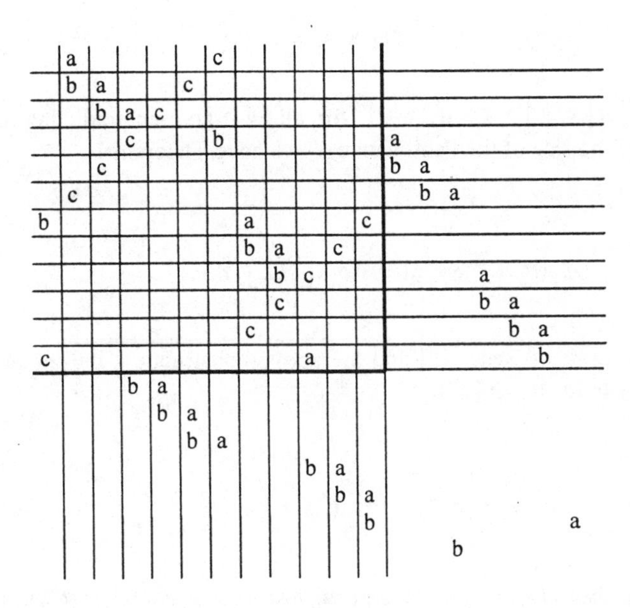

Fig. 7. $k = 6$

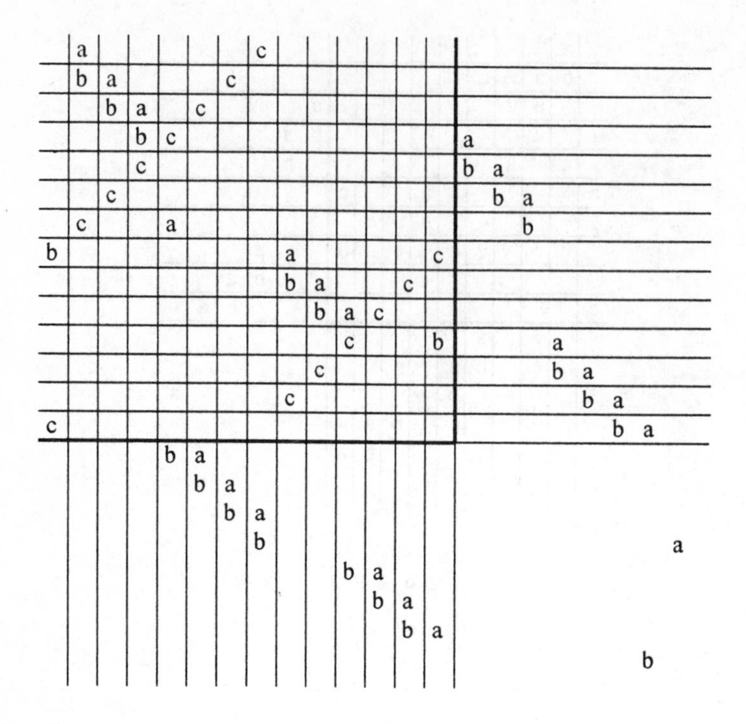

Fig. 8. $k = 7$

Note that $3 \times \lfloor s/2 \rfloor \geq s + \sqrt{2s} + 1$ for all $s \geq 16$, contradicting Bassalygo's claim. Consider the size of the Paull's matrix, we have Theorem 1.

Theorem 1:

$$\varphi_1(n,n,r;s) \geq \min(3 \times \lfloor s/2 \rfloor, r-1) \tag{7}$$

Proof: By the result of Benes [2], and the above construction, the inequality holds, and the equality holds if $3 \times \lfloor s/2 \rfloor \geq r-1$.

Theorem 2:

$$\varphi_1(n,n,r;s) \leq \min(2s+1, r-1) \tag{8}$$

Proof: Assume that (I_i, O_j) is the next request, row i contains a but not b, and column j contains b but not a. Define $D = (\text{first } s \text{ rows}) \cup (\text{first } s \text{ columns})$. Then we can find two rerouting paths with null second parts and starting with the above a and b respectively. Since there are at most $4s$ a's and b's in D, one of the two paths must get out of D in $2s+1$ steps. Then we can change the last letter to c and interchange a's and b's in the path. Thus the request can be routed in $2s+1$ rearrangements.

4 Two-path algorithm

We may also route a request by a *two-path algorithm*. For example, we can route the request (I_1, O_1) for the network state in Fig. 6 under a one-path algorithm, which requires $3 \times \lfloor 10/2 \rfloor = 15$ rearrangements if $r \geq 16$. But we can rearrange the connections lying on the two paths as shown in Fig. 9, and then route the request through c after only 4 rearrangements.

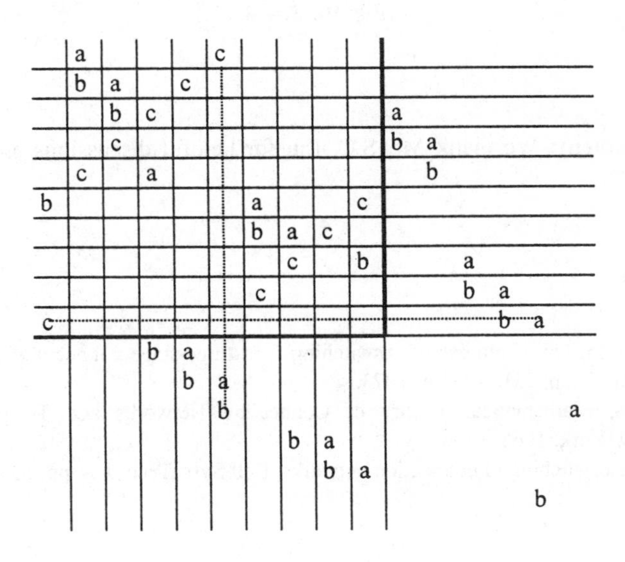

Fig. 9.

However, the number of rearrangements under a two-path algorithm is not always fewer than $\varphi_1(n,n,r;s)$. For example, we can route the request (I_1, O_1) for the network state in Fig. 4 by rearranging the connections lying on the two paths ($cbcb$ and $caca$) as in Fig. 10, but 8 rearrangements are needed. Clearly, the request can be routed after 6 rearrangements under a one-path algorithm. Moreover, this is a counterexample for $\varphi(n,n,r;s) = s+1$.

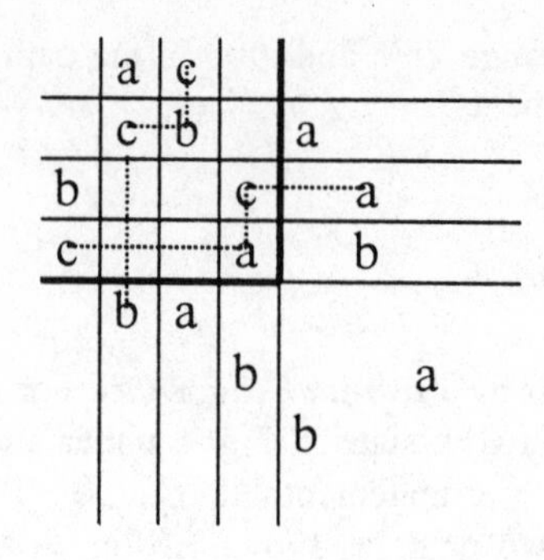

Fig. 10. $s = 4$

Acknowledgement: We thank Mr. S.C. Liu for helpful discussions and lament his untimely death.

References

1. L.A. Bassalygo, On a number of reswitching in a three-stage connecting network, Inter. Teletraffic Cong. 7, pp. 231/1-231/4, 1973.
2. V.E. Benes, Mathematical Theory of Connecting Networks and Telephone Traffic, Academic, New York, 1965.
3. M.C. Paull, Reswitching of connection networks, Bell Syst. Tech. J. 4, pp. 833-855, 1962.

Lower Bounds for Wide-Sense Non-blocking Clos Network

Kuo-Hui Tsai and Da-Wei Wang

Institute of Information Science, Academia Sinica, Nankang, Taipei 11529, Taiwan, ROC.

Abstract. The *3-stage* Clos network is generally considered the most basic multistage interconnecting network(MIN). The nonblocking property of such network has been extensively studied in the past. However there are only a few lower bound results regarding wide-sense nonblocking. We show that in the classical circuit switching environment, for large r to guarantee wide-sense nonblocking, $2n-1$ center switches are necessary where r is the number of input switches and n is the number of inlets of each input switch. For the multirate environment, we show that for large enough r any 3-stage Clos network needs at least $3n-2$ center switches to guarantee wide-sense nonblocking. Our proof works for even 2-rate environment.

1 Introduction

The *3-stage* Clos network is generally considered the most basic multistage interconnecting network(MIN). It is symmetric with respect to the center stage. The first stage, or the ***input stage***, has r $n \times m$ crossbar switches; the center stage has m $r \times r$ crossbar switches. The n inlets (outlets) on each input (output) switch are the ***inputs (outputs)*** of the network. There are exactly one link between every center switch and every input (output) switch. We use $C(n, m, r)$ to denote a 3-stage Clos network. An example of $C(3, 3, 4)$ is shown in figure 1.

In classical circuit switching, i.e. every link can only serve one connection request, three types of nonblocking properties have been extensively studied , they are ***strictly nonblocking***, ***wide-sense nonblocking*** and ***rearrangeably nonblocking***. The focus of this paper is to establish lower bounds for the number of center switches needed to guarantee wide-sense nonblocking. A network is wide-sense nonblocking (WSNB) if a new call is always routeable as long as all previous requests were routed according to a given routing algorithm.

In multirate environments, a request is a triple (u, v, w) where u is an inlet, v an outlet and w a weight which can be thought as the bandwidth requirement (rate) of that request. We normalize the weights so that $1 \geq w > 0$, and each link has capacity one; i.e., it can carry any number of calls as long as the sum of weights of these calls does not exceed one.

Clos [3] proved that for the classical model $2n-1$ center switches are necessary and sufficient to guarantee SNB for $C(n, m, r)$. Benes [1, 2] proved that $C(n, m, 2)$ is WSNB (using the packing routing) if and only if $m \geq 3n/2$, thus

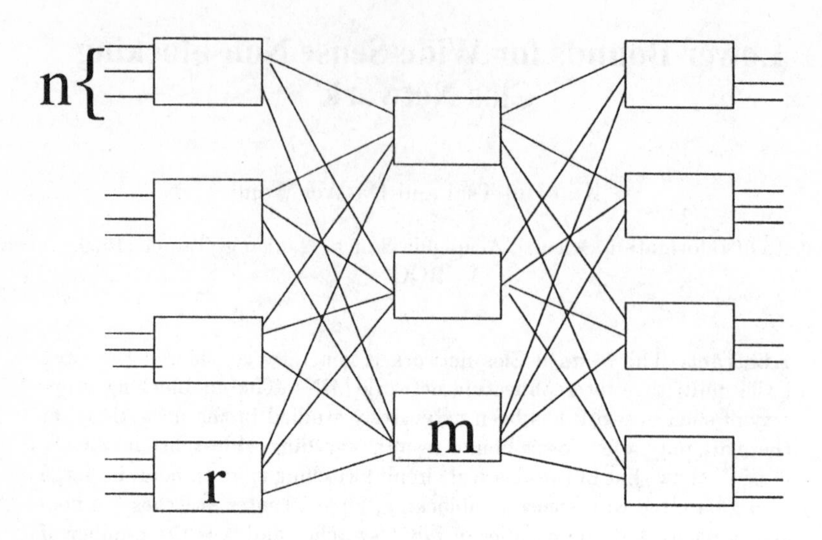

Fig. 1. C(3,3,4)

giving hope that WSNB can be achieved with fewer center switches than SNB in general. However recently, Du, Gao, Fishburn and Hwang [5] gave the surprising result that $C(n,m,r)$ for $r \geq 3$ is WSNB under the packing routing if and only if $m \geq 2n-1$; namely, it requires the same number of center switches as SNB. In this paper we further dash the hope by showing that for large r, $C(n,m,r)$ is WSNB (under any routing algorithm) if and only if $m \geq 2n-1$. While WSNB, as commented above, plays a very restrictive role in the classical model, the multirate environment provides a fertile playground. This is because we have a new dimension, the *rate*, to design routing algorithm. For example, Gao and Hwang [6]gave a routing algorithm such that $C(n,m,r)$ is WSNB if $m \geq 5.75n$. If there are only two different rates, then the requirement reduced to $4n$. In this paper we shown that $3n-2$ is a lower bound of WSNB under any routing algorithm, and this bound is obtained by using only two different rates. This lower bound provides a gauge to measure how good the algorithm of Gao and Hwangis , and how much room to improve. We will also talk about some impact on repackable algorithms, a relatively new type of nonblocking.

2 Main Results

We first study the classical model, where every link can serve only one connection request.

Theorem 1. *For* $r \geq (n-1)\binom{2n-2}{n-1}+2$, $C(n,m,r)$ *is WSNB if and only if* $m \geq 2n-1$.

Proof. Since SNB implies WSNB, it suffices to prove the "only if" part. Suppose $r = (n-1)\binom{2n-2}{n-1} + 2$ and $m = 2n-2$. Let the network contains a set of connections involving $n-1$ inputs from each input switches but does not involve the output switch O (easily verified to be a feasible state). Consider the $\binom{2n-2}{n-1}$ distinct subsets of $n-1$ center switches. The $n-1$ connections from each input switch are routed by one such subset. For the given r there must exist a subset Y which routes a set X of n input switches. Consider a new set of n requests $\{(x,o) : x \in X, o \in O\}$. Each of the n requests must be routed through a distinct center switch, which is not in Y. Hence at least

$$|X| + |Y| = 2n - 1$$

center switches are needed. (See figure 2) □

Next we consider the multirate model where each request has a rate requiremeet . We normalize the weights to be a number between zero and one, and each link has capacity one.

Theorem 2. *$C(n,m,r)$ is not WSNB for any two rates B, b satisfying $B+b>1$ if $m \leq \min k, n+2n-3$ and r is large enough, where $k = \lfloor 1/b \rfloor$.*

Proof. Without loss of generality, we may assume that $B = 1$ and $b = 1/k$ for some integer $k \geq 2$.
In phase 1, *all requests have weight b and come from the first inlet of each input switch* (output will be specified later if necessary).
Step 1, Each input generates one request. Since r is large enough, there exists a large set I_1 of input switches whose requests are all routed through the same middle switch M_1.
Step 2, Each input in I_1 generates a second request. Partition I_1 into size $k+1$ groups such that inputs in the same group all make the second request to the same output, this is possible since $k+1 \leq nk$, the capacity of an output switch. But the total weights of each group, $b(k+1)$, require at least a new center switch M_2 to carry. Therefore in each group at least one input whose second request is carried by some center switch other than M_1. Since I_1 is large there exists a large set I_2 of input switches whose requests are routed through the same set of middle switches M_1, M_2.
Step ℓ, Each input in $I_{\ell-1}$ generates a ℓth request. Partition $I_{\ell-1}$ into size $(\ell-1)k+1$ groups such that inputs in the same group all make the ℓth request to the same output, this is possible if $(\ell-1)k+1 \leq nk$. But the total weights of each group, $(\ell-1)bk+b$, is greater than the capacity of the $\ell-1$ center switches used before. Therefore in each group at least one input whose ℓ^{th} request is carried by some center switch other than $M_1, M_2, \ldots, M_{\ell-1}$. Since $I_{\ell-1}$ is large there exists a large set I_ℓ of input switches whose requests are routed through the same set of middle switches. (See figure 3)
Step k, Each input in I_{l-1} generates a kth request. Partition I_{k-1} into size $(k-1)k+1$ groups such that inputs in the same group all make the kth request

to the same output, this is possible if $(k-1)k+1 \leq nk$. But the total weights of each group, $(k-1)bk+b$, is greater than the capacity of the $k-1$ center switches used before. Therefore in each group at least one input whose k^{th} request is carried by some center switch other than $M_1, M_2, \ldots, M_{k-1}$. Since I_{k-1} is large there exists a large set I_k of input switches whose requests are routed through the same set of middle switches. Let M denotes the set of common center switches used by I_k.

In phase 2, Consider only the input switches in I_k. Let each of them generate $n-1$ weight B requests quests going to a set of clean output switches. Since $B+b>1$ none of these requests can be routed through M. By theorem 1, another set of $2n-2$ center switches is needed. Hence the total of $k+2n-3$ middle switches is necessary.

Corollary 3. *For $k \geq n$, then $m \geq 3n-3$ is impossible for WSNB when r is large.*

3 Conclusion

A network is *repackable* if existing calls can be rearranged at any moment a connection is deleted (e.g., a call hands up). Repacking algorithms have been studied for both the classical model [7] and the multirate model [8, 9] to show that they can help to reduce the number of center switches needed for $C(n,m,r)$ to be nonblocking. Theorem 1 and 2 showed that no repacking algorithm can be effective when r is large. since the request sequences we construct in these theorems have no deletion, they apply to all repacking algorithms. Thus our results solidly confirm a major difference between WSNB, Repackable and SNB, RNB, namely the numbers of center switches required for $C(n,m,r)$ are independent from r in the latter case. In other words, the hope of finding an effective WSNB or repackable algorithm is restricted to small r.

References

1. V.E. Benes, Mathematical Theory of Connecting Networks and Telephone Traffic. Academic Press, New York, 1965.
2. V.E. Benes, Blocking in the NAIU network, Bell Laboratories Tech. Memo. 1985
3. C. Clos, A study of nonblocking switching networks, Bell Syst. Tech. J., 32(1953) pp. 406–424.
4. D.Z. Du, B. Gao, F.K. Hwang, J.H. Kim, On rearrangeable multirate Clos networks. To appear in SIAM J. Comput.,
5. D.Z. Du, P.C. Fishburn, B. Gao, F.K.Hwang, Wide-sense nonblocking for 3-stage Clos networks, preprint.
6. B. Gao and Frank Hwang, Wide-sense Nonblocking for Multirate 3-stage Clos Networks, Theoretical Computer Science, 182(1997) pp. 171–182.

7. A. Jajszczyk, A new concept–repackable networks, IEEE Trans. Communications 41(1993), pp. 1232–1236.
8. F.K. Liptopoulos and S. Lhalaasani, Semi–rearrangably nonblocking operation of Clos networks in the multirate environment, To appear in IEEE Trans. Networking.
9. S. Ohta. A simple control algorithm for rearrangable switching networks with time division multiplexed links, IEEE J. Selected Areas of Comm., 5(1987), pp. 1302–1308.

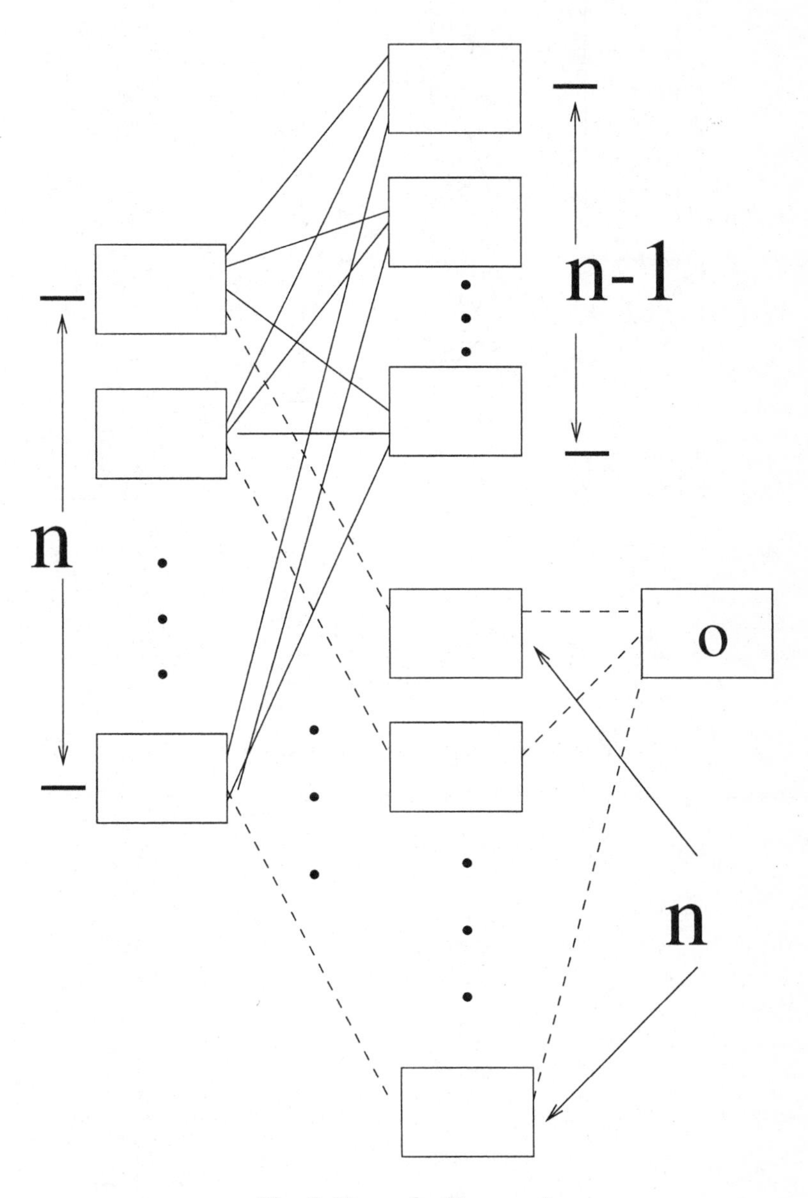

Fig. 2. Figure for theorem 1

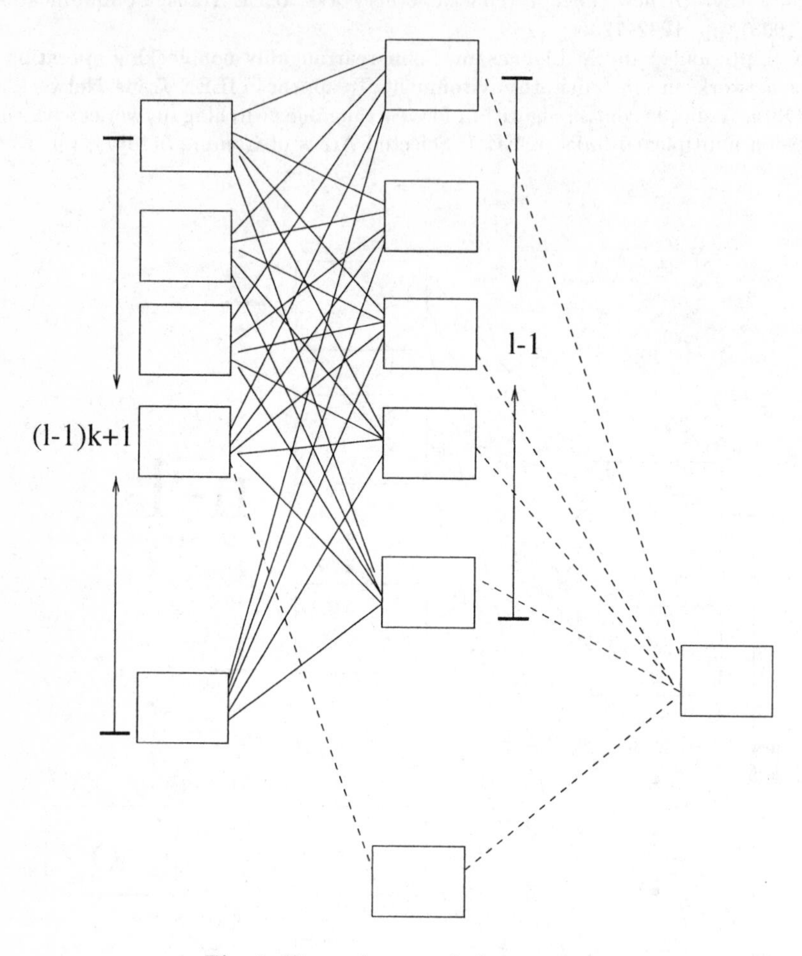

Fig. 3. Figure for step ℓ of theorem 2

Multirate Multicast Switching Networks

Dongsoo S. Kim[1] and Ding-Zhu Du[1,2,3]

[1] Department of Computer Science and Engineering, University of Minnesota
{dkim,dzd}@cs.umn.edu
[2] Institute of Applied Mathematics, Chinese Academy of Sciences
[3] Department of Computer Science, City University of Hong Kong

Abstract. This paper investigates multirate multicast Clos switching network which is nonblocking in a wide sense, where a compatible multicast request is guaranteed to be routed without disturbing the existing network if all requests have conformed to a given routing scheme. The routing strategy discovers $(2.875n - 1)\min(k + r^{1/k}) + 1$ middle switches sufficient for any multirate multicast requests, whereas strictly nonblocking multirate switching networks requires infinite number of middle switches if the range of weights can be widely distributed.
This paper also shows that Yang and Masson's nonblocking multicast Clos network for pure circuit switching is rearrangeable for multirate multicast communication if each weight is chosen from a given finite set of integer multiplicity. Note that a general rearrangeability of multirate Clos network even for point-to-point communications has not been known yet. In our work, the number of middle switches only depends on the configuration of the switch itself but not on the patterns of connection requests, which is critically advisable to construct large scale switching networks.

1 Introduction

A Clos network has been widely employed for telephone switching systems instead of crossbar switches because of its asymptotic advantage over the crossbars while providing nonblocking behavior [3]. The Clos switching network was first developed for a pure circuit switching, in which a connection request should set up a physical path between its source and destination, and the path should be dedicated to the single service during the entire conversation so that all links on the path cannot be used for any other communications.

Advances in digital technology have allowed several telephones and data terminals to share a single link if the total load on the link does not exceed its capacity [2, 6, 7]. In this *multirate* environment, each request takes a portion of a link bandwidth. A spectrum of network services have been introduced during the past decade, which need different connection characteristics. In those services, it is necessary to transmit their information to multiple destinations. To provide flexible communication environments and to obtain cost-effectiveness, there have been numerous efforts to integrate the dissimilar network services into a single network. The advent of asynchronous transfer mode(ATM) has further forced to combine the heterogeous communication networks. The current implementation of multicast switching devices, however, are based on blocking networks because nonblocking multicast switching networks were known to require unrealistic hardware complexities.

In this paper, we study routing schemes for multirate multicast Clos networks to obtain the minimum hardware complexity. The numbers of middle stage switches in the Clos networks depend only on the parameters of the network itself but not on the pattern of connection requests, which is highly desirable for constructing a real switching network.

2 Multistage Switching Networks

An interconnection network is represented as a directed weight graph $G = (V, E)$. $V_I \subset V$ is a set of external inlets, each of which has one outgoing edge and no incoming edge. $V_O \subset V$ is a set of external outlets, each of which has one incoming edge and no outgoing edge. For an n-stage network, V_I and V_O are said to be in stage 0 and $n+1$, respectively. Nodes in stage i have directed edges only to nodes in stage $i+1$ for $0 \leq i \leq n$, and there exists only one edge between any pair of nodes. We construct uniform 3-stage Clos networks, denoted as $C(n_1, r_1, n_2, r_2, m)$, forcing that each node in stage 1 (input stage) has n_1 incoming edges and m outgoing edges to every node in stage 2 (middle stage), in which each node has r_1 incoming edges and r_2 outgoing edges to every node in stage 3 (output stage). Each node in stage 3 has m incoming edges and n_2 outgoing edges. Each switch module is assumed to be a crossbar switch which has nonblocking multicast capability, even though it can be recursively constructed with smaller Clos subnetworks. A symmetric Clos network, denoted as $C(n, r, m)$, is induced from the asymmetric network with $n_1 = n_2$ and $r_1 = r_2$.

A *connection request* for point-to-point connection is a triple (x, y, ω), where $x \in V_I$, $y \in V_O$ and ω is a normalized bandwidth requirement (weight) of the connection request. A set of connection requests are said to be *compatible* if the sum of all weights passing each external link does not exceed its capacity. A connection request is compatible to the existing network if adding the request does not cause capacity overflows for any external links. A multicast connection request is defined as a triple (x, Y, ω), where $Y \subseteq V_O$ denotes a set of output ports. A point-to-point request can easily be represented with a multicast notation by imposing $|Y| = 1$. A *route* is a tree connecting an input port to a set of output ports through middle stage switches. A *configuration* is a set of all routes. A configuration is said to be *satisfied* if we can find all routes in such a way that the sum of weights on each $e \in E$ is not larger than its capacity.

A switching network is *strictly nonblocking*, or simply nonblocking, if a new connection request can be routed without disturbing the existing network no matter how the previous calls were routed. It is well-known that a symmetric Clos network, $C(n, r, 2n-1)$, is strictly nonblocking in circuit switching point-to-point communications [3]. Melen and Turner [6] found a sufficient condition of a nonblocking Clos network for multirate point-to-point communications. They took an advantage of higher speed of internal links over externals to reduce hardware requirements. Chung and Ross [2] determined necessary and sufficient conditions for multirate interconnection networks to be nonblocking for both discrete and continuous weights, especially when the external speed is same as the internal one.

A network is said to be *nonblocking in a wide-sense* if a new request can be satisfied without interfering the existing network configuration under a condition that all requests comply to a given routing algorithm. Beneš [1] proved a Clos network $C(n, 2, m)$ is wide-sense nonblocking in circuit switching if $m = \lfloor 3n/2 \rfloor$. Melen and Turner [6] devised a Clos network $C(n, r, 8n-2)$ to be wide-sense nonblocking in a multirate environment by assembling two nonblocking Clos networks in parallel, each of them has $4n-1$ middles stage switches. All connections with weights more than $1/2$ are routed through only one of subnetworks, and all requests with weights no more than $1/2$ are routed through the other subnetwork.

Hwang [5] had given conditions $m = O(nr)$ of rearrangeable multi-connection 3-stage Clos networks, which is a generalization of interconnection networks such that a set of input ports are able to connect to a set of output ports. Yang and Masson [9] showed that a Clos network in circuit switching is nonblocking for multicast requests when $m > \min(n-1)(k + r^{1/k}) = O(n \ln r / \ln \ln r)$, where $1 \leq k \leq \min(n-1, r)$. Yang [8] extended her previous result to obtain multirate nonblocking multicast networks with $m > \min(\lfloor 1/b \rfloor (n-1)(k + r^{1/k}))$. A weakness of the result is that the network requires an unbounded number of middle stage switches when b goes to 0.

3 Preliminaries

We can denote a set of destinations as a set of output switches instead of output ports because an output switch module can fan out to as many as outlets once it receives a compatible request. We use a vector notation as a set of destinations in such a way that the j-th element of a vector is 1 if the destinations contain output switch j, 0 otherwise. A multirate multicast request is denoted as $C_i = (x_i, \boldsymbol{y}_i, \omega_i)$, where $x_i \in \{1, \ldots, nr\}$ is an input port, $\boldsymbol{y}_i$ is a vector of size r denoting a set of output switches in a bit-vector format, and ω_i is a required weight. For the connection request, we can define a *connection vector* as $\boldsymbol{I}_i = \omega_i \cdot \boldsymbol{y}_i$. Figure 1 shows four multicast connection requests $C_1 = (2, (1, 1, 1, 1), .2)$, $C_2 = (6, (0, 0, 1, 0), .3)$, $C_3 = (8, (1, 1, 1, 1), .4)$ and $C_4 = (11, (0, 1, 0, 0), .5)$. Their connection vectors are $\boldsymbol{I}_1 = (.2, .2, .2, .2)$, $\boldsymbol{I}_2 = (0, 0, .3, 0)$, $\boldsymbol{I}_3 = (.4, .4, .4, .4)$ and $\boldsymbol{I}_4 = (0, .5, 0, 0)$.

Definition 1 (relational operators of vector). *For given two vectors, $\boldsymbol{x}$ and $\boldsymbol{y}$ of length n, $\boldsymbol{x} \leq \boldsymbol{y}$ if and only if $x_i \leq y_i$ for all $i = 1, 2, \cdots, n$. $\boldsymbol{x} < \boldsymbol{y}$ if $\boldsymbol{x} \leq \boldsymbol{y}$ and $\boldsymbol{x} \neq \boldsymbol{y}$. Similarly, we can define $\geq$ and $>$ operators.*

For example, if $\boldsymbol{x} = (.3, .2, .5)$, $\boldsymbol{y} = (.4, .3, .9)$ and $\boldsymbol{z} = (.4, .5, .3)$ then $\boldsymbol{x} \leq \boldsymbol{y}$ but $\boldsymbol{x} \not\leq \boldsymbol{z}$ because $x_3 \not\leq z_3$. The k-th element of a connection vector $\boldsymbol{I}$ denotes

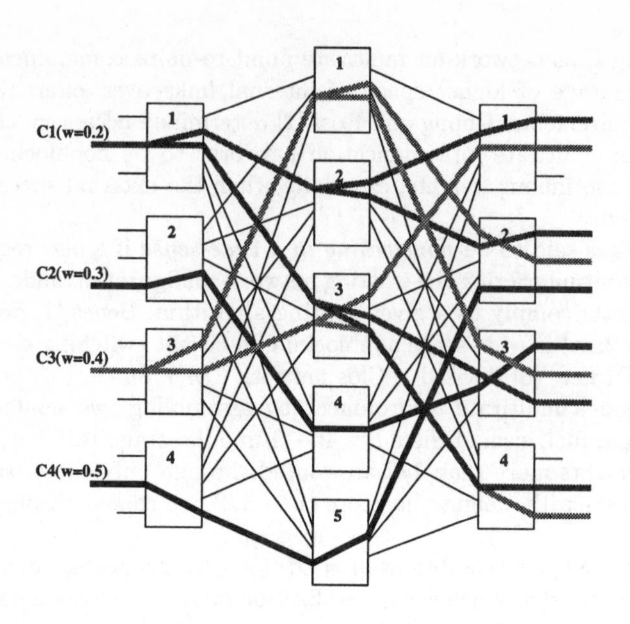

Fig. 1. $\mathcal{C}(3,4,5)$ Clos Network. Connection requests $C_1 = (2,(1,1,1,1),.2)$, $C_2 = (6,(0,0,1,0),.3)$, $C_3 = (8,(1,1,1,1),.4)$ and $C_4 = (11,(0,1,0,0),.5)$.

the weight on output switch k so that the sum of connection vectors, $\sum_{all\ i} \boldsymbol{I}_i$, represents the sum of weights loaded on each output switch.

To describe the configuration of links between middle stage and output stage, we define *destination vector* $\boldsymbol{M}_j$ for each middle switch j such that $\boldsymbol{M}_j(k)$ is the sum of weights loaded on link between middle switch j and output switch k. In Figure 1, for example, $\boldsymbol{M}_1 = (0,.4,.4,0)$, $\boldsymbol{M}_2 = (.2,.2,0,0)$, $\boldsymbol{M}_3 = (.4,0,.2,.6)$, $\boldsymbol{M}_4 = (0,0,.3,0)$ and $\boldsymbol{M}_5 = (0,.5,0,0)$. Like connection vectors, the sum of destination vectors, $\sum_{all\ j} \boldsymbol{M}_j$, represents the sum of weights loaded on each output switch. Because each output switch has n output ports, its load is at most n. We can easily obtain the relationship between connection vectors and destination vectors after routing all requests as follows:

$$\sum_{all\ i} \boldsymbol{I}_i = \sum_{all\ j} \boldsymbol{M}_j \leq n$$

Definition 2 (min and max of vectors). *For two vectors,* $\boldsymbol{x} = (x_1, x_2, \cdots, x_n)$ *and* $\boldsymbol{y} = (y_1, y_2, \cdots, y_n)$*, operators* min *and* max *are defined as follows:*

$$\mathsf{min}(\boldsymbol{x}, \boldsymbol{y}) = (\min(x_1, y_1), \min(x_2, y_2), \cdots, \min(x_n, y_n))$$
$$\mathsf{max}(\boldsymbol{x}, \boldsymbol{y}) = (\max(x_1, y_1), \max(x_2, y_2), \cdots, \max(x_n, y_n))$$

Generally, for k vectors, $\boldsymbol{x}_j = (x_{j,1}, x_{j,2}, \ldots, x_{j,n})$ *for* $1 \le j \le k$,

$$\min_{1\le j\le k}(\boldsymbol{x}_j) = (\min_{1\le j\le k}(x_{j,1}), \min_{1\le j\le k}(x_{j,2}), \ldots, \min_{1\le j\le k}(x_{j,n}))$$
$$\max_{1\le j\le k}(\boldsymbol{x}_j) = (\max_{1\le j\le k}(x_{j,1}), \max_{1\le j\le k}(x_{j,2}), \ldots, \max_{1\le j\le k}(x_{j,n}))$$

For the previous destination vectors, $\min_{1\le j\le 5}(\boldsymbol{M}_j) = (0,0,0,0)$ and $\max_{1\le j\le 5}(\boldsymbol{M}_j) = (.4, .5, .4, .6)$.

4 Wide-sense Nonblocking Multirate Multicast Networks

We consider a special network in which the weight of a new multicast request is no more than $\frac{1}{p+1}$ for some positive integer p.

Theorem 1. *A Clos network* $C(n, r, m)$ *is nonblocking for multirate multicast when* $C_{new} = (x, \boldsymbol{y}, \omega_S)$ *where* $\omega_S \le \frac{1}{p+1}$ *if each request uses at most k middle switches and*

$$m > \frac{(\beta n - \omega_S)(p+1)}{p} \min(k + r^{1/k})$$

Proof. $\omega_S \le \frac{1}{p+1}$ implies $1 - \omega_S \ge \frac{p}{p+1}$. Let m' be the number of middle switches blocking the new multicast request ω_S from x's input switch. We obtain that,

$$m'(1-\omega_S) \le (\beta n - \omega_S)k \qquad \text{and} \qquad m' \le \frac{(\beta n - \omega_S)(p+1)}{p}k \tag{1}$$

Consider $m'' \times r$ destination matrix by discarding rat most m' rows whose corresponding middle switches are blocking the new request from the input switch. Suppose that $1 - \min_{1\le i\le k}(\boldsymbol{M}_{j_i}) \not\ge \boldsymbol{I}_{new}$, which means that any k middle switches can not satisfy the new multicast. Let $c_1(j)$ be the number of elements in the j-th row whose values are greater than $1 - \omega_S$.

$$m'' c_1 \frac{p}{p+1} \le m'' c_1 (1-\omega_S) \le \sum_{j=1}^{m''} c_1(j)(1-\omega_S) \le (\beta n - \omega_S) r$$

where $c_1 = \min_{1\le j\le m''}\{c_1(j)\}$ and $c_1 \ne 0$. Hence, we obtain that

$$m'' \le \frac{(\beta n - \omega_S)(p+1)}{p} \frac{r}{c_1} \tag{2}$$

Without loss of generality, assume the h-th row has the minimum. We can route a part of the request to $r - c_1$ output switches by using the h-th row and delete those $r - c_1$ columns from destination matrix for finding the next middle switch to route the remaining destinations. Generally, assume there are only c_{i-1} output switches which are needed to be routed through by using $m'' \times c_{i-1}$ destination matrix $\mathbf{M}^{i-1}$ for $i < k$. Let $c_i(j)$ be the number of elements in the j-th rows whose values are greater than $1 - \omega_S$ and c_i be the minimum of $c_i(j)$ for all j. Then,

$$m'' c_i \frac{p}{p+1} \leq m'' c_i (1 - \omega_S) \leq \sum_{j=1}^{m''} c_i(j)(1 - \omega_S) \leq (\beta n - \omega_S) c_{i-1}$$

$$m'' \leq \frac{(\beta n - \omega_S)(p+1)}{p} \frac{c_{i-1}}{c_i} \tag{3}$$

where $c_i \neq 0$ for $i < k$. Otherwise, it is a contradiction to our assumption that any k middle switches can not satisfy the new multicast request. When $i = k$, each row vector has at least one element whose value is greater than $1 - \omega_S$. Therefore,

$$m'' \frac{p}{p+1} \leq m''(1 - \omega_S) \leq \sum_{j=1}^{m''} (1 - \omega_S) \leq (\beta n - \omega_S) c_{k-1}$$

$$m'' \leq \frac{(\beta n - \omega_S)(p+1)}{p} c_{k-1} \tag{4}$$

A geometric mean is not less than the minimum of a sequence so that the minimum m'' can be obtained from (2), (3) and (4) as

$$m'' \leq \frac{(\beta n - \omega_S)(p+1)}{p} r^{1/k}$$

To provide a general multirate multicast switching networks, we extend a routing algorithm called a *quota scheme* from Gao and Hwang [4]. Connection requests are partitioned into *large* calls (C_L) and *small* calls (C_S) by their weights. In addition, the set of middle switches(M) are also assumed to be partitioned to M_S and M_L whose sizes are m_S and m_L, respectively. The algorithm forces a C_S to use only M_S but allows a C_L to use not only M_L but also M_S. Theorem 1, a multicast request with $\omega_S \leq \frac{1}{p+1}$ for some positive p is classified to a small call and the one with $\omega_L > \frac{1}{p+1}$ be a large call. Of course, ω_L should not be greater than B because it is the upper bound of all connection requests. For simplicity, let us define $f(r) = \min(k + r^{1/k})$ which was known to have an approximate minimum of $O(\ln r / \ln \ln r)$ [9].

Theorem 2. *The multirate 3-stage Clos networks $C(n, r, m)$, in which weights are within the range of $[b, B]$ and external links can operate at β, is nonblocking in a wide-sense if*

$$m > \begin{cases} \dfrac{\beta n(p+1)(Bp + B + p - 1)}{p^2} f(r) & \text{for } B < 23/32 \\ (\dfrac{15\beta n}{8} + n - 1) f(r) & \text{for } B \geq 23/32 \end{cases}$$

where $p = \lfloor 1/B \rfloor$.

Proof. Assume a large call $C_l = (x, y, \omega_L)$ with $\omega_L > \frac{1}{p+1}$ is compatible to the existing configuration. Let M'_S be a subset of M_S blocking the large call from x's input switch and M'_L be a subset of M_L blocking the request from the input switch by carrying exactly p calls and their sizes are m'_S and m'_L, respectively. Each request is supposed to multiplicate its message at most k times at the input switch. Because of the compatibility, the maximum weights going to the middle stage out of the input switch is at most $(\beta n - \omega_L)k$. Therefore

$$m'_L \frac{p}{p+1} + m'_S(1-B) \leq m'_L \frac{p}{p+1} + m'_S(1-\omega_L) \leq (\beta n - \omega_L)k \tag{5}$$

Let $M''_S = M_S \setminus M'_S$ and $M''_L = M_L \setminus M'_L$ be the subset of M_S and M_L which are available for the large call, respectively. Their sizes are denoted as m''_S and m''_L. To find out the maximum number of blocking links to output switches, let us consider $(m''_S + m''_L) \times r$ destination matrix $\mathbf{M}$. Suppose that any k middle switches from $m''_S + m''_L$ can not satisfy the request. We will use the same notation for $c_i(j)$ and c_i as Theorem 1 but $c_i = \min_{j \in M''_L \cup M''_S} \{c_i(j)\}$.

$$\sum_{j \in M''_L} c_1(j) \frac{p}{p+1} + \sum_{j \in M''_S} c_1(j)(1-B) \leq (\beta n - \omega_L) r$$

$$m''_L \frac{p}{p+1} + m''_S(1-B) \leq (\beta n - \omega_L) \frac{r}{c_1} \tag{6}$$

We apply the similar method as before to contract the destination matrix and obtain the minimum number of middle switches as,

$$m''_L \frac{p}{p+1} + m''_S(1-B) \leq (\beta n - \omega_L) \frac{c_{i-1}}{c_i} \quad \text{for } i < k \tag{7}$$

$$m''_L \frac{p}{p+1} + m''_S(1-B) \leq (\beta n - \omega_L) c_{k-1} \quad \text{for } i = k \tag{8}$$

From (6), (7) and (8), we get

$$m''_L \frac{p}{p+1} + m''_S(1-B) \leq (\beta n - \omega_L) r^{1/k} \tag{9}$$

Because $m'_L + m''_L = m_L$ and $m'_S + m''_S = m_S$, we obtain the following by summing up (5) and (9):

$$m_L \leq \frac{[(\beta n - \omega_L) f(r) - (1-B) m_S](p+1)}{p}$$

From Theorem 1, the maximum numbers of blocking middles m^*_S and m^*_L in M_S and M_L are obtained by $\omega_S \to 0$ and $\omega_L \to 0$.

$$m^*_S = \frac{\beta n(p+1)}{p} f(r)$$

$$m^*_L = \frac{\beta n(Bp + B - 1)(p+1)}{p^2} f(r)$$

Hence, the network is wide-sense nonblocking if one more middle switch is provided in addition to $m_L^* + m_S^*$

$$m > \frac{\beta n(Bp + B + p - 1)(p + 1)}{p^2} f(r) \tag{10}$$

When $B = 1/2$ so as $p = 2$, a Clos network $C(n, r, m_1)$ is nonblocking if $m_1 > \frac{15\beta n}{8} f(r)$. For the other case of $\omega > 1/2$, a Clos network $C(n, r, m_2)$ is nonblocking if $m_2 > (n-1)f(r)$ because a link can not carry more than one request anyhow. We can combine these two Clos networks in parallel to construct a general wide-sense nonblocking Clos network $C(n, r, m)$ for multirate multicast in which $m > (\frac{15\beta n}{8} + n - 1)f(r)$. In this network, all multicast requests with $\omega \leq 1/2$ are routed through the first sub-network, and all requests with $\omega > 1/2$ are routed through the other sub-network.

Let us compare $g_1(B) = \beta n(Bp+B+p-1)(p+1)f(r)/p^2$ with $g_2 = (15\beta n/8 + n - 1)f(r)$. It is easy to verify $g_1(B)$ is an increasing function on B and it is always smaller than g_2 for $B \leq 1/2$ because $\beta \leq 1$. For $B > 1/2$, i.e. $p = 1$, $g_1(B)$ is approximately equal to g_2 at $B = 23/32$.

5 Rearrangeable Multirate Multicast

Yang and Masson [9] gave a nonblocking 3-stage multicast Clos network $C(n, r, m)$ for pure circuit switching if the number of middle stage switches is larger than $(n-1)\min(k + r^{1/k})$. In this section, we show the Clos network is rearrangeable for multirate multicast communications of some special discrete bandwidth cases. Each multicast request is assumed to have a normalized weight from a given finite set $\{p_1, p_2, \cdots, p_h\}$, where $p_3|p_2, \cdots, p_h|p_{h-1}$, and $1 \geq p_1 > 1/2 \geq p_2 > \cdots > p_h > 0$. It is called *integer multiplicity of discrete bandwidths for* p_2 *to* p_h. The rearrangement algorithm orders the requests by their weights and routes the heaviest request first, each of them is restricted to use at most k middle switches. To route the next heaviest request, the algorithm would not disturb the heavier requests which were already routed and route them by using at most k middle switches. It continues to route other requests until the lightest requests are successfully routed.

For a new multicast request, let us consider how many middle stage switches are needed to satisfy the request. Because the maximum fan-out is limited to r for a symmetric Clos network $C(n, r, m)$ as section 3, we use at most r middle switches. We are also able to discover that $n - 1$ is another upper bound a compatible multicast requests.

Theorem 3. *A symmetric Clos network* $C(n, r, m)$ *is multirate rearrangeable when each connection has a weight chosen from a given finite set* $\{p_1, p_2, \cdots, p_h\}$, *where* $p_3|p_2, \cdots, p_h|p_{h-1}$, *and* $1 \geq p_1 > 1/2 \geq p_2 > \cdots > p_h > 0$ *if*

$$m > (n-1) \min_{1 \leq k \leq \min(n-1, r)} (k + r^{1/k}) \tag{11}$$

Proof. We will prove this theorem by induction on h. For $h = 1$, each link can carry no more than one call due to $p_1 > 1/2$ so the Clos network is nonblocking and rearrangeable. Assume that the Clos network is rearrangeable for $h = h' - 1$. Consider two integers u and v such that $p_1 + up_{h'} \leq 1 < p_1 + (u+1)p_{h'}$ and $vp_{h'} \leq 1 < (v+1)p_{h'}$. If a link blocks a new connection request of weight $p_{h'}$, the blocking link is carrying either one p_1-call and weights of u $p_{h'}$-calls(U-blocking), or v $p_{h'}$-calls(V-blocking). Let us assume more than $(n-1)k$ middle switches are blocking the new $p_{h'}$-call from input stage. Because all connection requests were able to duplicate their messages at most k times at the input switch, at least n input ports should have carry full weights that are $p_1 + up_{h'}$ or $vp_{h'}$. This is a contradiction to our assumption for the compatible new $p_{h'}$-call.

Suppose that any k middle switches among m' can not satisfy the new $p_{h'}$-call. Then, we can obtain $m' \leq (n-1)r^{1/k}$. The total number of blocking links between middle stage and output stage is no more than $(n-1)r$ because each output switch have at most $(n-1)$ output ports which are either U-blocking or V-blocking for the new call. By using the similar approach as the previous section, we can obtain,

$$m' \leq (n-1)r/c_1 \tag{12}$$

$$m' \leq (n-1)c_{i-1}/c_i \qquad \text{for } i < k \tag{13}$$

$$m' \leq (n-1)c_{k-1} \qquad \text{for } i = k \tag{14}$$

A minimum of a sequence is not larger than its geometric mean so that, from (12), (13) and 14, we can obtain,

$$\begin{aligned} m' &\leq \left[(n-1)\frac{r}{c_1} \cdot (n-1)\frac{c_1}{c_2} \cdots (n-1)\frac{c_{k-2}}{c_{k-1}} \cdot (n-1)c_{k-1}\right]^{1/k} \\ &= (n-1)r^{1/k} \end{aligned}$$

We showed that the nonblocking multicast Clos network for pure circuit switching is also rearrangeable for multirate multicast communications if each request has a weight chosen from integer multiplicity of discrete bandwidth for p_2 to p_h and $1 \geq p_1 > 1/2 \geq p_2 > \cdots > p_h$. In the following, we observe other special cases that provide more flexibility to the finite set of weights.

Corollary 1. *A symmetric Clos network $C(n,r,m)$ is multirate rearrangeable when each connection has a weight chosen from a given finite set $\{p_1, p_2, \cdots, p_h\}$, where $p_{i+1}|p_i, p_{i+2}|p_{i+1}, \cdots, p_h|p_{h-1}$, and $1 \geq p_1 > p_2 > \cdots > p_{i-1} > 1/2 \geq p_i > p_{i+1} > \cdots > p_h > 0$ if*

$$m > (n-1) \min_{1 \leq k \leq \min(n-1,r)} (k + r^{1/k}) \tag{15}$$

Proof. We apply the same idea as Theorem 3, but consider i integers $u_1, u_2, \cdots, u_{i-1}$ and v such that $p_j + u_j p_{h'} \leq 1 < p_j + (u_j+1)p_{h'}$ for $j \leq i-1$ and $vp_{h'} \leq 1 < (v+1)p_{h'}$.

Corollary 2. *A symmetric Clos network $C(n, r, m)$ is multirate rearrangeable when each connection has a weight chosen from a given finite set $\{p_1, p_2, \cdots, p_h\}$, where $p_2|p_1, p_3|p_2, \cdots, p_h|p_{h-1}$, and $1 \geq p_1 > p_2 > \cdots > p_h > 0$ if*

$$m > (n-1) \min_{1 \leq k \leq \min(n-1,r)} (k + r^{1/k}) \tag{16}$$

Proof. Consider an integer u such that $up_i \leq 1 < (u+1)p_i$ for each iteration.

6 Conclusions

We have studied the construction of multirate 3-stage Clos switching networks which were nonblocking in a wide sense for multicast communications. To overcome the obstacle of other nonblocking multirate Clos networks requiring infinite number of switch elements at the worst case, the middle stage switches were partitioned into two or three subsets and the routing algorithm allowed connection requests to utilize one of two subset according to their normalized weights. The hardware complexities of the networks were determined only by the configurations of the networks themselves but not by the patterns of connection requests, which is extremely advisable to build a real large-scale switching network. The nonblocking circuit switching multicast Clos network was also shown to be rearrangeable for some discrete multirate multicast communications. The rearrangeable routing algorithm sorted connection requests by their normalized weights and routed the heaviest requests first and then the next heaviest and so forth in nonblocking fashion.

References

1. V. E. Benes. *Mathematical Theory of Connecting Networks and Telephone Traffice.* Academic Press, New York, 1965.
2. Shun-Ping Chung and Keith W. Ross. On nonblocking multirate interconnection networks. *SIAM J. on Comp.*, 20(4):726–726, August 1991.
3. C. Clos. A study of non-blocking switching networks. *Bell Syst. Tech. J.*, 32:406–424, March 1953.
4. Biao Gao and Frank K. Hwang. Wide-sense nonblocking for multirate 3-stage Clos networks. *to appear in Theoretical Computer Science.*
5. F. K. Hwang. Rearrangeability of multiconnection three-stage networks. *Networks*, 2:301–306, 1972.
6. Riccardo Melen and Jonathan S. Turner. Nonblocking multirate networks. *SIAM J. on Comp.*, 18(2):301–313, April 1989.
7. I. Svinnset. Nonblocking ATM switching networks. *IEEE Trans. on Comm.*, 42(2-4):1352–1358, Feb.-Apr. 1994.
8. Yuanyuan Yang. An analysis model on nonblocking multirate broadcast networks. In *International Conference on Supercomputing*, pages 256–263, 1994.
9. Yuanyuan Yang and Gerald M. Masson. Nonblocking broadcast switching networks. *IEEE Trans. on Comp.*, 40(9):1005–1015, September 1991.

Efficient Randomized Routing Algorithms on the Two-Dimensional Mesh of Buses

Kazuo IWAMA[1] and Eiji MIYANO[2] and Satoshi TAJIMA[3] and Hisao TAMAKI[4]

[1] Department of Information Science, Kyoto University, Kyoto 606-8501, JAPAN
iwama@kuis.kyoto-u.ac.jp

[2] Kyushu Institute of Design, Fukuoka 815-8540, JAPAN
miyano@kyushu-id.ac.jp

[3] Systems and Software Research Laboratories, Research and Development Center, Toshiba Corporation, Kawasaki 210-8501, JAPAN
tajima@ssel.toshiba.co.jp

[4] Department of Computer Science, Meiji University, Kawasaki 214-8571, JAPAN
tamaki@cs.meiji.ac.jp

Abstract. The mesh of buses (MBUS) is a parallel computation model which consists of $n \times n$ processors, n row buses and n column buses but no local connections between two neighboring processors. As for deterministic (permutation) routing on MBUSs, the known $1.5n$ upper bound appears to be hard to improve. Also, the information theoretic lower bound for any type of MBUS routing is $1.0n$. In this paper, we present two randomized algorithms for MBUS routing. One of them runs in $1.4375n+o(n)$ steps with high probability. The other runs $1.25n+o(n)$ steps also with high probability but needs more local computation.

1 Introduction

The two dimensional mesh is widely considered to be a promising parallel architecture in its scalability [Lei92,MS96]. In this architecture, processors are naturally placed at intersections of horizontal and vertical grids, while there can be two different types of communication links: The first type is shown in Figure 1-(1). Each processor is connected to its four neighbors and such a system is called a *mesh-connected computer* (an *MC* for short). Figure 1-(2) shows the second type: Each processor is connected to a couple of (row and column) buses. The system is then called a *mesh of buses* (an *MBUS* for short).

Permutation routing (simply *routing* in this paper) is apparently a basic form of communication among the processors: The input is given by n^2 packets that are initially held by the $n \times n$ processors, one by each. Routing requires that all n^2 such packets be moved to their destinations that are mutually distinct. In the case of MCs, a $2n-2$ lower bound comes from a fundamental nature of the model, i.e., the physical distance between the farthest two processors. Also, the same $2n-2$ upper bound can be achieved by an elementary algorithm based on the *dimension-order* strategy [Lei92,Tom94]. Thus there remains little for further research in the case of MCs. (This is not true for limited buffer-size as mentioned later.) In the case of MBUSs, on the other hand, there is a wide margin between the known upper and lower bounds. First of all, unlike the case of MCs, the dimension-order strategy only gives us a poor algorithm which takes trivial $2n$ steps. The best upper and lower bounds known are $1.5n$ and $(1-\varepsilon)n$, respectively [IMK96]. The $1.5n$ bound appears to be hard to improve; it is also known to be a (tight) lower bound if we impose the so-called "source-oblivious" condition [IM97a].

The main purpose of this paper is to decrease this $1.5n$ upper bound by allowing randomization. Two randomized algorithms are given: One of them runs in $1.4375n + o(n)$ steps with high probability. The other runs in $1.25n + o(n)$ steps but needs more local computation. The idea is an efficient use of the buses and a reduction of packet collisions. Consider, for example, the (deterministic) dimension-order routing where each packet first moves horizontally (in the order of original position) using the first n steps and then moves vertically (in the order of destination position) in the second n steps. One can see that column buses are completely idle in the first n steps and so are row buses afterwards in this algorithm.

A simple attempt to avoid this inefficiency is to try to move the first n packets vertically immediately after they move horizontally. If those n packets go to all different n columns, i.e., they have all different column destinations, then we can do this in a single step without collision. If collision happens in some column, then we can use randomization techniques to resolve the collision. If few collisions occur, then we might achieve an approximately $1.0n$ upper bound. Unfortunately, however, this observation is too optimistic. Some experiments show that a lot of collisions occur for even random permutation. It seems that this approach gives us no better bounds than $3n$; it is much worse than the deterministic version. Thus, an efficient use of buses tends to imply more collisions. Our first algorithm avoids this difficulty in a tricky way. The second one is based on a novel use of the technique that allows us to generate many pseudo-random numbers deterministically from a few random numbers.

Research on mesh routing has a long history and has a huge literature. Nevertheless there still remain a lot of unknowns. For example, our knowledge on the 3-dimensional (3-D) mesh is much weaker than the 2-D mesh. Recently, it is shown [IM97b] that minimum-bending oblivious routing on the 3-D mesh needs $O(N^{2/3})$ steps that is much *more* than $O(N^{1/2})$ for the 2-D mesh (N is the total number of processors). It is not known either whether we can improve $O(n^2)$ upper bounds substantially for 2-D obvious routing on MCs with *constant buffer-size* [CLT96,Kri91]. There is also a gap between the known upper and lower bounds, $(1+\varepsilon)n + o(n)$ and $0.691n$, respectively, for 2-D routing on the mesh equipped with both buses and local connections [CL93,LS94].

2 Models and Problems

An MBUS consists of n^2 processors, $P_{i,j}$, $1 \leq i, j \leq n$, and n row and n column buses. $P_{i,j}$ is connected to the ith row bus and the jth column bus. The problem of *permutation routing* on the MBUS is defined as follows: The input is given by n^2 *packets* that are initially held by the n^2 processors, one by each. Each packet, (s, d, σ), consists of three portions; s is a *source* address that shows the initial position of the packet, d is a *destination* address that specifies the processor to which the packet should be moved, and σ is a data portion that is not important in this paper. No two packets have the same destination address. Routing requires that all n^2 such packets be moved to their correct destinations.

Our discussion throughout this paper is based on the following four rules on the model: (i) We follow the common practice on how to measure the running time of MBUSs: One-step computation of each processor P consists of (a) reading the current data on both row and column buses P is connected to, (b) executing arbitrarily complicated instructions using the local memory and (c) if necessary, writing data to the row and/or column buses. The written data will be read in the next step. (ii) The queue size is not bounded, namely, an arbitrary number of packets can stay on a single processor temporarily. (iii) What can be written on the buses by the processor P must be the packet originally given to P as its input packet or one of the packets that have been read so far by P from its row or column bus. (Nothing other than packets can be written.) This means that any kind of data compression is not allowed. (iv) We allow the simultaneous write. However, if two or more packets are written on the same bus simultaneously, then a special value flows on the bus, which has no information other than collision.

As mentioned in the previous section, the $2n$-step dimension-order routing moves horizontally the leftmost n packets initially placed on $P_{1,1}, P_{2,1}, \cdots, P_{n,1}$ in step 1, $P_{1,2}, P_{2,2}, \cdots, P_{n,2}$ in step 2 and so on. Namely packets are moved in their "source-order" in this first stage. In the second stage, n packets whose destinations are the uppermost n processors, $P_{1,1}, P_{1,2}, \cdots, P_{1,n}$, are moved vertically in step 1, then, $P_{2,1}, P_{2,2}, \cdots, P_{2,n}$, and so on. Thus they are moved in the "destination-order" regardless of their current positions. It should be noted that this destination-order transmission can only be used after all the packets have moved horizontally. That is why column buses are completely idle in the first stage. If we do not wait, then we have to give up the destination-order transmission and encounter the more serious problem, i.e., packet collisions, as described in Section 1.

The $1.5n$-step algorithm, called *DR4* from now on, reduces the number of first-stage steps from n to $0.5n$ as follows: The whole $n \times n$ plane is divided into four $0.5n \times 0.5n$ subplanes. Packets in the upper-left $0.5n \times 0.5n$ and the lower-right $0.5n \times 0.5n$ subplanes are moved horizontally and those in the upper-right and the lower-left subplanes vertically, both in the source-order. Thus, all the buses are used in the first stage, which reduces the computation time one half. The second stage is almost the same as before.

3 $1.4375n + o(n)$ Randomized Algorithm

Note that there are n^2 packets, $2n$ buses and each packet has to ride on a bus twice (in general). Thus $2 \times n^2/2n = n$ steps are needed even if we have no idle buses. In the $1.5n$-step algorithm *DR4*, Stage 1 has no idle buses. However, it is impossible to improve Stage 2 since we can create an instance as an "adversary" which leaves n packets on a single bus after Stage 1. (Detours might help but it seems difficult to design an algorithm that exploits the possibility.)

Our first randomized algorithm, *RR*, is based on *DR4*. The basic idea is as follows: (i) We should avoid, for any instance, the bad case where n packets are gathered on a single bus after Stage 1. (ii) In other words, we should distribute packets evenly so that each single bus has approximately $0.5n$ packets at Stage 2. (iii) Then it is not so hard to design a randomized algorithm for Stage 2 that needs more than optimal $0.5n$ steps but some $0.75n$ steps are enough. (iv) In order to accomplish the even distribution in (ii), we now have to give up the very efficient Stage 1 of *DR4*. Some loss of performance is inevitable, but if we can keep it less than $0.75n$ steps then the $1.5n$ bound in total can be improved.

Algorithm: *RR*

Stage 1. The whole plane is divided into four subplanes as *DR4*. This stage consists of Stages 1-1 and 1-2.

Stage 1-1. For a while, we only look at the upper-left subplane. The $0.5n$ processors in each row are divided into $0.125n$ *blocks*; each block includes four consecutive processors. For example, $(P_{1,1}, P_{1,2}, P_{1,3}, P_{1,4})$ is the first block of row 1, $(P_{1,5}, P_{1,6}, P_{1,7}, P_{1,8})$ is the second block and so on. Now from $j = 1, 2, \cdots$, through $0.125n$, namely for each block from left to right, the following Phases 1 and 2 are executed. The same operation is executed on each row bus in parallel; the description below is for row i:

Phase 1 : The first two processors of block j, i.e., $P_{i,4j-3}$ and $P_{i,4j-2}$, write their initial packets on the row bus with probability 1/2.

Phase 2 : One of the following four operations is selected due to the result of Phase 1. Note that all the processors on the bus can figure out which case occurred.

1. If the packet whose source address is $(i, 4j-3)$ was on the bus, i.e., if only $P_{i,4j-3}$ wrote the packet, then $P_{i,4j-1}$ writes its initial packet on the row bus. Go to the next block. (See Figure 2-(1).)
2. If only $P_{i,4j-2}$ wrote the packet, then $P_{i,4j-1}$ writes its initial packet on the row bus. Go to the next block. (See Figure 2-(2).)
3. If collision occurred, i.e., if both $P_{i,4j-3}$ and $P_{i,4j-2}$ wrote the packets, then $P_{i,4j-3}$ again writes its packet and then $P_{i,4j}$ writes its packet on the row bus. Thus we need three steps in this (and next) case. Go to the next block. (See Figure 2-(3).)
4. If the bus was idle, i.e., if neither $P_{i,4j-3}$ nor $P_{i,4j-2}$ wrote the packets, then $P_{i,4j-2}$ writes its packet and then $P_{i,4j}$ writes its packet on the row bus. Go to the next block. (See Figure 2-(4).)

Thus exactly two packets out of four ones in each block are moved horizontally. The remaining two packets, called *m-packets*, are moved vertically in Stage 1-2.

Stage1-2. Now the $0.5n$ processors in each *column* are divided into $0.25n$ *blocks*, i.e., each block has two processors. From $i = 1, 2, \cdots$, through $0.25n$, namely for each block from top to bottom, Phases 1 and 2 are executed:

Phase 1 : Each of the two processors, $P_{2i-1,j}$ and $P_{2i,j}$, in block i writes its m-packet (if any).

Phase 2 : Again one of the four operations is selected:

1. If only $P_{2i-1,j}$ wrote the packet, then go to the next block. (See Figure 3-(1).)

2. If only $P_{2i,j}$ wrote the packet, then go to the next block. (See Figure 3-(2).)
3. If neither $P_{2i-1,j}$ nor $P_{2i,j}$ wrote the packet, then go to the next block. (See Figure 3-(3).)
4. If both $P_{2i-1,j}$ and $P_{2i,j}$ wrote the packets, then $P_{2i-1,j}$ writes its packet and then $P_{2i,j}$ writes its packet on the column bus. Go to the next block. (See Figure 3-(4).)

That concludes Stage 1 for the upper-left subplane. The algorithm is exactly the same for the lower-right subplane. As for the upper-right and the lower-left subplanes, we exchange rows and columns, i.e., Stage 1-1 uses columns and Stage 1-2 rows.

Stage 2. Every packet has already moved to its row or column destination. Thus the situation is the same as at the beginning of Stage 2 of *DR4*. The difference is that the number of packets held by processors on each single bus is evenly distributed, i.e., is about $0.5n$ (proof is given later). Let us look at some single bus, say, row 1. The basic idea of Stage 2 is to use the destination-order counterpart of Stage 1-2. Namely, at the first step, if a processor has a packet whose destination is $(1,1)$ or $(1,2)$ then it writes that packet on the row bus. If no collision occurs, i.e., if at most one of the two packets exists on this row, then we move forward. Otherwise, two more steps are used for sending each of those collided packets.

Unfortunately, this algorithm does not work. The reason is that the two packets whose destinations are $(1,1)$ and $(1,2)$ may be held by some single processor. If this is the case, that processor puts either one on the bus but other processors following the above algorithm assume that there is only one packet on this row whose destination is in the first block.

To solve this problem, we introduce a "special packet" (*SP* in short) whose purpose is to broadcast special information. (If one does not like to use a packet for such a special purpose, then we can give another algorithm whose performance is a little bit worse than the current one. See the end of this section.) As an *SP*, we use the packet whose destination is (n,n). At the beginning of Stage 2, we introduce two extra steps to let all the processors know this *SP*. Now from $i = 1, 2, \cdots$, through $0.5n$, i.e., for each block from left to right, the following two phases are executed:

Phase 1 : Let Q_1 and Q_2 be packets whose destinations are in block i. For each processor P on this row, if P holds either Q_1 or Q_2 then P writes that packet on the bus. If P holds both Q_1 and Q_2, then P writes the *SP*.

Phase 2 : One of the following four operations is selected:

1. If Q_1 flowed on the bus in Phase 1, then all the processors move on to the next iteration, namely, for $i+1$.
2. If Q_2 flowed, then all the processors move on to the next iteration also.
3. If nothing flowed then all the processors move on to the next iteration also.
4. If *SP* flowed or collision occurred, then we need two more steps; the processor holding Q_1 writes it first and then the processor holding Q_2 follows.

We have not yet stated when Stage 2 should be started. Our design of *RR* determined to start Stage 2 at some fixed step that is no later than $0.6875n+o(n)$.

At this moment all processors have finished Stage 1 with high probability (see the proof of Theorem 1). However, there is a slight chance that some processor is still executing Stage 1. If that is the case, then some unexpected data-collision will happen in Stage 2 and the algorithm fails. To improve it so as not to fail is possible but needs some technical details. For this purpose, we can use the *SP* again, which can also be avoided at the expense of extra $o(n)$ steps (omitted).

Theorem 1. *With high probability, RR can rout any instance within* $1.4375n + o(n)$ *steps.*

Proof. We first calculate the expected number of steps each stage takes. Then it is proved that the probability that *RR* takes essentially more steps than the average is very low.

Stage 1-1. See Phase 2. The probabilities for the cases 1 to 4 are all the same, i.e., 1/4. Cases 1 and 2 take two steps and 3 and 4 take three steps. Therefore, it takes 2.5 steps for each block on average or $0.3125n$ steps for $0.125n$ blocks.

Stage 1-2. Consider an arbitrary column in the upper-left subplane. It is not hard to see that each processor on this column holds an m-packet with probability 1/2 and furthermore that this occurs independently between any pair of processors on this column. In Phase 2, the cases 1 to 3 take one step and the case 4 three steps. Hence we need 1.5 steps on average per block or $0.375n$ steps for $0.25n$ blocks.

Stage 2. Let us calculate the probability of the cases 1 to 4 in Phase 2 where we have to be a bit more careful than before.

Case a : Q_1 and Q_2 come from different blocks of Stage 1-1. Then one can see easily that the probabilities for 1 to 4 are the same, i.e., 1/4 for each.

Case b : Q_1 and Q_2 come from the same block. Then we should check many different possibilities, such as coming from the top two positions, from the middle two positions, and so on. However, it turns out that in any possibility, the probability for the case 4 is at most 1/4 (1/4 or 0, in fact).

Recall that the cases 1 to 3 need one step and the case 4 needs three steps. Hence we need at most 1.5 steps on average per block or at most $0.75n$ steps for $0.5n$ blocks.

Now we shall evaluate the probability for bad behaviors using the Chernoff bound [Che52]:

Lemma 1. *Let* $X_1, X_2, ..., X_n$ *be independent Bernoulli trials having binomial distribution* $B(n,p), 0 < p < 1$. *Let* $X = \sum_{i=1}^{n} X_i$, $\mu = \mathrm{E}[X]$. *Then, for any* $0 < \varepsilon < 1$,

$$\Pr[\, X \; > \; (1+\varepsilon)\mu \,] \; < \; \exp(\frac{-\varepsilon^2\mu}{3}).$$

Stage 1-1. For a time being, we only consider some single bus. For block i, let $X_i = 2$ when the case 1 or 2 occurs and $X_i = 3$ otherwise. The $X = \sum X_i$ has a binomial distribution $B(0.125n, 1/2)$, and $\mu = \mathrm{E}[X] = 0.3125n$. Apply the Chernoff bound with $\varepsilon = c_1\sqrt{n \ln n}/\mu$ for $c_1 > 0$. Then,

$$\Pr[\, X \; > \; 0.3125n + c_1\sqrt{n \ln n}\,] \; < \; \exp(-\frac{16}{15}c_1^2 \ln n) = n^{-d_1}$$

for some $d_1 > 0$. Namely, the number of steps for Stage 1-1 is at most $0.3125n + c_1\sqrt{n \ln n}$ with probability $1 - n^{-d_1}$.

Stage 1-2. Our analysis is almost the same. Let $X_i = 1$ when one step is needed and $X_i = 3$ when three steps are needed, then the number of steps for this stage is at most $0.375n + c_2\sqrt{n \ln n}$ with probability $1 - n^{-d_2}$. Thus Stage 1 takes at most $0.6875n + o(n)$ steps with high probability.

Stage 2. Let $Y_i = 3$ when the case 4 occurs (i.e., three steps are needed) and $Y_i = 1$ otherwise. This time, however, $Y_1, Y_2, \cdots, Y_{0.5n}$ may not be totally independent. For example, if two packets heading for block 1 come from the top and third positions of some block (on the upper-right or the lower-left subplane) in Stage 1-1, then there is no possibility that two packets heading for block 2 come from the second and fourth positions of the same block. Namely, if $Y_1 = 3$ (with probability 1/4) then Y_2 must be 1. However, we can show that the sum of those non-independent random variables does not deviate from its expected value with high probability using the following lemma [MR95]:

Lemma 2. *Let $X_0, X_1, X_2, \cdots$ be a martingale sequence such that for each k,*

$$|X_k - X_{k-1}| \leq c$$

where c is some constant independent of k. Then, for all $t \geq 0$ and any $\lambda > 0$,

$$\Pr[\ |X_t - X_0| \geq \lambda c\sqrt{t}\ ] \leq 2\exp(-\lambda^2/2).$$

Let $Y = \sum_{i=1}^{0.5n} Y_i$ and define a martingale sequence $X_0, X_1, \cdots, X_{0.5n}$ by setting $X_0 = \mathrm{E}[Y]$ and for $1 \leq i \leq 0.5n$ $X_i = \mathrm{E}[Y|Y_1, Y_2, \cdots, Y_i]$. Now we shall evaluate the value $|X_i - X_{i-1}|$ for $1 \leq i \leq 0.5n$. Note that the difference between X_i and X_{i-1} depends only on the value of Y_i, which is determined by the behavior of two packets heading for the ith block. The behavior of each of those two packets affects the behavior of at most three other packets in its block of Stage 1 and hence at most three other random variables Y_j's. Therefore, conditioning the value of Y_i affects at most seven Y_j's (including Y_i itself) and the difference between X_i and X_{i-1} is bounded by some constant $c_3 \leq 3 \times 7 = 21$ since $Y_i \leq 3$. By applying Lemma 2 with $t = n/2$ and $\lambda = \sqrt{2d_3 \ln n}$ for some $d_3 > 0$, one can conclude that the number of steps for Stage 2 is at most $0.75n + c_3\sqrt{d_3 n \ln n}$ with probability $1 - 2n^{-d_3}$.

We have $2n$ buses. So, the probability that the bad behavior occurs in at least one bus can be as large as $2n$ times. However, since its probability can be written as n^{-d} for a large enough constant d, we do not have to worry about that. As a result, the whole algorithm takes at most $1.4375n + o(n)$ with high probability. □

Remark 1. We can also design Stage 2 without using an *SP* : *Phase 1* : If P holds both Q_1 and Q_2, then P writes Q_1 and then writes Q_2 at the next step. *Phase 2* : 1 to 3 are the same. 4. If collision occurred, then we examine which case happened in the previous block. If it is the case 1, then the collision might be caused by the packet Q_2 of the previous block. So, we insert an extra step to send this Q_2 (but it may be idle if the collision is caused by two packets of the current block). Although details are omitted, the algorithm runs in at most $1.46875n + o(n)$ steps with high probability, which is a little worth than Theorem 1 but is still better than $1.5n$.

Remark 2. Local computation in each step is very simple in RR. That is obviously a good point of this algorithm compared to the next algorithm.

4 $1.25n + o(n)$ Randomized Algorithm

Recall that there are n processors on each bus and roughly one half of these n processors put their packets on its bus in the previous algorithm RR. However, it takes much more time than $0.5n$ to finish the transmission. An obvious reason is a lot of packet collisions: If each processor P would know whether or not each other processor on the same bus is now trying to write its packet, then P could calculate a proper time-slot at which P should write its own packet without collision or waste of the bus.

This is in fact possible if P would know all the random numbers the other processors have generated. To this goal, one can use the technique of generating many pseudo-random numbers deterministically from a few random numbers. Some preliminaries are needed: Let $X_1, X_2, \cdots, X_n$ be discrete random variables defined on the same probability space. Such a set of random variables is said to be *pairwise independent* if for all $i \neq j$,

$$\Pr[\, X_i = x \mid X_j = y \,] = \Pr[\, X_i = x \,].$$

This pairwise independence is naturally extended to the *k-wise independence*: A set of similarly defined random variables $X_1, \cdots, X_n$ are said to be k-wise *independent* if for all different $i_1, i_2, \cdots, i_k$,

$$\Pr[\, X_{i_1} = x_{i_1} \mid X_{i_2} = x_{i_2}, X_{i_3} = x_{i_3}, \cdots, X_{i_k} = x_{i_k} \,] = \Pr[\, X_{i_1} = x_{i_1} \,].$$

Lemma 3 (see [Jof74]). *Let m be a prime number and Z_m denote the field of integers modulo m. Then the set of $n < m$ random variables $X_1, \cdots, X_n$ calculated by the following equation from the k numbers, $a_1, \cdots, a_k$, which are randomly chosen from Z_m are k-wise independent:*

$$X_i = a_1 i^{k-1} + a_2 i^{k-2} + \cdots + a_{k-1} i + a_k \mod m. \tag{1}$$

Consider for example the n processors $P_1, \cdots, P_n$ on the first row. The left-most processor P_1 generates k (truly) random numbers $a_1, \cdots, a_k$ and transmits them to P_2 through P_n. Then each P_i can generate its own random number X_i by equation (1). The set of X_i's are guaranteed to be k-wise independent. The degree of randomness for each X_i is smaller than before, but we can show that it is enough for our purpose when $k = 6$.

We have another technical problem; how to transmit $a_1, \cdots, a_k$. Recall that our rule is that only packets can be transmitted on the bus. Fortunately, the amount of information carried by $a_1, \cdots, a_k$ is not too large since k is a constant.

Moreover, we can set m to be nearly equal to n^2, so the number of total bits of $a_1, \cdots, a_k$ is $O(\log n)$. Consequently the following simple algorithm works: (i) P_1 creates $a_1, \cdots, a_k$. (ii) Suppose that the bit sequence of $a_1, \cdots, a_k$ (may be encoded) is $b_1, b_2, b_3, \cdots$. Then P_1 puts its original packet repeatedly on the bus at time-slot i if $b_i = 1$ and puts nothing if $b_i = 0$. This takes only $O(\log n)$ steps.

Now we are ready to give our new algorithm RR_k which consists of two stages as before. In the first stage, about one half of the whole packets move horizontally to their column destinations, and the rest to their row destinations. The second stage is exactly the same as the previous algorithm RR.

Algorithm: RR_k

Stage 1. Choose any prime number $m > n^2$. Let $Z_m = \{1, 2, \cdots, m-1\}$, $Z_m^0 = \{1, \cdots, \frac{m}{2}\}$ and $Z_m^1 = \{\frac{m}{2}+1, \cdots, m-1\}$. Note that $|Z_m| = m-1$ and $|Z_m^0| = |Z_m^1| = (m-1)/2$ ($m-1$ is even since m is a prime number).

Phase 1 : $P_{1,1}$ generates k prime numbers $a_1, \cdots, a_k \in Z_m$ (k will be set to six when the probability of success is calculated).

Phase 2 : $P_{1,1}$ transmits $a_1, \cdots, a_k$ to all the processors on the first column in the way described above.

Phase 3 : $P_{i,1}$ ($1 \le i \le n$) transmits $a_1, \cdots, a_k$ to all the processors on the ith row in the same way.

Phase 4 : Now each processor $P_{i,j}$ has $a_1, \cdots, a_k$, from which it computes $f(i,j) = a_1\{n(i-1)+j\}^{k-1} + a_2\{n(i-1)+j\}^{k-2} + \cdots + a_k \bmod m$. Then set $X_{i,j} = 1$ if $f(i,j) \in Z_m^1$ and $X_{i,j} = 0$ if $f(i,j) \in Z_m^0$.

Phase 5 : If $X_{i,j} = 1$, then $P_{i,j}$ puts its packet on the row bus first in the following way: $P_{i,j}$ computes how many processors among $P_{i,1}, \cdots, P_{i,j-1}$ also write to the row bus first by simulating Phase 4 of each processor (we need local computation proportional to n here). If that number is t, then $P_{i,j}$ writes to the row bus at step $t+1$. If $X_{i,j} = 0$, then $P_{i,j}$ puts its packet on the column bus first. It uses the similar calculation to decide when it should do so.

Stage 2. Note that all the processors can calculate when the first stage ends (by calculating the last processor which accesses to the row or column bus at Stage 1). After the first stage is finished, all the processors enter Stage 2 that is exactly the same as Stage 2 of RR. One might think why we cannot use the same technique as Stage 1. Recall that there are approximately $0.5n$ packets on each bus at the beginning of this stage. If each processor knows the current positions or the destinations of all those packets on the bus, then we can use the same technique as before. Unfortunately, neither is known.

Theorem 2. *For $k = 6$, RR_k halts within $1.25n + o(n)$ steps with high probability.*

Proof. Note that Phases 1 and 4 include only local computation. Also, as mentioned before, Phases 2 and 3 take only $O(\log n)$ steps. Therefore what we have to prove is that the number of packets written to a row bus first (and to a column bus first also) is sufficiently close to $0.5n$ and furthermore, at the beginning of Stage 2, the number of packets to move on each single bus is also close to $0.5n$.

Fix some row, say row j. Let P_i be the ith processor on this row and X_i be the random variable whose 1 or 0 is determined at Phase 4. Let $X = \sum_{i=1}^{n} X_i$. Then it turns out [MR95] that the expected value of X is $0.5n$. What we wish to know is the probability that X differs from $0.5n$ by a certain amount of value. Note that X is a random variable that is a sum of (not necessarily independent) random variables X_i, for which we cannot use Chernoff bound. Instead we use the (generalized) Chebyshev bound: For an integer $k \geq 2$, let $\mu_X^k = \mathrm{E}[(X - 0.5n)^k]$. (This is called *kth central moment*, which does not exist for some probability space. It obviously exists in the present case.)

Lemma 4 (see [MR95] for example). *For any* $t > 0$,

$$\Pr[\ |X - 0.5n| > t\sqrt[k]{\mu_X^k}\ ] \leq \frac{1}{t^k}.$$

In order to prove this theorem, it is enough to consider only the case where $k = 3$ (the reason will be described later). Let us evaluate the third central moment $\mu_U^3 = \mathrm{E}[(U - \mathrm{E}[U])^3]$, where $U = \sum U_i$ is the sum of n 3-wise independent binary random variables. Expand $\mu_U^3 = \mathrm{E}[(\sum U_i - \sum \mathrm{E}[U_i])^3]$, and consider each term. Such a term involves up to three variables from U_i's. However, we can claim that terms involving more than one variable cancel each other. To see this, let T be a term of the expansion that involves more than one variable. Then, T contains a variable U_i that appears in T exactly once. Thus, T can be written in the form $\mathrm{E}[T_1 U_i]$ or $\mathrm{E}[T_2 \mathrm{E}[U_i]]$, where T_1 or T_2 does not contain U_i. Note that the terms in these two forms (for fixed U_i) are in one to one correspondence, each $\mathrm{E}[T_1 U_i]$ corresponding to $\mathrm{E}[-T_1 \mathrm{E}[U_i]]$. Due to the 3-wise independence we can write $\mathrm{E}[T_1 U_i]$ as $\mathrm{E}[T_1]\mathrm{E}[U_i]$ and $\mathrm{E}[-T_1 \mathrm{E}[U_i]]$ as $-\mathrm{E}[T_1]\mathrm{E}[U_i]$ and thus cancel them out. The only remaining terms are of the form $\mathrm{E}[U_i^3]$, $\mathrm{E}[U_i^2]\mathrm{E}[U_i]$, or $\mathrm{E}[U_i]^3$, involving only one variable, with some constant coefficients. The contribution of these terms is a constant per variable and thus $O(n)$ in total.

Note that the 6-wise independent random variables X_i's satisfy 3-wise independency. Using Lemma 4 with $\mu_X^3 = O(n)$,

$$\Pr[\ |X - 0.5n| > t\sqrt[3]{O(n)}\ ] \leq \frac{1}{t^3}.$$

If we set $t = n^{c - \frac{1}{3}}$ for a positive constant c, then

$$\Pr[\ |X - 0.5n| > O(n^c)\ ] \leq \frac{1}{n^{3c-1}}.$$

This means $X \leq 0.5n + o(n)$ with probability at least $1 - n^{-(3c-1)}$. Then it follows that the number of steps needed in Stage 1 for the fixed single bus is at most $0.5n + o(n) + O(\log n)$ with at least the same probability. This holds for all other buses.

For the analysis of Stage 2, we can apply the same argument as above since random variables Y_i's are 3-wise independent. By the above calculation we can get $\mu_Y^3 = O(n)$, i.e.,

$$\Pr[\ |Y - 0.75n| > O(n^c)\ ] \leq \frac{1}{n^{3c-1}},$$

which says Stage 2 takes at most $0.75n + o(n)$ steps with probability at least $1 - n^{-(3c-1)}$ per bus.

Since there are $2n$ buses, the unsuccessful probability can be up to $2n$ times. However, we can still get a sufficiently small probability by setting $\frac{2}{3} < c < 1$. As a result, RR_k requires $1.25n + o(n)$ steps with high probability. □

5 Concluding Remarks

It is known that routing can be done in $1.0n$ steps if all the processors know the source and destination addresses of all the packets (so-called off-line routing) [IMK96]. This $1.0n$ is also an absolute lower bound. The question is how close we can be to this bound in the normal routing. Further improvements may be possible for Stage 2 of both RR and RR_k and for the local computation in Stage 1 of RR_k. Also an interesting question is whether we can apply our randomization technique to improve the upper bound for the mesh equipped with both buses and local links.

References

[Che52] H. Chernoff, "A Measure of asymptotic efficiency for tests of a hypothesis based on the sum of observations," *Annals of Mathematical Statistics*, 23 (1952) 493-507.

[CL93] S. Cheung and F.C.M. Lau, "A lower bound for permutation routing on two-dimensional bused meshes," *Information Processing Letters*, 45 (1993) 225-228.

[CLT96] D.D. Chinn, T. Leighton and M. Tompa, "Minimal adaptive routing on the mesh with bounded queue size," *J. Parallel and Distributed Computing* 34 (1996) 154-170.

[IMK96] K. Iwama, E. Miyano, and Y. Kambayashi, "Routing problems on the mesh of buses," *J. Algorithms*, 20 (1996) 613-631.

[IM97a] K. Iwama and E. Miyano, "Oblivious routing algorithms on the mesh of buses," In *Proc. International Parallel Processing Symposium* (1997) 721-727.

[IM97b] K. Iwama and E. Miyano, "Three-dimensional meshes are less powerful than two-dimensional ones in oblivious routing," In *Proc. European Symposium on Algorithms* (1997) 154-170.

[Jof74] A. Joffe, "On a set of almost deterministic k-independent random variables," *The Annals of Probability*, Vol.2 (1974) 161-162.

[Kri91] D. Krizanc, "Oblivious routing with limited buffer capacity," *J. Computer and System Sciences*, 43 (1991) 317-327.

[Lei92] F.T. Leighton, *Introduction to parallel algorithms and architectures: arrays, trees, hypercubes*, Morgan Kaufmann (1992).

[LS94] L.Y.T. Leung and S.M. Shende, "On multidimensional packet routing for meshes with buses," *J. Parallel and Distributed Computing*, 20 (1994) 187-197.

[MS96] R. Miller and Q.F. Stout, *Parallel algorithms for regular architectures: meshes and pyramids*, The MIT Press (1996).

[MR95] R. Motwani and P. Raghavan, *Randomized algorithms*, Cambridge University Press (1995).

[Tom94] M. Tompa, *Lecture notes on message routing in parallel machines*, Technical Report # 94-06-05, Department of Computer Science & Engineering, University of Washington (1994).

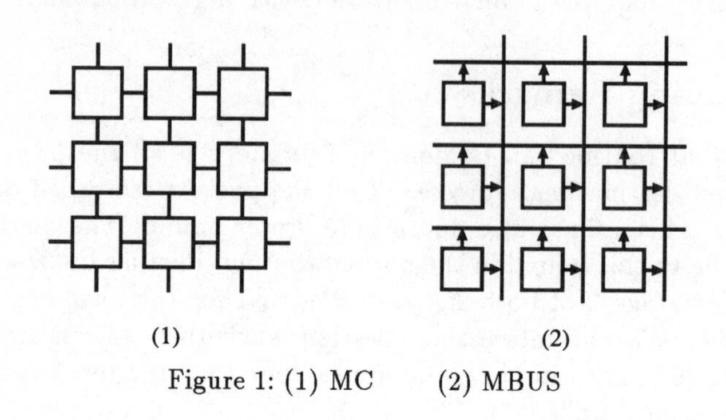

Figure 1: (1) MC (2) MBUS

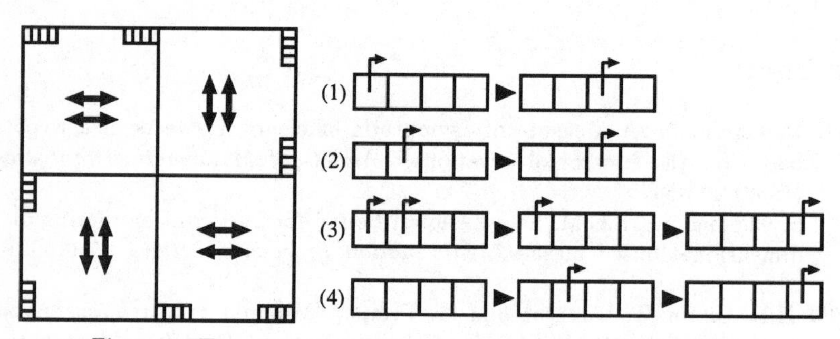

Figure 2: Two packets written on the row bus in Stage 1-1.

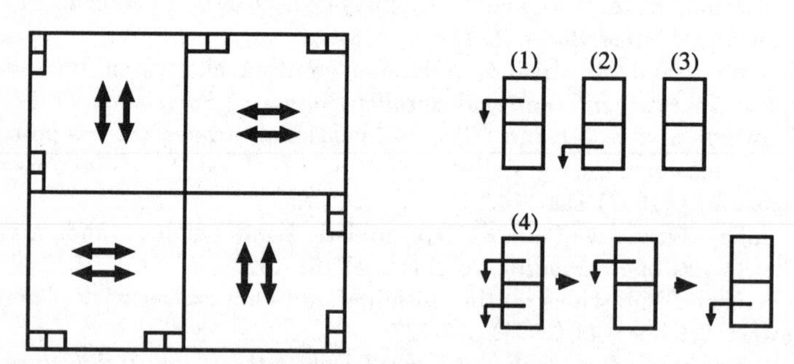

Figure 3: The remaining packets written on the column bus in Stage 1-2.

Executing Divisible Jobs on a Network with a Fixed Number of Processors (Extended Abstract)

Tsan-sheng Hsu[1*] and Dian Rae Lopez[2**]

[1] Institute of Information Science, Academia Sinica, Nankang, Taipei 11529, Taiwan, ROC
tshsu@iis.sinica.edu.tw

[2] Division of Science and Mathematics, University of Minnesota, Morris, Morris, Minnesota 56267, USA
lopezdr@cda.mrs.umn.edu

Abstract. In real practice, a job sometimes can be divided into s independent tasks to be distributed for execution on a network with a fixed number of processors. The overall finish time can vary widely depending on variables such as latency, data partitioning and/or data combining times, the individual execution times, the amount of data to be transferred, and the sending out of more tasks than needed. This paper studies the problem of finding an optimal task scheduling for a divisible job such that the overall finish time is minimized.
We first prove the studied problem is NP-complete and give a simple 3-OPT approximation algorithm. Then we develop a $(2 + \epsilon)$-OPT linear-time approximation algorithm by generalizing our simple algorithm, where ϵ is an arbitrarily small constant. A linear-time 2-OPT approximation algorithm is given when we divide the tasks evenly. Algorithms to find optimal solutions are then given for two special cases: 1) when the network has exactly two processors and 2) when the evenly divided tasks have symmetric behaviors. These cases happen frequently in real practice.

1 Introduction

This paper studies how to solve a problem that needs intensive computational time on a network with a fixed number of processors. Such a problem can often be divided into a set of tasks with precedence constraints and communication overheads [3, 13, 16]. This paper studies the case when the task graph is a directed graph with one source, one sink and with every other vertex having an incoming edge from the source and an outgoing edge to the sink. Jobs that can be represented using this type of task graph frequently happen in practice

* Research supported in part by NSC of Taiwan, ROC Grant 87-2213-E-001-022.
** Research supported in part by Academia Sinica, Taipei and the University of Minnesota, Morris.

and include matrix multiplication [17], and imaging, signal or pattern matching algorithms using hierarchical tree data structures (e.g., quad-trees) [6, 15, 19].

To model a network of processors, this paper uses a model proposed in [11]. This model is designed to represent a communication network of computers or workstations each with its own memory and microprocessor. A pipe-lined message sending (and receiving) cost with latency is used which is proportional to the message size. In such an environment, communication times are significant when tasks are located on different processors. The growth of such networks mandates more study into the efficient use of their parallel computing power. More importantly, the model we use is general enough to be used for any algorithm which can be represented as a set of tasks which communicate with each other and whose execution and communication costs are known or can be estimated. An example where such an algorithm would be helpful is a network of computers using PVM parallel software [8].

If the task precedence graph of the problem is represented as a one-level directed out- or in-tree, results are shown in [11, 12].

Previous work on this problem was based on different models of parallel computations and some of them were very theoretical in nature [1, 2, 4, 5, 14, 18]. This paper concentrates on solving problems on a network with a fixed number of processors. Previously, approximated solutions are given when the intermediate tasks have different execution times and for task graphs that are one-level trees [11, 12]. We improve both the approximation factor and the running time of the approximation algorithms presented in [12] by relating existing approximation results to ours. We further notice that in real applications the tasks fanning out have regular structures, i.e., their execution times are equal (e.g., the matrix multiplication problem). In many cases, the amount of data to be sent out and collected is also the same, which implies that the inter-processor communication times for tasks are likely to be the same. In this paper, we also show exact solutions under these constraints.

2 Preliminaries

2.1 Notation

In this paper, we use the following notation.

1. Let a job $\mathcal{J} = \{t_0, t_1, \ldots, t_n\}$ be a set of tasks whose precedence constraints form a one-level send-received graph $\mathrm{PC}(\mathcal{J})$.
2. Given a task t_i, let e_i be its execution time.
3. Given two tasks t_i and t_j with the precedence constraint that t_i must be executed before t_j can be executed, let $c_{i,j}$ be the communication time from t_i to t_j if they are not executed on the same processor. The communication time is 0 if t_i and t_j are executed on the same processor. All data streams are transmitted in a pipelined fashion, i.e., after t_i starts sending, all data arrives at t_j in $c_{i,j} + L$ units of time. If a task needs to send or receive two data elements at the same time, the two I/O operations must take place in sequence.

4. We schedule $\mathcal{J}$ on uniform processors $P_0, P_1, \ldots, P_r$, $r \leq n$, with the system I/O latency, L. We assume the job starts at P_0.

Given a one-level send-receive graph, the number of intermediate vertices is called its *fan-out*. We define the tasks represented by intermediate vertices *intermediate tasks*. The tasks represented by the starting vertex and the terminating vertex are called the *starting* task and the *terminating* task, respectively. We call a job *divisible* if its task precedence graph is a one-level send-receive task graph. A divisible job is *evenly divisible* if the execution times of all intermediate tasks are equal. Given two intermediate tasks in a one-level send-receive task graph, they are *symmetrical* if their execution times and communication times are equal. A one-level send-receive task graph is *symmetrical* if all tasks represented by intermediate vertices are symmetrical. A divisible job is *symmetrically divisible* if its task precedence graph is symmetrical. A scheduling S, for $\mathcal{J}$, is an assignment of tasks to processors.

A *legal realization* for S is the assignment of starting times for all tasks allocated to each processor such that it satisfies the precedence constraints and the I/O latency requirement. We also consider only the realization that if a processor has something waited to be done, then it will do it instead of idle. Note that a processor can be idle at some time if the processor is waiting for tasks allocated on other processors because of the precedence constraints. The *makespan* of a processor P_i for a realization is the time at which the processor P_i finishes the execution, including the idle time and the communication, of all tasks allocated to it. The makespan of a legal realization is the largest makespan among all processors. The makespan of a divisible job is always equal to the makespan of P_0. A legal realization with the smallest makespan is a *best* realization. The makespan of scheduling S is the makespan of its best realization and is denoted as $M(S)$. An *optimal scheduling* $\mathcal{J}$ is a scheduling with the smallest possible makespan. We further define $\mathrm{OPT}_k(\mathcal{J})$ to be a scheduling for $\mathcal{J}$ to be executed on k processors with the optimal makespan. Hence $M(\mathrm{OPT}_k(\mathcal{J}))$ is the *optimal makespan*. Note that t_0 and t_{n+1} are allocated to P_0.

An example of a realization for a scheduling on a divisible job $\mathcal{J} = \{t_0, \ldots, t_5\}$ is shown in Figure 1.

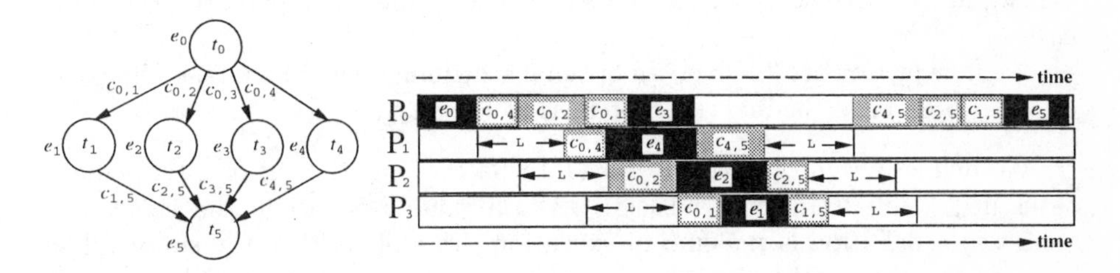

Fig. 1. A realization for a scheduling on the right for the send-receive task graph on the left. Assume that tasks t_0, t_3 and t_5 are allocated on P_0. Tasks t_1, t_2 and t_4 are allocated on processors other than P_0.

2.2 NP-complete Results

Lemma 1 *The problem of finding an optimal scheduling for a divisible job on a network with k processors is NP-complete if $k > 1$.*

Proof. Assume $L = 0$, $e_0 = e_{n+1} = 0$, and $c_{0,i} = c_{i,n+1} = 0$ for all $1 \leq i \leq n$. Then this problem is reduced to the problem of finding a best makespan on k identical processors without precedence constraints [7, Problem SS8], which is NP-complete even when $k = 2$.

Lemma 2 *[12] The problem of finding an optimal scheduling for an evenly divisible job on a network with k processors is NP-complete if $k > 2$.*

3 Divisible Jobs

This section presents approximation algorithms to find an optimal scheduling for a divisible job on a network with fixed k processors. The problem is NP-complete by Lemma 1 for any fixed $k \geq 2$. Previously, an $O(n \cdot \log n)$-time $(4 - 1/k)$-OPT algorithm is presented in [12].

3.1 A Simple 3-OPT Approximation Algorithm

Let $\mathcal{J}' = \{t_0, t_{n+1}\} \cup \{t_i \mid 1 \leq i \leq n \text{ and } c_{0,i} + c_{i,n+1} \geq e_i\}$. Let $\mathcal{J}'' = \{t_i \mid 1 \leq i \leq n \text{ and } c_{0,i} + c_{i,n+1} < e_i\}$. Hence $\mathcal{J}'$ and $\mathcal{J}''$ is a disjoint partition of $\mathcal{J}$.

Lemma 3

1. $M(\mathrm{OPT}_k(\mathcal{J})) \geq e_0 + e_{n+1} + \sum_{t_i \in \mathcal{J}''}(c_{0,i} + c_{i,n+1}) + \sum_{t_i \in \mathcal{J}'} e_i$.
2. *An optimal scheduling for $\mathcal{J}$ is to schedule all tasks on P_0 if and only if for all tasks $t_i \in \mathcal{J}''$, $\sum_{j=1}^{n} e_j \leq c_{i,n+1} + 2 \cdot L + e_i + c_{0,i}$.*
3. *Assume that executing all tasks on P_0 is not an optimal scheduling for $\mathcal{J}$. Then $M(\mathrm{OPT}_k(\mathcal{J})) \geq e_0 + e_{n+1} + 2 \cdot L + \min_{t_i \in \mathcal{J}''} e_i$.*

Proof. Part 2 is from [12]. We first prove part 1. In any optimal scheduling for $\mathcal{J}$, all tasks in $\mathcal{J}'$ must be allocated on P_0, since otherwise we could reschedule them on P_0 giving a smaller makespan. Note that $c_{0_i} + c_{i,n+1} < e_i$ if $t_i \in \mathcal{J}''$. Among all possible schedulings, the smallest possible makespan for P_0 is thus $e_0 + e_{n+1} + \sum_{t_i \in \mathcal{J}''} c_{0,i} + c_{i,n+1} + \sum_{t_i \in \mathcal{J}'} e_i$.

We now prove part 3. Given an optimal scheduling, let t_i be a task allocated on P_j, $j > 0$. Then the makespan of P_j is at least $e_0 + e_{n+1} + 2 \cdot L + e_i$.

We define a *valid k partition* for $\mathcal{J}$ to be the partitioning of intermediate tasks in $\mathcal{J}''$ into k disjoint subsets. Given a valid k partition, the *value* of a subset in a partition is the sum of execution times of tasks in the subset. The *largest value* of a valid k partition is the largest *value* of subsets in the partition. We also pick an arbitrary subset in the partition with the largest value to be its *largest subset.* We further define $U_{k,\mathcal{J}}$ to be the smallest largest value of all possible valid k partitions.

Lemma 4 $M(\mathrm{OPT}_k(\mathcal{J})) \geq U_{k,\mathcal{J}} \geq \max\{\max_{t_i \in \mathcal{J}''} e_i, \sum_{t_i \in \mathcal{J}''} e_i/k\}$.

Proof. In any optimal scheduling for $\mathcal{J}$, tasks in $\{t_0, t_{n+1}\} \cup \mathcal{J}'$ are executed on P_0. Tasks in $\mathcal{J}''$ are distributed among the k processors, which form a valid k partition. The makespan of any scheduling is hence at least the largest value of the corresponding valid k partition. Thus $M(\mathrm{OPT}_k(\mathcal{J})) \geq U_{k,\mathcal{J}}$. The proof that $U_{k,\mathcal{J}} \geq \max\{\max_{t_i \in \mathcal{J}''} e_i, \sum_{t_i \in \mathcal{J}''} e_i/k\}$ is straightforward.

We first prove two lemmas that are needed to derive our approximation algorithm.

Lemma 5 *There is a linear-time algorithm to find a a valid k partition for $\mathcal{J}$ whose largest value is at most $2{\cdot}U_{k,\mathcal{J}}$;*

Proof. Let $Q = \max\{\max_{t_i \in \mathcal{J}''} e_i, \sum_{t_i \in \mathcal{J}''} e_i/k\}$.

Algorithm VKP($\mathcal{J}$,Q) /* Finding a valid k partition. */

1. Let sets $S_1 = \cdots = S_{2k} = \emptyset$; $current := 1$, $full := Q$;
2. **for** $i = 1{\cdot\cdot}n$ **and** $t_i \in \mathcal{J}''$ **do**
   (a) **while** $e_i + \sum_{t_i \in S_{current}} e_i \geq full$ **do**
   $current := current + 1$;
   (b) $S_{current} := S_{current} \cup \{t_i\}$;
3. **return** $S_1 \cup S_2, S_3 \cup S_4, \ldots, S_{2k-1} \cup S_{2k}$ as a valid k partition.
4. **end**;

It is straightforward to see that the returned sets are a valid k partition and that the algorithm runs in linear time. If the value of *current* in Step 2a is at most $2{\cdot}k$, then the largest value of the returned valid k partition is at most $2{\cdot}Q$, which is at most $2{\cdot}U_{k,\mathcal{J}}$. We now prove the value of *current* in Step 2a is at most $2{\cdot}k$. Note that the value of $S_i \cup S_{i+1}$, $1 \leq i \leq current$, is at least Q because of Step 2a. Hence if $current > 2{\cdot}k$, then $\sum_{j=1}^{current} \sum_{t_i \in S_{2j-1} \cup S_{2j}} e_i > \sum_{t_i \in \mathcal{J}''} e_i$, which is a contradition. Hence the lemma is true.

Lemma 6 *In Algorithm* VKP, $2{\cdot}M(\mathrm{OPT}_k(\mathcal{J})) \geq 2{\cdot}L + \max_{j=2}^{k} \sum_{t_i \in S_{2j-1} \cup S_{2j}} e_i$.

Proof. Let S be an optimal scheduling whose corresponding valid k partition on tasks in $\mathcal{J}''$ is $\{R_0, \ldots, R_{k-1}\}$, where R_i is the set of tasks in $\mathcal{J}''$ that are allocated on P_i. Let V_i be the value of R_i. Let $W = \max_{i=1}^{k-1} V_i$. Note that $M(\mathrm{OPT}_k(\mathcal{J})) \geq \max\{2{\cdot}L + W, V_0\}$. By Lemma 5, $\max_{j=2}^{k} \sum_{t_i \in S_{2j-1} \cup S_{2j}} e_i \leq 2{\cdot}Q$ By the argument of the value of some subset in a partition must be at least its mean value, $W \geq Q$ if $V_0 \leq Q$. If $W \geq Q$, then our lemma follows from Lemma 3(3); otherwise, we have the following cases. Note that we assume $W < Q$. Hence $V_0 > Q$. We want to prove $(2{\cdot}L + 2{\cdot}Q)/\max\{2{\cdot}L + W, V_0\} \leq 2$. Hence it suffices to prove $(2{\cdot}L + 2{\cdot}Q)/(2{\cdot}L + W) \leq 2$ or $(2{\cdot}L + 2{\cdot}Q)/V_0 \leq 2$.
Case 1: $L \geq Q$. Hence $(2{\cdot}L + 2{\cdot}Q)/(2{\cdot}L + W) \leq 2$.
Case 2: $L < Q$ and $V_0 \geq 2{\cdot}Q$. Hence $(2{\cdot}L + 2{\cdot}Q)/V_0 \leq 2$.
Case 3: $L < Q$ and $V_0 = Q + \delta$, where $0 < \delta < Q$. Also by the argument

of the value of some subset in a partition must be at least its mean value, $W \geq Q - \delta/(k-1)$.

Case 3.1: $L \geq \delta/(k-1)$. Then $(2{\cdot}L + 2{\cdot}Q)/(2{\cdot}L + W) \leq 2$.

Case 3.2: $L < \delta/(k-1)$. Then $(2{\cdot}L + 2{\cdot}Q)/(Q + \delta) = (2{\cdot}L + 2{\cdot}Q)/V_0 \leq 2$.

Theorem 7 *There is a linear-time 3-OPT approximation algorithm for the problem of finding an optimal scheduling of a divisible job on a fixed k, $k > 1$, processors*

Proof. Our algorithm works as follows.

Algorithm KD($\mathcal{J}$)

1. Let S' be the scheduling with all tasks allocated on P_0;
   Let M' be the makespan of S';
2. (a) Find a valid k partition $\mathcal{P}$ for $\mathcal{J}$ whose largest value is at most $2{\cdot}U_{k,\mathcal{J}}$;
   (b) Let S'' be the scheduling by allocating tasks in $\{t_0, t_{n+1}\} \cup \mathcal{J}'$ and the largest subset in $\mathcal{P}$ on P_0, and the rest of the subsets one each on P_i, $1 \leq i < k$;
   (c) Let M'' be the makespan of an arbitrary realization of S'';
3. **if** $M' \geq M''$, **then return** S' **else return** S'';
4. **end**;

Using Lemma 3(2), Step 1 can be implemented in linear time. Let $Q = \max\{\max_{t_i \in \mathcal{J}''} e_i, \sum_{t_i \in \mathcal{J}''} e_i/k\}$. By Lemma 4, $Q \leq M(\mathrm{OPT}_k(\mathcal{J}))$. By Lemma 5, Step 2a can be done in linear time.

We then prove the approximation bound of Algorithm KD. We assume that S' is not an optimal scheduling; otherwise, our algorithm has found the optimal scheduling.

Assume *wlog* that $S_1 \cup S_2$ is the largest subset in $\mathcal{P}$. the makespan of P_0 is either one the the following:

$$e_0 + e_{n+1} + \sum_{t_i \in \mathcal{J}''} c_{0,i} + c_{i,n+1} + \sum_{t_i \in \mathcal{J}'} e_i + \sum_{t_i \in S_1 \cup S_2} e_i \tag{1}$$

$$e_0 + e_{n+1} + \sum_{t_i \in \mathcal{J}''} c_{0,i} + c_{i,n+1} + 2{\cdot}L + \max_{j=2}^{k} \sum_{t_i \in S_{2j-1} \cup S_{2j}} e_i \tag{2}$$

Note that $M(\mathrm{OPT}_k(\mathcal{J})) \geq U_{k,\mathcal{J}}$. By Lemma 5 and Lemma 4, $\sum_{t_i \in S_1 \cup S_2} e_i \leq 2{\cdot}M(\mathrm{OPT}_k(\mathcal{J}))$. By Lemma 3(1), $e_0 + e_{n+1} + \sum_{t_i \in \mathcal{J}''} c_{0,i} + c_{i,n+1} + \sum_{t_i \in \mathcal{J}'} e_i \leq M(\mathrm{OPT}_k(\mathcal{J}))$. Hence Equation 1 $\leq 3{\cdot}M(\mathrm{OPT}_k(\mathcal{J}))$. By Lemma 6, Equation 2 is also $\leq 3{\cdot}M(\mathrm{OPT}_k(\mathcal{J}))$.
Hence the theorem is true.

3.2 Generalizations

The problem of finding a valid k partition with the largest value $U_{k,\mathcal{J}}$ is equivalent to the problem of scheduling n independent tasks on k identical machines without any communication costs or any partitioning and combining computation. This problem is well-studied [2, 5].

It is well-known that computing $U_{k,\mathcal{J}}$ is NP-complete since if we set $k = 2$, $L = 2$, $e_0 = e_{n+1} = 0$, and all communication times to be zero, then it reduces to the partition problem. Hence we give a linear-time 3-OPT approximation algorithm in Theorem 7. We remark that the 2-OPT approximation algorithm to approximate $U_{k,\mathcal{J}}$ is first described in [9]. From the proof of Theorem 7, we observe the following.

Lemma 8 *Let $\mathcal{J}$ be a divisible job. If there is an approximation algorithm for finding a valid k partition whose largest value is within r_1 times $U_{k,\mathcal{J}}$ in $O(t_1)$ time, then there is an $O(t_1 + n)$-time algorithm to find a scheduling for $\mathcal{J}$ whose makespan is within $r_1 + 1$ times $M(\mathrm{OPT}_k(\mathcal{J}))$.*

Proof. Straightforward from the proof of Theorem 7, especially Lemma 6.

Note that Lemma 8 says that better approximation algorithms lead to a better approximation algorithm for our problem of finding an optimal scheduling for a divisible job on k processors.

Theorem 9 *There is linear-time approximation algorithm for finding a scheduling of a divisible job $\mathcal{J}$ on k processors whose makespan is arbitrarily close to $2{\cdot}M(\mathrm{OPT}_k(\mathcal{J}))$.*

Proof. From Lemma 8 and from the fact that there is an polynomial approximation scheme that runs in $O(c_\epsilon{\cdot}n)$time for finding a valid k partition whose largest value is within $(1+\epsilon){\cdot}U_{k,\mathcal{J}}$, where $0 < \epsilon \leq 1$ and $c_\epsilon = O(2^{2^{1/\epsilon}})$ [10].

4 Evenly and Symmetrically Divisible Jobs

4.1 Evenly Divisible Jobs

This subsection studies the optimal scheduling problem for an evenly divisible job on a fixed number k of processors. Lemma 2 shows that this problem is NP-complete if $k > 2$.

Algorithm for Two Processors We solve the problem of finding a best scheduling for an evenly divisible job on a network of two processors P_0 and P_1. Without loss of generality, assume that $c_{0,i} + c_{i,n+1} \leq c_{0,i+1} + c_{i+1,n+1}$, $1 \leq i < n-1$.

Lemma 10 *Let S_w be a scheduling of J with exactly w intermediate tasks scheduled on P_1. Given any w, among all possible selections of S_w, the one which schedules tasks $t_1, \ldots, t_w$ on P_1 has the least makespan.*

Proof. Given any S_w, let T_w be the set of tasks allocated on P_1 and let T'_w be the set of intermediate tasks allocated on P_0. The makespan of S_w is either $e_0 + \sum_{t_i \in T_w}(c_{0,i} + c_{i,n+1}) + (n-w)\cdot e + e_{n+1}$ or $e_0 + 2\cdot L + \sum_{t_i \in T_w}(c_{0,i} + c_{i,n+1}) + w\cdot e + e_{n+1}$. In both cases, the value $\sum_{i \in T_w}(c_{0,i} + c_{i,n+1})$ is minimized if $T_w = \{t_1, \ldots, t_w\}$. Hence the lemma holds.

Theorem 11 *There is a linear-time algorithm to find an optimal scheduling for an evenly divisible job on two processors when the total communication times of intermediate tasks are sorted.*

Proof. Our algorithm is based on Lemma 10. Let $T^*_w = \{t_1, \ldots, t_w\}$. Let S^*_w be the schedule realized by allocating the tasks in T^*_w on P_1 and everything else on P_0. We first find the makespan for S^*_n by allocating all tasks on P_1 except for task t_0. Then we incrementally obtain the makespan of S^*_{n-1} from what we have for computing the makespan of S^*_n. We continue removing the largest task from P_i and placing it on P_0 until we have obtained the makespan of S^*_0. An optimal scheduling is the minimum makespan of $S^*_i, 0 \leq i \leq n$, found above.

Algorithm for any $k > 2$ Processors

Theorem 12 *There is linear-time 2-OPT approximation algorithm to find a scheduling for an evenly divisible job $\mathcal{J}$ on a network with k processors whose makespan is at most $2\cdot M(\mathrm{OPT}_k(\mathcal{J}))$.*

Proof. Since the execution times of non-root tasks are all e, each processor must spend at least $\lceil n/k \rceil\cdot e$ time executing tasks. It is straightforward to find a valid k partition whose largest value is $\lceil n/k \rceil\cdot e$. Hence this theorem follows from lemma 8.

When the number of available processors in the network is fixed, the problem is NP-hard by [12]. We first give a 3-OPT approximation algorithm. Note that the difference d_i equals $e_i - (c_{0,i} + c_{i,n+1})$. We also assume for now that $d_i > 0$, $1 \leq i \leq n$.

4.2 Symmetrically Divisible Jobs

We first present needed notations. Let $\mathcal{S}$ be a scheduling for a symmetrically divisible job executed on a network of processors $P_0, \ldots, P_{k-1}$. Assume that the starting and terminating tasks are executed on P_0.

Lemma 13 *A best realization of $\mathcal{S}$ has the following properties.*

- *P_0 first executes the starting task, then spend time to send out data for tasks not allocated on P_0. Finally, P_0 executes intermediate tasks allocated on it, collects data sent from other processors, and then execute the terminating task.*

- *Processor P_i, $i > 0$, first receives data sent from P_0 and then immediately executes the first allocated task. It then sends output data back to P_0 and waits to receive data for the second allocated task sent from P_0. After the data is received, the second task is executed. The above continues until all tasks are executed.*

Proof. Straightforward from the fact that all intermediate tasks are symmetric.

Assume that w tasks are not allocated on P_0 and data for those tasks are sent out in the order of $t'_1, \ldots, t'_w$. Given a best realization for $\mathcal{S}$ as described in Lemma 13, we define its *ith snap shot*, $1 \leq i \leq w$, being an assignment of time to processors P_i, $i > 0$. The time assigned for each processor is the earliest time that this processor finishes executing tasks, including its communication time, in $\{t'_1, \ldots, t'_i\}$ that are allocated on it. The time assigned in the ith snap shot for processor P_j, $j > 0$, is called the *available time* for P_j.

Lemma 14 *There exists an optimal scheduling for a symmetrically divisible job whose best realization satisfies Lemma 13 and an intermediate task t'_i, $1 < i \leq w$, is allocated on the processor with the earliest available time in the $(i-1)$th snap shot.*

Proof. Let $\mathcal{S}$ be an optimal scheduling. Let r be the least integer larger than 1 such that t'_r is not allocated accordingly. Assume that t'_r is allocated on P_u whose available time is later than P_v. We obtain from $\mathcal{S}$ an scheduling $\mathcal{S}'$ such that the makespan of $\mathcal{S}'$ is no larger than the makespan of $\mathcal{S}$ using the following two rules. If no task in $\{t'_{r+1}, \ldots, t'_w\}$ is allocated on P_v, then $\mathcal{S}'$ is the same as $\mathcal{S}$ except that t'_r is allocated on P_v. Otherwise, let t'_s, $s > r$, be the first intermediate task that is allocated on P_v after the $(r-1)$th snap shot is taken. Then $\mathcal{S}'$ is the same as $\mathcal{S}$ except that t'_r is allocated on P_v and t'_s is allocated on P_u. Because all intermediate tasks are symmetrical, the makespan of $\mathcal{S}'$ is no larger than that of $\mathcal{S}$.

Theorem 15 *The optimal scheduling of a symmetrical one-level send-receive task graph to be executed on a network of $k > 1$ processors can be found in linear time.*

Proof. Since all intermediate tasks are symmetrical and by Lemmas 13 and 14, an optimal scheduling can be found if we know the the number of tasks that are not allocated on P_0. Assume that the set $\mathcal{W}$ of tasks not allocated on P_0 is $\{t'_1, \ldots, t'_w\}$. In order not to keep track of the available time of each processor in every snap shot, we use the following strategy. We allocate task t'_i, $1 \leq i \leq w$, on processor $P_{i \bmod (k-1)}$. Since all tasks in $\mathcal{W}$ are symmetrical, the available time for processor $P_{i \bmod (k-1)}$, $i > 1$, is the earliest in the $(i-1)$th snap shot. There are up to m possible values for w, where m is the number of intermediate tasks. However, we first assume $w = m$. Let p_i be the latest available times for the ith snap shot. Using p_i, we can compute the makespan on P_0 when there are exactly i intermediate tasks not allocated on P_0.

References

1. E. Bampis, F. Guinand, and D. Trystram. Some models for scheduling parallel programs with communication delays. *Discrete Applied Mathematics*, 72:5–24, 1997.
2. J. Błażewicz, K. H. Ecker, E. Pesch, G. Schmidt, and J. Węglarz. *Scheduling Computer and Manufacturing Processes.* Springer, 1996.
3. T. C. E. Cheng and C. C. S. Sin. A state-of-the-art review of parallel-machine scheduling research. *European J. Operational Research*, 47:271–292, 1990.
4. P. Chrétienne. Tree scheduling with communication delays. *Discrete Applied Mathematics*, 49:129–141, 1994.
5. P. Chrétienne, E. G. Coffman, Jr., J. K. Lenstra, and Z. Liu, editors. *Scheduling Theory and its Applications.* John Wiley & Sons Ltd, 1995.
6. Y. Cohen, M. S. Landy, and M. Pavel. Hierarchical coding of binary images. *IEEE Trans. on Pattern Analysis and Machine Intelligence*, PAMI-7(3):284–298, 1985.
7. M. R. Garey and D. S. Johnson. *COMPUTERS AND INTRACTABILITY A Guide to the Theory of* **NP**-*Completeness.* W. H. Freeman and Company, New York, 1979.
8. A. Geist, A. Beguelin, J. Dongarra, W. Jiang, R. Manchek, and V. Sunderam. *PVM 3 User's Guide and Reference Manual.* Oak Ridge National Laboratory, Oak Ridge, Tennessee 37831, USA, May 1993.
9. R. L. Graram. Bounds on multiprocessing timing anomalies. *SIAM Journal on Appl. Math.*, 17(2):416–429, 1969.
10. D. S. Hochbaum and D. B. Shmoys. A polynomial approximation scheme for scheduling on uniform processors: Using the dual approximatiom approach. *SIAM Journal on Computing*, 17(3):539–551, 1988.
11. L. Hollermann, T.-s. Hsu, D. R. Lopez, and K. Vertanen. Scheduling problems in a practical allocation model. *Journal of Combinatorial Optimization*, 1(2):129–149, 1997.
12. T.-s. Hsu and D. R. Lopez. Bounds and algorithms for a practical task allocation model (extended abstract). In *Proceedings of 7th International Symposium on Algorithms and Computation*, volume LNCS #1178, pages 397–406. Springer-Verlag, 1996.
13. V. M. Lo. *Task Assignment in Distributed Systems.* PhD thesis, Univ. of Illinois at Urbana-Champaign, USA, October 1983.
14. D. R. Lopez. *Models and Algorithms for Task Allocation in a Parallel Environment.* PhD thesis, Texas A&M University, Texas, USA, December 1992.
15. H. S. Malvar. Lapped transforms for efficient transformation/subband coding. *IEEE Trans. on Acoustics, Speech and Signal Processing*, 38(6):969–978, 1990.
16. M. G. Norman and P. Thanisch. Models of machines and computation for mapping in multicomputers. *ACM Computing Surveys*, 25(3):263–302, 1993.
17. S. A. Rees and J. P. Black. An experimental investigation of distributed matrix multiplication techniques. *Software – Practice and Experience*, 21(10):1041–1063, 1991.
18. H. S. Stone. Multiprocessor scheduling with the aid of network flow algorithms. *IEEE Trans. on Software Eng.*, SE-3(1):85–93, 1977.
19. P. Strobach. Tree-structured scene adaptive coder. *IEEE Trans. on Communications*, 38(4):477–486, 1990.

On the Ádám Conjecture on Circulant Graphs

Bernard Mans[1], Francesco Pappalardi[2] and Igor Shparlinski[1]

[1] Dept of Computing, MPCE, Macquarie University, Sydney, NSW 2109, Australia.
{bmans,igor}@mpce.mq.edu.au

[2] Departimento di Matematica, Terza Università delgi Studi, Roma, 00146, Italy.
pappa@mat.uniroma3.it

1 Introduction

In this paper we study isomorphism between circulant graphs. Such graphs have a vast number of applications to telecommunication network, VLSI design and distributed computation [4, 13, 15, 17]. By suitably choosing the length of the chord between two nodes of the network, one can achieve the appropriate property: e.g., low diameter, high connectivity, or implicit routing. A network that does provide labelled edges should be able to exploit the same properties as one with different labelling if the underlying graphs are isomorphic.

For general graph the problem is known to be in **NP**, not known to be in **P**, and probably is not **NP**-complete, see Section 6 of [3]. It has been conjectured by Ádám [1] that for circulant graphs there is a very simple rule to decide the isomorphism of two graphs. Although this rule is know to be false in general, even for undirected graphs [9], for several special cases it holds [5, 15, 18, 19, 20].

The *purpose* of this paper is to extend essentially the class of graphs having the Ádám property and even more general spectral Ádám property as well as to introduce a new technique, which is based on some number theoretic results about equations in roots of unity, which we believe can be applied to several other questions of graph theory. In particular, we believe that our method can be applied to study general *Cayley* graphs, see [20]. Indeed, at least in the case of Cayley graphs generated by an *abelian* group, the corresponding eigenvalues are linear combinations of group characters, that is they are linear combinations of roots of unity, see Section 3.12 of [3] or Lemma 9.2 of [12].

The new approach, which we propose here, is based on the combination of the spectral technique from [8, 9, 15] with some deep results of algebraic number theory on linear equations in roots of unity [6, 10, 16, 21, 22, 23]. In fact it can be extended to weighted circulant graphs as well. In particular, we settle the aforementioned Ádám conjecture [1] for a wide class of circulant graphs which are not covered by the previously known results.

We recall that an n-vertex *circulant graph* G is a graph whose adjacency matrix $A = (a_{ij})_{i,j=1}^{n}$ is a circulant. That is $a_{ij} = a_{1,j-i+1}$, $i, j = 1, \ldots, n$. Hereafter, subscripts are taken modulo n. We assume that $a_{ii} = 0$, $i = 1, \ldots, n$.

Therefore with every circulant graph one can associate a set $S \subseteq \mathbb{Z}_n$ of the positions of non-zero entries of the first row of the adjacency matrix of the graph. Respectively we denote by $\langle S \rangle_n$ the corresponding graph. We also recall that two graphs G_1, G_2 are *isomorphic*, and write $G_1 \simeq G_2$, if their matrices differ by a permutation of their rows and columns. We say that two sets $S, T \subseteq \mathbb{Z}_n$ are *proportional*, and write $S \sim T$, if for some integer l with $\gcd(l, n) = 1$, such that $S = lT$ where the multiplication is taken over $\mathbb{Z}_n$. Obviously, $S \sim T$ implies $\langle S \rangle_n \simeq \langle T \rangle_n$. For example ($S = \{1,5\}$, $T = \{1,9\}$, and $n = 23$), $\langle S \rangle_n \simeq \langle T \rangle_n$ since $S \sim T$ ($l = 5$).

Ádám [1] conjectured that the inverse statement is true as well. We say that a set $S \subseteq \mathbb{Z}_n$ has the *Ádám property* if for any other set $T \subseteq \mathbb{Z}_n$ the isomorphism $\langle S \rangle_n \simeq \langle T \rangle_n$ implies the proportionality $S \sim T$. Thus the *Ádám conjecture* is equivalent to the statement that all sets $S \subseteq \mathbb{Z}_n$ have Ádám property.

The discovered in [9] counterexample of 6-element sets $S_1 = \{\pm 1, \pm 2, \pm 7\} \subseteq \mathbb{Z}_{16}$, $S_2 = \{\pm 1, \pm 6, \pm 7\} \subseteq \mathbb{Z}_{16}$ shows that the Ádám conjecture is *false*. It is easy to verify that the isomorphism $\langle S_1 \rangle_{16} \simeq \langle S_2 \rangle_{16}$ is furnished by the permutation on $\mathbb{Z}_{16}$ given by

$$i \to \begin{cases} -5i, & \text{if } i \equiv 0 \pmod 2; \\ -5i - 4, & \text{if } i \not\equiv 0 \pmod 2; \end{cases} \qquad i \in \mathbb{Z}_{16},$$

but $S_1 \not\sim S_2$. In fact, counterexamples exist for any values of n except, maybe, n of the form $n = 2^\alpha 3^\beta m$, where $\alpha \in \{0,1,2,3\}$, $\beta \in \{0,1,2\}$, $\gcd(m, 6) = 1$ and m is squarefree [2].

Nevertheless, there are several very important families of circulant graphs for which the Ádám conjecture holds. For example, Muzychuk has shown that the Ádám conjecture is true for circulant graphs with a squarefree number of vertices [18] and with a twice squarefree number of vertices [19]. It also holds for 4-element sets S [7, 11], see also [14]. The corresponding graphs are known as *double loops* are have many applications to computer science. In fact it has been discovered in several paper that, under some additional restrictions, the isomorphism property of graphs can be replaced by the property of their isospectrality.

We recall that the spectrum $\operatorname{Spec} G$ of a graph G is the set of the eigenvalues of its adjacency matrix. In particular, isomorphic graphs have the same spectra (although the inverse statement is obviously false). Respectively, we say that a set $S \subseteq \mathbb{Z}_n$ has the *spectral Ádám property* if for any other set $T \subseteq \mathbb{Z}_n$ the isospectrality $\operatorname{Spec} S \simeq \operatorname{Spec} T$ implies the proportionality $S \sim T$.

Here we describe a general class of sets having the spectral Ádám property. For example, it is shown in [15] that any 4-element set $S = \{\pm 1, \pm d\} \subseteq \mathbb{Z}_n$ (an important sub-family of double loop circulant graphs), has the spectral Ádám property, provided that $2 \le d < \min\{n/4, \varphi(n)/2\}$, where $\varphi(n)$ is the Euler function. Here we settle the question almost completely, with only 5 possible exceptions, for this type of graphs; for all of them $d \ge n/6$. For more general sets of the form $S = \{\pm a, \pm b\}$ there are at most 12 possible exceptions. Moreover, if $\gcd(n, 12) < 3$ then there is no exceptions at all. We also show that for any fixed m the probability that a random m-element set $S \subseteq \mathbb{Z}_n$ does not have the spectral Ádám property, is $O(n^{-m/4})$.

2 Auxiliary Results

Let us consider the equation

$$a_0 + \sum_{j=1}^{k-1} a_j \zeta^{w_j} = 0, \qquad w_1, \dots, w_{k-1} \in \mathbb{Z}^{k-1}, \tag{1}$$

where $\zeta = \exp(2\pi i/n)$ and $a_0, \dots, a_{k-1}$ are nonzero integers.

We call a solution $(w_1, \dots, w_{k-1})$ of (1) *irreducible* if $\sum_{j \in J} a_j \zeta^{w_j} \neq 0$ for any *proper* subset $J \subset \{1, \dots, k-1\}$.

Such equations and their various generalizations have been studied in the literature a great deal [6, 10, 16, 21, 22, 23]. We summarize the results of [6, 16] in the following lemma.

Lemma 1. *For any irreducible solution of the equation (1), the fraction*

$$Q = \frac{n}{\gcd(n, w_1, \dots, w_{k-1})}$$

is squarefree and

$$\sum_{p|Q} (p-2) \leq k-2,$$

where the sum is taken over all prime divisors of Q.

In particular, one can see that $Q = 1$ if all prime divisors of n are greater than k and that $Q \leq 2$ if n is a power of 2. Let us denote

$$Q_k = \max \left\{ \prod_p p \mid \sum_p (p-2) \leq k-2 \right\}$$

where both, the product and the sum, are taken over distinct prime numbers. Thus for the quantity Q of Lemma 1 we have $Q \leq Q_k$.

From the known results on the distribution of prime numbers (see [6, 22, 23]) one easily derives that $Q_k \leq \exp\left((1 + o(1))\, k^{1/2} \log^{1/2} k\right)$.

It is easy to verify that for $S \subset \mathbb{Z}_n$, $\operatorname{Spec}\langle S \rangle_n = \left\{ \sum_{s \in S} \zeta^{ls} \mid l = 0, 1, \dots, n-1 \right\}$.

The following result is based on the previous representation and provides a connection between circulant graphs and equations roots of unity. It lies in the background of the approach of [15] (see the proof of Theorem 4 from [15]).

Lemma 2. *Let $S, T \subseteq \mathbb{Z}_n$ be such that* $\mathrm{Spec}\,\langle S\rangle_n = \mathrm{Spec}\,\langle T\rangle_n$ *but $S \not\sim T$. Then there exists l, $1 \le l \le n-1$, such that the polynomials*

$$F(X) = \sum_{s\in S} X^s - \sum_{t\in T} X^{lt}$$

is not identical to zero modulo $X^n - 1$ and $F(\zeta) = 0$.

We remark that in other words the polynomial does not vanish if one replaces the exponents lt, $t \in T$, by its smallest positive residues modulo n.

3 General Estimates

Let us denote

$$\rho_m = \min_{2\le k\le m} \max\left\{\frac{1}{Q_{2k}}, \frac{k-1}{Q_{m+k-\lceil m/k\rceil+1}}\right\}.$$

Obviously $\rho_m \ge 1/Q_{2m}$ thus we have the asymptotic inequality

$$\rho_m \ge \exp\left(-\left(2^{1/2}+o(1)\right) m^{1/2}\log^{1/2} m\right),$$

which apparently can be shown to be tight in the sense $\log\rho_m \sim -\log Q_{2m}$, $m \to \infty$. However, for smaller values of m one can obtain better than $\rho_m \ge 1/Q_{2m}$ numerical estimates. For example:

m	2	3	4	5	6	7	8	9	10
ρ_m	1/6	1/15	1/15	2/35	1/42	2/105	2/105	4/231	3/770
$1/Q_{2m}$	1/6	1/30	1/42	1/70	1/210	1/210	1/330	1/462	1/2310

Theorem 3. *Let $S = \{s_1, \ldots, s_m\} \subseteq \mathbb{Z}_n$ be an m-element set which does not satisfy the spectral Ádám property. Then*

$$\max_{1\le i<j\le m} |s_i - s_j| \ge \rho_m n.$$

Proof. From Lemma 2 we conclude that for some subsets $U \subseteq S$ and $V \subseteq T$ with $U \cap V = \emptyset$ we have

$$\sum_{u\in U}\zeta^u - \sum_{v\in V}\zeta^v = 0.$$

We split this equation into the maximal possible set of r, $m \ge r \ge 1$, subequations

$$\sum_{u\in U_\mu}\zeta^u - \sum_{v\in V_\mu}\zeta^v = 0, \qquad \mu = 1, \ldots, r,$$

where $U = \bigcup_{\mu=1}^r U_\mu$, $V = \bigcup_{\mu=1}^r V_\mu$ and $U_\mu \bigcap U_\nu = V_\mu \bigcap V_\nu = \emptyset$, $1 \le \mu < \nu \le r$.

Let R be the set of $\mu = 1, \ldots, r$ for which $\#U_\mu \geq 2$. We put

$$L = \#R \qquad \text{and} \qquad M = \sum_{\mu \in R} \#U_\mu.$$

Because $\#U_\mu = \#V_\mu = 1$ is not possible, $\mu = 1, \ldots, r$,we see that there are $r - L = m - M$ values of $\mu = 1, \ldots, r$ with $\mu \notin R$ and $\#V_\mu \geq 2$ for each of such μ. Therefore for any $\mu \in R$

$$\#V_\mu \leq m - L - 2(m - M) + 1 = 2M - m - L + 1 \tag{2}$$

Denote $\Delta = \max_{1 \leq i < j \leq m} |s_i - s_j|$.

First of all let us select a pair (U_μ, V_μ), $\mu \in R$, for which the total cardinality $N = \#U_\mu + \#V_\mu$ is minimal. Select two arbitrary distinct elements $u_1, u_2 \in U_\mu$ with $0 < |u_1 - u_2| \leq \Delta$. Dividing out the corresponding equation by u_1 and applying Lemma 1 we obtain $\Delta \geq n/Q_N$.

Now we select select a pair (U_μ, V_μ) for which the total cardinality $K = \#U_\mu$ is maximal. Then the selected subset $U_\mu \subseteq S$ contains at least one pair $u_1, u_2 \in U_\mu$ with $0 < |u_1 - u_2| \leq (\Delta - 1)/(K - 1)$. Dividing out the corresponding equation by u_1 and applying Lemma 1 we obtain that

$$\Delta \geq \frac{n(K-1)}{Q_{2M+K-m-L+1}} + 1.$$

We have $K \geq M/L$, thus $L \geq \lceil M/K \rceil$ and we derive

$$Q_{2M+K-m-L+1} \leq Q_{M+K-\lceil M/K \rceil+1} \leq Q_{m+K-\lceil m/K \rceil+1}.$$

We also remark that $N \leq 2M/L \leq 2K$. Combining the above inequalities we obtain the desired estimate. □

Similar arguments show that if the smallest prime divisor of n is greater than m then the spectral Ádám property holds for all m-element sets $S = \{s_1, \ldots, s_m\} \subseteq \mathbb{Z}_n$. It also easy to see that in the most interesting for applications case when n is a power of 2 sets $S = \{s_1, \ldots, s_m\} \subseteq \mathbb{Z}_n$, satisfy the Ádám property if

$$\max_{1 \leq i < j \leq m} |s_i - s_j| < n/2.$$

Denote by $A_m(n)$ the number of m-element sets $S = \{s_1, \ldots, s_m\} \subseteq \mathbb{Z}_n$ which do not satisfy the spectral Ádám property.

Theorem 4. *For any fixed m, the bound $A_m(n) = O(n^{3m/4})$ holds.*

Proof. Let $S = \{s_1, \dots, s_m\} \subseteq \mathbb{Z}_n$ be an m-element set which does not satisfy the spectral Ádám property. We use the the same notations as in the proof of Theorem 3. Then for every pair $u_1, u_2 \in U_\mu$ we have $\gcd(|u_1 - u_2|, n) \geq n/Q_{2m}$. Therefore there are at least

$$\sum_{\mu=1}^{r} \frac{\#V_\mu(\#V_\mu - 1)}{2} \geq \sum_{\mu=1}^{r} \frac{\#V_\mu}{2} = M/2 \geq m/4$$

pairs $u_1, u_2 \in S$, $u_1 \neq u_2$, with $\gcd(|u_1 - u_2|, n) \geq n/Q_{2m}$. It is easy to see that there are at most $O(n^{3m/4})$ m-element sets $S \subseteq \mathbb{Z}_n$ satisfying this condition. □

Obviously, this bound can be slightly improved as $A_m(n) = O\left(n^{m - \lceil m/4 \rceil}\right)$.

4 Double Loops

In this section we concentrate on double loop circulant graphs. That is consider circulant graphs generated by sets $S = \{\pm a, \pm b\} \in \mathbb{Z}_n$ with the condition that $1 \leq a < b < n/2$. We restrict ourselves with connected graphs which is equivalent to the solvability of the congruence $ax + by \equiv 1 \pmod{n}$, thus to the condition $\gcd(a, b, n) = 1$. We define the following 4 special families of graphs

$$W_e = \{\pm e, \pm(n/2 - e)\}, \qquad X_e = \{\pm e, \pm n/4\},$$
$$Y_e = \{\pm e, \pm(n/3 - e)\}, \qquad Z_e = \{\pm e, \pm(n/6 - e)\}.$$

First of all we need the following two simple lemmas which are based on elementary number theory

Lemma 5. *Let integers $1 \leq a < b < n/2$ be such that $\gcd(n, a, b) = 1$. If either $n/\gcd(n, a+b)$ or $n/\gcd(n, b-a)$ divides 6, then the set $\{\pm a, \pm b\}$ is proportional to one of the following 12 sets*

$$W_1; \qquad Y_e,\ e = \pm 1, 3; \qquad Z_e,\ e = \pm 1, \pm 2, \pm 3, \pm 6. \tag{3}$$

Lemma 6. *Let integers $1 \leq a < b < n/2$ be such that $\gcd(n, a, b) = 1$. If both $n/\gcd(n, 2a)$ and $n/\gcd(n, 2b)$ divides 6, then $n|12$.*

Theorem 7. *Any set $S = \{\pm a, \pm b\} \subseteq \mathbb{Z}_n$ with $\gcd(a, b, n) = 1$ except possibly one of the following 12 exceptional sets*

$$X_1; \qquad Y_e,\ e = \pm 1, 3; \qquad Z_e,\ e = \pm 1, \pm 2, \pm 3, \pm 6;$$

satisfies the spectral Ádám property.

Proof. First of all we prove the theorem assuming that S is neither one of the sets (3) nor one of the sets X_e, $e = 1, 2, 4$. Then we will rule out some of the remaining possibilities. The eigenvalues of $\langle\{\pm a, \pm b\}\rangle_n$ are

$$\begin{aligned}\lambda_k &= \zeta^{ka} + \zeta^{-ka} + \zeta^{kb} + \zeta^{-kb} \\ &= 2\left(\cos\left(\frac{2\pi ka}{n}\right) + \cos\left(\frac{2\pi kb}{n}\right)\right) = 4\cos\left(\frac{\pi k}{n}(a+b)\right)\cos\left(\frac{\pi k}{n}(a-b)\right),\end{aligned}$$

where $k = 1, \dots, n$.

Assume that $\langle\{\pm a, \pm b\}\rangle_n$ has the eigenvalue $\lambda_1 = 0$. This happens only when n is even and since $0 < b - a < n/2$ we must have $a + b = n/2$. By Lemma 5, $\{\pm a, \pm b\} \sim W_1$. From now on we suppose that $\zeta^a + \zeta^{-a} + \zeta^b + \zeta^{-b} \neq 0$. Note that if it happens that $\zeta^b + \zeta^{-b} = 0$ (respectively $\zeta^a + \zeta^{-a} = 0$) this forces $b = n/4$ (respectively $a = n/4$) and since $\gcd(a, n/4) = 1$, we see that $\{\pm a, \pm b\} \sim X_e$, with some $e = 1, 2, 4$.

From now on we suppose that $\operatorname{Spec}\langle S\rangle_n = \operatorname{Spec}\langle T\rangle_n$, where $T = \{\pm c, \pm d\} \subseteq \mathbb{Z}_n$, that $S \not\sim T$ and that neither S nor T is of the sets X_e, with some $e = 1, 2, 4$.

We conclude that there exists k, $1 \le k \le n$, such that

$$\zeta^a + \zeta^{-a} + \zeta^b + \zeta^{-b} - \zeta^{kc} - \zeta^{-kc} - \zeta^{kd} - \zeta^{-kd} = 0. \tag{4}$$

First of all we show that is the sum in (4) contains a subsum of length 2 that vanished that this is of the form $\zeta^t + \zeta^{-t} = 0$ with $t = n/4$ or $t = 3n/4$. Indeed, $\zeta^t + \zeta^s = 0$ is possible for $s = -t = n/2 \pm n/4$ only. If $\zeta^t - \zeta^s = 0$ for some t, s then $\{\pm a, \pm b\}$ and $\{\pm kc, \pm kd\}$ have at least two elements in common. It is now easy to deduce that if this is the case, then the two sets have to coincide and $(k, n) = 1$ in this case. Given the previous argument we have the following 5 possibilities:

i. The sum in (4) does not have any proper subsum that vanishes. Lemma 1 implies that $n/\gcd(n, 2a, b-a, b+a)$ has to be a factor of 210. An easy argument shows that $\gcd(n, 2a, b-a, b+a) = \gcd(n, 2a, 2b)$ which can be 1 or 2 since $\gcd(n, a, b) = 1$. Therefore n is a divisor of 420. It is proven in [18, 19] that for values of n that are squarefree or twice a squarefree number the spectral Ádám property holds.

ii. The sum in (4) splits as a vanishing sum with 6 terms plus a sum with 2 terms and no other subsums vanish. Because sets X_e, $e = 1, 2, 4$, are excluded we may assume that none of a and b equals $n/4$. Then the sum of length 6 contains $\zeta^b + \zeta^{-b} + \zeta^a + \zeta^{-a}$ as a subsum and we can proceed and in the case **i**, concluding that $n|60$. For these values of n the statement of the theorem can be verified by the exhaustive search.

iii. The sum in (4) splits as the sum of three terms each vanishing, one of length 2, and two of length 3. Once again we may assume that none of a and b equals $n/4$. By a similar argument as above and applying Lemma 6, we can assume that one of the two sums of length three is of the form $\zeta^a + \zeta^b - \zeta^r$

or $\zeta^a + \zeta^{-b} - \zeta^r$ with $r \in \{\pm kc, \pm kd\}$. By Lemma 1, we see that $n/\gcd(n, a \pm b)$ divides 6 and by Lemma 5 we have that $\{\pm a, \pm b\}$ is proportional to one of the exceptional sets (3).

vi. The sum in (4) splits as the sum of two terms of length four each vanishing and no other subsum vanishing. We can exclude all the sums that contain at the same time either ζ^a and ζ^{-a} or ζ^b and ζ^{-b} since in this case we would deduce that n divides 240. However if a and b or a and $-b$ belongs to the same sum of length 4 then in a similar way by Lemma 1 we see that $n/\gcd(n, a \pm b)$ divides 6 and by Lemma 5 we have that $\{\pm a, \pm b\}$ is proportional to one of the exceptional sets (3).

v. The sum in (4) splits as the sum of two terms, one of length 5 and one of length 3, each one vanishing. If the sum of length 5 contains at least three elements of $\{\pm a, \pm b\}$ then we immediately deduce that $n \mid 60$ and we have already ruled out these possibilities. If the sum of length 3 contains three elements of $\{\pm a, \pm b\}$, then we come to the same conclusion. Therefore we assume that both the sums of length 3 and the one of length 5 contains two elements of $\{\pm a, \pm b\}$. By a similar analysis as has been made in the case **iii**, we have that either $n/\gcd(n, a+b)$ or $n/\gcd(n, b-a)$ divide 6 or both $n/\gcd(n, 2a)$ and $n/\gcd(n, 2b)$ divide 6. By Lemma 5 and Lemma 6, we can conclude that $\{\pm a, \pm b\}$ is proportional to one of the exceptional sets (3).

So we have proved the statement of the theorem for all sets S except possible the sets (3) and X_e, $e = 1, 2, 4$. To reduce the number of exceptions let us define $\mu_\alpha(S)$ as the multiplicity of α as an eigenvalue of $\langle S \rangle_n$. So $\mu_\alpha(S) = 0$ if α is not an eigenvalue of $\langle S \rangle_n$.

It is also easy to see that $n/6 \le \mu_0(Z_e) \le n/6 + 6$, for $e = \pm 1, \pm 2, \pm 3, \pm 6$.

In the following table we present $\mu_\alpha(S)$ of other potentially exceptional graphs (together with the condition on n for which these graphs exist and are connected):

$\langle S \rangle_n$	Conditions on n	$\mu_0(\langle S \rangle_n)$
$\langle W_1 \rangle_n$	$2 \mid n$	$\begin{cases} n/2+2, \text{ if } n \equiv 0 \pmod 8; \\ n/2, \quad \text{if } n \not\equiv 0 \pmod 8. \end{cases}$
$\langle X_1 \rangle_n$	$n \equiv 0 \pmod 4$	$\begin{cases} 1, \text{ if } n \equiv 0 \pmod 8; \\ 2, \text{ if } n \not\equiv 0 \pmod 8. \end{cases}$
$\langle X_2 \rangle_n$	$n \equiv 4 \pmod 8$	1
$\langle X_4 \rangle_n$	$n \equiv 4 \pmod 8$	1
$\langle Y_1 \rangle_n$	$n \equiv 0 \pmod 3$	$\begin{cases} 6, \text{ if } n \equiv 24 \pmod{36}; \\ 2, \text{ if } n \equiv 0, 12 \pmod{36}; \\ 0, \text{ if } n \not\equiv 0 \pmod{12}. \end{cases}$
$\langle Y_{-1} \rangle_n$	$n \equiv 0 \pmod 3$	$\begin{cases} 6, \text{ if } n \equiv 12 \pmod{36}; \\ 2, \text{ if } n \equiv 0, 24 \pmod{36}; \\ 0, \text{ if } n \not\equiv 0 \pmod{12}. \end{cases}$
$\langle Y_3 \rangle_n$	$n \equiv 3 \pmod 9$	$\begin{cases} 2, \text{ if } n \equiv 0 \pmod{12}; \\ 0, \text{ if } n \not\equiv 0 \pmod{12}. \end{cases}$

For $n \leq 18$ the theorem can be proved by the exhaustive search. For $n > 18$ we see that the spectrum of $\langle W_1 \rangle_n$ is unique, so it cannot be an exceptional set. The sets X_2 and X_4 when $n \equiv 4 \pmod 8$ have both $\mu_0(X_2) = \mu_0(X_4) = 1$ (but not in the case when $\mu_0(X_1) = 1$). However $\mu_2(X_2) = 0$ while $\mu_2(X_4) = 2$. So the two graphs cannot be isomorphic and their spectra are distinct from spectra of all other possible exceptions. So X_2 and X_4 can be eliminated from the list of exceptions as well. □

So we have 12 sets of the form $\{\pm a, \pm b\}$ with $\gcd(a, b, n) = 1$ which do not satisfy the spectral Ádám property, and only 5 of them are of the form $\{\pm 1, \pm d\}$, the case of main interest.

5 Remarks

Naively speaking one can expect $12 \times 11/2 = 66$ 'suspicious' pairs $\text{Spec}\,\langle S \rangle_n = \text{Spec}\,\langle T \rangle_n$ for which $S \not\sim T$. However, by comparing the multiplicities of zero eigenvalues we see that sets Z_e, $e = \pm 1, \pm 2, \pm 3, \pm 6$, cannot form a suspicious pair with sets X_1, Y_e, $e = \pm 1, 3$. Hence there are only $4 \times 3/2 + 8 \times 7/2 = 34$ suspicious pairs. Similarly, we have only $3 \times 2/2 + 2 \times 1/2 = 4$ suspicious pairs of sets of the form $\{\pm 1, \pm d\}$. In fact, the number of suspicious pairs can be essentially reduced if one computes $\mu_0(Z_e)$ precisely for each $e = \pm 1, \pm 2, \pm 3, \pm 6$. Other eigenvalues may help as well. It is also easy to see that if $\gcd(n, 12) < 3$ then there is no exceptions at all.

Here instead of using the isomorphism property of graphs our approach is based on a weaker property of their spectral identity. Thus our results are more general than the original Ádám conjecture.

One can probably extend our results to the case of weighted circulant graphs. Indeed, in the case the question can be reduced to the equation of the form (1) as well however, instead of ± 1-coefficients it will have coefficient depending on the weights. For integer weights one can use Lemma 1 and obtain essentially the same results. For graphs with algebraic weights one can use more general results of [22, 23]. This case can also be considered with attracting any new ideas. For equations with roots of unity with complex coefficients very general and strong results applicable to equations with arbitrary complex coefficients are available [10, 21], however it is still now quite clear how to extract analogies of Theorems 3, 4 and 7.

One more possible generalization we can be tackled by our method is studying circulant graphs for which spectra have large intersection. It seems that for any $\varepsilon > 0$ one can obtain some non-trivial conclusions about sets $S, T \subseteq \mathbb{Z}_n$ such that the spectra of $\langle S \rangle_n$ and $\langle T \rangle_n$ have at least n^ε common elements, that is $\#\,\text{Spec}\,\langle S \rangle_n \bigcap \text{Spec}\,\langle T \rangle_n \geq n^\varepsilon$.

Finally we remark, that we hope that our approach will be useful for some other types of graphs, including Cayley graphs.

References

1. A. Ádám, 'Research problem 2-10', *J. Combinatorial Theory*, **3** (1967), 393.
2. B. Alspach and T. Parsons, 'Isomorphism of circulant graphs and digraphs', *Discrete Math.*, **25** (1979), 97–108.
3. L. Babai, 'Automorphism groups, isomorphism, reconstruction' *Handbook of Combinatorics*, Elsevier, Amsterdam, 1995, 1749–1783.
4. J.-C. Bermond, F. Comellas and D. F. Hsu, 'Distributed loop computer networks: A survey', *Journal of Parallel and Distributed Computing*, **24** (1995), 2–10.
5. F. Boesch and R. Tindell, 'Circulants and their connectivities', *J. of Graph Theory*, **8** (1984), 487–499.
6. J. H. Conway and A. J. Jones, 'Trigonometric diophantine equations (On vanishing sums of roots of unity)', *Acta Arithm.*, **30** (1976), 171–182.
7. C. Delorme, O. Favaron and M. Maheo, 'Isomorphisms of Cayley multigraphs of degree 4 on finite abelian groups', *European J. of Comb.*, **13** (1992), 59–61.
8. N. Deo and M. Krishnamoorthy, 'Toeplitz networks and their properties', *IEEE Trans. on Circuits and Sys.*, **36** (1989), 1089–1092.
9. B. Elspas and J. Turner, 'Graphs with circulant adjacency matrices', *J. Comb. Theory*, **9** (1970), 229–240.
10. J.-H. Evertse and H. P. Schlickewei, 'The absolute subspace theorem and linear equations with unknowns from a multiplicative group', *Preprint*, 1997, 1–22.
11. X. G. Fang and M. Y. Xu, ' On isomorphisms of Cayley graphs of small valency', *Algebra Colloquium*, **1** (1994), 67–76.
12. C. Godsil, *Algebraic Combinatorics*, Chapman and Hall, NY, 1993.
13. F. T. Leighton, *Introduction to parallel algorithms and architectures: Arrays, trees, hypercubes*, M. Kaufmann, 1992.
14. C.H. Li, 'The cyclic groups with the m-DCI property', *European J. of Comb.*, **18** (1997), 655-665.
15. B. Litow and B. Mans, 'A note on the Ádám conjecture for double loops', *Inf. Proc. Lett.*, (to appear).
16. H. B. Mann, 'On linear relations between roots of unity', *Mathematika*, **12** (1965), 171–182.
17. B. Mans, 'Optimal Distributed algorithms in unlabeled tori and chordal rings', *Journal on Parallel and Distributed Computing*, **46(1)** (1997), 80-90.
18. M. Muzychuk, 'Ádám's conjecture is true in the square free case', *J. of Combinatorial Theory, Series A*, **72** (1995), 118–134.
19. M. Muzychuk, 'On Ádám's conjecture for circulant graphs', *Discrete Math.*, **167/168** (1997), 497–510.
20. P. P. Pálfy, 'Isomorphism problem for relational structures with a cyclic automorphism', *Europ. J. Combin.*, **8** (1987), 35–43.
21. H.P. Schlickewei, 'Equations in roots of unity', *Acta Arithm.*, **76** (1996), 171–182.
22. U. Zannier, 'On the linear independence of roots of unity', *Acta Arithm.*, **50** (1989), 171–182.
23. U. Zannier, 'Vanishing sums of roots of unity', *Rend. Sem. Mat. Univ. Politec. Torino*, **53** (1995), 487–495.

Proof of Toft's Conjecture: Every Graph Containing No Fully Odd K_4 Is 3-Colorable (Extended Abstract)

Wenan Zang*
Department of Mathematics
University of Hong Kong
Hong Kong
wzang@maths.hku.hk

Abstract

The graph 3-coloring problem arises in connection with certain scheduling and partition problems. As is well known, this problem is NP-complete and therefore intractable in general unless $NP = P$. The present paper is devoted to the 3-coloring problem on a large class of graphs, namely, graphs containing no fully odd K_4, where a fully odd K_4 is a subdivision of K_4 such that each of the six edges of the K_4 is subdivided into a path of odd length. In 1974, Toft conjectured that every graph containing no fully odd K_4 can be vertex-colored with three colors. The purpose of this paper is to prove Toft's conjecture.

1 Introduction

A graph G is said to be *n-colorable* if there is an assignment of n colors, $1, 2, \ldots, n$, to the vertices of G so that no two adjacent vertices have the same color; the *chromatic number* of G is the minimum n for which G is n-colorable. The *graph-coloring problem* is to determine the chromatic number of a given graph. As is well known, this NP-hard problem has many potential real-world applications.

Graph coloring has a long history, and from the beginning it has been closely tied to the famous four-color conjecture which can be stated in terms of graph minors. A graph F is a *minor* of a graph G if F can be obtained from a subgraph of G by contracting edges; a graph G is a *subdivision* of a graph H if G can be obtained from H by replacing the edges of H with internally disjoint paths, each containing at least one edge. Hadwiger [5] conjectured that every graph with no minor isomorphic to K_{n+1} (the complete graph with $n+1$ vertices) is n-colorable; Hajós [6] further conjectured that every graph containing no subdivision of K_{n+1} is n-colorable. These two conjectures hold trivially for $n = 1$ and 2; they are equivalent when $n = 3$ and the validity of this case was established by Dirac [4]. Hadwiger's conjecture is equivalent to the four-color theorem for $n = 4$ and 5 [14, 1, 9, 10]; it remains unsolved for $n \geq 6$. Hajós' conjecture was disproved by Catlin [2] for $n \geq 6$; it is still open for $n = 4$ and 5; its validity for $n = 4$ would imply the four-color theorem.

*This work was supported in part by the Air Force Office of Scientific Research under grant F49620-93-1-0041 awarded to Rutgers Center for Operations Research, Rutgers University and by RGC grant 338/024/0009.

The problem of deciding if a graph is 3-colorable is called the *graph 3-coloring problem*, which is related to the four-color conjecture and arises in connection with certain scheduling and partition problems. Since this problem is NP-complete and therefore intractable in general unless $NP = P$, it would be interesting to consider some special classes of graphs.

A *fully odd* K_4 is a subdivision of K_4 such that each of the six edges of the K_4 is subdivided into a path of *odd* length. In 1974, Toft [13], with an attempt to strengthen Dirac's theorem, conjectured that every graph containing no fully odd K_4 can be vertex-colored with three colors. Motivated by Toft's conjecture, various authors have obtained several results in the past two decades. Let G be an arbitrary graph with chromatic number four, Catlin [2] proved that G contains a subdivision of K_4 such that each triangle of the K_4 is subdivided to form an odd cycle; Krusenstjerna-Hafstrøm and Toft [8] showed that G contains a subdivision of K_4 such that each of the three edges of a K_3 in the K_4 is subdivided into a path of odd length; Thomassen and Toft [12] verified that G contains a subdivision of K_4 such that three edges of an arbitrary spanning tree of the K_4 can be left undivided, corresponding to paths of length one. Jensen and Shepherd [7] confirmed Toft's conjecture for line graphs. The purpose of this paper is to prove Toft's conjecture.

THEOREM *Every graph containing no fully odd K_4 is 3-colorable.*

Our proof relies on decomposition methods, that is, we shall first decompose the graph in consideration into some nice smaller graphs, and then turn to color each smaller graph with three colors. By piecing together the colorings on these smaller graphs, we shall get a proper 3-coloring of the original graph. For ease of exposition, let us prove by contradiction.

2 Proof: Outline

Throughout, let G stand for a counterexample with the smallest number of vertices:

- G has no fully odd K_4,
- G is not 3-colorable,
- every graph with no fully odd K_4 and with fewer vertices than G is 3-colorable.

A *mixed* K_4 is a graph obtained from K_4 by inserting an odd number of new vertices of degree two into each of the three edges incident to a certain vertex and by inserting an even number of new vertices of degree two into each of the other edges.

Lemma 2.1 *G contains a mixed K_4.*

Mixed K_4's are important intermediate structures, which play crucial roles in our proof. We shall use repeatedly the following conventions of notation and terminology with regard to mixed K_4's obtained from a K_4 with vertex-set $\{v_0, v_1, v_2, v_3\}$.

Standard labeling of a mixed K_4:

- τ_1 denotes the path linking v_0 and v_1;
  this path has an even length; its vertices are $v_0, a_1, a_2, a_3, \ldots, v_1$;
  we set $A_1 = \{a_i : i \text{ is odd}\}$, $A_2 = \{a_i : i \text{ is even}\}$, and $A = A_1 \cup A_2$.

- τ_2 denotes the path linking v_0 and v_2;
  this path has an even length; its vertices are $v_0, b_1, b_2, b_3, \ldots, v_2$;
  we set $B_1 = \{b_i : i \text{ is odd}\}$, $B_2 = \{b_i : i \text{ is even}\}$, and $B = B_1 \cup B_2$.

- τ_3 denotes the path linking v_0 and v_3;
  this path has an even length; its vertices are $v_0, c_1, c_2, c_3, \ldots, v_3$;
  we set $C_1 = \{c_i : i \text{ is odd}\}$, $C_2 = \{c_i : i \text{ is even}\}$, and $C = C_1 \cup C_2$.

- π_1 denotes the path linking v_2 and v_3;
  this path has an odd length; its vertices are $v_2, x_1, x_2, x_3, \ldots, v_3$;
  we set $X_1 = \{x_i : i \text{ is odd}\}$, $X_2 = \{x_i : i \text{ is even}\}$, and $X = X_1 \cup X_2$.

- π_2 denotes the path linking v_1 and v_3;
  this path has an odd length; its vertices are $v_1, y_1, y_2, y_3, \ldots, v_3$;
  we set $Y_1 = \{y_i : i \text{ is odd}\}$, $Y_2 = \{y_i : i \text{ is even}\}$, and $Y = Y_1 \cup Y_2$.

- π_3 denotes the path linking v_1 and v_2;
  this path has an odd length; its vertices are $v_1, z_1, z_2, z_3, \ldots, v_2$;
  we set $Z_1 = \{z_i : i \text{ is odd}\}$, $Z_2 = \{z_i : i \text{ is even}\}$, and $Z = Z_1 \cup Z_2$.

Observe that G enjoys many nice properties in the presence of a mixed K_4, some of which are given below, where a path is called Σ-*external* if all the internal vertices of this path are outside Σ.

Lemma 2.2 *Let Σ be an arbitrary mixed K_4 with a standard labeling. Then the following statements hold:*

(i) Every Σ-external path linking any two of A_1, B_1, and C_1 is of odd length;

(ii) Every Σ-external path linking A_1 and $B_2 \cup C_2$ is of even length;

(iii) Every Σ-external path linking A_1 and $Y_1 \cup Z_1$ is of even length;

(iv) Every Σ-external path linking A_1 and π_1 is of even length;

(v) No component of $G - \Sigma$ is adjacent to all three of A_1, B_1, and $C \cup X_1 \cup Y_1 \cup \{v_3\}$;

(vi) If there is a Σ-external path linking any two of A_1, B_1, and C_1, then there is no Σ-external path linking any other two of A_1, B_1, and C_1.

To outline the proof, we still need one definition.

Definition 2.1 *A path τ_i in a mixed K_4 Σ (with a standard labeling) is said to meet the parity requirement with respect to Σ if τ_i satisfies the following two conditions:*

(a) For any two vertices u and v on τ_i such that $\tau_i[u,v]$ is of even length, u and v are nonadjacent in G;

(b) For any component H of $G-\Sigma$ which is adjacent to no vertex in $\Sigma-\tau_i$, every H-internal path linking any two vertices u and v on τ_i has the same parity as $\tau_i[u,v]$.

Lemma 2.3 *In every mixed K_4 Σ with a standard labeling, at least one of τ_1, τ_2, and τ_3 meets the parity requirement with respect to Σ.*

In completing our proof of the theorem, symmetry allows us to distinguish among three cases:

Case 1. There is a mixed K_4 Σ with a standard labeling such that A_1 and B_1 are linked by some Σ-external path and such that τ_3 meets the parity requirement with respect to Σ.

Case 2. For each mixed K_4 Σ with a standard labeling such that A_1 and B_1 are linked by some Σ-external path, τ_3 does not meet the parity requirement with respect to Σ.

Case 3. For each mixed K_4 Σ with a standard labeling, there is no Σ-external path linking any two of A_1, B_1, and C_1.

These three cases are the subjects of the next three sections of our paper. Using decomposition methods, we shall reach a contradiction in each case. Let us sketch the proof of the first case now.

3 Proof: Case 1

The assumption of this section: there is a mixed K_4 Σ with a standard labeling such that A_1 and B_1 are linked by some Σ-external path and such that τ_3 meets the parity requirement with respect to Σ.

Lemma 3.1 *There is no Σ-external path of odd length linking C_1 and $\Sigma-(C_2\cup\{v_0,v_3\})$.*

Since τ_3 meets the parity requirement with respect to Σ, by definition both C_1 and $C_2\cup\{v_0,v_3\}$ are stable sets.

Definition 3.1 *Let G^* denote the graph obtained from G by shrinking C_1 and shrinking $C_2\cup\{v_0,v_3\}$.*

We are going to prove that G^* is 3-colorable, and so (as 3-colorability of G^* implies 3-colorability of G) obtain the desired contradiction.

Definition 3.2 *Let Σ^* denote the graph obtained from Σ by shrinking C_1 and shrinking $C_2\cup\{v_0,v_3\}$.*

Lemma 3.2 *Let Q be a Σ-external path in G such that*

- *the length of Q is odd,*
- *one end, c_{2k-1}, of Q is in C_1,*
- *the other end, w, of Q is outside Σ.*

Then w is adjacent to at most two vertices of Σ^* other than C_1. Furthermore, if w is adjacent to precisely two vertices of Σ^* other than C_1 then these two vertices may be labeled u and v so that one of the following conditions is satisfied:

(i) u is in A_2, v is in B_2;
u is adjacent to no vertex in B_2 and v is adjacent to no vertex in A_2;

(ii) u is in X_1, v is in Y_1;
u is adjacent to no vertex in Y_1 and v is adjacent to no vertex in X_1.

Let us construct an equivalence relation on the set of vertices of Σ^* by the following algorithm:

Algorithm 3.1

Let each vertex of Σ^* form an equivalence class.
while *there is a path Q in G^* such that*

- *the length of Q is odd,*
- *one end of Q is C_1,*
- *all the other vertices are outside Σ^*,*
- *the end of Q that is outside Σ^* is adjacent to precisely two equivalence classes, U and V, other than C_1*

do *merge U and V.*

Lemma 3.3 *Each equivalence class produced by Algorithm 3.1 is a stable set in G.*

Definition 3.3 *Let G_0^* denote the graph obtained from G^* by shrinking each equivalence class produced by Algorithm 3.1.*

Our task is to prove that G^* is 3-colorable; we shall actually prove that G_0^* is 3-colorable; by Lemma 3.3, 3-colorability of G_0^* implies 3-colorability of G^*.

Definition 3.4 *Let $F_1, F_2, \ldots, F_k$ denote all the components of $G^* - \Sigma^*$ that are adjacent to C_1.*

Note that both C_1 and $C_2 \cup \{v_0, v_3\}$ are vertices of G_0^*. From Lemma 3.1, it follows that

$$C_2 \cup \{v_0, v_3\} \text{ is the only neighbor of } C_1 \text{ in } G_0^* - (F_1 \cup F_2 \cup \ldots \cup F_k). \tag{3.1}$$

Lemma 3.4 $G_0^* - (F_1 \cup F_2 \cup \ldots \cup F_k)$ *contains no fully odd K_4.*

By Lemma 3.4 and the assumption on G, $G_0^* - (F_1 \cup F_2 \cup \ldots \cup F_k)$ admits a 3-coloring φ; switching colors if necessary, we may assume that

$$\varphi(C_2 \cup \{v_0, v_3\}) = 1 \text{ and } \varphi(C_1) = 2. \tag{3.2}$$

We are going to extend φ into a 3-coloring of G^* by coloring each F_i with $1 \leq i \leq k$ separately.

Lemma 3.5 *For each i with $1 \leq i \leq k$, there are disjoint stable subsets D_1, D_2 of $V(F_i)$ and a partition of $V(F_i)$ into sets V_x ($x \in D_1 \cup D_2$) with the following properties:*

(i) V_x contains x, V_x is the union of some blocks of F_i if it has at least two vertices, and V_x is connected;

(ii) V_x and V_y are disjoint and $V_x - \{x\}$ and $V_y - \{y\}$ are nonadjacent whenever $x \neq y$;

(iii) $D_1 \cup D_2$ induces a connected bipartite subgraph, which is the union of some blocks of F_i;

(iv) $V(F_i) - D_2$ is nonadjacent to $\Sigma - (C_2 \cup \{v_0, v_3\})$;

(v) For each x in D_2, all the neighbors of x in $\Sigma - C_1$ have the same color in φ.

Now we are ready to construct a 3-coloring φ_i of F_i which is compatible with φ for each $1 \leq i \leq k$. We shall actually construct a 3-coloring φ_x of each subgraph of F_i induced by V_x so that the union of all φ_x's yields φ_i.

Description: Let c_{2k-1} be a vertex in C_1 which is adjacent to some vertex w in D_2.

To construct φ_x when $x \in D_1$, consider the graph G_x obtained from the subgraph of G induced by $V_x \cup \tau_3$ by removing all the vertices in $C_1 - \{c_{2k-1}\}$ and then identifying all the vertices in each of the four sets $\{v_0\} \cup \{c_{2i} : i < k\}$, $\{v_3\} \cup \{c_{2i} : i \geq k\}$, $\{v_0, v_3\}$, and $\{c_{2k-1}, x\}$. It can be shown that G_x contains no fully odd K_4. Since G_x has fewer vertices than G, G_x has a 3-coloring ϕ_x; trivially $\phi_x(x)$, $\phi_x(C_2 \cup \{v_0, v_3\})$ are distinct colors; without loss of generality, $\phi_x(x) = 2$, $\phi_x(C_2 \cup \{v_0, v_3\}) = 1$. The restriction of ϕ_x on V_x is φ_x.

To construct φ_x when $x \in D_2$, we distinguish between two cases.

Case 1.1. x has no neighbor in $\Sigma - \tau_3$ with color 1 or 3.

Consider the graph G_x obtained from the subgraph of G induced by $V_x \cup C_2 \cup \{v_0, v_3, c_{2k-1}\}$ by identifying all the vertices in each of the three sets $\{v_0\} \cup \{c_{2i} : i < k\}$, $\{v_3\} \cup \{c_{2i} : i \geq k\}$, and $\{v_0, v_3\}$ and then adding the edge xc_{2k-1}. It can be shown that G_x contains no fully odd K_4. Since G_x has fewer vertices than G, G_x has a 3-coloring ϕ_x; trivially $\phi_x(c_{2k-1})$ and $\phi_x(C_2 \cup \{v_0, v_3\})$ are distinct; without loss of generality, $\phi_x(C_2 \cup \{v_0, v_3\}) = 1$ and $\phi_x(c_{2k-1}) = 2$. The restriction of ϕ_x on V_x is φ_x.

Case 1.2. x has some neighbor u in $\Sigma - \tau_3$ with color 1 or 3.

In case $\varphi(u) = 1$ (resp. 3), consider the graph G_x obtained from the subgraph of G induced by $V_x \cup C_2 \cup \{v_0, v_3\}$ by identifying all the vertices in $C_2 \cup \{v_0, v_3\}$ and then adding the edge $v_0 x$

(resp. identifying v_0 and x). It can be shown that G_x contains no fully odd K_4. Since G_x has fewer vertices than G, G_x has a 3-coloring ϕ_x; without loss of generality, $\phi_x(C_2 \cup \{v_0, v_3\}) = 1$, $\phi_x(x) = 3$ (resp. $\phi_x(x) = 1$). The restriction of ϕ_x on V_x is φ_x.

Let us justify the validity of φ_x for each $x \in D_1 \cup D_2$. By Lemma 3.5(iv), $V_x - \{x\}$ is nonadjacent to $\Sigma - (C_2 \cup \{v_0, v_3\})$. For each $x \in D_1$, Lemma 3.5(iv) also implies that x is nonadjacent to $\Sigma - (C_2 \cup \{v_0, v_3\})$. Note that $\varphi_x(x) = 2$, from the construction it can be seen that φ_x is compatible with φ. For each $x \in D_2$, by Lemma 3.5(v) all the neighbors of x in $\Sigma - C_1$ has the same color in φ. Note that $\varphi_x(x) = 1$ or 3, from the construction it can be seen that φ_x is compatible with φ.

Since $\varphi_x(x) \neq \varphi_y(y)$ whenever $x \in D_1$ and $y \in D_2$, by Lemma 3.5(i)-(iii), the union of all φ_x's yields a proper 3-coloring, φ_i, of F_i which is compatible with φ. □

4 Remarks

The idea underlying the proof of case 1 can be used to handle both case 2 and case 3, although the latter two cases are technically more complicated.

A slight modification of our proof yields a polynomial time algorithm for coloring graphs containing no fully odd K_4 with three colors provided there is a polynomial time algorithm for recognizing graphs with no fully odd K_4. Sewell and Trotter [11] affirmatively settled a question of Chvátal [3] and proved that every connected, stability critical graph that is neither K_2 nor an odd cycle contains a fully odd K_4. This leads to a polynomial time algorithm for finding maximum stable sets in graphs with no fully odd K_4. Since both the 3-coloring problem and the maximum stable set problem are hard in general, it would be nice and interesting to have a polynomial time algorithm for recognizing graphs containing no fully odd K_4. We close with a conjecture that such an algorithm exists.

Acknowledgments

This work was carried out while I was a graduate student at Rutgers Center for Operations Research, Rutgers University; I am very much indebted to Professor Vašek Chvátal for his invaluable guidance and for his generous help in writing this paper. I am also grateful to Professor Paul Seymour for his stimulating suggestions.

References

[1] K. Appel and W. Haken, Every Planar Map Is Four Colorable, *A.M.S. Contemp. Math.* **98** (1989).

[2] P. Catlin, Hajós Graph-Coloring Conjecture: Variations and Counterexamples, *J. Combin. Theory Ser. B* **26** (1979), 268-274.

[3] V. Chvátal, On Certain Polytopes Associated with Graphs, *J. Combin. Theory Ser. B* **18** (1975), 138-154.

[4] G. A. Dirac, A Property of 4-Chromatic Graphs and Some Remarks on Critical Graphs, *J. London Math. Soc.* **27** (1952), 85-92.

[5] H. Hadwiger, Über eine Klassifikation der Streckenkomplexe, *Vierteljahrsch. Naturforsch. Ges. Zürich* **88** (1943), 133-142.

[6] G. Hajós, Über eine Konstruktion nicht n-färbbarer Graphen, *Wiss. Z. Martin-Luther-Univ. Halle-Wittenberg Math.-Natur. Reihe* **10** (1961), 116-117.

[7] T. R. Jensen and F. B. Shepherd, Note on a Conjecture of Toft, *Combinatorica* **15** (1995), 373-377.

[8] U. Krusenstjerna-Hafstrøm and B. Toft, Special Subdivisions of K_4 and 4-Chromatic Graphs, *Monatsh. Math.* **89** (1980), 101-110.

[9] N. Robertson, D. Sanders, P. Seymour and R. Thomas, The Four-Colour Theorem, *J. Combin. Theory Ser. B* **70** (1997), 2-44.

[10] N. Robertson, P. Seymour and R. Thomas, Hadwiger's Conjecture for K_6-Free Graphs, *Combinatorica* **13** (1993), 279-361.

[11] E. C. Sewell and L. E. Trotter, Stability Critical Graphs and Even Subdivisions of K_4, *J. Combin. Theory Ser. B* **59** (1993), 74-84.

[12] C. Thomassen and B. Toft, Non-separating Induced Cycles in Graphs, *J. Combin. Theory Ser. B* **31** (1981), 199-224.

[13] B. Toft, "Problem 10", in: *Recent Advances in Graph Theory, Proc. of the Symposium held in Prague, June 1974* (M. Fiedler, ed.), Academia Praha, 1975, pp. 543-544.

[14] K. Wagner, Über eine Eigenschaft der ebene Komplexe, *Math. Ann.* **114** (1937), 570-590.

A New Family of Optimal 1-Hamiltonian Graphs with Small Diameter *

Jeng-Jung Wang[1], Ting-Yi Sung[2], Lih-Hsing Hsu[1], Men-Yang Lin[3]

[1] Department of Computer and Information Science, National Chiao Tung University, Hsinchu, Taiwan, R.O.C.
[2] Institute of Information Science, Academia Sinica, Taipei, Taiwan, R.O.C.
[3] Department of Information Management, National Taichung Institute of Commerce, Taichung, Taiwan, R.O.C.

Abstract. In this paper, we construct a family of graphs denoted by $Eye(s)$ that are 3-regualr, 3-connected, planar, hamiltonian, edge hamiltonian, and also optimal 1-hamiltonian. Furthermore, the diameter of $Eye(s)$ is $O(\log n)$, where n is the number of vertices in the graph and to be precise, $n = 6(2^s - 1)$ vertices.

1 Introduction

Given a graph $G = (V, E)$, $V(G) = V$ and $E(G) = E$ denote the vertex set and the edge set of G, respectively. All graphs considered in this paper are undirected graphs. A *simple path* (or *path* for short) is a sequence of adjacent edges (v_1, v_2), (v_2, v_3), $\ldots$, (v_{m-2}, v_{m-1}), (v_{m-1}, v_m), written as $\langle v_1, v_2, v_3, \ldots, v_m\rangle$, in which all of the vertices $v_1, v_2, v_3, \ldots, v_m$ are distinct except possibly $v_1 = v_m$. The path $\langle v_1, v_2, v_3, \ldots, v_m\rangle$ is also called a *cycle* if $v_1 = v_m$ and $m \geq 3$. A cycle that traverses every vertex in the graph exactly once is called a *hamiltonian cycle.* A graph that contains a hamiltonian cycle is called a *hamiltonian graph* or said to be *hamiltonian.* A graph is *edge hamiltonian* if each edge in the graph is incident with some hamiltonian cycle in the graph. The *diameter* of graph G is the maximum *distance* among all pairs of vertices in G, where *distance* means the length of a shortest path joining vertices u, v.

For $V' \subset V$ and $E' \subset E$, $G-V'-E'$ denotes the graph obtained by removing all of the vertices in V' from V and removing the edges incident with at least one vertex in V' and also all of the edges in E' from E. Let k be a positive integer. A graph G is *k-edge hamiltonian* if $G - E'$ is hamiltonian for any $E' \subset E$ with $|E'| = k$; G^* is said to be *optimal k-edge hamiltonian* if G^* contains the least number of edges among all k-edge hamiltonian graphs having the same number of vertices as G^*. A graph G is *k-vertex hamiltonian* if $G - V'$ is hamiltonian for any $V' \subset V$ with $|V'| = k$; G^* is said to be *optimal k-vertex hamiltonian* if G^* contains the least number of edges among all k-vertex hamiltonian graphs having the same number of vertices as G^*. A graph G is *k-hamiltonian* if $G - V' - E'$

* This work was supported in part by the National Science Council of the Republic of China under contract NSC86-2213-E009-020.

is hamiltonian for any set $V' \subset V$ and $E' \subset E$ with $|V'| + |E'| \leq k$. It is clear that every k-hamiltonian graph has at least $k+3$ vertices. Moreover, the degree of every vertex in a k-hamiltonian graph is at least $k+2$. A k-hamiltonian graph having n vertices is said to be *optimal* if it contains the least number of edges among all k-hamiltonian graphs having n vertices.

Mukhopadhyaya and Sinha [6] proposed a family of optimal *1-hamiltonian* graphs which are also planar. These graphs are hamiltonian and have diameter of $\lfloor \frac{n}{6} \rfloor + 2$ if n is even, and $\lfloor \frac{n}{8} \rfloor + 3$ if n is odd. For any positive integer k, Harary and Hayes presented families of optimal k-edge hamiltonian graphs and optimal k-vertex hamiltonian graphs in [3] and [4], respectively. In particular, the family of optimal 1-edge hamiltonian graphs proposed in [3] are identical to the family of optimal 1-edge hamiltonian graphs proposed in [4]. Hence this family of graphs are 1-hamiltonian. Furthermore, each graph is planar, hamiltonian, and of diameter $\lfloor \frac{n+2}{4} \rfloor + ((\frac{n}{2} + 1) \textbf{ mod } 2)$ if n is even, $\lfloor \frac{n+1}{4} \rfloor + ((\lfloor \frac{n}{2} \rfloor + 1) \textbf{ mod } 2)$ if n is odd, where n is the number of vertices in the graph. Wang, Hung, and Hsu [9] presented another family of optimal 1-hamiltonian graphs, each of which is planar, hamiltonian, 3-regular, and of diameter $O(\sqrt{n})$ with n vertices in the graph. We are interested in finding more families of optimal 1-hamiltonian graphs that are also hamiltonian.

The three families of optimal 1-hamiltonian graphs presented in [4], [6] and [9] are planar, 3-regular and hamiltonian. It is natural to ask whether we can find such graphs with smaller diameter. This problem relates to the famous (n, d, D) problem in which we want to construct a graph of n vertices with maximum degree d such that the diameter D is minimized. When d and n are given, the lower bound on diameter D, called the *Moore bound* (on diameter), is given by $D \geq \log_{d-1} n - \frac{2}{d}$ [2]. In this paper, we propose a family of optimal 1-hamiltonian graphs that are hamiltonian, edge hamiltonian, planar, 3-regular and 3-connected. Furthermore, the diameter of our graphs is no more than $4 \log_2 \frac{n+6}{6}$, i.e., less than 4 times of Moore bound.

2 Definitions

Let $N_0 = 3$ and $N_k = 9 \cdot 2^{k-1}$ for any positive integer k. Let $[i]_m$ denote i **mod** m. An *eye* network $Eye(s)$, $s \geq 1$, is a graph with $s+1$ layers of concentric cycles. These $s+1$ cycles are denoted by $I_0, I_1, I_2, \ldots, I_{s-1}$, and O_s. In particular, O_s is the outermost cycle. The vertex set $V(Eye(s))$ is given by $\bigcup_{k=0}^{s-1} V(I_k) \cup V(O_s)$ where

$$V(I_k) = \{(k, j) \mid 0 \leq j \leq N_k - 1\} \text{ for } 0 \leq k \leq s-1 \text{ and}$$
$$V(O_s) = \{(s, j) \mid 0 \leq j \leq N_s - 1 \text{ and } [j]_3 = 0\}.$$

For vertex (k, j), k and j are referred to as the first and the second coordinate, respectively. Throughout this paper all computation on the second coordinate of a vertex at the kth concentric cycle are carried out with modulo N_k. Graph $Eye(s)$ contains two types of edges, i.e., *cycle edges*, denoted by $e_{k,j}$, and *inter-*

cycle edges, denoted by $e_{k,j}^{k+1}$, which are given as follows:

$$e_{k,j} = \begin{cases} ((k,j),(k,j+1)) \text{ for } 0 \le k \le s-1 \text{ and } 0 \le j \le N_k - 1, \\ ((k,j),(k,j+3)) \text{ for } k = s, 0 \le j \le N_s - 1 \text{ and } [j]_3 = 0, \end{cases}$$

$$e_{k,j}^{k+1} = \begin{cases} ((0,j),(1,2j+[j]_3)) & \text{for } k = 0 \text{ and } j = 0,1,2, \{v\}space1mm \\ ((k,j),(k+1,2j+[j]_3)) & \text{for } 1 \le k \le s-1,\ 0 \le j \le N_k - 1 \text{ and } [j]_3 \ne 0. \end{cases}$$

The set $\{e_{k,j} | 0 \le j \le N_k - 1\}$ is denoted by $E(I_k)$ if $0 \le k \le s-1$, and denoted by $E(O_s)$ if $k = s$. We use E_k^{k+1} to denote the set $\{e_{k,j}^{k+1} | 0 \le j \le N_k - 1\}$. The edge set of $Eye(s)$ is defined as $E(Eye(s)) = \bigcup_{k=0}^{s-1} E(I_k) \cup E(O_s) \cup (\bigcup_{k=0}^{s-1} E_k^{k+1})$. We illustrate $Eye(s)$ for $1 \le i \le 4$ in Figure 1. In all figures, the first coordinate of each vertex is omitted. Graph $Eye(s)$ is 3-regular and contains $3 + 9\sum_{k=1}^{s-1} 2^{k-1} + 3 \cdot 2^{s-1} = 6(2^s - 1)$ vertices and $9(2^s - 1)$ edges.

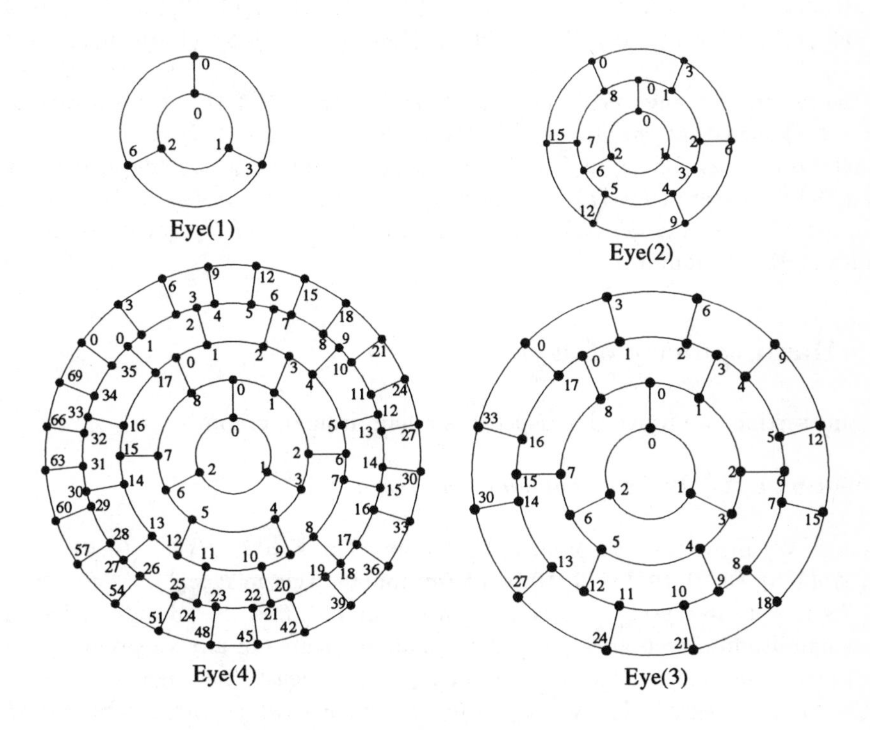

Fig. 1. Examples of eye networks

On the other hand, graphs $Eye(s+1)$, $s \ge 1$, can be recursively drawn (or constructed) from $Eye(s)$ by performing the following two steps:

Step_SUBDIVISION. Subdivide each edge $e_{s,j}$ of $O_s, 0 \le j \le N_s - 1$ and $[j]_3 = 0$, into a path of length 3, i.e., replace $e_{s,j}$ with a path $\langle (s,j), (s,j+1), (s,j+2), (s,j+3) \rangle$ to connect its two ends (s,j) and $(s,j+3)$. Rename O_s as I_s.

Step_EXTENSION. Construct graph O_{s+1} as a concentric cycle outside I_s and join every vertex (s,j) in I_s with vertex $(s+1, 2j+[j]_3)$ in O_{s+1} for $1 \leq j \leq N_s - 1$ and $[j]_3 = 1$ and 2.

The above recursive construction also shows that $Eye(s)$ is a planar graph. All vertices in each cycle can be drawn at equal distance, and all intercycle edges drawn in the normal directions of the corresponding cycles as shown in Figure 1. Therefore, $Eye(s)$ is invariant under the rotation of 120°. Furthermore, we can use a specified subgraph to obtain other isomorphic subgraphs by proper rotation, which we mean rotation of 120° or 240°. To be specific, each vertex (k,i) is relabelled by $(k, i+\delta N_k/3)$, where $\delta = 1$ for 120° rotation and $\delta = 2$ for 240° rotation. For example, consider a cycle in $Eye(2)$ $\mathcal{H}_2 = \langle (0,0), (1,0), (1,1), (2,3), (2,0), (1,8), (1,7), (2,15), (2,12), (2,9), (2,6), (1,2), (1,3), (1,4), (1,5), (1,6), (0,2), (0,1), (0,0)\rangle$. Then we can obtain another cycle $\mathcal{C}_2$ by 120° rotation of $\mathcal{H}_2$, which is given by $\mathcal{C}_2 = \langle (0,1), (1,3), (1,4), (2,9), (2,6), (1,2), (1,1), (2,3), (2,0), (2,15), (2,12), (1,5), (1,6), (1,7), (1,8), (1,0), (0,0), (0,2), (0,1)\rangle$.

Note that for any vertex (k,i) in $Eye(s)$ with $k \geq 1$ there exists a vertex $(k-1, j)$ such that the distance between (k,i) and $(k-1,j)$ is at most 2. In other words, each vertex in I_k can reach I_{k-1} with at most two edges. It follows that the diameter of $Eye(s)$ is at most $2(2(s-1)+1)+1 = 4s-1$. Because the number of vertices in $Eye(s)$ is $6(2^s-1)$, the diameter of $Eye(s)$ is less than 4 times of Moore bound.

3 Hamiltonicity of *Eye*(*s*)

In this section, we prove that $Eye(s)$ is a hamiltonian graph.

Theorem 1. *$Eye(s)$ is hamiltonian for every $s \geq 1$.*

Proof: We prove this theorem by induction. For $s = 1$, $\mathcal{H}_1 = \langle (0,0), (1,0), (1,3), (1,6), (0,2), (0,1), (0,0)\rangle$ is a hamiltonian cycle in $Eye(1)$.

Assume that $Eye(k)$ is hamiltonian for all $1 \leq k \leq s$ and $s \geq 1$. Let $\mathcal{H}_s$ be a hamiltonian cycle in $Eye(s)$. Note that the degree of any vertex in $Eye(s)$ is three, at least one edge of O_s is in $\mathcal{H}_s$ and at least one edge of O_s is not in $\mathcal{H}_s$. Since $Eye(s+1)$ can be obtained from $Eye(s)$ by Step_SUBDIVISION and Step_EXTENSION, we can construct a cycle $\mathcal{H}'_s$ in $Eye(s+1)$ from $\mathcal{H}_s$ by subdividing the edges in $E(O_s) \cap E(\mathcal{H}_s)$. Let U_s denote the collection of vertices in $V(I_s)$ which are not in $\mathcal{H}'_s$. It follows that $\mathcal{H}'_s$ is a hamiltonian cycle in the graph $Eye(s+1) - V(O_{s+1}) - U_s$.

Observe that vertex $(s+1, j)$ in O_{s+1} is adjacent to $(s, \lceil \frac{j}{2} \rceil - 1)$ in I_s. In order to augment O_{s+1} to include all of the vertices in U_s, we replace edge $e_{s+1,2m+1}$ in O_{s+1} with a path $\langle (s+1, 2m+1), (s,m), (s,m+1), (s+1, 2m+4)\rangle$ if $(s,m) \in U_s$ and $[m]_3 = 1$. In this way, we attain a cycle $\mathcal{O}'_{s+1}$ that traverses each vertex of $U_s \cup V(O_{s+1})$ exactly once.

Cycles $\mathcal{H}'_s$ and $\mathcal{O}'_{s+1}$ are two disjoint cycles that traverse every vertex of $Eye(s+1)$ exactly once. To combine these two cycles into a hamiltonian cycle, we arbitrarily choose an edge $e_{s,j}$ in $\mathcal{H}'_s$ with $[j]_3 = 1$ and then define a cycle $\mathcal{H}_{s+1}$ given by

$$(\mathcal{H}'_s - e_{s,j}) \cup (\mathcal{O}'_{s+1} - e_{s+1,2j+1}) \cup \{e^{s+1}_{s,j}, e^{s+1}_{s,j+1}\}. \quad (1)$$

Thus the theorem follows. □

Henceforth, we write "the (construction) scheme (1)" to mean the construction scheme presented in the proof of Theorem 1. We use $\mathcal{H}'_s$ and $\mathcal{O}'_{s+1}$ to denote the two cycles as specified in the scheme (1). Given $\mathcal{H}'_s$ and an edge $e_{s,j}$ of $\mathcal{H}'_s$ with $[j]_3 = 1$, the hamiltonian cycle $\mathcal{H}_{s+1}$ is uniquely defined. Since $Eye(s+1)$ is a 3-regular graph for $s \geq 1$, no two consecutive edges in O_{s+1} can be excluded from $\mathcal{H}_{s+1}$. The following properties hold when deleting edges $e_{s,j}$ in $\mathcal{H}'_s$ and $e_{s+1,2j+1}$ in $\mathcal{O}'_{s+1}$ with $[j]_3 = 1$ to construct $\mathcal{H}_{s+1}$:

(P1) Edges $e_{s+1,2j-2}$ and $e_{s+1,2j+4}$ are included in $\mathcal{H}_{s+1}$.
(P2) Edge $e_{s,i} \in \mathcal{H}_s$ with $i \neq j-1$ if and only if $\langle(s+1, 2i), (s+1, 2i+3), (s+1, 2i+6), (s+1, 2i+9)\rangle$ is included as a subpath of $\mathcal{H}_{s+1}$.
(P3) Any edge $e_{s,i} \notin \mathcal{H}_s$ implies $e_{s,i} \notin \mathcal{H}_{s+1}$, $e_{s,i+1} \in \mathcal{H}_{s+1}$, $e_{s,i+2} \notin \mathcal{H}_{s+1}$ and $e_{s+1,2i+3} \notin \mathcal{H}_{s+1}$.
(P4) Each edge in $O_{s+1} - \mathcal{H}_{s+1}$ is given by $e_{s+1,m}$ with $[m]_6 = 3$, and on the other hand, $\mathcal{H}_{s+1}$ includes all of the edges $e_{s+1,l}$ with $[l]_6 = 0$.

(P1) and (P2) can be easily verified. (P3) follows from the definition of $\mathcal{O}'_{s+1}$. The chosen edge in $\mathcal{O}'_{s+1}$ to be deleted to form $\mathcal{H}_{s+1}$ has the form $e_{s+1,2j+1}$ with $[j]_3 = 1$. Therefore, (P4) follows from (P1) and (P3).

For $s \geq 2$, a hamiltonian cycle $\mathcal{H}_s$ constructed by scheme (1) is called the *fundamental hamiltonian cycle* of $Eye(s)$ if

(i) $\mathcal{H}_1 = \langle(0,0), (1,0), (1,3), (1,6), (0,2), (0,1), (0,0)\rangle$, and
(ii) edge $e_{i,1}$ in $\mathcal{H}'_i$ is chosen to be deleted in the recursive construction of $\mathcal{H}_{i+1}$ for all $1 \leq i \leq s-1$.

In particular, we use $\mathcal{FH}_s$ to denote the fundamental hamiltonian cycle in $Eye(s)$. An example of the fundamental cycle of $Eye(4)$ is illustrated in Figure 2. Since $O_1 - \mathcal{H}_1$ consists of only one edge and we delete one edge from $\mathcal{O}'_{i+1}$ in the recursive construction of $\mathcal{H}_{i+1}$, $\mathcal{FH}_s \cap O_s$ is a set of s disjoint paths which are denoted by $P_0, P_1, \cdots, P_{s-1}$. To be precise,

$$\begin{aligned}
P_0 &= \langle(s,0), (s,3)\rangle, \\
P_i &= \langle(s, 3\cdot 2^i), (s, 3\cdot 2^i + 3), (s, 3\cdot 2^i + 6), \cdots, (s, 3\cdot 2^{i+1} - 3)\rangle \text{ for } 1 \leq i \leq s-2, \\
P_{s-1} &= \langle(s, 3\cdot 2^{s-1}), (s, 3\cdot 2^{s-1} + 3), (s, 3\cdot 2^{s-1} + 6), \cdots, (s, N_s - 3)\rangle.
\end{aligned}$$

Path P_i has $2^i - 1$ edges for $1 \leq i \leq s-2$, and $2^s - 1$ edges for $i = s-1$. Furthermore, there is a set of $2^s - 2$ intercycle edges, given by $T_s = \{e^s_{s-1,3\cdot 2^{s-2}+1}, e^s_{s-1,3\cdot 2^{s-2}+2}, e^s_{s-1,3\cdot 2^{s-2}+4}, \cdots, e^s_{s-1,N_{s-1}-4}\}$, which are not included in $\mathcal{FH}_s$.

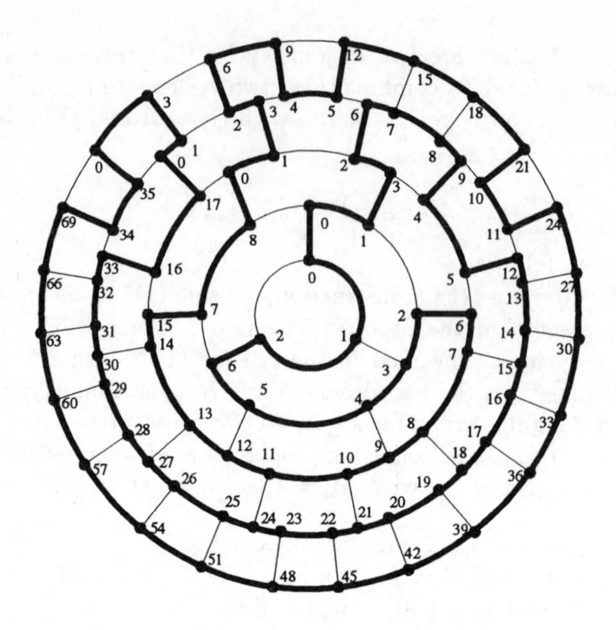

Fig. 2. The fundamental cycle of $Eye(4)$

Remark 1: Since $\frac{1}{2} \leq \frac{|P_{s-1}|}{|E(O_s)|} < \frac{2}{3}$ for $s \geq 2$ where $|X|$ denotes the number of edges in X, we can always obtain a hamiltonian cycle in $Eye(s)$ containing any specific edge $e \in E(O_s)$ by proper rotation of $\mathcal{FH}_s$. Moreover, since $\frac{1}{3} \leq \frac{|T_s|}{|E^s_{s-1}|} < \frac{2}{3}$, we can always obtain a hamiltonian cycle in $Eye(s) - e$ by rotating $\mathcal{FH}_s$ properly for any edge $e \in E^s_{s-1}$ and $s \geq 2$.

4 1-Edge Hamiltonicity of *Eye*(*s*)

The hamiltonian cycle $\mathcal{H}_1 = \langle(0,0), (1,0), (1,3), (1,6), (0,2), (0,1), (0,0)\rangle$ of $Eye(1)$ does not include edges $e_{1,6}$, $e_{0,2}$ and $e^1_{0,1}$. By proper rotation of $\mathcal{H}_1$, we can obtain other two distinct hamiltonian cycles $\mathcal{H}^1_1 = \langle(0,1), (1,3), (1,6), (1,0), (0,0), (0,2), (0,1)\rangle$ and $\mathcal{H}^2_1 = \langle(0,2), (1,6), (1,0), (1,3), (0,1), (0,0), (0,2)\rangle$ such that edges $e_{1,0}$, $e_{0,0}$ and $e^1_{0,2}$ are not in $\mathcal{H}^1_1$ and edges $e_{1,3}$, $e_{0,1}$ and $e^1_{0,0}$ not in $\mathcal{H}^2_1$. Thus, $Eye(1)$ is a 1-edge hamiltonian graph. For $Eye(2)$, it can be verified by exhaustive construction that $Eye(2) - e_{2,0}$ is not hamiltonian. Hence, $Eye(2)$ is not 1-edge hamiltonian. Furthermore, $Eye(2) - e$ is not hamiltonian if $e \in \{e_{2,0}, e_{2,6}, e_{2,12}\}$, and is hamiltonian otherwise.

Let $E^1_s = \{e_{s,m} \mid [m]_6 = 0,\ 0 \leq m < N_s\}$ and $E^2_s = \{e_{k,m} \mid [m]_6 = 0 \text{ or } 2, 0 \leq m < N(k),\ 2 \leq k \leq s-1\}$ be two subsets of $E(Eye(s))$. By definition, we have $E^2_1 = E^2_2 = \emptyset$.

Lemma 2. *$Eye(s) - e$ is hamiltonian for all $e \in E(Eye(s)) - (E^1_s \cup E^2_s)$.*

The following two lemmas are provided with proofs omitted.

Lemma 3. *$Eye(s) - e$ is hamiltonian for $e \in E_s^1$ and $s \geq 3$.*

Lemma 4. *$Eye(s) - e$ is hamiltonian for $e \in E_s^2$ and $s \geq 3$.*

We illustrate the special cases of Eye(3) in Lemma 2 and Lemma 3 in Figures 3(a)–3(d). Lemmas 1, 2, 3 can immediately lead to the following theorem.

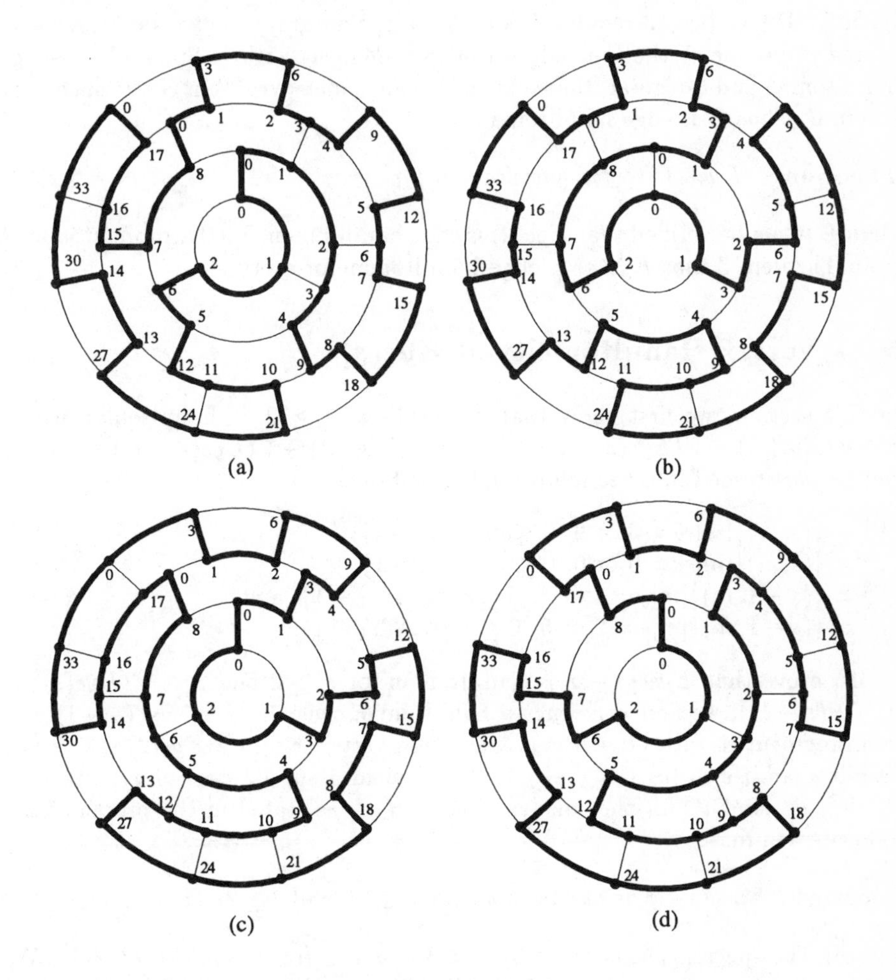

Fig. 3. The hamiltonian cycle of $Eye(3)$ does not traverse the edges $e_{3,m}$ where $m = 0, 6, 12, 18$ in (a) and $m = 24, 30, 0, 6$ in (b). The hamiltonian cycle of $Eye(3)$ does not contain the edges $e_{2,m}$ where $m = 0, 2, 6, 8$ in (c) and $m = 0, 2, 12, 14$ in (d).

Theorem 5. *$Eye(s)$ is 1-edge hamiltonian for every $s \geq 1$ and $s \neq 2$.*

Because the degree of every vertex in $Eye(s)$ is three, we have the following corollary:

Corollary 6. *$Eye(s)$ is optimal 1-edge hamiltonian for $s \neq 2$.*

To verify the relationship between 1-edge hamiltonicity and edge hamiltonicity, we have the following theorem.

Theorem 7. *Any 3-regular 1-edge hamiltonian graph is edge hamiltonian.*

Proof: Let G be a 3-regular 1-edge hamiltonian graph. Let e be an edge in G and e' be one of the four edges that are incident with e. Since G is 1-edge hamiltonian and 3-regular, there exists a hamiltonian cycle $\mathcal{C}$ in $G - e'$ such that e is in $\mathcal{C}$. Thus G is edge hamiltonian. □

Theorem 8. *$Eye(s)$ is edge hamiltonian for every s.*

Proof: It can be verified that $Eye(2)$ is edge hamiltonian. Furthermore, it follows from Theorem 3 that $Eye(s)$ is edge hamiltonian for every s. □

5 1-Vertex Hamiltonicity of $Eye(s)$

In this section, we first show that $Eye(s) - x$, $s \geq 2$, is hamiltonian where $x \in V(Eye(s)) - V(Eye(s-1))$. The set $V(Eye(s)) - V(Eye(s-1))$ for $s \geq 2$ can be partitioned into the following four subsets:

$$\mathcal{V}_s^1 = \{(s,m) \,|\, [m]_{12} = 3 \text{ or } 9,\ 0 \leq m \leq N_s - 1\},$$
$$\mathcal{V}_s^2 = \{(s,m) \,|\, [m]_{12} = 0 \text{ or } 6,\ 0 \leq m \leq N_s - 1\},$$
$$\mathcal{V}_s^3 = \{(s-1,m) \,|\, [m]_6 = 1 \text{ or } 4,\ 0 \leq m \leq N_{s-1} - 1\}, \text{ and}$$
$$\mathcal{V}_s^4 = \{(s-1,m) \,|\, [m]_6 = 2 \text{ or } 5,\ 0 \leq m \leq N_{s-1} - 1\}.$$

To prove that $Eye(s) - x$ is hamiltonian for $s \geq 2$ and $x \in V(Eye(s)) - V(Eye(s-1))$, we first construct a hamiltonian cycle $\mathcal{H}_{s-1}$ in $Eye(s-1)$ and then augment it with O_s without traversing vertex x. Since $Eye(2) - e$ is not hamiltonian if $e \in \{e_{2,0}, e_{2,6}, e_{2,12}\}$, it will yield a special case of x in each of $\mathcal{V}_s^1$, $\mathcal{V}_s^2$, $\mathcal{V}_s^3$, and $\mathcal{V}_s^4$ for the construction of $\mathcal{H}_{s-1}$ as stated in the proofs of the following lemmas.

Lemma 9. *$Eye(s) - x$ is hamiltonian for $x \in \mathcal{V}_s^1$ and $s \geq 2$.*

Proof: The special case is given by $s = 3$ and $x \in \{(3,3), (3,15), (3,27)\}$. We can construct a hamiltonian cycle in $Eye(3) - x$ (the construction is omitted here).

For the cases of $s = 2$ and $x = (s,m)$, we define $\mathcal{H}_{s-1}$ as $\mathcal{H}_1^1$ for $m = 3$, as $\mathcal{H}_1^2$ for $m = 9$, and as $\mathcal{H}_1$ for $m = 15$, where $\mathcal{H}_1$, $\mathcal{H}_1^1$, and $\mathcal{H}_1^2$ are defined in the first paragraph of Section 4. For the remaining cases of s and $x = (s,m)$, let $\mathcal{H}_{s-1}$ denote a hamiltonian cycle of $Eye(s-1) - e_{s-1,\lceil\frac{m}{2}\rceil-2}$ obtained by applying the construction scheme (1) if $e_{s-1,\lceil\frac{m}{2}\rceil-2} \notin E_{s-1}^1$, and by applying the construction presented in the proof of Lemma 2 if $e_{s-1,\lceil\frac{m}{2}\rceil-2} \in E_{s-1}^1$. Since deleting vertex x from $Eye(s)$ reduces the degree of each vertex of $(s,m-3)$, $(s,m+3)$ and

$(s-1, \lceil\frac{m}{2}\rceil-1)$ by one, it follows that any hamiltonian cycle in $Eye(s)-x$ must traverse edges $e_{s,m-6}$, $e^s_{s-1,\lceil\frac{m}{2}\rceil-3}$, $e^s_{s-1,\lceil\frac{m}{2}\rceil}$, $e_{s,m+3}$, $e_{s-1,\lceil\frac{m}{2}\rceil-2}$, and $e_{s-1,\lceil\frac{m}{2}\rceil-1}$. Let $\mathcal{P}$ denote the path $(\mathcal{O}'_s - e_{s,m-3} - e^s_{s-1,\lceil\frac{m}{2}\rceil-1}) \cup \{e_{s-1,\lceil\frac{m}{2}\rceil-2}, e^s_{s-1,\lceil\frac{m}{2}\rceil-3}\}$. Then $\mathcal{P}$ contains all vertices in $(V(Eye(s)) - V(\mathcal{H}'_{s-1}) - x) \cup \{(s-1, \lceil\frac{m}{2}\rceil - 2), (s-1, \lceil\frac{m}{2}\rceil-3)\}$. Furthermore, $\mathcal{P}$ and $\mathcal{H}'_{s-1} - e_{s-1,\lceil\frac{m}{2}\rceil-3}$ are two disjoint paths with common end vertices $(s-1, \lceil\frac{m}{2}\rceil-2)$, $(s-1, \lceil\frac{m}{2}\rceil-3)$ and contain all vertices in $Eye(s)-x$. Thus, $\mathcal{P} \cup (\mathcal{H}'_{s-1} - e_{s-1,\lceil\frac{m}{2}\rceil-3})$ defines a hamiltonian cycle of $Eye(s)-x$. □

The following three lemmas are provided with proofs ommitted.

Lemma 10. *$Eye(s)-x$ is hamiltonian for $x \in \mathcal{V}^2_s$ and $s \geq 2$.*

Lemma 11. *$Eye(s)-x$ is hamiltonian for $x \in \mathcal{V}^3_s$ and $s \geq 2$.*

Lemma 12. *$Eye(s)-x$ is hamiltonian for $x \in \mathcal{V}^4_s$ and $s \geq 2$.*

Theorem 13. *$Eye(s)$ is 1-vertex hamiltonian for every $s \geq 1$.*

Proof: For $Eye(1)$, the cycle $\langle(0,0), (0,1), (1,3), (1,6), (0,2), (0,0)\rangle$ does not include vertex $(1,0)$. Since this cycle is unique of length five up to isomorphism, it follows that $Eye(1)$ is 1-vertex hamiltonian. Now consider $s \geq 2$. Let $x = (k,m) \in Eye(s)$, where $0 \leq k \leq s$. If $k = s$, it follows from Lemmas 4, 5, 6, and 7 that $Eye(s)-x$ is hamiltonian. If $k \leq s-1$, then we distinguish the following four cases of x.

case 1: Let $k=0$. Since $Eye(1)$ is 1-vertex hamiltonian, we can obtain a hamiltonian cycle, denoted by $\mathcal{NH}^1_1$, in $Eye(1)-x$.

case 2: Let $k=1$ and $[m]_3 = 0$. Similarly, we can obtain a hamiltonian cycle, denoted by $\mathcal{NH}_2$, in $Eye(1)-x$.

case 3: Let $k \geq 2$ and $[m]_3 = 0$. Since $x \in \mathcal{V}^1_k \cup \mathcal{V}^2_k$, it follows from Lemma 4 and Lemma 5 that we can obtain a hamiltonian cycle, denoted by $\mathcal{NH}_k$, in $Eye(k)-x$.

case 4: Let $k \geq 1$ and $[m]_3 \neq 0$. Since $x \in \mathcal{V}^3_{k+1} \cup \mathcal{V}^4_{k+1}$, it follows from Lemma 6 and Lemma 7 that we also obtain a hamiltonian cycle, denoted by $\mathcal{NH}^4_{k+1}$, in $Eye(k+1)-x$.

In case 3, $\mathcal{NH}_k$ is a cycle that traverses all vertices in $Eye(k)$ except x exactly once. Let $\mathcal{NH}'_k$ denote the cycle obtained from $\mathcal{NH}_k$ by subdividing all edges in $(O_k - x) \cap \mathcal{NH}_k$. By performing Step_SUBDIVISION and Step_EXTENSION on $Eye(k)$, $\mathcal{NH}'_k$ will not include vertices $(k, m-2)$, $(k, m-1)$, (k,m), $(k, m+1)$ and $(k, m+2)$. To construct a cycle, denoted by $\mathcal{NH}^3_{k+1}$, in $Eye(k+1)$ to traverse all vertices excluding x exactly once, we use the scheme (1) and replace the path $\langle(k+1, 2m-3), (k+1, 2m), (k+1, 2m+3), (k+1, 2m+6)\rangle$ in $\mathcal{O}'_{k+1}$ with path $\langle(k+1, 2m-3), (k, m-2), (k, m-1), (k+1, 2m), (k+1, 2m+3), (k, m+1), (k, m+2), (k+1, 2m+6)\rangle$. In case 2, we can obtain a hamiltonian cycle in $Eye(2)-x$, denoted by $\mathcal{NH}^2_2$, using the same approach as in case 3.

We then apply the scheme (1) on $\mathcal{NH}_1^1$, $\mathcal{NH}_2^2$, $\mathcal{NH}_{k+1}^3$ or $\mathcal{NH}_{k+1}^4$, respectively, to construct a hamiltonian cycle in $Eye(s) - x$. □

Since $Eye(s)$ is 1-vertex hamiltonian and 3-regular, the following corollaries hold.

Corollary 14. *$Eye(s)$ is 3-connected for every s.*

Corollary 15. *$Eye(s)$ is optimal 1-vertex hamiltonian for every s.*

Corollary 1 and Corollary 3 immediately lead to the following theorem.

Theorem 16. *$Eye(s)$ is optimal 1-hamiltonian for $s \neq 2$.*

6 Concluding Remarks

In summary, $Eye(s)$ is optimal 1-edge hamiltonian for $s \neq 2$, optimal 1-vertex hamiltonian for every s, and optimal 1-hamiltonian for $s \neq 2$. Furthermore, $Eye(s)$ is edge hamiltonian for every s. $Eye(s)$ indeed has nice properties on hamiltonicity and is planar. This graph has diameter of $O(\log n)$ and to be precise, bounded by $4 \log \frac{n+6}{6} - 1$, where n is the number of vertices in $Eye(s)$. It would be interesting to construct a family of planar optimal 1-hamiltonian graphs which have nice properties as $Eye(s)$ and have even smaller diameter than $Eye(s)$.

REFERENCES

[1] J.A. Bondy and U.S.R. Murty, *Graph Theory with Applications*, North-Holland, New York, (1980).

[2] F.R.K. Chung, Diameters of graphs: Old problems and new results, *Proc. 18th South-Eastern Conf. Combinatorics, Graph Theory, and Computing, Congressus Numerantium*, 60(1987), 298–319.

[3] F. Harary and J.P. Hayes, Edge fault tolerance in graphs, *Networks*, 23(1993), 135–142.

[4] F. Harary and J.P. Hayes, Node fault tolerance in graphs, *Networks*, 27(1996), 19–23.

[5] P. Horák and J. Širáň, On a construction of Thomassen, *Graphs and Combinatorics*, 2(1986), 347–350.

[6] K. Mukhopadhyaya and B.P. Sinha, Hamiltonian graphs with minimum number of edges for fault-tolerant topologies, *Information Processing Letters*, 44(1992), 95–99.

[7] L. Stacho, Maximally non-hamiltonian graphs of girth 7, *Graphs and Combinatorics*, 12(1996), 361–371.

[8] C. Thomassen, Hypohamiltonian and hypotraceable graphs, *Discrete Math.*, 9(1974), 91–96.

[9] J.J. Wang, C.N. Hung, and L.H. Hsu, Optimal 1-hamiltonian graphs, accepted by *Information Processing Letters*.

A Linear-Time Algorithm for Constructing an Optimal Node-Search Strategy of a Tree

Sheng-Lung Peng[1], Chin-Wen Ho[3], Tsan-sheng Hsu[2], Ming-Tat Ko[2], and Chuan Yi Tang[1]

[1] National Tsing Hua University,Taiwan
[2] Academia Sinica,Taiwan
[3] National Central University, Taiwan

Abstract. Ellis *et al.*, proposed algorithms (in terms of vertex separation) to compute the node-search number of an n-vertex tree T in $O(n)$ time and to construct an optimal node-search strategy of T in $O(n \log n)$ time. An open problem is whether the latter can also be done in linear time. In this paper, we solve this open problem by exploring fundamental graph theoretical properties.

1 Introduction

The graph searching problem was first proposed by Parsons [22]. In the original version, it is called *edge searching*, an undirected graph is considered as a system of tunnels in which all the tunnels are initially contaminated by a gas. There are three kinds of moves in edge searching: (1) placing a searcher on a vertex; (2) removing a searcher from a vertex; and (3) moving a searcher from one vertex to another along an edge. A contaminated edge is *cleared* if move (3) is applied. Another version is called *node searching*, which was proposed by Kirousis and Papadimitriou [16]. In node searching, the move (3) is not allowed and a contaminated edge is *cleared* if both its two endpoints simultaneously contain searchers. The objective of graph searching problem is to obtain a state of the graph in which all the edges are simultaneously cleared by a sequence of allowed moves using the least number of searchers. The graph searching problem is not only interesting theoretically, but also have applications on combinatorial problems [2, 12, 15, 17, 20, 25].

A *search strategy* is a sequence of allowed moves to clear the initial contaminated graph. A search strategy is *optimal* if it uses the minimum number of searchers among all possible search strategies. The number of searchers needed in an optimal search strategy of G is called the search number of G. In node searching, we call it the *node-search number* of G and denote it as $ns(G)$. A cleared edge may be recontaminated if there is a path from a contaminated edge to the cleared edge without any searcher on its vertices. It has been shown in [16] and [3] that there always exists an optimal node-search strategy for a graph that does not recontaminate any edge. In the rest of paper, we only consider the node-search strategies which do not recontaminate any edge.

The node searching problem attracted the attention of the theoretical computer science community because of its equivalences with different seemingly unrelated problems. For example, the following numbers coincide: the node-search number of G, the interval thickness of G [15], the pathwidth of G plus one [25, 20], and the vertex separation number of G plus one [16, 12]. Several excellent surveys for these problems can be found in [2, 4, 10, 20]. The node searching problem is NP-complete on planar graphs with vertex degree at most three [21], chordal graphs [11], bipartite graphs [13], cobipartite graphs [1], and bipartite distance-hereditary graphs [14]. For some special classes of graphs, it can be solved in polynomial time [5, 6, 7, 8, 9, 11, 20, 24, 26].

For trees, the following results are known. Ellis *et al.* [9] proposed a linear-time algorithm (in terms of vertex separation) to find the node-search number of a tree T. However, this algorithm cannot be used to construct, in linear time, an optimal node-search strategy of T. Scheffler [26] presented a fairly complicate linear-time algorithm (in terms of pathwidth) to compute the node-search number of T and she claimed that a linear-time algorithm for constructing an optimal node-search strategy of T easily followed. Unfortunately, few details are recorded in the literature.

Megiddo *et al.* [19] proposed a linear-time algorithm to find the edge-search number of a tree T. They also gave an algorithm to construct an optimal edge-search strategy of T. However, this algorithm runs in $O(n \log n)$ time where n denotes the number of vertices of T. Peng *et al.* [23] showed that an optimal edge-search strategy of T can be obtained in linear time if an optimal node-search strategy of T is given.

Peng *et al.* [23] proposed the concept of the avenue on a tree for node searching. In this paper we extend this avenue concept of a tree to an extended avenue system and an avenue tree. Based on these structures, we design a linear-time algorithm to construct an optimal node-search strategy of a tree.

2 The Extended avenue system and the avenue tree

2.1 The Extended avenue system

Let T be an unrooted tree, and let $V(T)$ and $E(T)$ denote the vertex and edge sets of T, respectively. For a vertex $t \in V(T)$, a connected component of $T \backslash \{t\}$ is called a *branch* of T at t. Let v be adjacent to t in T. The branch of T at t containing v is denoted as T_{tv}.

Let P be a path in T. Given a vertex $x \in V(P)$ and an edge $(x, y) \in E(T)$, T_{xy} is a *path branch at* P if $y \in V(P)$ and a *nonpath branch at* P if $y \notin V(P)$. If $|V(T)| > 1$, then $ns(T) \geq 2$. We define $ns(T) = 1$ if $|V(T)| = 1$. Thus, $ns(T) \geq 2$ if and only if there exists a vertex $t \in V(T)$ with at least one branch.

Lemma 1. *[9, 26] For any tree T, $ns(T) \geq k + 1$ for $k \geq 2$ if and only if there exists a vertex $t \in V(T)$ with at least three branches T_{tu}, T_{tv}, and T_{tw} whose node-search numbers are all at least k. For any tree T, $ns(T) \geq 2$ if and only if there exists a vertex $t \in V(T)$ with at least one branch.*

Using Lemma 1, Peng *et al.* [23] proposed the concept of an avenue of a tree in node searching. A path P is an *avenue* of T, if the following conditions hold:

1. for every path branch T' at P, $ns(T') = ns(T)$; and
2. for every nonpath branch T' at P, $ns(T') < ns(T)$.

Note that if $|V(P)| = 1$, then P is also called a *hub*. When T has a hub as an avenue, it may have another vertex to be a hub. In general, a hub is not unique. Peng *et al.* showed that for any tree T, T has either a hub or a unique avenue [23]. In fact, if T has an avenue having at least two vertices, then the avenue is induced by edges (x, y) such that $ns(T_{xy}) = ns(T_{yx}) = ns(T)$. Now we extend the concept of an avenue as follows. A path P is an *extended avenue* of a tree T, if the following conditions hold:

1. for every path branch T' at P, $ns(T') \leq ns(T)$; and
2. for every nonpath branch T' at P, $ns(T') < ns(T)$.

Note that an avenue of T is also an extended avenue of T. Let P be an extended avenue of T. Then P must contain an avenue of T. Path branches (respectively, nonpath branches) at P are also called the *avenue branches* (respectively, *nonavenue branches*) at P. In the following, we define tag values on vertices of T, $\mathcal{A}(T)$, and $\mathcal{F}(\mathcal{A}(T)$ recursively as follows.

1. If $V(T) = \{v\}$, then
   (a) $tag(v) = 1$;
   (b) $\mathcal{A}(T) = \{[v]\}$;
   (c) $\mathcal{F}(\mathcal{A}(T)) = \{T\}$.
2. If $|V(T)| > 1$, let P be an extended avenue of T. Let $\mathcal{T}_P$ denote the set $\{T' | T'$ is a connected component after removing P from $T\}$, then
   (a) for all $v \in V(P)$, $tag(v) = ns(T)$;
   (b) $\mathcal{A}(T) = \{P\} \cup (\cup_{T' \in \mathcal{T}_P} \mathcal{A}(T'))$;
   (c) $\mathcal{F}(\mathcal{A}(T)) = \{T\} \cup (\cup_{T' \in \mathcal{T}_P} \mathcal{F}(\mathcal{A}(T')))$.

As defined above, $\mathcal{A}(T)$ is a set of vertex disjoint paths of T. We call $\mathcal{A}(T)$ an *extended avenue system of* T. The set $\mathcal{F}(\mathcal{A}(T))$ is corresponding to $\mathcal{A}(T)$. By definition, for each path $P \in \mathcal{A}(T)$, there exists a tree $T' \in \mathcal{F}(\mathcal{A}(T))$ such that P is an extended avenue of T'. Hence $|\mathcal{A}(T)| = |\mathcal{F}(\mathcal{A}(T))|$. For convenience, we denote T' by B_P in the rest of paper. Let P be any maximal path induced by vertices with the same tag value in T. Then $P \in \mathcal{A}(T)$. In other words, $\mathcal{A}(T)$ can also be determined according to the tag values of all vertices of T. Note that once $\mathcal{A}(T)$ is defined, the tag value of every vertex of T and $\mathcal{F}(\mathcal{A}(T))$ are then well defined and we will refer them without specifying. Since an extended avenue of T is not unique, $\mathcal{A}(T)$ is also not unique.

By definition, all the vertices in a path $P \in \mathcal{A}(T)$ are with the same tag value. Thus we may define *the tag value of* P in $\mathcal{A}(T)$ being the tag value of a vertex of P, i.e., $tag(P) = tag(v)$ for $v \in V(P)$. Note that if $tag(P) = 1$, then $|V(P)| = 1$. For simplicity, all the paths in $\mathcal{A}(T)$ are called *avenues*. Especially, the path in $\mathcal{A}(T)$ with the tag value $ns(T)$ is called the *main avenue* of T with respect to $\mathcal{A}(T)$. Note that if $P \in \mathcal{A}(T)$ is the main avenue of T, then $B_P = T$.

For any two avenues P and Q in $\mathcal{A}(T)$, let C_{PQ} denote the *connecting path* of P and Q which is the path $[u, w_1, w_2, \ldots, w_s, v]$ in T such that $u \in V(P)$, $v \in V(Q)$, and $w_i \notin V(P) \cup V(Q)$ for all i. Every vertex w_i, $1 \leq i \leq s$, is called an *internal vertex* of C_{PQ}. Since T is a tree, there is exactly one connecting path that connects any two avenues of $\mathcal{A}(T)$. By definition of $\mathcal{A}(T)$, It is not hard to see that for any two avenues with the same tag value s in $\mathcal{A}(T)$, $s < ns(T)$, their connecting path contains at least one vertex whose tag value is greater than s.

For each $P \in \mathcal{A}(T)$, let $\mathcal{A}_P = \{Q \in \mathcal{A}(T) | tag(P) < tag(Q)$ and for each internal vertex $v \in V(C_{PQ})$, $tag(v) < tag(P)\}$. Note that if P is the main avenue of T, then $\mathcal{A}_P = \emptyset$; otherwise $\mathcal{A}_P \neq \emptyset$. It is not hard to show that for any avenue $P \in \mathcal{A}(T)$ which is not the main avenue of T, all the tag values of the avenues in $\mathcal{A}_P$ are distinct.

2.2 The Avenue tree

We use the notation "$v \to u$" in a rooted tree to denote the parent of v is u. With respect to an extended avenue system $\mathcal{A}(T)$, an avenue tree $\mathcal{L}(T)$, which is a rooted tree, is defined as follows.

1. If T is an isolated vertex v, then v itself is the root of $\mathcal{L}(T)$.
2. If T is not an isolated vertex, then let $P = [v_1, v_2, \ldots, v_r] \in \mathcal{A}(T)$ be the main avenue of T and let $\mathcal{F}_P$ be the set of all the nonavenue branches at P. We construct $\mathcal{L}(T)$ as follows.
   (a) We arbitrarily choose a vertex v_i from $V(P)$ as the root of $\mathcal{L}(T)$.
   (b) Let $v_j \to v_{j+1}$ for $1 \leq j \leq i-1$.
   (c) Let $v_j \to v_{j-1}$ for $i+1 \leq j \leq r$.
   (d) For each $T' \in \mathcal{F}_P$, let u be the root of $\mathcal{L}(T')$ and assume T' is a nonavenue branch at $v_j \in V(P)$; then let $u \to v_j$.

Since the root of $\mathcal{L}(T)$ is arbitrarily chosen, $\mathcal{L}(T)$ is not unique. A rooted tree is called a *rooted path* if the underlying unrooted tree is a path. From the definition of $\mathcal{L}(T)$, every avenue in $\mathcal{A}(T)$ is a rooted path in $\mathcal{L}(T)$. In $\mathcal{L}(T)$, the tag value of every internal vertex is greater than 1 and every vertex with tag value 1 is a leaf. For any avenue $P \in \mathcal{A}(T)$, $tag(P) > 1$, and a vertex $v \in V(P)$, if v is a leaf of $\mathcal{L}(T)$ then (i) v is not the root of P in $\mathcal{L}(T)$; (ii) v is an endpoint of P; and (iii) there is no nonavenue branch at v in B_P. For $u, v \in V(\mathcal{L}(T))$ and $v \to u$ in $\mathcal{L}(T)$, let $P, Q \in \mathcal{A}(T)$ be the avenues containing v and u, respectively. In the case of $tag(v) < tag(u)$, we call Q the *parent* of P in $\mathcal{L}(T)$. We define $\min(\mathcal{A}_P)$ to be the avenue in $\mathcal{A}_P$ with the smallest tag value.

Lemma 2. *For $Q \in \mathcal{A}(T)$, B_P is a nonavenue branch at Q in B_Q if and only if $Q = \min(\mathcal{A}_P)$.*

Lemma 3. *Let $P \in \mathcal{A}(T)$ which is not the main avenue of T. Then in any $\mathcal{L}(T)$, the parent of P is Q if and only if $Q = \min(\mathcal{A}_P)$.*

Lemma 4. *Let $\mathcal{A}(T)$ be an extended avenue system of T and T^* be a rooted tree with $V(T^*) = V(T)$, T^* is an avenue tree of T with respect to $\mathcal{A}(T)$ if T^* satisfies the following conditions.*

1. For each $P \in \mathcal{A}(T)$, $V(P)$ induces a rooted path in T^*.

2. For each $P \in \mathcal{A}(T)$, $tag(P) < ns(T)$, let $Q = \min(\mathcal{A}_P)$ and B_P be a nonavenue branch at $v \in V(Q)$. Then v is the parent of the root of P in T^*.

Given an $\mathcal{A}(T)$ and its one corresponding avenue tree $\mathcal{L}(T)$, we design the following algorithm to construct an optimal node-search strategy of T.

```
procedure SEARCH(v); {v is the root of L(T)}
  Let (v1, v2, ..., vr) be the sequence such that tag(vi) = tag(v);
  for i := 1 to r do begin
    place a searcher on vi;
    if i > 1 then remove the searcher from v(i-1);
    for all children y of vi with tag(y) < tag(v) do SEARCH(y)
  end;
  remove the searcher from vr
end SEARCH;
```

Theorem 5. *Given an avenue tree $\mathcal{L}(T)$ rooted at v, Algorithm $SEARCH(v)$ constructs an optimal node-search strategy of T in linear time.*

3 Constructing an optimal node-search strategy

In the following, T is a rooted tree. The node-search number of a rooted tree is the same as the node-search number of its underlying unrooted tree. Let $T[u]$ denote the subtree of T rooted at u. Let $T[u, v_1, v_2, \ldots, v_i]$ denote the resulting tree of removing the subtrees rooted at v_1 through v_i from $T[u]$. A vertex x is k-*critical* in a rooted tree T if $ns(T[x]) = k$ and there are two children y and z of x such that $ns(T[y]) = ns(T[z]) = k$.

For each vertex $u \in V(T)$, the *label* of u is a list of integers $(a_1, \ldots, a_p)$, $a_1 > a_2 > \cdots > a_p \geq 1$, and each a_i is associated with a vertex u_i, $1 \leq i \leq p$, such that the following conditions hold.

1. $ns(T[u]) = a_1$.
2. For $1 \leq i < p$, $ns(T[u, u_1, \ldots, u_i]) = a_{i+1}$.
3. For $1 \leq i < p$, u_i is an a_i-critical vertex in $T[u, u_1, \ldots, u_{i-1}]$.
4. u_p is u. If a_p is marked with a prime "$'$" then there is no a_p-critical vertex in $T[u, u_1, \ldots, u_{p-1}]$. If a_p is not marked with a prime, then u_p is an a_p-critical vertex in $T[u, u_1, \ldots, u_{p-1}]$.

A *critical element* is an element of the label that is associated with a critical vertex. Note that the prime marker is used only on the last element, since all the others are necessarily critical.

Let $I = [a, b]$, where $a \geq b$ and a, b are integers, represent the list of consecutive integers $(a, a-1, \ldots, b)$. Then the label $\lambda = (a_1, \ldots, a_p)$ can be represented as a list of intervals $(I_1, I_2, \ldots, I_r)$, where I_i represents a maximal set of consecutive integers in λ for all $1 \leq i \leq r$ [9]. In λ's integer representation $(a_1, \ldots, a_p)$, let u_i be the vertex associated with a_i for $1 \le i \le p$. In λ's interval representation $(I_1, I_2, \ldots, I_r)$, only the vertices associated with the endpoints of intervals are recorded.

For convenience, a label is also treated as a set. For a label λ and an integer k, in the interval representation, we say $k \in \lambda$ if there exists an interval $[a, b] \in \lambda$ such that $a \ge k \ge b$. Let $cut(\lambda, k)$ be the resulting label by deleting all elements which are no greater than k from λ. Let $\min(\lambda)$ (respectively, $\max(\lambda)$) be the smallest (respectively, largest) element of label λ. For any two trees $T_1 = (V_1, E_1)$ and $T_2 = (V_2, E_2)$, $T_1 \cup T_2 = (V_1 \cup V_2, E_1 \cup E_2)$.

For a subtree $T[u]$ of T, we first compute the label λ of u and then decide the tag value of u according to λ. By the tag value of u and the labels of its children, an avenue tree $\mathcal{L}(T[u])$ can be constructed. Let "&" denote the concatenating operation in label combination. For simplicity, we only define the concatenating operation in integer representation. In our algorithm, the concatenation $\lambda = \lambda_1 \& \lambda_2$ only occurs when $\min(\lambda_1) \ge \max(\lambda_2)$. If $\min(\lambda_1) > \max(\lambda_2)$, then $\lambda = \lambda_1 \cup \lambda_2$; otherwise we recursively compute $\lambda = cut(\lambda_1, \max(\lambda_2)) \& (\max(\lambda_2) + 1)$. The details of our algorithm are as follows.

```
function AVTREE(T: unrooted tree): rooted tree;
    Initially, V(T*) = V(T) and for each vertex v ∈ V(T*), v → v in T*;
    Choose some vertex u in T and make u the root of T;
    λ := (); /* λ is the label of u and, initially, it is empty */
    BUILD_AVTREE(T, u, λ, T*);
    AVTREE := T*
end AVTREE;
procedure BUILD_AVTREE(T: tree; u: vertex; λ: label; T*: tree);
    if u is the only vertex in the tree T[u] then
      λ := ([1', 1']) and ntag(u) = 1
    else
      for all vertices v_i, 1 ≤ i ≤ d, the d children of u, do
        Let T*_i be the subforest of T* induced by V(T[v_i]);
        λ_i := (); /* λ_i is the label of v_i */
        BUILD_AVTREE(T, v_i, T*_i, λ_i)
      endfor;
      λ := COMBINE_LABELS(λ_1, λ_2, ..., λ_d);
      ntag(u) := min(λ);
      T* := COMBINE_AVTREES(u, λ_1, ..., λ_d, T*_1, ..., T*_d)
    endif
end BUILD_AVTREE;
function COMBINE_LABELS(λ_1, λ_2, ..., λ_d : label): label;
    if there is one or more label containing 1 then λ := ([2', 2']) else λ := ([1', 1']);
    Let m be the second largest element in {max(λ_1), max(λ_2), ..., max(λ_d)};
    for k := 2 to m do
      Let n be the number of labels containing an element k;
      Case 1: {n ≥ 3} λ := ([k + 1', k + 1']);
      Case 2: {n = 2 and at least one element k is critical} λ := ([k + 1'.k + 1']);
      Case 3: {n = 2 and neither element k is critical} λ := ([k, k]);
      Case 4: {n = 1 and element k is critical and k ∈ λ} λ := ([k + 1', k + 1']);
      Case 5: {n = 1 and element k is critical and not (k ∈ λ)} λ := ([k, k])&λ;
      Case 6: {n = 1 and element k is not critical} λ := ([k', k'])
    endfor;
    Let λ_s be the label with max(λ_s) > m and μ = cut(λ_s, m);
    COMBINE_LABELS := μ&λ
end COMBINE_LABELS;
function COMBINE_AVTREES(u: vertex; λ_1, ..., λ_d: labels; T*_1, ..., T*_d: trees): tree;
    T* := (∪_{i=1}^{d} T*_i) ∪ {the rooted tree consists of the single vertex u};
    Let m̃ be the second largest element in {max(λ_1), max(λ_2), ..., max(λ_d)};
```

```
    for k = 1 to m do
        for each λ_i with k ∈ λ_i do
          Case 1: {k < ntag(u) and k is a left endpoint of λ_i}
            /* this case decides the children of u */
            Let v be the vertex associated with k and v → w;
            if ntag(w) > ntag(u) or w = v then v → u; u' = u;
          Case 2: {k = ntag(u) and k is an endpoint of λ_i}
            /* this case decides the rooted path (u is the root) */
            Let v be the vertex associated with k and v → w;
            Case 2.1: {k is a left endpoint} v → u;
            Case 2.2: {k is a right endpoint} v → u; u → w;
          Case 3: {k > ntag(u) and k is an endpoint of λ_i}
            /* this case decides the ancestors of u */
            Let v be the vertex associated with k;
            Case 3.1: {k is a left and right endpoint} u' → v; u' = v;
            Case 3.2: {k is a right endpoint} u' → v;
            Case 3.3: {k is a left endpoint} u' = v
        endfor
    endfor;
    Let λ_s be the label with max(λ_s) > m and μ = (..., I_{r-1}, I_r) = cut(λ_s, m);
A:  if min(μ) is a right endpoint of λ_s then
      /* the remaining work of Case 3 */
      Let v be the vertex associated with min(μ);
      u' → v
B:  else /* m ∈ λ_s, this implies ntag(u) > m */
      /* decide the children of u */
      Let v be the vertex associated with the left endpoint of I_r;
      v → u;
C:    if r > 1 then
        Let w be the vertex associated with the right endpoint of I_{r-1};
        u → w
      endif
    endif;
    COMBINE_AVTREES := T*
end COMBINE_AVTREES;
```

The function *COMBINE_LABELS* is presented in [9] which computes the labels correctly. As proved in [9], using the interval representation of labels, $ns(T)$ can be computed in linear time.

Lemma 6. *For any vertex $u \in V(T)$, let $(a_1, \ldots, a_p)$ be the label of u and u_i be the vertex associated with a_i for $1 \le i \le p$. Let T^0 denote the tree $T[u]$ and T^i denote the subtree $T[u, u_1, \ldots, u_i]$ for $1 \le i < p$. Then the following statements hold.*

1. *For each critical u_i, $1 \le i \le p$, u_i is in any extended avenue of T^{i-1}.*
2. *T^i is a nonavenue branch at u_i in T^{i-1}, for $1 \le i < p$.*
3. *Let $B^i = T^{i-1} \backslash T^i$, $1 \le i < p$. Then $\mathcal{A}(T^0) = (\cup_{i=1}^{p-1} \mathcal{A}(B^i)) \cup \mathcal{A}(T^{p-1})$.*

Lemma 7. *Let u be a vertex in the rooted tree T with the label λ. Let $a_p = \min(\lambda)$. Let v_i with the label λ_i, $1 \le i \le d$, be the d children of u. For each i, $1 \le i \le d$, let $\pi_i = cut(\lambda_i, a_p)$. Then the following statements hold.*

1. *π_i, $1 \le i \le d$, are mutually disjoint.*
2. *$cut(\lambda, a_p) = \cup_{i=1}^{d} \pi_i$.*

Lemma 8. *The ntag values computed by Algorithm $AVTREE(T)$ determine an extended avenue system of T.*

Lemma 9. *Let $\mathcal{A}(T)$ be the avenue system of T determined by $AVTREE(T)$ and u be the root of T. Let the label of u be $\lambda = (a_1, \ldots, a_p)$ and u_i be the vertex associated with a_i for $1 \leq i \leq p$. Let $P_i \in \mathcal{A}(T)$ be the avenue containing u_i. Then $P_{i-1} = \min(\mathcal{A}_{P_i})$.*

Lemma 10. *Algorithm $AVTREE(T)$ constructs an avenue tree T^* of T.*

Proof: It suffices to prove that T^* constructed by $AVTREE(T)$ satisfies the conditions in Lemma 4. We consider T as a rooted tree with root at u. We prove this lemma by induction on the height of T. If T is of height 1, then it is a star graph. In $AVTREE(T)$, since $ntag(u) = 2$ and every child of u has label $([1', 1'])$, only Case 1 of *COMBINE_AVTREES* is executed. Thus every child v of u, $v \rightarrow u$. It is easy to verify that T^* satisfies the conditions of Lemma 4. Assume that for all rooted tree T of height less than k, the rooted tree T^* constructed by $AVTREE(T)$ satisfies the conditions of Lemma 4. Now we consider a tree T of height k. Let the label of u be $\lambda = (a_1, \ldots, a_p)$ and u_i be the vertex associated with a_i for $1 \leq i \leq p$. Let v_i with the label $\lambda_i = (a_{i,1}, \ldots, a_{i,p_i})$, $1 \leq i \leq d$, be the children of u. Let $u_{i,j}$ be the vertex associated with $a_{i,j}$ for all $1 \leq i \leq d$ and $1 \leq j \leq p_i$.

By Lemma 8, the *ntag* values computed by $AVTREE(T)$ determine an extended avenue system $\mathcal{A}(T)$ of T. Let $Q \in \mathcal{A}(T)$ be the avenue containing u. Let $P \in \mathcal{A}(T)$ be any avenue not containing u. Thus, P is in $T[v_t]$ for some t, $1 \leq t \leq d$ and P must be in $T[v_t, u_{t,1}, \ldots, u_{t,p_t-1}]$ or $T[u_{t,s}]$ for some s, $1 \leq s \leq p_t$. By induction hypothesis on $T[v_t]$, P is a rooted path in $T[v_t]^*$. Hence P is also a rooted path in T^*. That is, P satisfies Lemma 4(1) in T^*. In the following, we show that P satisfies Lemma 4(2) in T^*.

We first consider the case that P does not contain any $u_{t,s}$, $1 \leq s \leq p_t$. Let P be contained in $T[u_{t,s}]$ for some s. Since $ntag(P) < a_{t,s}$, all the avenues in $\mathcal{A}_P$ is contained in $T[u_{t,s}]$. That is, $\min(\mathcal{A}_P)$ is also in $T[u_{t,s}]$. By the induction hypothesis, P satisfies Lemma 4(2) in $T[v_t]^*$. Since the pointer of the root of P does not change in *COMBINE_AVTREES*, P satisfies Lemma 4(2) in T^*.

Next, we consider the case that P contains $u_{t,s}$ for some s. Let $P_{t,s}$ denote the avenue in $\mathcal{A}(T)$ containing $u_{t,s}$.

Case 1: $ntag(P_{t,s}) < ntag(u)$. In this case, $\mathcal{A}_{P_{t,s}}$ contains Q and every avenue in $\mathcal{A}_{P_{t,s}}$ except Q is contained in $T[v_t]$. Let $a_{t,s'}$ be the largest element in λ_t such that $a_{t,s'} < tag(Q)$. Then $\min(\mathcal{A}_{P_{t,s}})$, $s \neq s'$, is contained in $T[v_t]$. For $P_{t,s}$, $s \neq s'$, by the induction hypothesis on $T[v_t]$, $P_{t,s}$ satisfies Lemma 4(2) in $T[v_i]^*$. The root of $P_{t,s}$ has pointed to the right vertex of $\min(\mathcal{A}_{P_{t,s}})$ in $T[v_t]^*$. Since the pointer does not change in *COMBINE_AVTREES*, $P_{t,s}$ satisfies Lemma 4(2) in T^*. For $P_{t,s'}$, since $a_{t,s'-1} > ntag(Q)$, $Q = \min(\mathcal{A}_{P_{t,s'}})$. In Case 1 or B: of *COMBINE_AVTREES*, the parent of the root of $P_{t,s'}$ is assigned to be u. Hence $P_{t,s'}$ satisfies Lemma 4(2) in T^*.

Case 2: $ntag(P_{t,s}) > ntag(u)$. Since $ntag(u) = a_p$, $u_{t,s} \in \{u_1, \ldots, u_{p-1}\}$ by Lemma 7. Let $u_{t,s} = u_k$. By Lemma 9, the parent of u_k is u_{k-1}. If a_k (i.e., $a_{t,s}$) is not a left endpoint in λ_t, then $a_{k-1}(= a_k + 1) \in \lambda_t$. That is, u_{k-1} is also contained in $T[v_t]$. By Lemma 9, the avenue contains u_{k-1} is $\min(\mathcal{A}_{P_{t,s}})$. By induction hypothesis on $T[v_t]$, $P_{t,s}$ satisfies Lemma 4(2) in $T[v_t]^*$. The root of

$P_{t,s}$ has pointed to the right vertex of $\min(\mathcal{A}_{P_{t,s}})$ in $T[v_t]^*$. Since the pointer does not change in *COMBINE_AVTREES*, $P_{t,s}$ satisfies Lemma 4(2) in T^*. If a_k is a left endpoint in λ_t, then by Lemma 7 a_{k-1} must be a right endpoint in some λ_l, $1 \leq l \neq t \leq d$. Then $a_{t,s}$ is determined in Cases 3.1 or 3.3 of *COMBINE_AVTREES* and $a_{l,s'}$ (i.e., a_{k-1}) is determined in Cases 3.1, 3.2 or A: of *COMBINE_AVTREES*. The pointer $u_{t,s} \rightarrow u_{l,s'}$ is assigned in Cases 3.1, 3.2 or A: of *COMBINE_AVTREES*. Hence $P_{t,s}$ satisfies Lemma 4(2) in T^*.

Finally we consider the avenue Q. There are six cases in determining a_p of λ in *COMBINE_LABELS*. Case 5 can not happen. In Cases 1, 2, and 4, since there is no child of u whose label contains a_p, $Q = [u]$ and therefore Q is a rooted path. In Case 3, there are exact two children of u, say v_i and v_j, such that $a_p \in \lambda_i$, $a_p \in \lambda_j$, and a_p is not critical in λ_i and λ_j. Let Q_i be the avenue containing v_i in $\mathcal{A}(T[v_i])$. Then $V(Q) = V(Q_i) \cup V(Q_j) \cup \{u\}$. By the induction hypothesis, the root of Q_i is v_i. The Case 2 of *COMBINE_AVTREES* makes Q be a rooted path with root u in T^*. In Case 6, there is exact one child v_t of u, say v_t, such that $a_p \in \lambda_t$ and a_p is not critical in λ_t. Let Q_t be the avenue containing v_t in $\mathcal{A}(T[v_t])$. Then $V(Q) = V(Q_t) \cup \{u\}$. By induction hypothesis on $T[v_t]$, v_t is the root of Q_t in $T[v_t]^*$. In Case 2 of *COMBINE_AVTREES*, the pointer of v_t is assigned to u. Hence Q satisfies Lemma 4(1). Next, we prove that Q satisfies Lemma 4(2). We consider P being the avenue which contains u_{p-1}. By Lemma 9, $P = \min(\mathcal{A}_Q)$. By Lemma 7, only one child of u whose label contains a_{p-1}. If both a_p and a_{p-1} are contained in λ_t for some t, then the pointer $u \rightarrow u_{p-1}$ in T^* is done in Case 2.2 of *COMBINE_AVTREES* since u is the root of Q. Otherwise, a_{p-1} is a right endpoint in λ_t. The pointer $u \rightarrow u_{p-1}$ in T^* is done in one of the Cases 3.1, 3.2, A:, and C: of *COMBINE_AVTREES*. Hence Q satisfies Lemma 4(2) in T^*. Thus, our lemma follows. □

It can be verified that the time complexities of *COMBINE_AVTREES* and *COMBINE_LABELS* have the same order. Since *COMBINE_AVTREES* only uses the information of interval representation of labels, the time complexity of Algorithm $SEARCH(v)$ is linear. Together with Theorem 5, we have the following theorem.

Theorem 11. *An optimal node-search strategy of a tree can be constructed in linear time.*

References

1. S. Arnborg, D.G. Corneil, and A. Proskurowski, Complexity of finding embeddings in a k-tree, *SIAM J. Alg. Disc. Meth.*, **8**(1987), 277-284.
2. D. Bienstock, Graph searching, path-width, tree-width and related problems (a survey), in: F. Roberts, F. Hwang and C. Monma, eds., *Reliability of Computer and Communication Networks*, DIMACS series in Disc. Math. and Theoretical Comp. Scie., Vol **5**, American Math. Society, 1991, 33-49.
3. D. Bienstock and P. Seymour, Monotonicity in graph searching, *J. Algorithms*, **12**(1991), 239-245.

4. H.L. Bodlaender, *A partial k-arboretum of graphs with bounded treewidth*, Technical Report UU-CS-1996-02, Department of Computer Science, Utrecht University, Utrecht, the Netherlands, 1996.
5. H.L. Bodlaender and T. Kloks, Efficient and constructive algorithms for the pathwidth and treewidth of graphs, *J. Algorithms*, **21**(1996), 358-402.
6. H.L. Bodlaender, T. Kloks, and D. Kratsch, Treewidth and pathwidth of permutation graphs, *SIAM J. Disc. Math.*, **8**(1995), 606-616.
7. H.L. Bodlaender, T. Kloks, D. Kratsch, and H. Muller, *Treewidth and minimum fill-in on d-trapezoid graphs*, Technical Report UU-CS-1995-34, Department of Computer Science, Utrecht University, Utrecht, the Netherlands, 1995.
8. H.L. Bodlaender and R.H. Möhring, The pathwidth and treewidth of cographs, *SIAM J. Disc. Math.*, **6**(1993), 181-188.
9. J.A. Ellis, I.H. Sudborough, and J.S. Turner, The vertex separation and search number of a graph, *Information and Computation*, **113**(1994), 50-79.
10. F.V. Fomin and N.N. Petrov, Pursuit-evasion and search problems on graphs, In *Nonlinear and Game-Theoretic Control Synthesis*, Euler Int. Math. Institute, St. Petersburg, 13-26 March, 1995, Update May 1996.
11. J. Gustedt, On the pathwidth of chordal graphs, *Disc. Appl. Math.*, **45**(1993), 233-248.
12. N.G. Kinnersley, The vertex separation number of a graph equals its path-width, *Inform. Process. Lett.*, **42**(1992), 345-350.
13. T. Kloks, *Treewidth, Computations and Applications*, LNCS **842**, Springer, 1994.
14. T. Kloks, H. Bodlaender, H. Muller, and D. Kratsch, Computing treewidth and minimum fill-in: all you need are the minimal separators, *ESA'93*, LNCS **726**, 260-271, 1993. Erratum: *ESA'94*, LNCS **855**, pp. 508, 1994.
15. L.M. Kirousis and C.H. Papadimitriou, Interval graph and searching, *Disc. Math.*, **55**(1985), 181-184.
16. L.M. Kirousis and C.H. Papadimitriou, Searching and pebbling, *Theoretical Comput. Scie.*, **47**(1986), 205-218.
17. A. Kornai and Z. Tuza, Narrowness, pathwidth, and their application in natural language processing, *Disc. Appl. Math.*, **36**(1992), 87-92.
18. A.S. LaPaugh, Recontamination does not help to search a graph, *J. Assoc. Comput. Mach.*, **40**(1993), 224-245.
19. N. Megiddo, S.L. Hakimi, M.R. Garey, D.S. Johnson, and C.H. Papadimitriou, The complexity of searching a graph, *J. Assoc. Comput. Mach.*, **35**(1988), 18-44.
20. R.H. Möhring, Graph problems related to gate matrix layout and PLA folding, in: G. Tinnhofer et al., eds., *Computational Graph Theory*, Springer, 1990, 17-32.
21. B. Monien and I.H. Sudborough, Min cut is NP- complete for edge weighted trees, *Theoretical Comput. Sci.*, **58**(1988), 209-229.
22. T.D. Parsons, Pursuit-evasion in a graph, in Y. Alavi and D.R. Lick, eds., *Theory and applications of graphs*, Springer-Verlag, New York, 1976, 426-441.
23. S.L. Peng, C.W. Ho, T.-s. Hsu, M.T. Ko, and C.Y. Tang, Edge and node searching problems on trees, *COCOON'97*, LNCS **1276**, 284-293, 1997.
24. S.L. Peng, M.T. Ko, C.W. Ho, T.-s. Hsu, and C.Y. Tang, Graph searching on chordal graphs, *ISAAC'96*, LNCS **1178**, 156-165, 1996.
25. N. Robertson and P.D. Seymour, Graph minors I. Excluding a forest, *J. Comb. Theory Ser. B*, **35**(1983), 39-61.
26. P. Scheffler, A linear algorithm for the pathwidth of trees, in: R. Bodendiek and R. Henn, eds., *Topics in Combinatorics and Graph Theory* (Physica-Verlag, Heidelberg, 1990), 613-620.

A Graph Optimization Problem in Virtual Colonoscopy

Jie Wang[1]* and Yaorong Ge[2]**

[1] The University of North Carolina at Greensboro
[2] Wake Forest University

Abstract. This paper studies a graph optimization problem occurring in virtual colonoscopy, which concerns finding the central path of a colon model created from helical computed tomography (CT) image data. The central path is an essential aid for navigating through complex anatomy such as colon. Recently, Ge et al. [GSZ+] devised an efficient method for finding the central path of a colon. The method first generates colon data from a helical CT data volume by image segmentation. It then generates a 3D skeleton of the colon. In the ideal situation, namely, if the skeleton does not contain branches, the skeleton will be the desired central path. However, almost always the skeleton contains extra branches caused by holes in the colon model, which are artifacts produced during image segmentation. To remove false branches, we formulate a graph optimization problem and justify that the solution of the optimization problem represents the accurate central path of a colon. We then provide an efficient algorithm for solving the problem.

1 Introduction

Virtual endoscopy is a new medical technology that allows physicians to examine computer simulations of patients' anatomy rendered from CT scans. It combines medicine, clinical experience, radiology, image processing, computer algorithms, and applied mathematics to provide the public with alternative medical procedures that are less painful, less costly, and less risky compared to conventional endoscopic procedures. For instance, medical research has shown that small colon polyps, the precursor to colon cancer, can be detected with virtual endoscopy [VGB94,VS94].

When applying virtual endoscopy to complex anatomy such as colon, users often find it difficult to keep track of their position and orientation inside the

* Department of Mathematical Sciences, The University of North Carolina at Greensboro, Greensboro, NC 27402, USA. Email: wang@uncg.edu. Supported in part by NSF under grant CCR-9424164.

** Department of Mathematics and Computer Science, and Department of Medical Engineering of the Division of Radiological Sciences, Bowman Gray School of Medicine, Wake Forest University, Box 7388, Winston-Salem, NC 27109, USA. Email: gey@mthcsc.wfu.edu. Supported in part by NSF under grant BES-9520388 and by ARPA under grant F41624-96-2-001.

complex colon image. As a consequence, a part of colon lumen may be left without inspection. Hence, one would like to have a tour guide for traveling through a virtual colon. Finding the central path through the lumen of a colon provides a natural solution. The central path can be used, for example, to create a movie that displays the internal views of the colon lumen generated automatically along the path; and to guide the user to walk through, by using a computer mouse, the colon lumen without getting lost in the virtual space. In addition to aiding navigation, the central path also provides a vital component for more advanced processing and visualization of the virtual anatomy. For example, one can slice the virtual colon into segments of similar length and split each segment into two halves based on a curvilinear cutting plane that passes through the central path, which provides a clear visualization with a single view [VSH+96].

Several methods have been proposed to determine the central path of a virtual anatomy. One approach requires that users manually select a number of points along the pathway for constructing a central path [VS94,HJR+96,SAN+96]. Using this method, users often need to spend considerable amount of time to explore and understand the data volume prior to placing the points, which makes the method less desirable for routine clinical applications. Another approach offers automatic methods [HKW95,LJK97], but these methods can only find a central path (or a path closed to the center) in the colon lumen for limited cases. These methods fail when the colons have complex shapes. In some cases, a part of the small bowel may also be included in the colon data, which makes the case even more difficult to analyze. To overcome these obstacles, Ge et al. [GSZ+] recently devised an efficient method based on the concepts of skeletons. A *skeleton* of a 3D object is the locus of the centers of the largest balls that can fit inside the object; such balls are referred to as *maximally inscribed balls* [Blu73].

Ge et al.'s method consists of four steps. First, it generates a 3D image data volume of a colon from helical CT scans by image segmentation. Second, it generates a 3D skeleton of the colon image by using an improved thinning algorithm based on Li et al.'s thinning algorithm [LKC94]. The thinning process preserves the topological constraints and the geometric constraints of the original object. (The geometric constraints are the two endpoints of the colon provided by the user.) This means that if the skeleton does not contain branches, then it is the desired central path. However, almost always the skeleton contains extra branches caused by holes in the object that are artifacts produced during image segmentation. The number of holes may be reduced by a finer segmentation; but no segmentation, however fine, seems likely to eliminate holes completely. Hence, false branches need to be removed, which is the task of the third step of the method. Note that the image segmentation process may also produce cavities, which corresponds to an isolated point on the skeleton, and hence can be removed easily. After the false branches have been pruned, the remaining skeleton provides the accurate central path. However, the path contains many abrupt direction changes due to the discrete nature of image data. The last step of the method computes a smooth representation of the central path by approximating the final skeleton with B-splines.

To remove the false branches of the skeleton in Step 3, the skeleton is first converted to a connected graph with positive weights, referred to as a *skeletal graph.* Here by connected graph we mean that the graph does not contain isolated vertices. In the skeletal graph, a vertex corresponds to a point on the skeleton where either three or more branches join, or only one branch joins (in this case, the point is an endpoint of the colon), and an edge corresponds to a branch. The weight of each edge is determined as follows. Note that each point on a skeleton branch is associated with a maximally inscribed ball. It has been observed that holes are almost always near the surface [GSZ+], and so the extra branches that are induced by these holes usually pass through narrow segments. In other words, the desired central path lies in the center of the colon lumen with a relatively large diameter. Imagine that each point on the skeleton corresponds to a pipe whose size is the maximally inscribed ball. A skeletal branch can then be viewed as a connected sequence of pipes in which balls can roll back and forth. The smallest pipe along the branch determines the largest ball that can pass through this branch. Thus, the radius of the smallest ball along a skeleton branch is used as the weight of its corresponding edge in the graph. In case there are multiple edges between two vertices (this occurs when a skeletal branch splits into two to go around a hole), the edges with smaller weights are removed.

The skeletal graph of a colon has two endpoints. We want to find a path from one endpoint to the other endpoint in the skeletal graph such that its minimum weight is the maximum among all paths, and it avoids passing through any narrow passage whenever it is possible. Such a path corresponds to the true central path of the colon. Ge et al. [GSZ+] formulated this optimization problem as a problem of finding a path with the maximum *flow*, where the flow of a path is defined as the minimum weight of all edges on the path. However, in some cases, there may be more than one path with the maximum flow. In such a case, a heuristic was used in [GSZ+] for finding a central path among all paths with the maximum flow. We refine this formulation in this paper and fully justify that the solution of the refined optimization problem represents the accurate central path. These results are given in Section 2. In Section 3, we present a $O(n \log n)$-time algorithm for solving the optimization problem on skeletal graphs, as well as mathematical analysis of the algorithm, where n is the number of vertices in a graph. In Appendix, we present a running example of the algorithm on a complex colon case.

2 Finding the Central Path

We consider connected graphs with positive weights. Since an undirected graph can be simulated by a directed graph, where each edge is represented by two arcs traveling in opposite directions, we assume that all graphs are directed. Denote by $w(u, v)$ the weight of an edge from u to v.

Let G be a skeletal graph (i.e., the weighted graph converted from a skeleton as described in the second last paragraph in Section 1). Let s and t be the two endpoints in G. In what follows, we will fix the usage of s and t to denote the two endpoints of the colon image. To search for the central path from s to t, we should avoid selecting branches with small weights whenever it is possible. There are two subtle issues in formulating a mathematical definition to capture the essential features of the central path, and we discuss them below.

Issue 1: False branches of large weights. In searching for the central path from s to t, one may want to use the following greedy strategy; namely, at a current vertex u, select the next vertex v with the largest $w(u,v)$. This strategy works fine if all false branches from vertex u have strictly smaller weights than the weight of the true branch from u (here by a true branch it means the branch on the central path). But this strategy fails if a false branch actually has a larger weight than that of the true branch on a particular segment, which may occur due to limitations of segmentation. Although such cases do not always occur, we have encountered such cases during our experiments. Hence, if this greedy strategy is used, then a false branch will be selected, which will then lead us to other false branches.

One may then perhaps suggest that, if such a case happens, one should try to run a finer segmentation and start the whole process over again. However, the segmentation process and the thinning algorithm are extremely time consuming because a large data volume is involved: Each helical CT volume may contain up to 200MB of data. Hence, this approach is not desirable. Compared to the large data volume of the original image, the size of the skeletal graph is substantially smaller; naturally we should avoid re-running the whole process. Note that selecting a false branch with a large weight may lead to another false branch, which may lead to a false path with a smaller flow than the maximum flow. Hence, we want to find a path with the maximum flow. We note that there may be two or more such paths. If x is a common vertex occurred on such paths, the one that has the largest weight from s to x should be selected.

Let $p = \langle v_0, v_1, \ldots, v_k \rangle$ be a simple path. Denote by $f(p)$ the flow of p; namely,

$$f(p) = \min\{w(v_{i-1}, v_i) : 1 \le i \le k\}.$$

We use $v_i \overset{p}{\rightsquigarrow} v_j$ $(j > i)$ to denote the portion of p from v_i to v_j, i.e., the path $\langle v_i, \ldots, v_j \rangle$.

Definition 1. *Let G be a weighted graph. Let u and v be two vertices. A simple path p from u to v in G is a* largest path *if for any other path p' from u to v, the following two conditions hold.*

1. $f(p') \le f(p)$.
2. *If x is a vertex shared by both p and p', then* $f(u \overset{p'}{\rightsquigarrow} x) \le f(u \overset{p}{\rightsquigarrow} x)$.

Issue 2: True branches of fluctuating weights. If there is only one largest path from s to t, then that path represents the accurate central path. However, in some cases, there are two or more largest paths. For example, imagine that a colon image contains some segments that are narrower than the other segments. Then the narrowest segment may determine the flow of the central path. If after this narrowest segment the colon image (obtained from image segmentation) becomes substantially larger, then a false path may also have the same flow as the flow of the central path, because the maximum flow of all paths is already small. Several heuristics may be used to help identify the accurate central path among the largest paths. For example, one may suggest to find a largest path p from s to t such that for any other largest path p' from s to t, $f(p'_i) \leq f(p_i)$ for all i, where p_i (respectively, p'_i) represents the portion on p from the ith vertex to t. This heuristic works for some cases, but it may fail if the colon image has a narrow segment that is followed by a wide segment, and then followed by a narrow segment.

Let e_1 and e_2 be two edges on a path. Denote by $e_1 < e_2$ if e_1 is reached first before e_2 is reached. If e_1, e_2, and e_3 are three edges on a path with $e_1 < e_2 < e_3$ such that $w(e_1) < w(e_2)$, and $w(e_2) > w(e_3)$, then we say that the path has *fluctuating weights.*

The following is a possible heuristic to handle fluctuating weights; namely, find a largest path p from s to t such that for any largest path p' from s to t, if there is an edge e_1 on p and an edge e'_1 on p' with $w(e'_1) > w(e_1)$, then p must contain an edge $e_2 > e_1$ such that for some edge $e'_2 \geq e'_1$ on p', $w(e_2) > w(e'_2)$. This heuristic works for some cases, but it may fail in some other cases. For example, assume that e_1 and e_2 are the two edges at the end on the true central path, and e is an edge at the end of a false path such that $w(e) > w(e_1)$ and $w(e) > w(e_2)$, then the true central path will not be selected following this strategy.

To overcome these obstacles, let us imagine that the skeleton of a colon represents the density of the colon. In the ideal situation, namely, if at any point, all false branches have smaller weights than the true branch, then the central path has the heaviest *average weight.* An average weight of a path is the total weight of all edges divided by the number of edges on the path. Note that we cannot use the maximum total weight as a criterion because a false branch may be long and hence its total weight may be large. In a skeletal graph, almost all false branches have small weights. So even if on some segments of a colon, a false branch has a larger weight than that of the true branch, the central path must still have the heaviest average weight.

Let $p = \langle v_0, v_1, \ldots, v_k \rangle$ be a simple path. Denote by $W(p)$ the average weight of p; namely,

$$W(p) = \sum_{i=0}^{k-1} \frac{w(v_i, v_{i+1})}{k}.$$

Definition 2. *Let G be a weighted graph. Let u and v be two vertices. A simple path from u to v in G is a* heaviest path *if for any other path p' from u to v, $W(p') \leq W(p)$.*

In some cases a path with the heaviest weight may imply that it is also a largest path. But it is not always true because a heaviest path may contain an edge with very small weight. Hence, we want to find a path that is the heaviest among the largest paths. This gives rise to the following definition.

Definition 3. *Let G be a weighted graph. Let u and v be two vertices. A simple path p from u to v in G is a* critical path *if p is a largest path from u to v, and for any largest path p' from u to v, $W(p') \leq W(p)$.*

It is easy to see that if there is a path from u to v, then there must be a critical path from u to v. Recall that for any vertex u in a skeletal graph G, almost always false branches from u have strictly smaller weights than the true branch from u. This implies that if a path is the heaviest among the largest paths, then it represents the accurate central path. Any reasonable segmentation guarantees that such a path is unique. (But we note that in a general weighted graph, there may be more than one critical path.) In practical terms, the central path can be found by solving the following optimization problem.

CENTRAL PATH PROBLEM
Input: A weighted graph G and two vertices s and t.
Output: A critical path from s to t.

We present a fast algorithm for solving this problem in the next section.

3 A Fast Algorithm

Let $G = (V, E)$ be a connected graph with positive weights. Let s and t be two vertices. We want to find a critical path from s to t. Let

$$\Delta(u,v) = \begin{cases} \max\{f(p) : u \overset{p}{\rightsquigarrow} v\}, & \text{if there is a path from } u \text{ to } v \\ -1, & \text{otherwise} \end{cases}$$

$$\Gamma(u,v) = \begin{cases} \max\{W(p) : u \overset{p}{\rightsquigarrow} v \text{ and } f(p) = \Delta(s,u)\}, & \text{if there is a path from } u \text{ to } v \\ 0, & \text{otherwise} \end{cases}$$

Lemma 1. *Let G be a weighted graph. Let u and v be two vertices. Then a path p from u to v is a critical path if and only if for every vertex x on p, $f(u \overset{p}{\rightsquigarrow} x) = \Delta(u,x)$ and $W(u \overset{p}{\rightsquigarrow} x) = \Gamma(u,x)$.*

We observe that in a skeletal graph, the number of edges is in the same asymptotic order of the number of vertices. Hence, we use an adjacency list *Adj* to represent G for saving memory space. Note that although G does not contain isolated vertices, it does not mean that for any pair of vertices u and v, there always exists a path from u to v. But in a skeletal graph G, there is always a path from s to u for any vertex u. We present an algorithm that can also be used to handle non-connected graphs. For each vertex $u \in V$, we maintain four attributes $d[u]$, $\pi[u]$, $l[u]$, and $a[u]$ in the algorithm, where $d[u]$ represents the flow from s to u, $\pi[u]$ returns the predecessor of u on the current path, $l[u]$ represents the number of edges from s to u on the current path through the π attributes, and $a[u]$ represents the average weight of the current path from s to u.

CRITICAL_PATH(G, s, t)

1. **for** each vertex $u \in V$ **do**
   $d[u] \leftarrow -1$, $\pi[u] \leftarrow$ NIL, $l[u] \leftarrow 0$, $a[u] \leftarrow 0$
   endfor
   $d[s] \leftarrow 0$, $S \leftarrow \emptyset$, $Q \leftarrow V$
2. **while** $Q \neq \emptyset$ **do**
   (a) $u \leftarrow$ EXTRACT_MAX(Q)
   if $d[u] \neq -1$ **then** $S \leftarrow S \cup \{u\}$ **else** goto Step 3
   (b) **for** each vertex $v \in Adj[u]$ **do**
   if $(d[v] < \min\{d[u], w(u,v)\})$ **then**
   $d[v] \leftarrow \min\{d[u], w(u,v)\}$
   $\pi[v] \leftarrow u$, $l[v] \leftarrow l[u] + 1$
   $a[v] \leftarrow (l[u] \cdot a[u] + w(u,v))/l[v]$
   if $(\pi[v] \neq$ NIL **and** $\min\{d[u], w(u,v)\} \geq d[v])$ **then**
   if $((l[u] \cdot a[u] + w(u,v))/(l[u]+1) > a[v])$ **then**
   $\pi[v] \leftarrow u$, $l[v] \leftarrow l[u] + 1$
   $a[v] \leftarrow (l[u] \cdot a[u] + w(u,v))/l[v]$
   endfor
   endwhile
3. **if** $t \in S$ **then** output the path from s to t using the π attributes
   else there is no path from s to t

Part 1 of CRITICAL_PATH(G, s, t) is for initialization. The procedure EXTRACT_MAX(Q) in 2(a) finds an element $u \in Q$ that has the largest $d[u]$. For each vertex v in $Adj[u]$, if v has not been visited, namely, $\pi[v] =$ NIL, then the first **if**-statement of 2(b) updates $d[v]$. If v has been visited, then the second **if**-statement of 2(b) checks whether the current path has a flow at least as large as the flow of the old path. If the answer is yes, the third **if**-statement of 2(b) checks whether the current path has a strictly larger average weight than that of the old path; if the answer is yes, the old path is flipped over to the current path by changing the π attribute. The formula $(l[u] \cdot a[u] + w(u,v))/(l[u]+1)$ calculates the average weight of the current path from s to v via u.

Lemma 2. *When the algorithm* CRITICAL_PATH(G, s, t) *terminates, we have* $d[u] = \Delta(s, u)$ *for all vertices* $u \in V - \{s\}$*. Moreover, if* $\pi[u] \neq$ NIL*, the path* p *from* s *to* u *obtained from the* π *attributes is a largest path.*

Lemma 3. *When the algorithm* CRITICAL_PATH(G, s, t) *terminates, we have* $a[u] = \Gamma(s, u)$ *for all vertices* $u \in V - \{s\}$*. Moreover, if* $\pi[u] \neq$ NIL*, the path* p *from* s *to* u *obtained from the* π *attributes is a critical path.*

It follows from Lemma 2 that the algorithm CRITICAL_PATH(G, s, t) returns a critical path from s to t. Namely, we have proven the following theorem.

Theorem 1. *Assume that there is a path from* s *to* t *in* G*, then when the algorithm* CRITICAL_PATH(G, s, t) *terminates, the path from* s *to* t *obtained from the* π *attributes is a critical path.*

Next, we analyze the time complexity of the algorithm.

Theorem 2. *Depending on how the* EXTRACT_MAX *operation is implemented, the algorithm* CRITICAL_PATH(G, s, t) *runs in time* $O(n^2)$ *or* $O((n + |E|) \log n)$*, where* $|V| = n$*. Hence, when* $|E| = O(n)$*, the algorithm can be run in* $O(n \log n)$ *time.*

Since in a skeletal graph G, $|E| = O(|V|)$, it follows from Theorems 1 and 2 that finding the central path of the colon lumen can be carried out in $O(n \log n)$ time, where $|V| = n$.

Acknowledgment. We thank Dr. David Vining at Wake Forest University School of Medicine for providing the clinical context of this research and for suppling the data sets to verify our results. We also thank Xiangjun Wu and Xiangliang Zha for implementing part of the algorithm presented in this paper.

References

[Blu73] H. Blum. Biological shape and visual science: part I. *Journal of Theoretical Biology*, 38:205–287, 1973.

[GSZ+] Y. Ge, D. Stelts, X. Zha, J. Wang, and D. Vining. Computing the central path of colon lumen in Helical CT images. In *SPIE International Symposium on Medical Imaging 1998: Image Processing* (in press).

[HJR+96] A. Hara, C. Johnson, J. Reed, R. Ehman, and D. Ilstrup. Colorectal polyp detection with CT colography: two- versus three-dimensional techniques. *Radiology*, 200:49–54, 1996.

[HKW95] L. Hong, A. Kaufman, and Y.-C. Wei. 3D virtual colonoscopy. In *Proc. 1995 IEEE Biomedical Visualization Symposium*, pages 26–32, 1995.

[LKC94] T.-C. Lee, R. Kashyap, and C.-N. Chu. Building skeleton models via 3D medial surface/axis thinning algorithms. *CVGIP: Graphical Models and Image Processing*, 56:462–478, 1994.

[LJK97] W. Lorensen, F. Jolesz, and R. Kikinis. United states patent: Virtual internal cavity inspection system. Tech. Rep. 5611025, 1997.

[SAN+96] R. Shahidi, V. Argiro, S. Napel, L. Gray, H. McAdams, G. Rubin, C. Beaulieu, R. Jeffrey, and A. Johnson. Assessment of several virtual endoscopy techniques using computed tomography and perspective volume rendering. In *Lecture Notes in Computer Science: Visualization in Biomedical Computing*, Karl Heinz Hohne and Ron Kikinis, Editors, vol. 1131, pages 521–528. Springer-Verlag, 1996.

[VSH+96] D. Vining, D. Stelts, G. Hunt, D. Aho, Y. Ge, and P. Hemler. Technical improvement in virtual colonoscopy. *Radiology*, 201:524, 1996.

[VGB94] D. Vining, D. Gelfand, and R. Bechtold et al. Technical feasibility of colon imaging with helical CT and virtual reality. *American Journal of Roentgenology*, 162:104, 1994.

[VS94] D. Vining and R. Shifrin. Virtual reality imaging with helical CT. *American Journal of Roentgenology*, 162:188, 1994.

Appendix: Running Examples

We have tested our algorithm on a number of virtual endoscopy colon cases. Each helical CT volume consists of up to 500 images with 512×512 pixels per image. We present in this section a running example of our algorithm on a complex colon case.

Figure 1(a) shows a colon that has been segmented from a CT volume and rendered for visualization. Notice that the segmentation result is inexact. A large portion of small bowel has been segmented in addition to the colon. We use this example to demonstrate the robustness of our algorithms.

The skeleton resulting from the 3D thinning algorithm is shown in Figure 1(b), superimposed on the original colon rendering. Notice that multiple colon segments touch each other, and that a portion of the small bowel touches the transverse colon. These touching segments result in segmentation artifacts known as holes and causes many extra branches in the skeleton.

Figure 1(c) shows the true central path extracted from the initial skeleton (dotted line) and its B-spline approximation (solid line).

Figure 2 shows two internal renderings of the colon lumen from two positions on the central path. Note that these are views of the colon surface from inside. The small holes that we see in these views are not the holes in the segmented colon object. Rather, they are artificial tunnels connecting colon segments that create holes visible from external views.These tunnels are where the extra skeletal branches pass through in order to go around the holes in the object.

The time required to execute our algorithms varies with the type of platform, the size of input volume, and the complexity of the colon itself. On an SGI System with R10000 (Silicon Graphics, Inc., Mountain View, CA), the time required to convert a skeleton to a graph and to search for the true central path never exceeded 15 seconds.

Fig. 1. (a) Rendering of the segmented colon. Touching segments of colon and inclusion of a portion of small bowel are due to limitations in the segmentation process. (b) The initial skeleton preserves the original colon topology. The many extra branches that deviate from the center of the colon are caused by holes in the original object. (c) The central path and its smooth B-spline approximation. The dotted line represents the central path from the original skeleton and the solid line is its B-spline approximation.

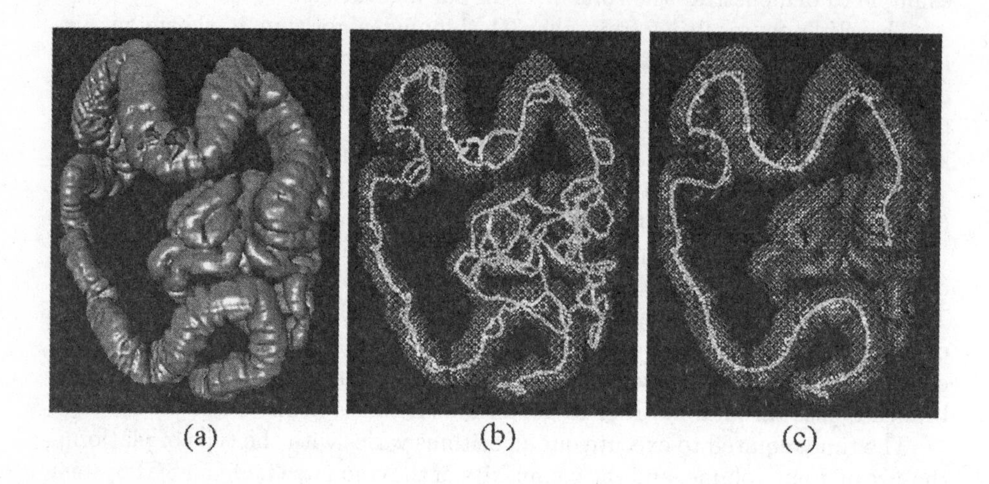

Fig. 2. Two internal renderings of the colon lumen as seen from positions along the central path. The central path is projected into the internal views for illustration.

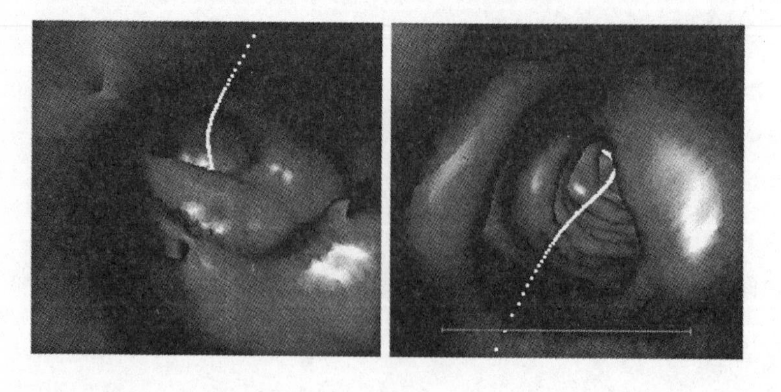

Approximation and Exact Algorithms for Constructing Minimum Ultrametric Trees from Distance Matrices

Bang Ye Wu[1], Kun–Mao Chao[2], and Chuan Yi Tang[1]

[1] Dept. of Computer Science, National Tsing Hua University, Hsinchu, Taiwan, R.O.C. { dr838305, cytang}@cs.nthu.edu.tw

[2] Dept. of Computer Science and Information Management, Providence University. Shalu, Taiwan, R.O.C. kmchao@csim.pu.edu.tw

Abstract. Constructing minimum ultrametric trees from distance matrices is an important problem in computational biology. In this paper, we examine its computational complexity and approximability. When the distances satisfy the triangle inequalities, we show that the minimum ultrametric tree problem can be approximated in polynomial time with error ratio $1.5(1 + \lceil \log n \rceil)$, where n is the number of species. We also developed an efficient branch and bound algorithm for constructing the minimum ultrametric tree for both metric and nonmetric inputs. The experimental results show that it can find an optimal solution for 25 species within reasonable time, while, to the best of our knowledge, there is no report of algorithms solving the problem even for 12 species.
Keywords: computational biology, ultrametric trees, approximation algorithms, branch and bound.

1 Introduction

Constructing evolutionary trees from distances is an important problem in biology and in taxonomy and there are many different models to define the problems [11, 14]. Most of the optimization problems of evolutionary tree construction have been shown to be NP-hard [3–6, 8, 9, 13]. An important model is to assume the rate of evolution is constant. With this assumption, the evolutionary tree will be an ultrametric tree [11, 14]. An ultrametric tree is a rooted, leaf labeled, and edge weighted binary tree in which every internal node has the same path length to all the leaves in its subtree. We only need to consider binary trees since a general ultrametric tree can be converted to a binary tree by replacing any vertex of degree larger than 3 (except the root) with a number of vertices of degree 3 connected by edges of weight 0 [12]. Some results about ultrametric trees had been studied in [1, 2, 6, 13]. Because of the high computational complexity, biologists usually construct the trees by heuristic algorithms. For example, the UPGMA (Unweighted Pair Group Method with Arithmetic mean, see [14] for introduction) is a popular heuristic algorithm to construct ultrametric trees.

In this paper, we examine the complexity and the approximability of the minimum ultrametric tree construction problem, and develop an efficient branch and bound algorithm which finds an optimal solution for moderate data instance in reasonable time. With the rapid development of the technique of DNA sequencing and alignment, the dissimilarities between species are often obtained by DNA (or RNA, protein) sequence alignment. Depending on the scoring scheme of alignment, the distance matrix may or may not obey the triangle inequality, that is, it may be either a metric or not. In [6], the problem of constructing a minimum ultrametric tree from a nonmetric distance matrix was shown to be NP-complete and cannot be approximated within n^{ε} in polynomial time for some $\varepsilon > 0$ if NP$\neq$P. In this paper, we show that the problem remains to be NP-hard if the distance matrix is a metric, but can be approximated within ratio $1.5(1 + \lceil \log n \rceil)$.

In the view of practice, the number of species is often not so large in many cases. So, it seems possible to compute an optimal tree by exhaustive search, that is, by checking all the possible trees. However, for n species, the number of unrooted, leaf labeled binary trees is $1 \times 3 \times 5 \times ..(2n-5)$ [12]. In the rooted case, for any unrooted tree, we can locate the root at any edge of the tree. Therefore, the number of rooted, leaf labeled binary trees is $A(n) = 1 \times 3 \times 5 \times ..(2n-3)$. The function A grows very rapidly. For example, $A(10) > 3 \times 10^7$, $A(15) > 2 \times 10^{14}$, and $A(20) > 8 \times 10^{21}$. Apparently, it becomes impossible to exhaustively search a minimal tree even when n is moderate.

The "Branch and Bound" is a strategy to avoid exhaustive search. Theoretically, a branch and bound algorithm cannot ensure a polynomial time complexity in the worst case. But it has been successfully used to solve some NP-hard problems. In addition, a branch and bound algorithm can often find the near optimal solutions as well as an optimal one. This may be an important feature in the evolutionary tree problem since the correct tree is not necessarily a minimal one. In [12], a branch and bound algorithm was designed to construct a minimum evolutionary tree of 11 species, and there is no report of algorithms for constructing minimum evolutionary trees of more species. In this paper, we present a branch and bound algorithm for constructing a minimum ultrametric tree from a metric or nonmetric distance matrix. The experimental results show that the algorithm can solve the problem in reasonable time for $n \leq 25$ if the input is a metric, or $n \leq 19$ for the nonmetric input case.

The rest of the paper is organized as follows: In Section 2, we give some preliminaries. We then show the complexity and the approximation algorithm in Section 3. The branch and bound algorithm will be presented in Section 4. Section 5 presents the experimental results. Finally, concluding remarks are given in Section 6.

2 Preliminaries

We first give some definitions as follows:

Definition 1. A metric M is an ultrametric iff $M[i,j] \leq \max\{M[i,k], M[j,k]\}$ $\forall i,j,k$ [1].

Definition 2. Let $T = (V, E, w)$ be an edge weighted tree and $i, j \in V$. The path length from i to j is denoted by $d(T, i, j)$. The weight of T, denoted by $w(T)$, is defined as $\sum_{e \in E} w(e)$.

Definition 3. Let T be a rooted tree and r be any node of T. T_r denotes the subtree rooted at r, and $L(T)$ denotes the leaf set of T.

Definition 4. An ultrametric tree T of $\{1..n\}$ is a rooted, edge weighted binary tree with $L(T) = \{1..n\}$, and for each node v of T, $d(T, i, v) = d(T, j, v)$ $\forall i, j \in L(T_v)$. So, we can define $height(v) = d(T, v, i)$ for any $i \in L(T_v)$.

It should be noted that an $n \times n$ metric is ultrametric if and only if there is an ultrametric tree T of $\{1..n\}$ such that $\mathrm{d}(T, i, j) = M[i, j]$ $\forall i, j$ [1].

Definition 5. *Minimum Ultrametric Tree from General distances* (MUTG): Given an $n \times n$ distance matrix M (not necessary a metric), find an ultrametric tree T such that $L(T) = \{1..n\}$, $d(T, i, j) \geq M[i, j]$ $\forall i, j$, and $w(T)$ is minimum. The *Minimum Ultrametric Tree from Metric distances* (MUTM) problem has the same definition except that M is a metric.

In the following paragraphs, unless specifically indicated, a tree is a rooted, nonnegative edge weighted binary tree. Given a tree $T = (V, E, w)$, the unweighted tree $P = (V, E)$ is called the topology of T. A tree with topology P and weight w is denoted as $P(w)$.

3 The complexity and approximability

The MUTG problem was defined and was shown NP-hard in [6]. They also proved that it is hard to approximated. Precisely, it had been shown that there is an $\varepsilon > 0$ such that the MUTG problem cannot be approximated in polynomial time within ratio n^{ε} unless NP=P. However, the complexity of MUTM has not been found. In this section, we shall show MUTM is NP-hard, but has different approximability. We now show some properties which will be used in the proofs and in the branch and bound algorithm.

Definition 6. *Min Ultrametric Tree with a given Topology* (MUTT) problem: Given any distance matrix M, a topology P with $L(P) = \{1..n\}$, find a nonnegative edge weight function w, such that $T = P(w)$ is an ultrametric tree, $d(T, i, j) \leq M[i, j] \forall i, j$ and $w(T)$ is minimum.

Definition 7. Let P and M be the tree topology and distance matrix of an MUTT problem, and s be any node of P. $h(s) = \max\{M[i, j] | i, j \in L(P_s)\}/2$.

Lemma 1. Let $T = P(w)$ be the solution of an MUTT problem with input M and P. For any internal node s of P, $height(s) = h(s)$.

Proof. It is easy to see that $height(s) \geq h(s)$, otherwise there exist two leaves $i, j \in T_s$ such that $d(T, i, j) = 2height(s) < M[i, j]$. Assume there is an internal node s such that $height(s) = h(s) + \delta > h(s)$, and $height(v) = h(v) \forall v \in T_s$. Let x and y be the sons of s, then $w(s, x) \geq \delta$ and $w(s, y) \geq \delta$. If s is the root, we can set $w_2(s, x) = w(s, x) - \delta$ and $w_2(s, y) = w(s, y) - \delta$, and result in a feasible solution with less tree size. If s is not the root and z is the father of s, we can set $w_2(s, x) = w(s, x) - \delta$, $w_2(s, y) = w(s, y) - \delta$, and $w_2(z, s) = w(z, s) + \delta$, and also result in a feasible solution with less tree size, which again contradicts to the assumption that $w(T)$ is minimal. □

From Lemma 1 and the property of ultrametric tree, we have the following corollaries:

Corollary 2. Let $T = P(w)$ be the solution of an MUTT problem with input M and P, r be any internal node of P, and u, v be the two sons of r. $w(T_r) = w(T_u) + w(T_v) + 2h(r) - h(u) - h(v)$.

Corollary 3. Let $T = P(w)$ be the solution of an MUTT problem with input M and P, $w(T) = height(r) + \sum_{s \in T} height(s)$, where r is the root of T.

Based on Lemma 1 and Corollary 2, it is not hard to compute the edge weight of the MUTT problem using a postorder traversal of the tree. We list the result but omit the algorithm in this abstract.

Theorem 4. Given the distance matrix and the topology of the ultrametric tree, the MUTT problem can be solved in linear time.

We now show that the MUTM problem is NP-hard. The proof is similar as the one for MUTG in [6]. Given a graph $G = (V, E)$, we construct a matrix M in which $M[i, j] = 4$ if $(i, j) \in E$, and $M[i, j] = 2$ otherwise. Obviously, M is a metric. We can show that G is k-colorable if and only if there is an ultrametric tree T of M where $w(T) = k$. Since the Graph k-colorability problem is NP-complete[10], we conclude that the MUTM problem is NP-hard.

Theorem 5. MUTM is NP-hard.

Unlike MUTG was shown to be hardly approximated, we shall show MUTM can be approximated within ratio $1.5(1 + \lceil \log n \rceil)$ by giving an approximation algorithm.

Definition 8. Let M be an $n \times n$ distance matrix and $G = (V, E, w)$ be the corresponding complete graph of M, in which $V = \{1..n\}$, $w(i, j) = M[i, j]$. $TSP(M)$ denotes the length of a minimal Hamiltonian cycle on G, which is the solution of Travelling Salesman Problem on G.

Lemma 6. Let T be a minimum ultrametric tree of M. $w(T) \geq TSP(M)/2$.

Proof. Let C be an Euler tour on T. Then, the length of the tour C, $w(C) = 2w(T)$. Without loss of generality, assume C visit the leaves in the order 1,2,3,..,n. We have

$$w(C) = \sum_{i=2}^{n} d(T, i-1, i) + d(T, n, 1) \geq \sum_{i=2}^{n} M[i-1, i] + M[n, 1]$$

Since $TSP(M)$ is the minimum length among all Hamiltonian cycles, we have $w(C) \geq TSP(M)$. It follows $w(T) \geq TSP(M)/2$. □

Definition 9. Let $i \leq j$. $CBTT(i,j)$ is a binary tree topology with leaf set $\{i, i+1, .., j\}$. If $i = j$, it contains only one vertex i. Otherwise, if r is the root and x,y are the two sons of r, then, the two subtrees rooted at x and y are $T_x = CBTT(i, \lfloor (i+j-1)/2 \rfloor)$ and $T_y = CBTT(\lfloor (i+j-1)/2 \rfloor + 1, j)$. We also define that the level of root is 0, and a node has level $i+1$ if its father has level i. Obviously, the leaves of $CBTT(1,n)$ have level $\lceil \log n \rceil$ or $\lceil \log n \rceil - 1$.

Our algorithm is based on the 1.5-approximation algorithm of the Travelling Salesman Problem with metric input [10]. The approximation algorithm is listed as follows:

Algorithm Approx_MUTM
Input: a metric M.
Output: an ultrametric tree T with $w(T) \leq 1.5(1 + \lceil \log n \rceil)OPT$,
where OPT is the weight of minimum ultrametric tree of M.
Step1: Run 1.5-approximation algorithm of TSP problem.
Relabel the leaves such that the solution is (1,2,..,n), that is, $\sum_{i=2}^{n} M[i-1, i] + M[n, 1] \leq 1.5TSP(M)$.
Step 2: Construct a topology $P = CBTT(1, n)$.
Step 3: Find the minimal edge weight function w by solving the MUTT problem, and output $T = P(w)$ as an approximation solution.

Lemma 7. Let T be the ultrametric tree constructed by algorithm Approx_MUTM with input M. $w(T) \leq 0.75\,(\lceil \log n \rceil + 1)\,TSP(M)$.

Proof. First, for any internal node s which is the root of $CBTT(a,b)$,

$$height(s) = \max\{M[i,j] \| a \leq i < j \leq b\}/2 \leq \sum_{i=a+1}^{b} M[i-1, i]/2$$

by the triangle inequality. Let s_1,s_2,..,s_k be the nodes of level i, and s_j is the root of $CBTT(a_j, b_j)$. Obviously, $\{a_1..b_1\}$, $\{a_2..b_2\}$,..,$\{a_k..b_k\}$ is a partition of $\{1..n\}$. So,

$$\sum_{j=1}^{k} height(s_j) \leq \sum_{i=2}^{n} M[i-1, i]/2 \leq 0.75TSP(M)$$

Since the tree has $(\lceil \log n \rceil + 1)$ levels and the nodes at the last level has height 0,

$$\sum_{s \in T} height(s) \leq 0.75 \lceil \log n \rceil TSP(M)$$

Let r be the root of T. From Corollary 3,

$$w(T) = height(r) + \sum_{s \in T} height(s) \leq 0.75\left(\lceil \log n \rceil + 1\right) TSP(M)$$

□

Theorem 8 directly comes from Lemma 7 and 6.

Theorem 8. MUTM problem can be approximated in polynomial time within ratio $1.5(1 + \lceil \log n \rceil)$, where n is the number of species.

4 A branch and bound algorithm

Before presenting the branch and bound algorithm, we first show some useful properties:

4.1 Some properties

Definition 10. Let M be a matrix. $\max(M)$ denotes $\max_{i,j}\{M[i,j]\}$.

Definition 11. Let P be a topology, and $a, b \in L(P)$. $LCA(a,b)$ denotes the lowest common ancestor of a and b. Let x,y be two nodes of P, we define $x \to y$ if and only if x is a ancestor of y.

The following lemma can be proved from Lemma 1, and we omit the proof in this abstract.

Lemma 9. If $M[u,v] = \max(M)$, there exists a minimum ultrametric tree T such that u and v are in the two subtrees of root r.

Definition 12. Let P be a topology. A relation $R(P)$ is defined to be $\{(a,b,c) | a,b,c \in L(P), LCA(a,c) = LCA(b,c) \to LCA(a,b)\}$.

Definition 13. Let P, Q be topologies. $P \subset Q$ if and only if $R(P) \subset R(Q)$.

It should be noted that $P \subset Q$ means that for all leaves of P, the topological relation are the same in P and Q. It also means that we can obtain Q by inserting the leaves in $L(Q) - L(P)$ into P. The insertion operation is defined below:

Definition 14. Let $P = (V, E)$ be a topology, $e = (a,b) \in E$. $Q = Insert(P, e, x)$ is a topology obtained by inserting a new leaf x into P at e. The insertion is to replace e with two edges (a,s) and (s,b), and insert an edge (s,x), in which s is a new internal node.

It is easy to see that $P \subset Q$ if $Q = Insert(P, e, x)$. For any topology P, P can be obtained by a sequence of insertions with any specified sequence of $L(P)$. For developing the branch and bound algorithm, we find a lower bound function which holds for both metric and nonmetric input. We list the result but omit the proof in this abstract.

Lemma 10. Let $T(i)$ be a minimum ultrametric tree of M with leaf set $\{1..i\}$ and topology P, and $T(n)$ be a minimum ultrametric tree of M with leaf set $\{1..n\}$ and contains $T(i)$. Then
$w(T(n)) \geq w(T(i)) + \sum_{i<j\leq n} \min\{M[k,j] \| \forall k < j\}/2$.

Lemma 10 is true for any permutation of species. To make the algorithm more efficient, we hope the lower bound is as large as possible, and use a special permutation of species. The permutation is defined as follows, and the efficiency will be shown in Section 5.

Definition 15. Let M be an n by n distance matrix. A permutation $(a_1, a_2, .., a_n)$ of $\{1..n\}$ is called as a **maxmin** permutation if $M[a_1, a_2] = max(M)$ and $\min_{k<i}\{M[a_i, a_k]\} \geq \min_{k<i}\{M[a_j, a_k]\}$ $\forall 1 < i < j$.

Given any distance matrix M, it is not hard to find a **maxmin** permutation in linear time. We omit the algorithm in this abstract.

4.2 The algorithm

The branch and bound algorithm is a tree search algorithm. The algorithm repeat searching the Branch and Bound Tree (BBT) for better solutions until an optimal solution is found. Instead of listing the algorithm, we describe the key points of the algorithm in the following:

- The algorithm first finds a maxmin permutation. After relabelling, assume $\{1..n\}$ is a maxmin permutation.
- The root (level 0) of BBT represents the minimum ultrametric tree of 1,2. It is a tree with root r, and leaves 1 and 2 are the sons of r, and $w(1,r) = w(2,r) = M[1,2]/2$. According to Lemma 9, there is an optimal solution contains this topology. A son of a node with level i will be said to have level $i+1$.
- A node v at level i represents the topology of an ultrametric tree T of $\{1..i+2\}$. $C(v)$ is the minimum tree size for this topology.
- The branching rule: Any topology, which represented by a node at level i, has $i+2$ leaves and $2+2i$ edges. Therefore, there are $2i+2$ ways to insert leaf $i+3$ into the topology, and each results in a different topology. A node at level i has $2i+2$ sons.
- The upper bound: Initially, the algorithm uses a heuristic algorithm UPGMM (Unweighted Pair Group Method with Maximum) to find a feasible solution. While the algorithm running, it retains the best solution so far, and uses it as an upper bound. UPGMM is modified from the popular heuristic

algorithm UPGMA. The heuristic is described as follows: Initially, there are n ultrametric trees, and each contains only one species. It finds two trees A,B such that $\max\{M[x,y] | x \in L(A), y \in L(B)\}$ is minimum, and merges A and B into one ultrametric tree by creating a common root, and then constructs the tree recursively on the $n-1$ trees.

- The lower bound: According to Lemma 10, for a node v at level i, we can set the lower bound by $LB(v) = C(v) + \sum_{j=i+3}^{n} \min\{M[k,j] | \forall k < j\}/2$.
- When the upper bound updated, that is, a better solution found, all nodes will be checked. Any node whose lower bound is no less than the upper bound can be deleted (bounding) safely since any tree containing this topology is of tree size no less than the upper bound.
- Search strategy: Two strategies are often used: **Depth-First** and **Best-First**. It is believed that the **Depth-First** search uses less space while the **Best-First** search often achieves better time efficiency. In our algorithm, we use both strategies to get a balance between time and space.
- Data structure for a node of BBT: Instead of storing the ultrametric tree in a node, we only store the attached edge where the new leaf inserted. Furthermore, the edge can be represented by a code sequence of 0,1. Starting from the root and setting the code empty, if the edge is in the left (right) subtree, we append a code 0 (1) and go down to the left (right) son. Continue this process and we can get the 0-1 code for each edge. Under this data structure, we need to reconstruct the tree every time we select a node to branch. For the sake of efficiency, we modify the **Best-First** search strategy. When a node is selected, our **Best-First** search only select the local best node among its children. Until the last level reached or all the branching of a node bounded, it selects the global best node again. It should be noted that the tree topology in a BBT node is uniquely determined by the codes of it and all its ancestors.

5 The experimental results

5.1 The environment and data instances

We have implemented the above methods in C language on PCs(Pentium-90) running MS- DOS. The data instances are generated randomly. Since the distances are even integers between 2 and 100, there is no rounding error for the tree size. Both metric and nonmetric data were tested.

5.2 Results of running time

For metric input, we ran experiments for the number of species ranging from 12 to 25. The results are shown in Table 1. In the table, n is the number of species, and the second row indicates how many data instances we perfromed for each n. For each n, the worst and the average running time are listed below. The last row presents the median of the time of the instances. For example, if we ran the program with 5 (different) data instances for 10 species, and the running time are 1, 1, 2, 6, 10 respectively. The worst case is 10 (sec.), the average is 4, and the median is 2. For nonmetric input, we performed experiments for the number of species ranging between 9 to 19, and the results are shown in Table 2.

Table 1: Running time(seconds) for metric input

n	12	13	14	15	16	17	18	19	20	24	25
# of instances	100	100	100	50	50	20	20	20	20	5	5
worst case	8	15	33	63	333	2730	6047	6786	1374	53126	23273
average	1	2	3	6	16	210	409	702	317	13724	8644
median	1	1	1	2	3	12	33	195	101	3492	5584

Table 2: Running time (seconds) for nonmetric input

n	9	10	11	12	13	14	15	16	17	18	19
# of instances	100	100	100	100	100	100	50	50	10	5	5
worst case	2	6	8	134	368	1617	2433	12420	26154	44627	29895
average	1	1	2	6	33	114	293	1288	8933	13852	18862
median	1	1	1	2	6	62	165	586	3160	5206	24469

5.3 The benefit of maxmin permutation

To show the benefit of maxmin permutation, we compare the running time of algorithms with and without using maxmin permutation. The comparison is shown in Table 3, and the experiments are for metric input. Without maxmin permutation, it takes more than one day for n=21 while it completes within one day for n=25 with maxmin permutation. We have also performed some tests for nonmetric cases (not listed in the table). Without maxmin permutation, it failed to complete a test for n=16 within one day. It took no more than one day for n=19 if the maxmin permutation was used.

Table 3: The efficiency of maxmin permutation (time in secs)

n	11	12	13	14	15	16	17	18	19	20
with maxmin	2	8	15	33	63	333	2730	6043	6786	1374
without maxmin	2	13	61	898	624	1356	43001	31705	40742	41367

5.4 The performance of UPGMM

We also recorded the ratio that UPGMM can find the optimal solution. The results are shown in Table 4. For metric cases, UPGMM finds the optimal solution with high probability only when $n < 10$, and for nonmetric cases, it can hardly find the optimal even when n=8. But in another point of view, we found that the error ratio between optimal and the tree size found by UPGMM is very small. In our tests, it never exceeds 5%. So, it is a very good upper bound for the branch and bound algorithm, and it indeed makes our algorithm very time efficient.

Table 4: The ratio of the data which optimum found by UPGMM														
n	8	9	10	11	12	13	14	15	16	17	18	19	20	24	25
metric input	.84	.78	.68	.62	.61	.44	.45	.40	.32	.45	.15	.35	.20	0	.20
general input	.40	.25	.24	.13	.11	.09	.10	.02	.04	0	0	0	-	-	-

6 Conclusion remarks

The branch and bound algorithm has another two important features: (1) It can be easily modified to generate all near optimal solutions. (2) It can be easily parallelized. We close this paper by mentioning a few open problems. Theoretically, the approximability of the problem with metric input is still unknown. In this paper, we only show it is not as hard as the nonmetric problem. Another important future work is to examine the complexity and approximability of the additive tree construction problem, and to develop a similar branch and bound algorithm for it.

References

1. H.J. Bandelt, Recognition of tree metrics, *SIAM Journal on Discrete Mathematics.*, 3(1)1-6,1990.
2. E.Dahlhaus, Fast parallel recognition of ultrametrics and tree metrics, *SIAM Journal on Discrete Mathematics*, 6(4):523-532, 1993.
3. W.H.E. Day, Computationally difficult parsimony problems in phylogenetic systematics, *Journal of theoretic Biology*, 103:429-438, 1983.
4. W.H.E. Day, D.S. Johnson and D. Sankoff, The computational complexity of inferring rooted phylogenies by parsimony, *Mathematical Biosciences* 81:33-42, 1986.
5. W.H.E. Day, Computational complexity of inferring phylogenies from dissimilarity matrices, *Bulletin of Mathematical Biology*, 49(4): 461-467,1987.
6. M. Farach, S. Kannan, and T. Warnow, A robust model for finding optimal evolutionary trees, *Algorithmica*, 13:155-179, 1995.
7. W.M. Fitch, A non-sequential method for constructing trees and hierarchical classifications, *Journal of Molecular Evolution*, 18:30-37, 1981.
8. L.R. Foulds and R.L. Graham, The Steiner problem in phylogeny is NP-complete, *Advances in Applied Mathematics*, 3:43-49, 1982.
9. L.R. Foulds, Maximum savings in the Steiner problem in phylogeny, *Journal of theoretic Biology*, 107:471-474, 1984.
10. M.R. Garey and D.S. Johnson, *Computers and Intractability: A guide to the theory of NP-Completeness*, W.H.Freeman and Company, San Fransisco, 1979.
11. D. Gusfield, *Algorithms on Strings, Trees, and Sequences, computer science and computational biology*, Cambridge University Press, 1997.
12. M.D. Hendy and D. Penny, Branch and bound algorithms to determine minimal evolutionary trees, *Mathematical Biosciences*, 59:277-290, 1982.
13. M. Krivanek, The complexity of ultrametric partitions on graph, *Information Processing Letter*, 27(5):265-270, 1988.
14. W.H. Li and D. Graur, *Fundamentals of Molecular Evolution*, Sinauer Associates, INC. 1991.

An Optimal Algorithm for Finding the Minimum Cardinality Dominating Set on Permutation Graphs

H. S. Chao[1], F. R. Hsu[2], and R. C. T. Lee[3]

[1] Dept. of Computer Science, National Tsing Hua University, Hsinchu, Taiwan
hschao@cs.nthu.edu.tw
[2] Department of Accounting, Providence University, Shalu, Taichung Hsien, Taiwan
frhsu@simon.pu.edu.tw
[3] Office of the President, Providence University, Shalu, Taichung Hsien, Taiwan
rctlee@simon.pu.edu.tw

Abstract. A dominating set D of an undirected graph G is a set of vertices such that every vertex not in D is adjacent to at least one vertex in D. Given a undirected graph G, the *minimal cardinality dominating set* problem is to find a dominating set of G with minimum number of vertices. The minimal cardinality dominating set problem is NP-hard for general graphs. For permutation graphs, the best-known algorithm ran in $O(n \log\log n)$ time, where n is the number of vertices. In this paper, we present an optimal $O(n)$ algorithm.

1 Introduction

Let $\pi = [\pi_1, \pi_2, \ldots, \pi_n]$ be a permutation of the numbers $1, 2, \ldots, n$. A *permutation* graph $G[\pi] = (V, E)$ with respect to π is defined as follows [3]:

$$V = \{1, 2, \ldots, n\}$$

and

$$(i, j) \in E \text{ iff } (i - j)(\pi_i^{-1} - \pi_j^{-1}) < 0,$$

where π_k^{-1} is the position of k in π. A graph G is a permutation graph if and only if G is isomorphic to some $G[\pi]$. Fig. 1 shows the permutation graph of a given permutation $\pi = [3, 1, 5, 7, 4, 2, 6]$. In this paper, we assume that the input is a permutation $\pi = [\pi_1, \pi_2, \ldots, \pi_n]$.

For an undirected graph $G = (V, E)$, a vertex $i \in V$ is said to *dominate* another vertex $j \in V$ if $(i, j) \in E$. For any two sets S_1 and S_2, the set $S_2 \backslash S_1$ is the set of all elements which belong to S_2 but not belong to S_1. For an undirected graph $G = (V, E)$, a vertex set S_1 is said to *dominate* another vertex set S_2 if every vertex in $S_2 \backslash S_1$ is dominated by at least one vertex in S_1. Let $S_1 \rhd S_2$ and $S_1 \not\rhd S_2$ denote that S_1 dominates S_2 and S_1 does not dominate S_2 respectively. A vertex set S is said to be a *dominating set* for G if $S \rhd V$. The *minimum*

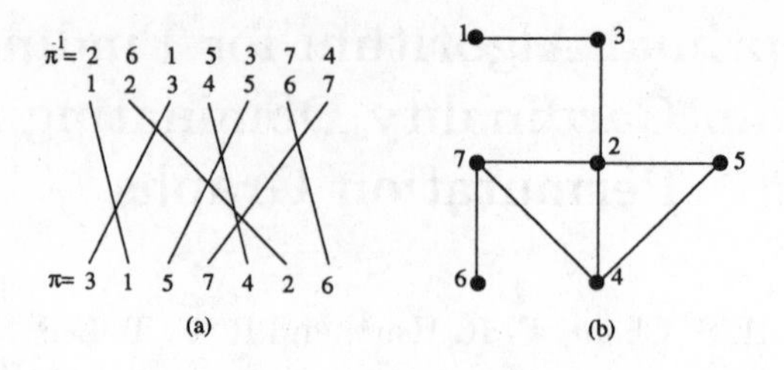

Fig. 1. A permutation graph $G[\pi]$ for $\pi = [3, 1, 5, 7, 4, 2, 6]$

cardinality dominating set (MCDS) problem and the *minimum weighted dominating set* (MWDS) problem are to find a dominating set S for G where the number and the total weight of vertices of S are minimized respectively.

Both of the MCDS and the MWDS problems are *NP*-hard for general graphs [2]. On permutation graphs, Farber and Keil [1] proposed $O(n^2)$ and $O(n^3)$ algorithms for the MCDS and the MWDS problems respectively based upon the dynamic programming method. Later, by utilizing the monotone ordering among the intermediate terms of the recursive formula in [1], Tsai and Hsu [6] improved the time-complexities to $O(n \log \log n)$ and $O(n^2 \log^2 n)$ respectively. For the MWDS problem, Liang *el al.* [4] proposed an $O(n(n+m))$ algorithm, where m is the number of edges. Later, they reduced the time-complexity to $O(n+m)$ time [5]. In this paper, we propose an optimal $O(n)$ algorithm for solving the MCDS problem on permutation graphs. Our algorithm is based on a new recursive formula by using the dynamic programming method, which is different from the formula in [1]. There is also a monotone ordering among the intermediate terms of our recursive formula. Then, we propose the new updating rules so that we can design an optimal linear time algorithm.

2 The Dynamic Programming Approach

Consider a permutation graph $G[\pi]$ defined by a permutation $\pi = [\pi_1, \pi_2, \ldots, \pi_n]$. Throughout this paper, we assume that the permutation π is given. Following the notations defined in [1, 6], define $V_i = \{\pi_1, \pi_2, \ldots, \pi_i\}$ and $V_{i,j} = V_i \cap \{1, 2, \ldots, j\}$. For each i, $1 \le i \le n$, we define π_i^* to be the minimum number over the suffix $\pi_i, \pi_{i+1}, \ldots, \pi_n$. Define $V_i^* = V_i \cup \{\pi_i^*\}$. For the example in Fig. 1, $V_4 = \{3, 1, 5, 7\}$, $V_{4,3} = \{1, 3\}$ and $V_4^* = \{3, 1, 5, 7, 2\}$.

For any vertex set S, define $\max(S)$ to be the maximum number in S. For each i and j, $1 \le i, j \le n$, define $D_{i,j}$ as follows:

1. $D_{i,j}$ is a minimum cardinality subset of V_i^* dominating $V_{i,j}$.
2. $\max(D_{i,j})$ is as large as possible.

Obviously, $D_{n,n}$ is a desired minimum cardinality dominating set for G.

Let X be a set of subsets of V. Let S be a non-empty element in X such that S is with the minimum cardinality among all elements in X and $\max(S)$ is is as large as possible, if $X \ne \phi$ and $X \ne \{\phi\}$. Then, define set_min(X) as follows: set_min(X) $= \phi$ if $X = \phi$ or $X = \{\phi\}$, and set_min(X) $= S$ if otherwise. set_min(X) may not be unique. If there are more than one candidate for set_min(X), select any one to be set_min(X). It is easy to prove that if X, Y and Z are sets of subsets of V and $X = Y \cup Z$, we have set_min(X) = set_min({set_min(Y), set_min(Z)}).

According to the definitions of $D_{i,j}$ and set_min, we have the following:

$$D_{i,j} = \begin{cases} \phi & \text{if } V_{i,j} = \phi, \\ \text{set\_min}(\{S \mid S \subset V_i^* \text{ and } S \rhd V_{i,j}\}) & \text{if otherwise.} \end{cases}$$

Consider the case where $i = 1$. We have the following rule:

$$D_{1,j} = \begin{cases} \phi & \text{if } j < \pi_1, \\ \{\pi_1\} & \text{if otherwise.} \end{cases}$$

Let $D_{\pi_i^*} = D_{i-1,\pi_i^*} \cup \{\pi_i^*\}$, $D_{\pi_i} = D_{i-1,\pi_i} \cup \{\pi_i\}$ and $D_{max} = D_{i-1,j} \cup \{\max(V_i)\}$.

Lemma 1 *For each i_1, i_2 and j, $1 \le i_1 < i_2 \le n$ and $1 \le j \le n$, $V_{i_1,j} \subset V_{i_2,j}$ and $V_{i_1}^* \subset V_{i_2}^*$.*

Proof. The proof is straightforward and omitted. □

Lemma 2 *For each i and j, $1 < i \le n$ and $1 \le j \le n$,*

$$\{D_{\pi_i^*}, D_{\pi_i}, D_{max}\} \subset \{S | S \subset V_i^* \text{ and } S \rhd V_{i,j}\}.$$

Proof. $D_{i-1,\pi_i^*} \subset V_{i-1}^* \subset V_i^*$ by Lemma 1. Hence (1) $D_{i-1,\pi_i^*} \cup \{\pi_i^*\} \subset V_i^*$. Since $\{\pi_i^*\} \rhd V_{i,j} \setminus V_{i-1,\pi_i^*}$ and $D_{i-1,\pi_i^*} \rhd V_{i-1,\pi_i^*}$. Hence (2) $V_{i,j}$ is dominated by $D_{i-1,\pi_i^*} \cup \{\pi_i^*\}$. According to (1) and (2), we have $D_{\pi_i^*} \in \{S | S \subset V_i^* \text{ and } S \rhd V_{i,j}\}$. The proofs for D_{π_i} and D_{max} are similar and omitted. □

Lemma 3 *For each i and j, $1 < i \le n$ and $\pi_i^* \le j \le n$, if $Y \subset \{S | \pi_i^* \in S, S \subset V_i^* \text{ and } S \rhd V_{i,j}\}$, then set_min($Y \cup \{D_{\pi_i^*}\}) = D_{\pi_i^*}$.*

Proof. Let X denote $\{S | \pi_i^* \in S, S \subset V_i^* \text{ and } S \rhd V_{i,j}\}$. Suppose $Y \subset X$ and Y is not empty. Consider any $A \in Y$. Since $V_{i,j} \supset V_{i-1,\pi_i^*}$ and $A \rhd V_{i,j}$, $A \rhd V_{i-1,\pi_i^*}$. Since no vertex in V_{i-1,π_i^*} is dominated by π_i or π_i^*, $A \setminus \{\pi_i, \pi_i^*\} \rhd V_{i-1,\pi_i^*}$. Since $A \subset V_i^* = V_{i-1} \cup \{\pi_i, \pi_i^*\}$, $A \setminus \{\pi_i, \pi_i^*\} \subset V_{i-1} \subset V_{i-1}^*$. Therefore, $A \setminus \{\pi_i, \pi_i^*\} \in \{S \mid S \subset V_{i-1}^* \text{ and } S \rhd V_{i-1,\pi_i^*}\}$. Note that $\pi_i^* \in A$. Then, $|D_{i-1,\pi_i^*}| \le |A| - 1$ and $|D_{i-1,\pi_i^*}| = |A| - 1$ only if $\pi_i^* = \pi_i$ or $\pi_i \notin A$. Hence, $|D_{\pi_i^*}| \le |A|$ and if $|D_{\pi_i^*}| = |A|$, $\max(D_{\pi_i^*}) \ge \max(A)$. Therefore, set_min($Y \cup \{D_{\pi_i^*}\}) = D_{\pi_i^*}$ □

The proofs of the following three lemmas are similar to Lemma 3 and omitted.

Lemma 4 *For each i and j, $1 < i \le n$ and $\pi_i \le j \le n$, if $Y \subset \{S | \pi_i \in S, \pi_i^* \notin S, S \subset V_i^* \text{ and } S \rhd V_{i,j}\}$, then set_min($Y \cup \{D_{\pi_i}\}) = D_{\pi_i}$.* □

Lemma 5 *For each i and j, $1 < i \le n$ and $\pi_i \le j \le n$, if $\max(D_{i-1,j}) < \pi_i$ and $Y \subset \{S | \pi_i \notin S, \pi_i^* \notin S, S \subset V_i^* \text{ and } S \rhd V_{i,j}\}$, then set_min($Y \cup \{D_{max}\}) = D_{max}$.* □

Lemma 6 *For each i and j, $1 < i \leq n$ and $1 \leq j \leq n$, if $Y \subset \{S|\pi_i \notin S, \pi_i^* \notin S, S \subset V_i^*$ and $S \rhd V_{i,j}\}$, set_min$(Y \cup \{D_{i-1,j}\}) = D_{i-1,j}$.* □

Theorem 1 *The following recursive formula correctly computes $D_{i,j}$, where $1 < i \leq n$ and $1 \leq j \leq n$.*

$$D_{i,j} = \begin{cases} \textit{set\_min}(\{D_{\pi_i^*}, D_{\pi_i}, D_{max}\}) & \textit{if } j \geq \pi_i \textit{ and } \max(D_{i-1,j}) < \pi_i, \\ \textit{set\_min}(\{D_{i-1,j}, D_{\pi_i^*}, D_{\pi_i}\}) & \textit{if otherwise} \end{cases}$$

Proof. Let $X = \{S \mid S \subset V_i^* \text{ and } S \rhd V_{i,j}\}$, $X_1 = \{S|\pi_i^* \in S, S \subset V_i^* \text{ and } S \rhd V_{i,j}\}$, $X_2 = \{S|\pi_i \in S, \pi_i^* \notin S, S \subset V_i^* \text{ and } S \rhd V_{i,j}\}$ and $X_3 = \{S \mid \pi_i \notin S, \pi_i^* \notin S, S \subset V_i^* \text{ and } S \rhd V_{i,j}\}$. It is obvious that $X = X_1 \cup X_2 \cup X_3$. According to Lemma 2, $\{D_{\pi_i^*}, D_{\pi_i}, D_{max}\} \subset X$. Hence,

$$\begin{aligned} \text{set\_min}(X) &= \text{set\_min}(X \cup \{D_{\pi_i^*}, D_{\pi_i}, D_{max}\}) \\ &= \text{set\_min}(\{\text{set\_min}(X_1 \cup \{D_{\pi_i^*}\}), \\ &\quad \text{set\_min}(X_2 \cup \{D_{\pi_i}\}), \text{set\_min}(X_3 \cup \{D_{max}\})\}). \end{aligned}$$

Furthermore, if $D_{i-1,j} \in X$, since $|D_{i-1,j}| < |D_{max}|$, then

$$\begin{aligned} \text{set\_min}(X) &= \text{set\_min}(X \cup \{D_{\pi_i^*}, D_{\pi_i}, D_{max}, D_{i-1,j}\}) \\ &= \text{set\_min}(\{\text{set\_min}(X_1 \cup \{D_{\pi_i^*}\}), \\ &\quad \text{set\_min}(X_2 \cup \{D_{\pi_i}\}), \text{set\_min}(X_3 \cup \{D_{i-1,j}\})\}). \end{aligned}$$

Case 1. $j \geq \pi_i$ and $\max(D_{i-1,j}) < \pi_i$:
According to Lemmas 3, 4 and 5, we have

$$\text{set\_min}(X) = \text{set\_min}(\{D_{\pi_i^*}, D_{\pi_i}, D_{max}\}).$$

Case 2. $j \geq \pi_i$ and $\max(D_{i-1,j}) \geq \pi_i$:
It can be easily shown that $D_{i-1,j} \in X$. By Lemmas 3, 4 and 6, we have

$$\text{set\_min}(X) = \text{set\_min}(\{D_{\pi_i^*}, D_{\pi_i}, D_{i-1,j}\}).$$

Case 3. $\pi_i^* \leq j < \pi_i$:
It can be easily shown that $D_{i-1,j} \in X$. Furthermore, since $j < \pi_i$, no vertex in $V_{i,j}$ is dominated by π_i. Hence π_i must not belong to $D_{i,j}$. Therefore, X_2 and D_{π_i} need not be considered for $D_{i,j}$. Then, by Lemma 3 and 6,

$$\begin{aligned} \text{set\_min}(X) &= \text{set\_min}(\{\text{set\_min}(X_1 \cup \{D_{\pi_i^*}\}), \text{set\_min}(X_3 \cup \{D_{i-1,j}\})\}) \\ &= \text{set\_min}(\{D_{\pi_i^*}, D_{i-1,j}\}) \\ &= \text{set\_min}(\{D_{\pi_i^*}, D_{\pi_i}, D_{i-1,j}\}). \end{aligned}$$

Case 4. $j < \pi_i^*$:
It can be easily shown that $D_{i-1,j} \in X$. Furthermore, since $j < \pi_i^* \leq \pi_i$,

neither π_i^* nor π_i belongs to $D_{i,j}$. We need not consider X_1, X_2, $D_{\pi_i^*}$ and D_{π_i} for finding $D_{i,j}$. By Lemma 6, we have

$$\text{set\_min}(X) = \text{set\_min}(X_3 \cup \{D_{i-1,j}\}) = D_{i-1,j}.$$

Since both D_{π_i} and $D_{\pi_i^*}$ belong to X, the following equality holds

$$\text{set\_min}(X) = D_{i-1,j} = \text{set\_min}(\{D_{\pi_i^*}, D_{\pi_i}, D_{i-1,j}\}).$$

Note that all cases are considered. Hence, the proof is complete. □

3 The New Updating Rules.

For each i and j, define $d_{i,j} = |D_{i,j}|$ and $m_{i,j} = \max(D_{i,j})$.

Lemma 7 (The Monotone Lemma) *For each i, if $j_1 < j_2$, then $d_{i,j_1} \leq d_{i,j_2}$. Furthermore, if $j_1 < j_2$ and $d_{i,j_1} = d_{i,j_2}$, then $m_{i,j_1} \geq m_{i,j_2}$.*

Proof. Suppose $j_1 < j_2$. We have $V_{i,j_1} \subset V_{i,j_2}$ and $D_{i,j_2} \rhd V_{i,j_1}$. Note that $D_{i,j_2} \subset V_i^*$. Hence, $D_{i,j_2} \in \{S|S \subset V_i^* \text{ and } S \rhd V_{i,j_1}\}$. By the definition of D_{i,j_1}, the proof is complete. □

Lemma 8 *For each i and j, $1 < i \leq n$ and $1 \leq j \leq n$, $d_{i-1,j} \leq d_{i-1,\pi_i^*} + 1$. Furthermore, if $d_{i-1,j} = d_{i-1,\pi_i^*} + 1$, then $m_{i-1,j} \geq m_{i-1,\pi_i^*}$.*

Proof. Consider the set $V_{i-1,j} \setminus V_{i-1,\pi_i^*}$. If $V_{i-1,j} \setminus V_{i-1,\pi_i^*}$ is an empty set, we have $D_{i-1,\pi_i^*} \rhd V_{i-1,j}$. Hence, $d_{i-1,j} \leq d_{i-1,\pi_i^*}$. Suppose $V_{i-1,j} \setminus V_{i-1,\pi_i^*}$ is not empty. For any vertex $v \in V_{i-1,j} \setminus V_{i-1,\pi_i^*}$, we have v is dominated by π_{i-1}^*. Hence, $D_{i-1,\pi_i^*} \cup \{\pi_{i-1}^*\} \in \{S|S \subset V_{i-1}^* \text{ and } S \rhd V_{i-1,j}\}$. Therefore, $d_{i-1,j} \leq d_{i-1,\pi_i^*} + 1$. Furthermore, if $d_{i-1,j} = d_{i-1,\pi_i^*} + 1$, $m_{i-1,j} \geq m_{i-1,\pi_i^*}$. □

Define $m^* = \max(D_{i-1,\pi_i^*} \cup \{\pi_i^*\})$. Obviously, $m^* = \max(\{m_{i-1,\pi_i^*}, \pi_i^*\})$. Our new updating rules are listed in Table 1.

Theorem 2 *The updating rules listed in Table 1 correctly compute $d_{i,j}$ and $m_{i,j}$ for $1 < i \leq n$ and $1 \leq j \leq n$.*

Proof. By Lemma 8, $d_{i-1,j} \leq d_{i-1,\pi_i^*} + 1$. Specially, $d_{i-1,\pi_i} \leq d_{i-1,\pi_i^*} + 1$. Since $\pi_i \geq \pi_i^*$, $d_{i-1,\pi_i} \geq d_{i-1,\pi_i^*}$ by Lemma 7. Hence, there are only two cases which should be considered: (1). $d_{i-1,\pi_i} = d_{i-1,\pi_i^*}$ and (2). $d_{i-1,\pi_i} = d_{i-1,\pi_i^*} + 1$.

Case (1). $d_{i-1,\pi_i} = d_{i-1,\pi_i^*}$:

Since $d_{i-1,\pi_i} = d_{i-1,\pi_i^*}$ and $\pi_i^* \leq \pi_i$, we have $m_{i-1,\pi_i^*} \geq m_{i-1,\pi_i}$.

(1a). $d_{i-1,j} = d_{i-1,\pi_i^*} + 1$:

Since $d_{i-1,j} > d_{i-1,\pi_i}$, we have $j > \pi_i$. If $m_{i-1,j} < \pi_i$, by Theorem 1, $D_{i,j} = \text{set\_min}(\{D_{\pi_i^*}, D_{\pi_i}, D_{max}\})$. Since $d_{i-1,j} = d_{i-1,\pi_i^*} + 1$, we have $\pi_i > m_{i-1,j} \geq m_{i-1,\pi_i^*} \geq m_{i-1,\pi_i}$, by Lemma 8. Hence, $\max(D_{\pi_i}) \geq \max(D_{\pi_i^*})$. Note that $|D_{\pi_i^*}| = |D_{\pi_i}| = d_{i-1,\pi_i^*} + 1$ and $|D_{max}| = d_{i-1,j} + 1 = d_{i-1,\pi_i^*} + 2$. Therefore, $D_{i,j} = D_{\pi_i}$. Otherwise, i.e. if $m_{i-1,j} \geq \pi_i$, by Theorem 1, $D_{i,j} = \text{set\_min}(\{D_{i-1,j}, D_{\pi_i^*}, D_{\pi_i}\})$. Note that $|D_{i-1,j}| = |D_{\pi_i^*}| = |D_{\pi_i}|$. Furthermore, we have $m_{i-1,j} \geq \pi_i \geq \pi_i^*$ and $m_{i-1,\pi_i^*} \geq m_{i-1,\pi_i}$ and by Lemma 8, $m_{i-1,j} \geq m_{i-1,\pi_i^*}$. Hence, $D_{i,j} = D_{i-1,j}$.

Case (1). $d_{i-1,\pi_i} = d_{i-1,\pi_i^*}$:

(1a) If $d_{i-1,j} = d_{i-1,\pi_i^*} + 1$, then

$$(d_{i,j}, m_{i,j}) = \begin{cases} (d_{i-1,\pi_i^*} + 1, \pi_i) & \text{if } m_{i-1,j} < \pi_i, \\ (d_{i-1,j}, m_{i-1,j}) & \text{if otherwise.} \end{cases}$$

(1b) If $d_{i-1,j} = d_{i-1,\pi_i^*}$, then

$$(d_{i,j}, m_{i,j}) = \begin{cases} (d_{i-1,\pi_i^*} + 1, \max(V_i)) & \text{if } m_{i-1,j} < \pi_i \text{ and } j \geq \pi_i, \\ (d_{i-1,j}, m_{i-1,j}) & \text{if otherwise.} \end{cases}$$

(1c) If $d_{i-1,j} < d_{i-1,\pi_i^*}$, then

$$(d_{i,j}, m_{i,j}) = (d_{i-1,j}, m_{i-1,j}).$$

Case (2). $d_{i-1,\pi_i} = d_{i-1,\pi_i^*} + 1$:

(2a) If $d_{i-1,j} = d_{i-1,\pi_i^*} + 1$, then

$$(d_{i,j}, m_{i,j}) = \begin{cases} (d_{i-1,\pi_i^*} + 1, m^*) & \text{if } (m_{i-1,j} < \pi_i \text{ and } j \geq \pi_i) \\ & \quad \text{or } (m_{i-1,j} < \pi_i^*), \\ (d_{i-1,j}, m_{i-1,j}) & \text{if otherwise.} \end{cases}$$

(2b) If $d_{i-1,j} \leq d_{i-1,\pi_i^*}$, then

$$(d_{i,j}, m_{i,j}) = (d_{i-1,j}, m_{i-1,j}).$$

Table 1. The New Updating Rules.

(1b). $d_{i-1,j} = d_{i-1,\pi_i^*}$:
If $j \geq \pi_i$ and $m_{i-1,j} < \pi_i$, $D_{i,j} = \text{set\_min}(\{D_{\pi_i^*}, D_{\pi_i}, D_{max}\})$ by Theorem 1. Note that $|D_{\pi_i^*}| = |D_{\pi_i}| = |D_{max}| = d_{i-1,\pi_i^*} + 1$. Hence, $D_{i,j} = D_{max}$. Otherwise, by Theorem 1, $D_{i,j} = \text{set\_min}(\{D_{i-1,j}, D_{\pi_i^*}, D_{\pi_i}\})$. Note that $|D_{i-1,j}| = d_{i-1,\pi_i^*} < |D_{\pi_i^*}| = |D_{\pi_i}|$. Hence, $D_{i,j} = D_{i-1,j}$.

(1c). $d_{i-1,j} < d_{i-1,\pi_i^*}$:
We have $j < \pi_i$. By Theorem 1, $D_{i,j} = \text{set\_min}(\{D_{i-1,j}, D_{\pi_i^*}, D_{\pi_i}\})$. Note that $|D_{i-1,j}| < |D_{\pi_i^*}| = |D_{\pi_i}|$. Hence, $D_{i,j} = D_{i-1,j}$.

Case (2). $d_{i-1,\pi_i} = d_{i-1,\pi_i^*} + 1$: In this case, we have $|D_{\pi_i}| = d_{i-1,\pi_i^*} + 2$.

(2a). $d_{i-1,j} = d_{i-1,\pi_i^*} + 1$:
If $m_{i-1,j} < \pi_i$ and $j \geq \pi_i$, $D_{i,j} = \text{set\_min}(\{D_{\pi_i^*}, D_{\pi_i}, D_{max}\})$ by Theorem 1. Note that $|D_{\pi_i^*}| < |D_{\pi_i}| = |D_{max}|$. Hence $D_{i,j} = D_{\pi_i^*}$. Otherwise, $D_{i,j} = \text{set\_min}(\{D_{i-1,j}, D_{\pi_i^*}, D_{\pi_i}\})$. Note that $|D_{i-1,j}| = |D_{\pi_i^*}| < |D_{\pi_i}|$. Hence, if $m_{i-1,j} < m^*$, $D_{i,j} = D_{\pi_i^*}$. Otherwise, we have $D_{i,j} = D_{i-1,j}$.

(2b). $d_{i-1,j} \leq d_{i-1,\pi_i^*}$:
Since $d_{i-1,j} \leq d_{i-1,\pi_i^*}$, $j < \pi_i$. Hence, $D_{i,j} = \text{set\_min}(\{D_{i-1,j}, D_{\pi_i^*}, D_{\pi_i}\})$. Since $|D_{i-1,j}| < |D_{\pi_i^*}| < |D_{\pi_i}|$, $D_{i,j} = D_{i-1,j}$. □

For example, consider the permutation graph shown in Fig. 1. The updating of $(d_{i,j}, m_{i,j})$'s by using Theorem 2 is shown in Table 2.

4 The Linear Time Algorithm

We now use Theorem 2 and show that we can obtain an $O(n)$ time algorithm.

<table>
<tr><th>j =</th><th>1</th><th>2</th><th>3</th><th>4</th><th>5</th><th>6</th><th>7</th></tr>
<tr><td>i = 1</td><td colspan="2">(0,0)</td><td colspan="5">(1,3)</td></tr>
<tr><td>i = 2</td><td colspan="7">(1,3)</td></tr>
<tr><td>i = 3</td><td colspan="4">(1,3)</td><td colspan="3">(2,5)</td></tr>
<tr><td>i = 4</td><td colspan="4">(1,3)</td><td colspan="2">(2,5)</td><td>(2,3)</td></tr>
<tr><td>i = 5</td><td colspan="3">(1,3)</td><td>(2,7)</td><td colspan="2">(2,5)</td><td>(2,4)</td></tr>
<tr><td>i = 6</td><td colspan="3">(1,3)</td><td>(2,7)</td><td colspan="2">(2,5)</td><td>(2,4)</td></tr>
<tr><td>i = 7</td><td colspan="3">(1,3)</td><td>(2,7)</td><td colspan="2">(2,5)</td><td>(3,7)</td></tr>
</table>

Table 2. The values of $(d_{i,j}, m_{i,j})$'s for the permutation graph in Fig. 1.

From Lemma 7, we know that if $j_1 \leq j \leq j_2$ and $(d_{i,j_1}, m_{i,j_1}) = (d_{i,j_2}, m_{i,j_2})$, then $(d_{i,j_1}, m_{i,j_1}) = (d_{i,j}, m_{i,j}) = (d_{i,j_2}, m_{i,j_2})$, Thus we do not have to compute all $d_{i,j}$'s and $m_{i,j}$'s. We only have to compute the range of j's such that $(d_{i,j}, m_{i,j})$ is the same in this range.

Consider the case when $i = 1$.

According to the discussion in Section 2, there are only two cases.

(1) $1 \leq j < \pi_1$, $D_{1,j} = \phi, d_{1,j} = 0$ and $m_{1,j} = 0$.

(2) $\pi_1 \leq j \leq n$, $D_{1,j} = \{\pi_1\}, d_{1,j} = 1$ and $m_{1,j} = \pi_1$.

For each i, $1 \leq i \leq n$, let us divide the range of j into blocks. In each block, the 2-tuple $(d_{i,j}, m_{i,j})$ is the same for that range of j's. Each such largest block is called a *dm-block*. For instance, when $i = 1$, there are two dm-blocks, as in Fig. 2. For each dm-block, it is associated with a 4-tuple $[j_1, j_2, d, m]$ where variables j_1 and j_2, $j_1 \leq j_2$, define the range of this dm-block and $(d_{i,j}, m_{i,j}) = (d, m)$ for each j, $j_1 \leq j \leq j_2$. We define an *α-group* to be all of the dm-blocks $[j_1, j_2, d, m]$'s, where $d = \alpha$. According to Lemma 7, all of the dm-blocks in an α-group occupy a consecutive range of j's and the rightmost (resp. leftmost) dm-block $[j_1, j_2, d, m]$ is associated with a smallest (resp. largest) value of m.

<table>
<tr><td>$j =$</td><td>1</td><td>2</td><td>$\cdots$</td><td>$\pi_i - 1$</td><td>π_i</td><td>$\cdots$</td><td>n</td></tr>
<tr><td>$(d_{1,j}, m_{1,j}) =$</td><td colspan="4">$(0,0)$</td><td colspan="3">$(1, \pi_i)$</td></tr>
</table>

Fig. 2. The dm-blocks for $i = 1$.

The following is our algorithm to solve the problem. We shall discuss the data structure needed to implement the algorithm later.

Algorithm A. Finding a MCDS on a Permutation Graph.
Input: A permutation $\pi = [\pi_1, \pi_2, \ldots, \pi_n]$.
Output: A minimum cardinality dominating set of the graph $G[\pi]$.
Step 1: Compute π_i^* and $\max(V_i)$ for all i's.

Step 2: For $i = 1$, add $[1, \pi_1 - 1, 0, 0]$ into 0-group and add $[\pi_1, n, 1, \pi_1]$ into 1-group.

Step 3: $i \leftarrow i + 1$.

If $(i > n)$, then goto Step 7.

Find $d_{i-1,\pi_i}, d_{i-1,\pi_i^*}$ and m_{i-1,π_i^*}.

$m^* \leftarrow \max(\{\pi_i^*, m_{i-1,\pi_i^*}\})$.

If $d_{i-1,\pi_i} = d_{i-1,\pi_i^*}$, then goto Step 4 else goto Step 6.

Step 4: /* Case 1a */

For the $(d_{i-1,1_i^*} + 1)$-group, find the rightmost dm-block B.

$[j_1, j_2, d, m] \leftarrow B$.

$j_2' \leftarrow j_2$.

While $(m < \pi_i)$ Do

Mark this dm-block.

$j_1' \leftarrow j_1$.

Find the left dm-block B' of the current dm-block in this group.

$[j_1, j_2, d, m] \leftarrow B'$.

EndOfWhile

Delete all marked dm-blocks in this group.

Insert $[j_1', j_2', d_{i-1,\pi_i^*} + 1, \pi_i]$ to the right end of $(d_{i-1,\pi_i^*} + 1)$-group.

Step 5: /* Case 1b */

For the (d_{i-1,π_i^*})-group, find the rightmost dm-block B.

$[j_1, j_2, d, m] \leftarrow B$.

$j_2' \leftarrow j_2$.

While $(m < \pi_i)$ and $(j_1 \geq \pi_i)$ do

Mark this dm-block.

$j_1' \leftarrow j_1$.

Find the left dm-block B' of the current dm-block in this group.

$[j_1, j_2, d, m] \leftarrow B'$.

EndOfWhile

Delete all marked dm-blocks in this group.

If $(m < \pi_i)$ and $(j_2 \geq \pi_i)$, then

Delete this block.

Insert dm-block $[j_1, \pi_i - 1, d, m]$ to the right end of this group.

$j_1' \leftarrow \pi_i$.

EndOfIf

Find the leftmost dm-block B'' of $(d_{i-1,\pi_i^*} + 1)$-group.

$[j_1, j_2, d, m] \leftarrow B''$.

If $m = \max(V_i)$, then $j_2' \leftarrow j_2$ and delete this dm-block B''.

Insert dm-block $[j_1', j_2', d_{i-1,\pi_i^*} + 1, \max(V_i)]$ to the left end of $(d_{i-1,\pi_i^*} + 1)$-group.

Goto Step 3.

Step 6: /* Case 2a */

For the (d_{i-1,π_i^*})-group, find the rightmost dm-block B.

$[j_1, j_2, d, m] \leftarrow B$.

$j_2' \leftarrow j_2$.

While $((m < \pi_i)$ and $(j_1 \geq \pi_i))$ or $(m < \pi_i^*)$
  Mark this dm-block.
  $j_1' \leftarrow j_1$.
  Find the left dm-block B' of the current dm-block in this group.
  $[j_1, j_2, d, m] \leftarrow B'$.
EndOfWhile
Delete all marked dm-blocks in this group.
If $(m < \pi_i)$ and $(j_2 \geq \pi_i)$ then
  Delete this block.
  Insert dm-block $[j_1, \pi_i - 1, d, m]$ to the right end of this group.
  $j_1' \leftarrow \pi_i$.
EndOfIf
Insert dm-block $[j_1', j_2', d_{i-1,\pi_i^*} + 1, m^*]$ to the right end of this group.
Goto Step 3.

Step 7:

Find a MCDS by backtracking.

The End of Algorithm A.

In the above algorithm, m_{i-1,π_i^*} has to be found. This is one of the most critical steps. A straightforward way is to scan all of the dm-blocks. This will be time-consuming. We shall use the fact that $\pi_{i-1}^* \leq \pi_i^*$. Using this fact, we shall only start from the dm-block $[j_1, j_2, d, m]$ in which $j_1 \leq \pi_{i-1}^* \leq j_2$ and move to the right dm-block until we reach the dm-block $[j_1', j_2', d', m']$ in which $j_1' \leq \pi_i^* \leq j_2'$. Then, m_{i-1,π_i^*} is equal to m'. Thus, in the amortized sense, only a linear scan is used for obtaining all of the m_{i-1,π_i^*}'s.

In the following, we shall discuss the data structure needed to implement the above algorithm.

First of all, for each α-group, we have α is an integer and $0 \leq \alpha \leq n$. Thus, we may have an array of size $n + 1$. Whenever a α-group is created, we shall create two pointers inside the α-th cell. The first and the second pointers always point to the leftmost and the rightmost dm-blocks of this group respectively.

Secondly, within each α-group, there are several dm-blocks. They will be linked by a doubly-linked list so that the locating of the leftmost and the rightmost dm-blocks can be done in $O(1)$ time. Since in our algorithm, insertions and deletions occur at the ends of the α-group, each insertion or deletion can be done in $O(1)$ time with this doubly-linked list. Furthermore, as we indicated previously, we need to find m_{i-1,π_i^*}'s. With the array of α-groups and all double-linked lists, the finding can be done by traversing the lists from the dm-block $[j_1, j_2, d, m]$ in which $j_1 \leq \pi_{i-1}^* \leq j_2$, where this dm-block was found in the $(i-1)$-th iteration.

For time-complexity analysis, note the following facts:

1. In Step 1, the computation of all π_i^*'s is a suffix-minima computation and it can be done in $O(n)$ time by scanning the permutation array from right to left. Similarly, the computation of all $\max(V_i)$'s is a prefix-maxima computation and it can be done in $O(n)$ time by scanning the permutation array from left to right.

2. In Step 2, there are 2 insertions.
3. Once Step 3 is executed, the algorithm goes either through Steps 4 and 5 or through Step 6. For the path going through Steps 3, 4 and 5 (Steps 3 and 6), at most 3 (2) insertions are executed. Thus totally at most $3n + 2$ insertions are executed.
4. The total number of deletions is less than the total number of insertions. Thus there are at most $3n + 1$ deletions.
5. The time needed to find all m_{i-1,π_i^*} is $O(n)$ time because we always start from the previous dm-block where we find m_{i-2,π_{i-1}^*} and traverse the linked lists to the right.

From the above discussion, we conclude that the time-complexity of Algorithm A is $O(n)$ in amortized sense.

5 Conclusions

In this paper, we presented an algorithm for finding a minimum cardinality dominating set for a permutation graph G. This algorithm exploits some subtle properties of this problem and runs in $O(n)$ time in amortized sense. Thus it is optimal. Our algorithm is easy to implement because doubly linked lists can be used.

References

1. M. Farber and J. M. Keil. Domination in permutation graphs. *Journal of Algorithms*, 6:309–321, 1985.
2. M. R. Garey and D. S. Johnson. *Computers and Intractability, A Guide to the Theory of NP-Completeness.* W. H. Freeman, San Francisco, CA, 1979.
3. M. C. Golumbic. *Algorithmic Graph Theory and Perfect Graphs.* Academic Press, New York, 1980.
4. Y. D. Liang, C. Rhee, S. K. Dhall, and S. Lakshmivarahan. A new approach for the domination problem on permutation graphs. *Information Processing Letters*, 37:219–224, 1991.
5. C. Rhee, Y. D. Liang, S. K. Dhall, and S. Lakshmivarahan. An $O(m+n)$ algorithm for finding minimum weight dominating set in permutation graphs. *SIAM Journal on Computing*, 25(2):404–419, 1996.
6. K. H. Tsai and W. L. Hsu. Fast algorithms for the dominating set problem on permutation graphs. *Algorithmica*, 9:601–614, 1993.

Similarity in Two-Dimensional Strings

Ricardo A. Baeza-Yates

Depto. de Ciencias de la Computación
Universidad de Chile
Casilla 2777, Santiago, Chile
E-mail: rbaeza@dcc.uchile.cl

Abstract. In this paper we discuss how to compute the edit distance (or similarity) between two images. We present new similarity measures and how to compute them. They can be used to perform more general two-dimensional approximate pattern matching. Previous work on two-dimensional approximate string matching either work with only substitutions or a restricted edit distance that allows only some type of errors.

1 Introduction

A number of important problems related to string processing lead to algorithms for approximate string matching: text searching, pattern recognition, computational biology, audio processing, etc. Two-dimensional pattern matching with errors has applications on computer vision.

The *edit distance* between two strings a and b, $ed(a, b)$, is defined as the minimum number of *edit operations* that must be carried out to make them equal. The allowed operations are insertion, deletion and substitution of characters in a or b. The problem of *approximate string matching* is defined as follows: given a *text* of length n, and a *pattern* of length m, both being sequences over an alphabet Σ of size σ, find all segments (or "occurrences") in *text* whose edit distance to *pattern* is at most k, where $0 < k < m$. The classical solution is $O(mn)$ time and involves dynamic programming [14].

Krithivasan and Sitalakshmi (KS) [11] proposed the following extension of edit distance for two dimensions. Given two images of the same shape, the edit distance is the sum of the edit distance of the corresponding row images. This definition is justified when the images are transmitted row by row and there are not too many communication errors. However, for many other problems, this distance does not reflect well simple cases of approximate matching in different settings. For example, we could have a match that only has the middle row of the pattern missing. In the definition above, the edit distance would be $O(m^2)$ if all pattern rows are different. Intuitively, the right answer should be at most $2m$, because only m characters were deleted in the pattern and m characters are inserted at the bottom. In this paper we extend the edit distance to two dimensions lifting the problem just mentioned and also extending the edit distance to images of different shapes.

This paper is organized as follows. First we discuss previous work on two-dimensional pattern matching with errors and image similarity. Next, we introduce new notions of similarity between two-dimensional strings or images. As for the one-dimensional counterpart, we consider first the comparison of two images and we also discuss how to compute efficiently these new similarity measures.

Then, we look at the problem of finding approximate matches with at most k errors of a rectangular pattern image of size $m \times m$ in a larger rectangular image (the text) of size $n \times n$. We present a $O(m^3n^2)$ worst case algorithm, and a $O(n^2k \log_\sigma m / m^2)$ for k up to $O(m^2/\log m)$ where σ denotes the size of the (finite) alphabet, for two of the new measures that we define. We end by discussing possible extensions and open problems. We denote by σ the size of the (finite) alphabet.

2 Previous Work

Two-dimensional approximate string matching usually considers only substitutions for rectangular patterns, which is much simpler than the general case with insertions and deletions (because in this case, rows and/or columns of the pattern can match pieces of the text of different length). For substitutions, the pattern shape matches the same shape in the text.

If we consider matching the pattern with at most k substitutions, one of the best results on the worst case is due to Amir and Landau [2] achieving $O((k+\log\sigma)n^2)$ time but using $O(n^2)$ space. A similar algorithm is presented in Crochemore and Rytter [5]. Ranka and Heywood [13], on the other hand, solve the problem in $O((k+m)n^2)$ time and $O(kn)$ space. Amir and Landau also present a different algorithm running in $O(n^2 \log n \log\log n \log m)$ time. On average, the best algorithm is due to Karkkäinen and Ukkonen [9], with its analysis and space usage improved by Park [12]. The expected time is $O((n^2k\log_\sigma m)/m^2)$ for

$$k \leq \left\lfloor \frac{m}{\lceil \log_\sigma(m^2) \rceil} \right\rfloor \frac{m}{2} - 1 \approx \frac{m^2}{4\log_\sigma m}$$

using $O(m^2)$ space ($O(k)$ space on average). This time result is optimal for the expected case.

Under the KS definition, Krithivasan [10] presents an $O(m(k+\log m)n^2)$ algorithm that uses $O(mn)$ space. This was improved (for $k < m$) by Amir and Landau [2] to $O(k^2n^2)$ worst case time using $O(n^2)$ space. Amir and Farach [1] also considered non-rectangular patterns achieving $O(k(k+\sqrt{m\log m}\sqrt{k\log k})n^2)$ time. This algorithm is very complicated and non-practical because it uses numerical convolutions.

Very recently, Baeza-Yates and Navarro [4] obtained the first fast algorithm on average for the KS model. They use a filter algorithm based in multiple approximate string matching, achieving $O(n^2k\log_\sigma m / m^2)$ average-case behavior for $k < m(m+1)/(5\log_\sigma m)$, and using $O(m^2)$ space. This time matches the best known result for the same problem allowing just substitutions and is optimal [9], being the upper bound on k only a bit smaller. For higher error levels, they present an algorithm with time complexity $O(n^2k/(w\sqrt{\sigma}))$ (where w is the size in bits of the computer word), which works for $k < m(m+1)(1-e/\sqrt{\sigma})$.

Another related problem is geometric matching, where we have to match a geometric figure or a set of points. In this case, the problem is in a continuous space rather than a discrete space and usually the Hausdorff measure [3] is used.

There are other approaches to matching images, which are very different to our approach (which belongs to what is called combinatorial pattern matching). Among them we can mention techniques used in pattern matching related to artificial intelligence (for example image processing and neural networks [15]) and techniques used in databases (extracting features of the image like color histograms [6, 7]).

3 Extending the Edit Distance

Let a and b be two images of size $nr \times nc$ and $mr \times mc$ respectively. In the sequel we use $row_i(a)$ to denote the i-th row of a and $col_i(a)$ to denote the i-th column of a. For example, the KS distance is given by

$$KS(a, b) = \sum_{i=1}^{nr} ed(row_i(a), row_i(b))$$

with the restriction that $nr = mr$. We also use the L-shape idea of Giancarlo [8] used for extending suffix trees to two dimensions. We denote by $LS_{i,j}(a)$ the L-shaped string consisting of the first (left) j elements of the i-th row and the first (top) $i - 1$ elements of the j-th column.

Because our main motivation is approximate matching, we assume that the pattern and a text subimage are compared from top to bottom and from left to right. That is, a sub-image can be enlarged by extending it the bottom or/and the right side. It can be argued that a pattern can match better fixing a different corner, but this does not make any difference, because that only changes the text position where the match will be reported, and still only one match is found. Another convention is that the text occurrence must have the same shape of the pattern. Otherwise, we may have occurrences that have at most k errors that basically do not count unmatched characters on the boundaries, which is not fair. Hence, although our similarity measures work for two images of different size, they will be used later for subimages in the text that have the same shape as the pattern.

First, we solve the limitation of the KS model to handle deletions or insertions of whole rows. The solution is simple, we just treat each row as a single string which is compared to other rows using the normal edit distance (that is, only one dimension). If $R_{i,j}$ is the distance between rows 1 and i of image a and rows 1 and j of image b, we have that

$$R_{i,j} = \min(R_{i-1,j} + nc, R_{i,j-1} + mc, R_{i-1,j-1} + ed(row_i(a), row_j(b)))$$

where the boundary conditions are $R_{i,0} = i \cdot nc$ and $R_{0,j} = j \cdot mc$, and the distance between the two images is given by $R(a, b) = R_{nr,mr}$.

In the example given in the introduction, the distance is reduced to less or equal than $2m$ instead of being $O(m^2)$ as in the KS model. Similarly, we could use columns instead of rows, obtaining $C(a,b)$. This model is much more fair than the KS model. Although we use rectangular images, this measure also works for any images where rows are connected and continuous.

Generalizing this idea to insertions and deletions at the same time in rows and/or columns is not as simple. Suppose that we have two subimages that we want to compare. One alternative is to decompose the border of a subimage in rows or columns. Then we can use the following decompositions:

1. removing one row or one column from one of the subimages or
2. removing one row or one column in the same side of each subimage and computing the edit distance between them.

We can apply dynamic programming to find the best possible decomposition. That is, if $RC_{i,j,k,\ell}$ is the distance between the left-top corner of a bounded by row i and column j and the left-top corner of b bounded by row k and column ℓ, we have that $RC_{i,j,k,\ell}$ is the minimum of the following values:

- $RC_{i-1,j,k,\ell} + j$, $RC_{i,j-1,k,\ell} + i$, $RC_{i,j,k-1,\ell} + \ell$, and $RC_{i,j,k,\ell-1} + k$, which corresponds to deleting one row or column in one sub-image; and
- $RC_{i-1,j,k-1,\ell} + ed(pref_j(row_i(a)), pref_\ell(row_k(b))))$ and $RC_{i,j-1,k,\ell-1} + ed(pref_i(col_j(a)), pref_k(col_\ell(b))))$ where $pref_i(s)$ denotes the first i characters of string s (a prefix) and which corresponds to comparing two rows at the bottom or two columns at the right.

The boundary conditions are $RC_{0,0,i,j} = RC_{i,j,0,0} = i \cdot j$. The distance $RC(a,b)$ is given by $R_{nr,nc,mr,mc}$. Figure 1 shows all these cases. This distance can also be applied to any convex image, for example circles or other regular polygons.

Nevertheless, this distance does not handle cases where we want to change at the same time a row and a column (for example, a motivation could be scaling). For that we use the L-shape mentioned earlier. So, we can also decompose the border of a subimage using L-shapes and we can have the same extensions as for rows or columns. To compare two L-shapes we see them as two one-dimensional strings. Then we have the following cases to find the minimal decomposed distance:

- $L_{i-1,j-1,k,\ell} + i + j - 1$ and $L_{i,j,k-1,\ell-1} + k + \ell - 1$ which corresponds to removing an L-shape in a subimage; and
- $L_{i-1,j-1,k-1,\ell-1} + ed(LS_{i,j}(a), LS_{k,\ell}(b)))$ which corresponds to comparing two L-shapes.

The boundary conditions are the same as the RC measure and the final distance is similarly given by $L(a,b) = L_{nr,nc,mr,mc}$. Figure 1 shows the decompositions associated to L. We will see later that this definition can be simplified by using the fact that one row and one column are considered at the same time when using L-shapes.

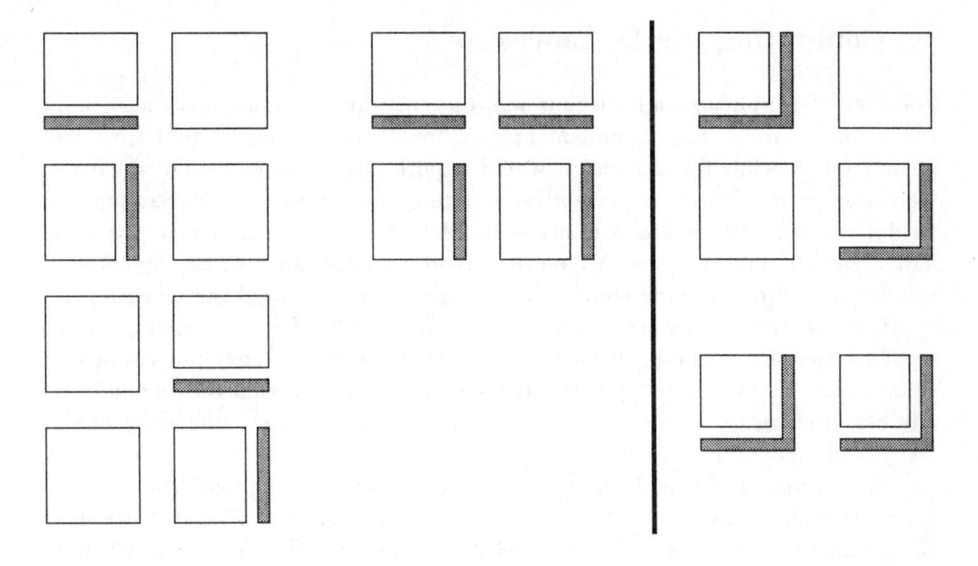

Fig. 1. Decomposition used in RC (left, 6 cases) and L (right, 3 cases).

Finally, we can have a general distance $All(a,b)$ that uses both decompositions at the same time (RC and L) computing the minimal value of all possible cases. It is easy to show that $KS(a,b) \geq R(a,b) \geq RC(a,b) \geq All(a,b)$ and that $L(a,b) \geq All(a,b)$ because each case is a subset of the next. On the other hand, there are cases where $RC(a,b)$ will be less than $L(a,b)$ and vice versa. In fact, Figure 2 this is shown together with other examples, where each color is a different symbol. The last example shows that combining RC and L can actually lead to a distance less than each separate case.

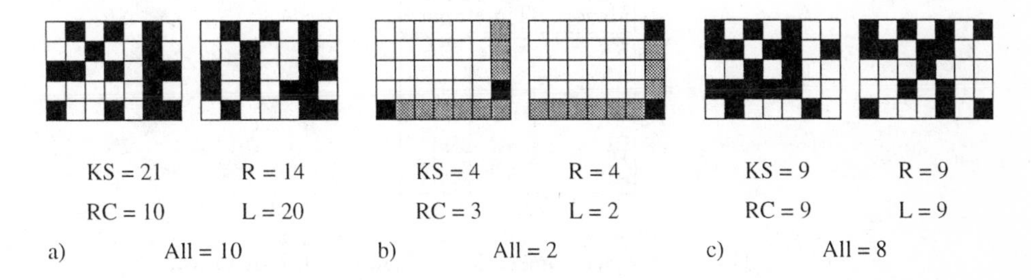

Fig. 2. Three examples for our new measures.

4 Computing the Distances

For sake of simplicity and without loss of generality assume that $nr = nc = mr = mc = m$. A direct implementation for R would take $O(m^4)$ time and $O(m^2)$ space, while for RC and L would require $O(m^6)$ time and $O(m^4)$ space. The later, in particular, is prohibitive even for small images. However, this can be done better. The space is easily reduced to $O(m)$ for R and $O(m^3)$ to RC and L by noticing that we only need to store the boundary of the matrices of the dynamic programming computation as they are computed incrementally.

However, the computation of L can be simplified further by noticing that to compute the best decomposition $i - j$ and $k - \ell$ are always constant (in fact, is equal to $nr - nc$, which for squares images is 0). This means that only a quadratic number of entries must be computed, which implies a matrix boundary of size $O(m)$ and a running time of $O(m^4)$.

We can also improve the computation of RC by noticing that the edit distances of prefixes involved can also be computed incrementally. That is, we store the distances between each two last prefixes computed and we use them when we extend the distance between two prefixes by one or two characters. This needs additional space which matches the improved space bound mentioned before. Therefore, each edit distance needed in the dynamic programming computation can be computed in constant time, reducing the total time to $O(m^4)$ for RC, and hence for *All* considering that L requires the same time. These optimizations allows to handle patterns of reasonable size (say up to 50×50).

Table 1 summarizes the space and time complexity obtained for all the measures, including KS. We can see that our measures need only one order of magnitude more time with respect to KS, using the same space, except for RC and *All*.

Measure	Time	Space
KS	m^3	m
R, C	m^4	m
L	m^4	m
RC	m^4	m^3
All	m^4	m^3

Table 1. Time and space complexity to compute the distance.

5 Two Dimensional Approximate Pattern Matching

Now we discuss approximate two-dimensional pattern matching. In this section, image a is the text, and image b is the pattern. We use k to denote the maximum number of errors allowed. For simplicity we use $nc = nr = n$ and $mc = mr = m$.

The straight forward technique to search an image in a larger image is by scanning sequentially the larger image (the text) and computing its distance to the pattern. For R, we will scan the text from top to bottom, using m columns each time, from column k to $k + mc - 1$, starting with $k = 1$ and ending with $k = n - m + 1$. For each k, we use the same dynamic programming formulas, changing the boundary conditions to allow the pattern to match in any possible row i. That is, we use $R_{i,0} = 0$ and $R_{0,j} = j \cdot mc$. We report a match whenever we have at most k errors and we have considered all the rows of the pattern. That is, when $R_{i,mr} \leq k$. For each scan, the time complexity is $O(nm^3)$ and we repeat that n times. Therefore, the total worst case time is $O(n^2m^3)$, using $O(m)$ extra space. This algorithm also works for C by exchanging rows and columns.

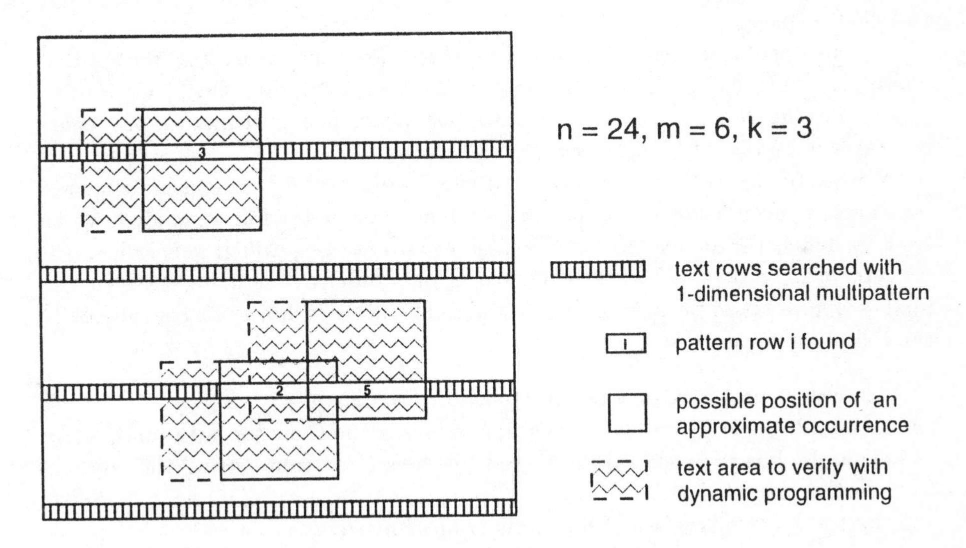

Fig. 3. Filtering algorithm to find potential matches.

Now, we consider a fast algorithm on average. For the R (or C) measures we can use the same fast expected time algorithm of Baeza-Yates and Navarro [4]. This algorithm uses a filter that searches all the pattern rows (or columns) with a multiple approximate string matching algorithm to find potential matching areas (see Figure 3). We only change the verification phase (each potential area found by the filter must be verified using the $O(n^2m^3)$ worst case searching algorithm described before) by using R (or C) to compute the distance in a fixed area (see Figure 3). Because this algorithm works up to certain threshold level up to where the verification time is negligible, the fact that computing R (or C) takes $O(m)$ the time of computing KS, will only change the maximal k for which the algorithm performs well. Because the maximal k depends on the inverse of the logarithm of the verification algorithm plus two, the maximal k changes from $m(m+1)/(5\log_\sigma m)$ to $k < m(m+1)/(7\log_\sigma m)$.

Therefore, we obtain the same average-case time bound of $O(n^2 k \log_\sigma m\ /m)$ and almost the same error level bound for which the expected time result is valid. That is, $k < m(m+1)/(7\log_\sigma m)$, using $O(m)$ extra space. This expected time is optimal [9]. How to extend these results to the other measures is still an open problem.

6 Concluding Remarks

Our measures can be easily extended to more dimensions. For d-dimensions we use $(d-1)$-dimensional strings for the decompositions. The only drawback is that the number of cases grows exponentially with the number of dimensions. Then, computing $All(a,b)$ for d-dimensional strings would require $O(2^d n^{2d})$ time and $O(n^d)$ space.

An open problem is to design optimal worst-case time algorithms for approximate searching using the new measures. That is, achieving $O(n^4)$ time complexity for the R measure. Also, finding fast filtering algorithms for the other measures is matter of future research.

Neither of the new measures defined can handle scaling transformations nor rotations. A more realistic distance can be defined using the following idea, which tries to define the *largest common image* of two images, which generalizes the concept of longest common subsequence of one-dimensional strings. Given two images, find a set of position pairs that match exactly in both images subject to the following restrictions:

1. The set of positions for the same pattern are disjoint;
2. a suitable order given by the position values is the same for both images (for example, image pixels can be sorted by their $i+j$ value, using the value of i in the case of ties); and
3. the total size of the set of positions is maximized.

For the edit distance, condition 3 has to be changed to:

3. Minimize the number of mismatches, insertions and deletions needed to obtain the set of matching positions.

This model may match a rotated pattern, because no corner is fixed. Figure 4 gives an example. All pieces of the pattern not in the text corresponds to deletions and mismatches and should be counted. In the text, black regions are not counted, because correspond to mismatches. All other pieces are insertions in the pattern. It is not clear that the minimal string editing solution gives the same answer as the largest common set of sub-images. Also, it could be argued

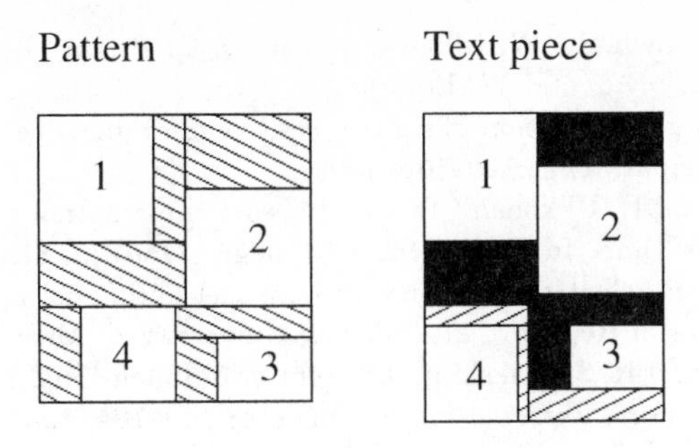

Fig. 4. Example of largest common image.

that characters inserted/deleted on external borders should not be counted as errors.

The approximate two-dimensional pattern matching problem can be stated as usual using the above definition as searching for all rectangular subimages of the text that have edit distance at most k with the pattern. An alternative definition would be to find all pieces of the text that have at least m^2-k matching positions with the pattern.

Acknowledgements

We would like to thanks the comments of Gonzalo Navarro, Ladan Rostami, Sven Schuerier (which in particular pointed out the simplification for computing the L measure), and the unknown referees. This work was supported by Fondecyt Grant 95-0622.

References

1. A. Amir and M. Farach. Efficient 2-dimensional approximate matching of non-rectangular figures. In *Proc. SODA'91*, pages 212–223, San Francisco, CA, Jan 1991.
2. A. Amir and G. Landau. Fast parallel and serial multidimensional approximate array matching. *Theoretical Computer Science*, 81:97–115, 1991. Also as report CS-TR-2288, Dept. of Computer Science, Univ. of Maryland, 1989.
3. M. Atallah. A linear time algorithm for the Hausdorff distance between convex polygons, *Information Processing Letters* 17, 207-209, 1983.
4. R. Baeza-Yates and G. Navarro. Fast two-dimensional approximate string matching. In *LATIN'98*, Campinas, Brazil, April 1998. Springer-Verlag LNCS 1380, edited by C. Lucchesi and A. Moura, 341-351.
5. M. Crochemore and W. Rytter. *Text Algorithms*. Oxford University Press, Oxford, UK, 1994.
6. C. Faloutsos, R. Barber, M. Flickner, J. Hafner, W. Niblack, D. Petkovic, and W. Equitz. Efficient and effective querying by image content. *J. of Intelligent Information Systems*, 3:231-262, 1994.

7. M. Flickner, H. Sawhney, W. Niblack, J. Ashley, Q. Huang, B. Dom, M. Gorkani, J. Hafner, D. Petkovic, and W. Equitz.
8. R. Giancarlo. A generalization of suffix trees to square matrices, with applications. *SIAM J. on Computing*, 24:520–562, 1995.
9. J. Karkkäinen and E. Ukkonen. Two and higher dimensional pattern matching in optimal expected time. In *Proc. SODA'94*, pages 715–723. SIAM, 1994.
10. K. Krithivasan. Efficient two-dimensional parallel and serial approximate pattern matching. Technical Report CAR-TR-259, University of Maryland, 1987.
11. K. Krithivasan and R. Sitalakshmi. Efficient two-dimensional pattern matching in the presence of errors. *Information Sciences*, 43:169–184, 1987.
12. K. Park. Analysis of two dimensional approximate pattern matching algorithms. In *Proc. CPM'96*, LNCS 1075, pages 335–347, 1996.
13. S. Ranka and T. Heywood. Two-dimensional pattern matching with k mismatches. *Pattern recognition*, 24(1):31–40, 1991.
14. P. Sellers. The theory and computation of evolutionary distances: pattern recognition. *J. of Algorithms*, 1:359–373, 1980.
15. P. Suetens, P. Fua, and A. Hanson. Computational Strategies for Object Recognition, *ACM Computing Surveys*, 24:5-62, 1992.

On Multi-dimensional Hilbert Indexings

Jochen Alber and Rolf Niedermeier*

Wilhelm-Schickard-Institut für Informatik, Universität Tübingen,
Sand 13, D-72076 Tübingen, Fed. Rep. of Germany,
alber/niedermr@informatik.uni-tuebingen.de

Abstract. Indexing schemes for grids based on space-filling curves (e.g., Hilbert indexings) find applications in numerous fields. Hilbert curves yield the most simple and popular scheme. We extend the concept of curves with Hilbert property to arbitrary dimensions and present first results concerning their structural analysis that also simplify their applicability. As we show, Hilbert indexings can be completely described and analyzed by "generating elements of order 1", thus, in comparison with previous work, reducing their structural complexity decisively.

1 Introduction

Discrete multi-dimensional spaces are of increasing importance. They appear in various settings such as combinatorial optimization, parallel processing, image processing, geographic information systems, data base systems, and data structures. In many applications it is necessary to number the points of a discrete multi-dimensional space (or, equivalently, a grid) by an indexing scheme mapping each point bijectively to a natural number in the range between 1 and the total number of points in the space. Often it is desirable that this indexing scheme preserves some kind of locality, that is, close-by points in the space are mapped to close-by numbers or vice versa. For this purpose, indexing schemes based on space-filling curves have shown to be of high value [4–9].

In this paper we study Hilbert indexings, perhaps the most popular space-filling indexing schemes. Properties of 2D and 3D Hilbert indexings have been extensively studied recently [4–10]. However, most of the work so far has focused on empirical studies. Up to now, little attention has been paid to the theoretical study of structural properties of multi-dimensional Hilbert curves, the focus of this paper. Whereas with "modulo symmetry" there is only one 2D Hilbert curve, there are many possibilities to define Hilbert curves in the 3D setting [4, 9]. The advantage of Hilbert curves is their (compared to other curves) simple structure.

Our results can shortly be sketched as follows. We generalize the notion of Hilbert indexings to arbitrary dimensions. We clarify the concept of Hilbert curves in multi-dimensional spaces by providing a natural and simple mathematical formalism that allows combinatorial studies of multi-dimensional Hilbert

* Work partially supported by a Feodor Lynen fellowship of the Alexander von Humboldt-Stiftung, Bonn, and the Center for Discrete Mathematics, Theoretical Computer Science, and Applications (DIMATIA), Prague.

indexings. For reasons of (geometrical) clearness, we base our formalism on permutations instead of e.g. matrices or other formalisms [2–4, 10]. So we obtain the following insight: Space-filling curves with Hilbert property can be completely described by simple generating elements and permutations operating on them. Structural questions for Hilbert curves in arbitrary dimensions can be decided by reducing them to basic generating elements. Putting it in catchy terms, one might say that for Hilbert indexings what holds "in the large" (i.e., for large side-length), can already be detected "in the small" (i.e., for side-length 2). In particular, this provides a basis for mechanized proofs of locality of curves with Hilbert property (cf. [9]). In addition, this observation allows the identification of seemingly different 3D Hilbert indexings [4], the generalization of a locality result of Gotsman and Lindenbaum [6] to a larger class of multi-dimensional indexing schemes, and the determination that there are exactly $6 \cdot 2^8 = 1536$ structurally different 3D Hilbert curves. The latter clearly generalizes and answers Sagan's quest for describing 3D Hilbert curves [10]. Finally, we provide an easy recursive formula for computing Hilbert indexings in arbitrary dimensions and sketch a recipe for how to construct an r-dimensional Hilbert curve for arbitrary r in an easy way from two $(r-1)$-dimensional ones. Some missing details and proofs can be found in the full version of the paper [1].

2 Preliminaries

We focus our attention on cubic grids, where, grid of side-length n. An r-dimensional *(discrete) curve* C is simply a bijective mapping $C : \{1, \ldots n^r\} \rightarrow \{1, \ldots, n\}^r$. Note that, by definition, we do not claim the continuity of a curve. A curve C is called *continuous* if it forms a Hamilton path through the n^r grid points. An r-dimensional cubic grid is said to be of *order* k if it has side-length 2^k. Analogously a curve C has order k if its range is a cubic grid of order k.

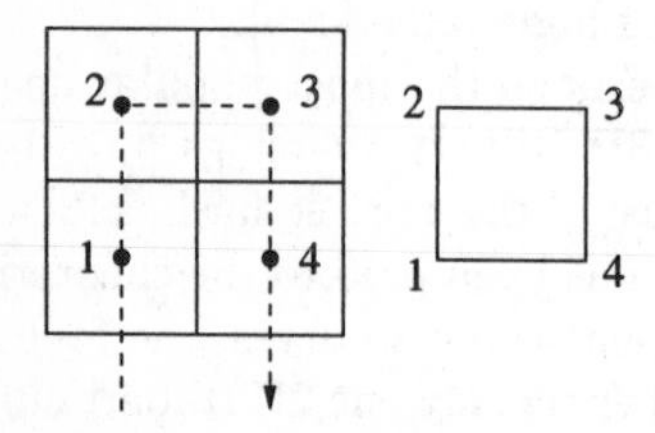

Fig. 1. The generator Hil_1^2 and its canonical corner-indexing $\widetilde{\mathrm{Hil}_1^2}$.

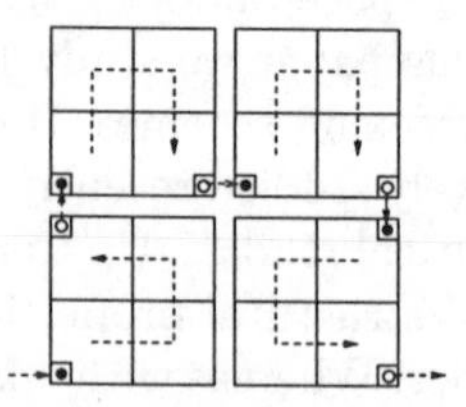

Fig. 2. Construction scheme for the 2D Hilbert indexing.

Fig. 1 shows the smallest 2D continuous curve indexing a grid of size 4. This curve can be found in Hilbert's original work (see [11]) as a constructing unit for a whole family of curves. Fig. 2 shows the general construction principle for these so-called Hilbert curves: For any $k \geq 1$ four Hilbert indexings of size 4^k

are combined into an indexing of size 4^{k+1} by rotating and reflecting them in such a way that concatenating the indexings yields a Hamilton path through the grid. One of the main features of the Hilbert curve is its "self-similarity". Here "self-similar" shall simply mean that the curve can be generated by putting together identical (basic construction) units, only applying rotation and reflection to these units. In a sense, the Hilbert curve is the "simplest" self-similar, recursive, locality-preserving indexing scheme for square meshes of size $2^k \times 2^k$.

3 Formalizing Hilbert curves in r dimensions

In this section, we generalize the construction principle of 2D Hilbert curves to arbitrary dimensions in a rigorous, mathematically precise way.

3.1 Classes of Self-Similar Curves and their generators

Let $V_r := \{x_1 x_2 \cdots x_{r-1} x_r \mid x_i \in \{0,1\}\}$ be the set of all 2^r corners of an r-dimensional cube coded in binary. Moreover, let $\mathcal{I} : V_r \longrightarrow \{1, \ldots, 2^r\}$ denote an arbitrary indexing of these corners. To describe the orientation of subcurves inside a curve of higher order, we want to use symmetry mappings, which can be expressed via suitable permutations operating on such corner-indexings. Observe that any r-dimensional curve C_1 of order 1 naturally induces an indexing of these corners (see Fig. 1 and Fig. 3). We call the obtained corner-indexing the *canonical* one and denote it by $\widetilde{C_1} : V_r \longrightarrow \{1, \ldots, 2^r\}$. Furthermore, let $W_{\mathcal{I}}$ denote the group of all permutations (operating on $\mathcal{I}$) that describe rotations and reflections of the r-dimensional cube. In other words, $W_{\mathcal{I}}$ is the set of all permutations that preserve the neighborhood-relations $n(i,j)$ of the corner indexing $\mathcal{I}$:

$$W_{\mathcal{I}} := \{\pi \in Sym(2^r) \ : \ n(i,j) = n(\pi(i), \pi(j)) \ \forall i,j \in \{1, \ldots, 2^r\}\}.$$

For a given permutation $\tau \in W_{\mathcal{I}}$, we sometimes write $(\tau : \mathcal{I})$ in order to emphasize that τ is operating on a cube with corner-indexing $\mathcal{I}$. The point here is that once we have fixed a certain corner-indexing $\mathcal{I}$, the set $W_{\mathcal{I}}$ will provide all necessary transformations to describe a construction principle of how to generate curves of higher order by piecing together a suitable curve of lower order. Obviously each permutation $(\tau : \mathcal{I})$ acting on a given corner-indexing $\mathcal{I}$ canonically induces a bijective mapping on a cubic grid of order k. Subsequently, we do not distinguish between a permutation and the corresponding mapping on a grid.

We partition an r-dimensional cubic grid of order k into 2^r subcubes of order $k-1$. For each $x_1 \cdots x_r \in V_r$ we therefore set

$$p^{(k)}_{(x_1 \cdots x_r)} := (x_1 \cdot 2^{k-1}, \ldots, x_r \cdot 2^{k-1}) \in \{0, \ldots, 2^k - 1\} \times \ldots \times \{0, \ldots, 2^k - 1\}$$

to be the "lower-left corner" of such a subcube. Let C_{k-1} be an r-dimensional curve of order $k-1$ ($k \geq 2$). Our goal is to define a "self-similar" curve C_k of order k by putting together 2^r pieces of type C_{k-1}. Let $\mathcal{I} : V_r \longrightarrow \{1, \ldots, 2^r\}$ be a corner-indexing. We intend to arrange the 2^r subcurves of type C_{k-1} "along"

$\mathcal{I}$. The position of the i'-th (where $i' \in \{1, \dots, 2^r\}$) subcurve inside C_k can formally be described with the help of the grid-points $p^{(k)}_{(x_1 \cdots x_r)}$. Bearing in mind the classical construction principle for the 2D Hilbert indexing, the orientation of the constructing curve C_{k-1} inside C_k can be expressed by using symmetric transformations (that is reflections and rotations). For any sequence of permutations $\tau_1, \dots, \tau_{2^r} \in W_{\mathcal{I}}$ we therefore define

$$C_k(i) := (\tau_{i'} : \mathcal{I}) \circ C_{k-1}\left(i \bmod (2^{k-1})^r\right) + p^{(k)}_{\mathcal{I}^{-1}(i')}, \tag{1}$$

where $i \in \{1, \dots, (2^k)^r\}$ and $i' = (i-1) \text{ div } (2^{k-1})^r + 1$. The geometric intuition behind is that the curve C_k can be partitioned into 2^r components of the form C_{k-1} (reflected or rotated in a suitable way). These subcurves are arranged inside C_k "along" the given corner-indexing $\mathcal{I}$. The orientation of the i'-th subcurve inside C_k is described by the effect of $\tau_{i'}$ operating on $\mathcal{I}$.

Definition 1. Whenever two r-dimensional curves C_{k-1} of order $k-1$ and C_k of order k satisfy equation (1) for a given sequence of permutations $\tau_1, \dots, \tau_{2^r} \in W_{\mathcal{I}}$ (operating on the corner-indexing $\mathcal{I} : V_r \longrightarrow \{1, \dots, 2^r\}$), we will write $C_{k-1} \overset{\mathcal{I}}{_{(\tau_1, \dots, \tau_{2^r})}\ll} C_k$ and call C_{k-1} the *constructor* of C_k.

Our final goal is to iterate this process starting with a curve C_1 of order 1. It's only natural and in our opinion "preserves the spirit of Hilbert" to fix the corner-indexing according to the structure of the defining curve C_1. Hence, in this situation we can specify our $\mathcal{I}$ to be the canonical corner-indexing $\widetilde{C_1}$. By successively repeating the construction principle in equation (1) k times, we obtain a curve of order k.

Definition 2. Let $\mathcal{C} = \{\, C_k \mid k \geq 1 \,\}$ be a family of r-dimensional curves of order k. We call $\mathcal{C}$ a *Class of Self-Similar Curves (CSSC)* if there exists a sequence of permutations $\tau_1, \dots, \tau_{2^r} \in W_{\widetilde{C_1}}$ (operating on the canonical corner-indexing $\widetilde{C_1}$) such that for each curve C_k it holds that

$$C_1 \overset{\widetilde{C_1}}{_{(\tau_1, \dots, \tau_{2^r})}\ll} C_2 \overset{\widetilde{C_1}}{_{(\tau_1, \dots, \tau_{2^r})}\ll} \cdots \overset{\widetilde{C_1}}{_{(\tau_1, \dots, \tau_{2^r})}\ll} C_{k-1} \overset{\widetilde{C_1}}{_{(\tau_1, \dots, \tau_{2^r})}\ll} C_k.$$

In this case, C_1 is called the *generator of the CSSC* $\mathcal{C}$ and we define the set $\mathcal{H}(C_1, (\tau_1, \dots, \tau_{2^r})) := \{\, C_k \mid k \geq 1 \,\}$ to be the CSSC generated by C_1 and $\tau_1, \dots, \tau_{2^r}$. A CSSC $\mathcal{C} = \{\, C_k \mid k \geq 1 \,\}$ is called *Class with Hilbert Property (CHP)* if all curves C_k are continuous.

Note that the CSSC $\mathcal{H}(C_1, (\tau_1, \dots, \tau_{2^r}))$ is well-defined, because any CSSC is uniquely determined by its generator C_1 and the choice of the permutations $\tau_1, \dots, \tau_{2^r} \in W_{\widetilde{C_1}}$. Our concept for multi-dimensional CHPs only makes use of the very essential tools which can be found in Hilbert's context (cf. [11]) as rotation and reflection. We deliberately avoid more complicated structures (e.g., the use of different sequences of permutations in each inductive step, or the use of several generators for the constructing principle) in order to maintain conceptual simplicity and ease of construction and analysis. However, the theory which we develop in this paper doesn't necessarily restrict to the continuous case. We end this subsection with an example. One easily checks that the classical 2D Hilbert indexing can be described via $\mathcal{H}(\,\mathrm{Hil}_1^2,\ ((2\,4),\mathrm{id},\mathrm{id},(1\,3))\,) = \{\,\mathrm{Hil}_k^2\ \mid\ k \geq 1\,\}$, where the generator Hil_1^2 is given in Fig. 1. As Theorem 2 will show, this is the only CHP of dimension 2 "modulo symmetry."

3.2 Disturbing the generator of a CSSC

In this subsection we analyze the effects of disturbing the generator of a CSSC by a symmetric mapping. We will see that any disturbance of the generator will be hereditary to the whole CSSC in a very canonical way. And also the other way round: if two different CSSCs show a certain similarity in one of their members, this similarity can already be found in the structure of the corresponding generators. We illustrate this by the following diagram. Given two CSSCs $\mathcal{H}(\,C_1,\,(\tau_1,\dots,\tau_{2^r})\,) = \{C_k \mid k \geq 1\}$ and $\mathcal{H}(\,D_1,\,(\tau_1,\dots,\tau_{2^r})\,) = \{D_k \mid k \geq 1\}$, respectively.[1] Suppose there is a similarity at a certain stage of the construction, i.e., for some k_0 the curves C_{k_0} and D_{k_0} can be obtained from each other by a similarity transformation Φ. The investigations in this section will show that the inner structure of CSSCs are strong enough to yield the same behavior at the stage of any order.

$$\begin{array}{ccccccc}
C_1 & {}_{(\tau_1,\dots,\tau_{2^r})}\!\overset{\widetilde{C_1}}{\ll} & C_2 & {}_{(\tau_1,\dots,\tau_{2^r})}\!\overset{\widetilde{C_1}}{\ll} & \cdots & {}_{(\tau_1,\dots,\tau_{2^r})}\!\overset{\widetilde{C_1}}{\ll} & C_k \\
\downarrow \Phi & \circlearrowleft & \downarrow \Phi & \circlearrowleft & \downarrow \Phi & \circlearrowleft & \downarrow \Phi \\
D_1 & {}_{(\tau_1,\dots,\tau_{2^r})}\!\overset{\widetilde{D_1}}{\ll} & D_2 & {}_{(\tau_1,\dots,\tau_{2^r})}\!\overset{\widetilde{D_1}}{\ll} & \cdots & {}_{(\tau_1,\dots,\tau_{2^r})}\!\overset{\widetilde{D_1}}{\ll} & D_k
\end{array}$$

Consequently, for issues like structural behavior, it will be sufficient to analyze the generating elements of a CSSC only, since we find all necessary information encoded here. We start with a simple observation concerning the behavior of the construction principle of Definition 1 under the "symmetric disturbance" of a constructor. We omit the proof.

Lemma 1. *Let C_{k-1} and C_k be curves of order $k-1$ and k, respectively. Suppose C_{k-1} is the constructor of C_k, i.e., $C_{k-1}\ {}_{(\tau_1,\dots,\tau_{2^r})}\!\overset{\mathcal{I}}{\ll}\ C_k$, for any sequence of permutations $\tau_1,\dots,\tau_{2^r} \in W_{\mathcal{I}}$ (acting on a given corner-indexing $\mathcal{I}$). Then for arbitrary $\phi \in W_{\mathcal{I}}$ we have*

$$(\phi : \mathcal{I}) \circ C_{k-1}\ {}_{(\tau_1\circ\phi^{-1},\dots,\tau_{2^r}\circ\phi^{-1})}\!\overset{\mathcal{I}}{\ll}\ C_k .$$

Whereas, by Lemma 1, we investigated the influence of disturbing the constructor, we now, in a second step, analyze how transforming the underlying

[1] Note that the τ's used in the definition of both CSSCs yield completely different automorphisms on the grid. Whereas in the first case they refer to the corner-indexing $\widetilde{C_1}$, in the second case they act on the corner-indexing $\widetilde{D_1}$, given by generator D_1.

corner-indexing influences the construction principle. We will need such a result, since two different CSSCs (by definition) come up with two different corner-indexings, each of which given by the underlying generator. Again we omit the proof.

Lemma 2. *Given the assumptions of Lemma 1 (that is: $C_{k-1} \;{}_{(\tau_1,\dots,\tau_{2^r})}\!\overset{\mathcal{I}}{\ll}\; C_k$ for two curves C_{k-1} and C_k of successive order), then for arbitrary $\phi \in W_{\mathcal{I}}$ and the modified corner-indexing $\mathcal{K} := \phi^{-1} \circ \mathcal{I}$ with $\Phi = (\phi : \mathcal{I}) = (\phi : \mathcal{K})$ we have*[2]

$$C_{k-1} \;{}_{(\tau_1\circ\phi,\dots,\tau_{2^r}\circ\phi)}\!\overset{\mathcal{K}}{\ll}\; \Phi \circ C_k \,.$$

Lemma 1 and 2 now allow the proof of the main result of this section. For its illustration we refer to the diagram at the beginning of this section. Do also recall the point made in the footnote there.

Theorem 1. *Let C_1 be the generator of the CSSC $\mathcal{H}(C_1, (\tau_1, \dots, \tau_{2^r})) = \{C_k \mid k \geq 1\}$ and D_1 the generator of the CSSC $\mathcal{H}(\, D_1, (\tau_1, \dots, \tau_{2^r})\,) = \{D_k \mid k \geq 1\}$. For an arbitrary permutation $\phi \in W_{\widetilde{C_1}}$ and the corresponding symmetric mapping $\Phi = (\phi : \widetilde{C_1}) = (\phi : \widetilde{D_1})$, the following statements are equivalent:*

(i) *$\Phi \circ C_{k_0} = D_{k_0}$ for some $k_0 \geq 1$.*
(ii) *$\Phi \circ C_k = D_k$ for all $k \geq 1$.*

Proof. (ii) $\Rightarrow$ (i) is trivial. For (i) $\Rightarrow$ (ii) we first show that statement (ii) is true for the generators C_1 and D_1: If $k_0 > 1$ we can divide the cubic grid of order k_0 into 2^r subgrids of order $k_0 - 1$. By the construction principle for CSSCs, the curves C_{k_0} and D_{k_0} traverse these subgrids "along" the canonical corner-indexings $\widetilde{C_1}$ resp. $\widetilde{D_1}$. Since, by assumption, $\Phi \circ C_{k_0} = D_{k_0}$, the corresponding relation also holds true for the corner-indexings $\widetilde{C_1}$ and $\widetilde{D_1}$, which finally yields the validity of the equation $\Phi \circ C_1 = D_1$, because of the isomorphisms $C_1 \simeq \widetilde{C_1}$ resp. $D_1 \simeq \widetilde{D_1}$. We proceed proving (ii) by induction on k. Assuming that $D_k = \Phi \circ C_k$ we show this relation for $k+1$ by applying Lemma 1 and Lemma 2. Since $\{C_k \mid k \geq 1\}$ is a CSSC, we get

$$C_k \;{}_{(\tau_1,\dots,\tau_{2^r})}\!\overset{\widetilde{C_1}}{\ll}\; C_{k+1} \overset{\text{Lemma 1}}{\Longrightarrow} \underbrace{\Phi \circ C_k}_{=D_k} \;{}_{(\tau_1\circ\phi^{-1},\dots,\tau_{2^r}\circ\phi^{-1})}\!\overset{\widetilde{C_1}}{\ll}\; C_{k+1}$$

$$\overset{\text{Lemma 2}}{\Longrightarrow} D_k \;{}_{(\tau_1,\dots,\tau_{2^r})}\!\overset{\widetilde{D_1}}{\ll}\; \Phi \circ C_{k+1}\,,$$

where the last relation makes use of $\widetilde{D_1} = \phi^{-1} \circ \widetilde{C_1}$, which we immediately obtain from the given equation $D_1 = \Phi \circ C_1$.[3] This implies $D_{k+1} = \Phi \circ C_{k+1}$ because of the CSSC-property of $\{D_k \mid k \geq 1\}$. □

In particular, the result of Theorem 1 implies that any questions concerning the structural similarity of two CSSCs can be reduced to the analysis of their generators.

---

[2] The fact that the corner-indexing is disturbed by ϕ^{-1} instead of ϕ is due to technical reasons only.

[3] A disturbance by Φ implies a transformation of the corner-indexings by ϕ^{-1}, which can be easily checked.

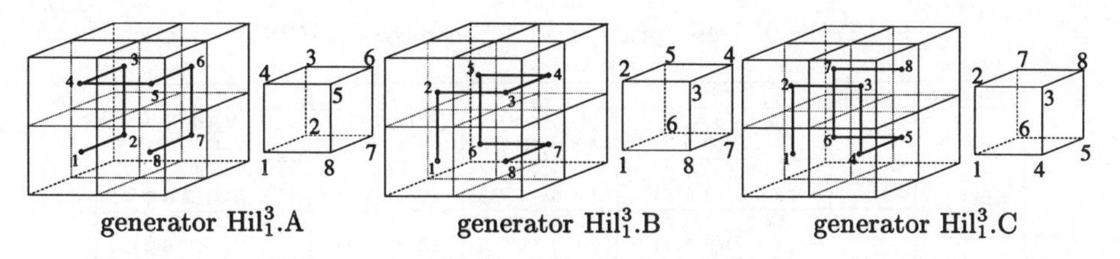

Fig. 3. Continuous 3D generators Hil^3_1.x and their canonical corner-indexings $\widetilde{\text{Hil}^3_1}$.x.

4 Applications: Computing and analyzing CHPs

First in this section, we attack a classification of all structurally different CHPs for higher dimensions. Whereas we can provide concrete combinatorial results for the 2D and 3D cases, the high-dimensional cases appear to be much more difficult. The basic tool for such an analysis, however, is given by Theorem 1. The following theorem justifies the naming "class with Hilbert property" (CHP).

Theorem 2. *The classical 2D Hilbert indexing* $\mathcal{H}(\,Hil^2_1,\ ((2\,4), id, id, (1\,3))\,)$ *is the only CHP of dimension 2 modulo symmetry.*

Proof. Due to Theorem 1 it suffices to show that Hil^2_1 is the only continuous 2D generator, which is obvious. In addition, we have to check whether there is another sequence of permutations such that 4 generators Hil^2_1 can be arranged in a grid of order 2 along the canonical corner-indexing $\widetilde{\text{Hil}^2_1}$ in a continuous way. A simple combinatorial consideration shows that no other sequence of permutations yields a continuous curve of order 2 whose starting- and endpoints are located at corners of the grid. However, any constructor for a continuous curve of higher order must have this property. □

What about the 3D case? The analysis of the "Simple Indexing Schemes" (which are related to our CHPs) in Chochia and Cole [4] already shows that the number of CHPs in the 3D case grows drastically compared to the 2D setting. However, by our analysis, lots of "Simple Indexing Schemes" in [4] now turn out to be identical modulo symmetry. We state the following classification-theorem, which treats the 3D case entirely. It also generalizes and answers work of Sagan [10].

Theorem 3. *For the 3D case there are* $6 \cdot 2^8 = 1536$ *structurally different (that is: not identical modulo reflection and rotation) CHPs. These types are listed in Table 1.*

Proof (Sketch). Theorem 1 says that we can restrict our attention to checking any continuous curves of order 1 which are different modulo symmetry. Given such a continuous generator C, the total amount of CHPs which can be constructed by C is given by all possibilities of piecing together 8 (rotated or reflected) versions of C ("subcurves") along its canonical corner-indexing $\widetilde{C}$. By exhaustive search, we get that there are 3 different (modulo symmetry) types

Table 1. Description of all 3-dimensional CHPS.

generator	version	τ_1	τ_2	τ_3	τ_4
Hil_1^3.A	(a)	(2 8)(3 5) / (2 4 8)(3 5 7)	(3 7)(4 8) / (2 8 4)(3 7 5)	(3 7)(4 8) / (2 8 4)(3 7 5)	(1 3)(6 8) / (1 3)(2 4)(5 7)(6 8)
	(b)	(2 8)(3 5) / (2 4 8)(3 5 7)	(3 7)(4 8) / (2 8 4)(3 7 5)	id / (2 4)(5 7)	(1 7 3)(4 6 8) / (1 7 5 3)(2 8 6 4)
	(c)	(2 8)(3 5) / (2 4 8)(3 5 7)	(3 7)(4 8) / (2 8 4)(3 7 5)	id / (2 4)(5 7)	(1 7)(4 6) / (1 7 5)(2 6 4)
	(d)	(2 8)(3 5) / (2 4 8)(3 5 7)	(3 7)(4 8) / (2 8 4)(3 7 5)	(3 7)(4 8) / (2 8 4)(3 7 5)	(1 3)(6 8) / (1 3)(2 4)(5 7)(6 8)
Hil_1^3.B	(a)	(2 8)(5 7) / (2 6 8)(3 5 7)	id / (2 6)(3 7)	(3 5)(6 8) / (2 8 6)(3 7 5)	(2 8)(5 7) / (2 6 8)(3 5 7)
	(b)	(2 8)(5 7) / (2 6 8)(3 5 7)	id / (2 6)(3 7)	(3 5)(6 8) / (2 8 6)(3 7 5)	(3 5)(6 8) / (2 8 6)(3 7 5)

generator	version	τ_5	τ_6	τ_7	τ_8
Hil_1^3.A	(a)	(1 3)(6 8) / (1 3)(2 4)(5 7)(6 8)	(1 5)(2 6) / (1 5 7)(2 4 6)	(1 5)(2 6) / (1 5 7)(2 4 6)	(1 7)(4 6) / (1 7 5)(2 6 4)
	(b)	(1 3 5)(2 6 8) / (1 3 5 7)(2 4 6 8)	id / (2 4)(5 7)	(1 5)(2 6) / (1 5 7)(2 4 6)	(1 7)(4 6) / (1 7 5)(2 6 4)
	(c)	(2 8)(3 5) / (2 4 8)(3 5 7)	id / (2 4)(5 7)	(1 5)(2 6) / (1 5 7)(2 4 6)	(1 7)(4 6) / (1 7 5)(2 6 4)
	(d)	(1 3 5)(2 6 8) / (1 3 5 7)(2 4 6 8)	id / (2 4)(5 7)	(1 5)(2 6) / (1 5 7)(2 4 6)	(1 7)(4 6) / (1 7 5)(2 6 4)
Hil_1^3.B	(a)	(1 3)(4 6) / (1 3 7)(2 4 6)	(1 3)(4 6) / (1 3 7)(2 4 6)	id / (2 6)(3 7)	(1 7)(2 4) / (1 7 3)(2 6 4)
	(b)	(1 7)(2 4) / (1 7 3)(2 6 4)	(1 3)(4 6) / (1 3 7)(2 4 6)	id / (2 6)(3 7)	(1 7)(2 4) / (1 7 3)(2 6 4)

of continuous generators, namely Hil_1^3.A, Hil_1^3.B and Hil_1^3.C (see Fig. 3). As described above, we now have to check whether there are continuous arrangements of these generators along their canonical corner-indexings. Beginning with type A, an exhaustive combinatorial search yields that there are 4 possible continuous formations of Hil_1^3.A along $\widetilde{\mathrm{Hil}_1^3}$.A. All possibilities are shown in Fig. 4, where the orientation of each subcube is given by the position of an edge (drawn in bold lines). For each subcube there are two symmetry mappings which yield possible arrangements for the generator within such a subgrid. The permutations expressing these mappings are listed in Table 1. Analogously, we find out the possible arrangements for generator type B. Note that there are no more than 2 different continuous arrangements of this generator along its canonical corner-indexing. Finally we easily check that Hil_1^3.C cannot even be the constructor of a continuous curve of order 2. Table 1 thus yields that there are exactly $4 \cdot 2^8 + 2 \cdot 2^8 = 6 \cdot 2^8$ structurally different CHPs. □

Construction of an r-dimensional Hilbert curve. Without giving an explicit proof here, we just indicate how the construction of a high-dimensional CHP can be done inductively: A continuous generator of dimension r can be derived simply by "joining together" two continuous generators of dimension $r-1$. As an example we give a CHP of dimension 4, whose generator Hil_1^4 is constructed by joining together two generators Hil_1^3, version (a) (cf. Figure 3). The generator Hil_1^4 and a suitable sequence of permutations are shown in Fig. 5. Note that this construction principle can be extended to obtain Hilbert indexings in arbitrary dimensions in an expressive and easy, constructive way: Following the

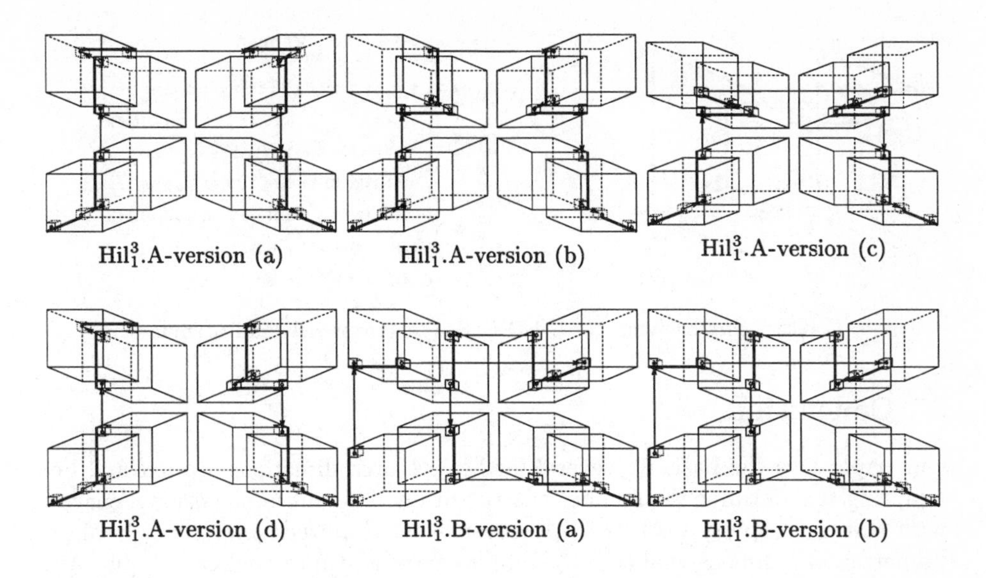

Fig. 4. Construction principles for CHPs with generators Hil_1^3.A and Hil_1^3.B.

construction principle of Hil_1^3, version (a), first pass through an $r-1$-dimensional structure, then in "two steps" do a change of dimension in the rth dimension, and finally again pass through an $r-1$-dimensional structure. This method applies to finding the generators as well as to finding the permutations.

Recursive computation of CSSCs. Note that whenever a CSSC $\mathcal{C} = \{C_k \mid k \geq 1\}$ is explicitly given by its generator and the sequence of permutations, we may use the recursive formula (1) of Subsection 3.1 to compute the curves C_k. In other words, the defining formula (1) itself provides a computation-scheme for CSSC, which is parameterized by the generating elements (generator and sequence of permutations).

Aspects of locality. The above mentioned parameterized formula might, for example, also be used to investigate locality properties of CSSCs by mechanical methods. The locality properties of Hilbert curves have already been studied in great detail. As an example for such investigations, we briefly note a result of Gotsman and Lindenbaum [6] for multi-dimensional Hilbert curves. In [6] they investigate a curve $C: \{1, \ldots, n^r\} \to \{1, \ldots, n\}^r$ with the help of their locality measure $L_2(C) = \max_{i,j \in \{1,\ldots,n^r\}} (d_2(C(i), C(j)))/|i-j|$, where d_2 denotes the Euclidean metric. In their Theorem 3 they claim the upper bound $L_1(H_k^r) \leq (r+3)^{\frac{r}{2}} 2^r$ for any r-dimensional Hilbert curve of order k, without precisely specifying what an r-dimensional Hilbert curve shall be. Since the proof of their result does not utilize the special Hilbert structure of the curve, this result can even be extended to arbitrary CSSCs.

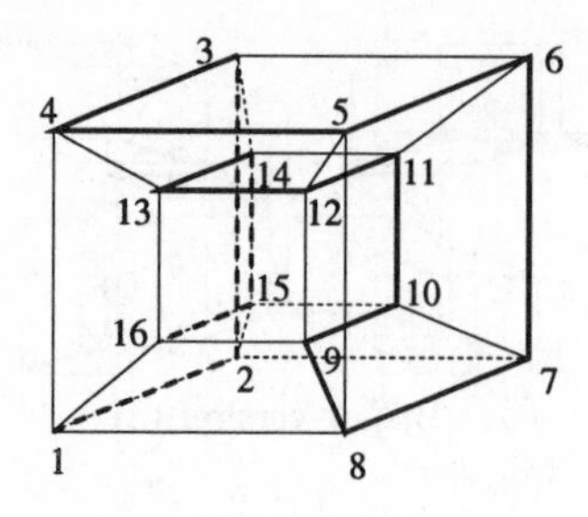

$$
\begin{aligned}
\tau_1 &= (2\ 16)(3\ 13)(6\ 12)(7\ 9)\\
\tau_2 &= (3\ 15)(4\ 16)(5\ 9)(6\ 10)\\
\tau_3 &= \tau_2\\
\tau_4 &= (1\ 3\ 13\ 11\ 9\ 7)(2\ 4\ 14\ 12\ 10\ 8)(5\ 15)(6\ 16)\\
\tau_5 &= \tau_4\\
\tau_6 &= (1\ 5\ 13\ 9)(2\ 6\ 14\ 10)(3\ 11\ 15\ 7)(4\ 12\ 16\ 8)\\
\tau_7 &= \tau_6\\
\tau_8 &= (1\ 7)(4\ 6)(10\ 16)(11\ 13)\\
\tau_9 &= \tau_8\\
\tau_{10} &= (1\ 9\ 13\ 5)(2\ 10\ 14\ 6)(3\ 7\ 15\ 11)(4\ 8\ 16\ 12)\\
\tau_{11} &= \tau_{10}\\
\tau_{12} &= (1\ 11)(2\ 12)(3\ 5\ 7\ 9\ 15\ 13)(4\ 6\ 8\ 10\ 16\ 14)\\
\tau_{13} &= \tau_{12}\\
\tau_{14} &= (1\ 13)(2\ 14)(7\ 11)(8\ 12)\\
\tau_{15} &= \tau_{14}\\
\tau_{16} &= (1\ 15)(4\ 14)(5\ 11)(8\ 10)
\end{aligned}
$$

Fig. 5. Constructing elements for a 4-D CHP (generator Hil_1^4 and permutations).

5 Conclusion

Our paper lays the basis for several further research directions. So it could be tempting to determine the number of structurally different r-dimensional curves with Hilbert property for $r > 3$. Moreover, a (mechanized) analysis of locality properties of r-dimensional ($r > 3$) Hilbert curves is still to be done (cf. [9]). An analysis of the construction of more complicated curves using more generators or different permutations for different levels remains open.

References

1. J. Alber and R. Niedermeier. On multi-dimensional Hilbert indexings. Technical Report 98-392, KAM-DIMATIA Series, Faculty of Mathematics and Physics, Charles University, Praha, Czech Republic, May 1998.
2. A. R. Butz. Space filling curves and mathematical programming. *Information and Control*, 12:314–330, 1968.
3. A. R. Butz. Convergence with Hilbert's space filling curve. *Journal of Computer and System Sciences*, 3:128–146, 1969.
4. G. Chochia and M. Cole. Recursive 3D mesh indexing with improved locality. Technical report, University of Edinburgh, 1997. Short version appeared in the Proceedings of HPCN'97, LNCS 1225.
5. G. Chochia, M. Cole, and T. Heywood. Implementing the Hierarchical PRAM on the 2D mesh: Analyses and experiments. In *Symposium on Parallel and Distributed Processing*, pages 587–595. IEEE Computer Science Press, October 1995.
6. C. Gotsman and M. Lindenbaum. On the metric properties of discrete space-filling curves. *IEEE Transactions on Image Processing*, 5(5):794–797, May 1996.
7. H. V. Jagadish. Linear clustering of objects with multiple attributes. *ACM SIGMOD Record*, 19(2):332–342, June 1990.
8. I. Kamel and C. Faloutsos. Hilbert R-tree: An improved R-tree using fractals. In *20th International Conference on Very Large Data Bases*, pages 500–509, 1994.
9. R. Niedermeier, K. Reinhardt, and P. Sanders. Towards optimal locality in mesh-indexings. In *Proceedings of the 11th Conference on Fundamentals of Computation Theory*, number 1279 in Lecture Notes in Computer Science, pages 364–375, Kraków, Poland, Sept. 1997. Springer-Verlag.
10. H. Sagan. A three-dimensional Hilbert curve. *Int. J. Math. Educ. Sci. Technol.*, 24(4):541–545, 1993.
11. H. Sagan. *Space-Filling Curves.* Universitext. Springer-Verlag, 1994.

Combinatorial Properties of Classes of Functions Hard to Compute in Constant Depth

Anna Bernasconi

Institut für Informatik, Technische Universität München (Germany).
e-mail: bernasco@informatik.tu-muenchen.de

Abstract. Any attempt to find connections between mathematical properties and complexity has a strong relevance to the field of Complexity Theory. This is due to the lack of mathematical techniques to prove lower bounds for general models of computation.
This work represents a step in this direction: we define a combinatorial property that makes Boolean functions "hard" to compute and show how the harmonic analysis on the hypercube can be applied to derive new lower bounds on the size complexity of previously unclassified Boolean functions.

1 Introduction

Any attempt to find connections between mathematical properties of functions and their computational complexity has a strong relevance to theory of computation. Indeed, there is the hope that developing new mathematical techniques could lead to discovering properties that might be responsible for lower bounds. The subject of this paper is related to the above general arguments, and in particular to showing how the Abstract Harmonic Analysis on the hypercube can provide some insight in our current understanding of Boolean function complexity. Our main result consists of new lower bounds on the size complexity of explicit functions, exactly derived by applying the above techniques.

One of the best-known results in Circuit Complexity is that constant depth circuits require exponential size to compute the parity function (see [2] and [3]). Here we generalize this result to a hierarchy of previously unclassified classes of functions.

This hierarchy is defined as follows. Let $0 < t \leq n$, and let $\mathcal{B}_n^{(t)}$ be the class of functions, depending on n variables, that take the value 1 on exactly 2^{n-t} input strings. We then divide $\mathcal{B}_n^{(t)}$ into *levels*, where the k-th level, which we denote by $B^{(t,k)}$, is defined as the subset of the functions $f \in \mathcal{B}_n^{(t)}$ such that any subfunction of f, depending on k ($k \geq t$) variables, takes the value 1 on 2^{k-t} input strings. These definitions are made precise below.

Our main result is that AC^0-circuits cannot compute functions in the k-th level of $\mathcal{B}_n^{(t)}$, whenever $k = n - (\log n)^{\omega(1)}$ and $t = (\log n)^{O(1)}$. More precisely, we prove that a circuit of constant depth d require size $\Omega\left(2^{\frac{(n-k)^{1/d}}{20}-t}\right)$ to compute any function in $B^{(t,k)}$, for any t and any k.

We also prove that nontrivial examples of functions exist for each level of this hierarchy if $k > \frac{t-1}{t} n$, and conjecture that this bound is not far from being asymptotically optimal.

The main tool of the lower bound proof is the harmonic analysis on the hypercube, that yields an interesting spectral characterization of the functions in the above hierarchy, together with a result proved in [6], stating that AC^0 functions have almost all of their *power spectrum* on the low-order coefficients.

Finally, notice that this paper generalizes results in [1], where it has been proven that AC^0-circuits cannot compute *strongly balanced functions.* Indeed, the class of strongly balanced functions coincides with the $[n - (\log n)^{\omega(1)}]$-th level of the class $\mathcal{B}_n^{(1)}$.

The rest of the paper is organized as follows. In Section 2 we provide some of the notation we use, and recall some basic definitions. In Section 3 we give the necessary background on Fourier transform on the hypercube, and review the results by Linial et al. (see [6]) about the spectral characterization of AC^0 functions. Section 4 is devoted to the definition of the classes $\mathcal{B}_n^{(t)}$ and of their levels $B^{(t,k)}$. In Section 5 we derive a spectral characterization of the functions in any level of $\mathcal{B}_n^{(t)}$, and in Section 6 we prove our main result stating that AC^0-circuits cannot compute functions in the level $B^{(t,k)}$, whenever $k = n-(\log n)^{\omega(1)}$ and $t = (\log n)^{O(1)}$. In Section 7 we provide nontrivial examples of functions in any level $B^{(t,k)}$ such that $k > \frac{t-1}{t} n$. Finally, in Section 8 we provide a framework for future research.

2 Basic Definitions

First of all, we provide some of the notation we use.

Given a Boolean function f on n binary variables, we will use different kinds of notation: the *classical notation*, where the input string is given by n binary variables; the *set notation*, based on the correspondence between the set $\{0,1\}^n$ and and the power set of $\{1,2,\ldots,n\}$; the 2^n-tuple *vector representation* $f = (f_0\, f_1 \ldots f_{2^n-1})$, where $f_i = f(x(i))$ and $x(i)$ is the binary expansion of i. Unless otherwise specified, the indexing of vectors and matrices starts from 0 rather than 1.

We will use the notation $|f|$ to denote the *cardinality* of f, that is the number of strings accepted f:

$$|f| = |\{w \in \{0,1\}^n \mid f(w) = 1\}| .$$

Given a binary string $w \in \{0,1\}^n$, we denote with $w^{(i)}$ the string obtained from w by flipping its i-th bit ($1 \le i \le n$), i.e. w and $w^{(i)}$ differ only on the i-th bit, and by $|w|$ the number of ones in w, which is sometimes called *cardinality* of the string because of the correspondence between sets of positive integers and strings over the alphabet $\{0,1\}$. If w and v are two binary strings of the same length, then $w \oplus v$ denotes the string obtained by computing the *exclusive or* of the bits of w and v. Finally, all the logarithms are to the base 2.

We now review some basic definitions.

AC^0 circuits

An AC^0 circuit consists of AND, OR and NOT gates, with inputs $x_1, \ldots, x_n$. Fan-in to the gates is unbounded. The size of the circuit (i.e. the number of the gates) is bounded by a polynomial in n, and its depth is bounded by a constant. Without loss of generality we can assume that negations occur only as negated input variables. If negations appear higher up in the circuit we can move them down to the inputs using De Morgan's laws which at most doubles the size of the circuit. Finally, observe that we have alternating levels of AND and OR gates, since two adjacent gates of the same type can be collapsed into one gate (for a more detailed description, see [3]).

Restriction

A *restriction* ρ is a mapping of the input variables to the set $\{0, 1, \star\}$, where

- $\rho(x_i) = 0$ means that we substitute the value 0 for x_i;
- $\rho(x_i) = 1$ means that we substitute the value 1 for x_i;
- $\rho(x_i) = \star$ means that x_i remains a variable.

Given a function f on n binary variables, we will denote by f_ρ the function obtained from f by applying the restriction ρ; f_ρ will be a function of the variables x_i for which $\rho(x_i) = \star$.

The *domain* of a restriction ρ, $dom(\rho)$, is the set of variables mapped to 0 or 1 by ρ. The *size* of a restriction ρ, $size(\rho)$, is defined as the number of variables which were given the value $\star$, i.e. $size(\rho) = n - |dom(\rho)|$.

3 Abstract Harmonic Analysis and AC^0 Functions

We give some background on abstract harmonic analysis on the hypercube. We refer to [5] for a more detailed exposition.

We consider Boolean functions as 0-1 valued real functions defined on the domain $\{0,1\}^n$. They are a vector space of dimension 2^n, and the set of functions $\{f_x(y) = (1 \text{ if and only if } x = y)\}$, where x ranges over $\{0,1\}^n$ is a basis. Another basis is given by the functions $\{g_S(x) = \sum_{i \in S} x_i\}$, where the sum is modulo 2. The Fourier coefficients of f are the coefficients of f in this basis.

More precisely, consider the space $\mathcal{F}$ of all the two-valued functions on $\{0,1\}^n$. The domain of $\mathcal{F}$ is a locally compact Abelian group and the elements of its range, i.e. 0 and 1, can be added and multiplied as complex numbers. The above properties allow one to analyze $\mathcal{F}$ by using tools from harmonic analysis. This means that it is possible to construct an orthogonal basis set of Fourier transform kernel functions for $\mathcal{F}$. The kernel functions of the Fourier transform are defined in terms of a group homomorphism from $\{0,1\}^n$ to the direct product of n copies of the multiplicative subgroup $\{\pm 1\}$ on the unit circle of the complex plane. The functions $Q_w(x) = (-1)^{w_1 x_1}(-1)^{w_2 x_2} \ldots (-1)^{w_n x_n} = (-1)^{w^T x}$ are known as *Fourier transform kernel functions*, and the set $\{Q_w | w \in \{0,1\}^n\}$ is an orthogonal basis for $\mathcal{F}$.

We can now define the *Abstract Fourier Transform* of a Boolean function f as the rational valued function f^* which defines the coordinates of f with respect to the basis $\{Q_w(x), w \in \{0,1\}^n\}$, i.e.,

$$f^*(w) = 2^{-n} \sum_x Q_w(x) f(x) .$$

Then

$$f(x) = \sum_w Q_w(x) f^*(w)$$

is the Fourier expansion of f.

It is interesting to note that the zero-order Fourier coefficient is equal to the probability that the function takes the value 1, while the other Fourier coefficients measure the correlation between the function and the parity of subsets of its input bits (see [6] for more details).

Using the binary 2^n-tuple representation for the functions f and f^*, and considering the natural ordering of the n-tuples x and w, one can derive a convenient matrix formulation for the transform pair. Let us consider a $2^n \times 2^n$ matrix H_n whose (i,j)-th entry h_{ij} satisfies $h_{ij} = (-1)^{x(i)^T x(j)}$, where $x(i)^T x(j)$ denotes the inner product of the binary expansions of i and j. If $f = [f_0\, f_1 \ldots f_{2^n-1}]^T$ and $f^* = [f_0^*\, f_1^* \ldots f_{2^n-1}^*]^T$, then, from the fact that $H_n^{-1} = 2^{-n} H_n$, we get $f = H_n f^*$ and $f^* = 2^{-n} H_n f$.

Note that the matrix H_n is the Hadamard symmetric transform matrix and can be recursively defined as

$$H_1 = \begin{pmatrix} 1 & 1 \\ 1 & -1 \end{pmatrix} , \qquad H_n = \begin{pmatrix} H_{n-1} & H_{n-1} \\ H_{n-1} & -H_{n-1} \end{pmatrix} .$$

We now present an interesting application of harmonic analysis to circuit complexity, due to Linial et al. (see [6]).

As we have already mentioned, one of the best known results in circuit complexity is that AC^0 circuits require exponential size to compute the parity function. More precisely, AC^0-circuits cannot even approximate the parity function. This fact has a direct consequence on the Fourier transform, because, as we have already mentioned, the Fourier coefficients measure the correlation between the function and the parity of subsets of its input bits. Consequently, each high order Fourier coefficient of an AC^0 function must be very small (where "high order" means coefficients corresponding to strings of large cardinality). By exploiting this fact, Linial et al. were able to prove that not only is each individual high order coefficient small, but in fact the sum of squares (i.e. the *power spectrum*) associated with all high Fourier coefficients is very small.

Lemma 1 ([6]) *Let f be a Boolean function on n variables computable by a Boolean circuit of depth d and size M, and let θ be any integer. Then*

$$\sum_{|w|>\theta} (f^*(w))^2 \leq \frac{1}{2} M\, 2^{-\frac{\theta^{1/d}}{20}} .$$

∎

4 The Classes $\mathcal{B}_n^{(t)}$ and their Levels $B^{(t,k)}$.

In this section we define classes of functions which generalize the notion of k-*balanced functions* introduced in [1]. Let t be a positive integer, $0 < t \leq n$.

Definition 1 $\mathcal{B}_n^{(t)}$ *is the class of Boolean functions depending on n variables that take the value 1 on exactly 2^{n-t} input strings.*

Making use of the notion of *restriction* (see Section 2), we organize the functions in each class $\mathcal{B}_n^{(t)}$ into a sequence of *levels* according to the following definition:

Definition 2 *For any $t \leq k \leq n$, $B^{(n,k)}$ is the subset of $\mathcal{B}_n^{(t)}$ consisting of all functions f such that, for any restriction ρ of size k, $f_\rho \in \mathcal{B}_k^{(t)}$. We call $B^{(n,k)}$ the* k-th level *of $\mathcal{B}_n^{(t)}$.*

In other words, $B^{(t,k)}$ consists of all functions f, of cardinality $|f| = 2^{n-t}$, such that any subfunction f_ρ depending on k variables has cardinality $|f_\rho| = 2^{k-t}$.
We now state some basic properties of the hierarchy of levels $B^{(t,k)}$. Let $t \leq k \leq n$.

- $B^{(t,k)} \subseteq B^{(t,k+1)}$.
- $B^{(t,n)} = \mathcal{B}_n^{(t)}$.
- The classes of k-balanced functions defined in [1] correspond to the k-th levels of $\mathcal{B}_n^{(1)}$.
- The parity function and its complement are the only two functions which belong to the first level of $\mathcal{B}_n^{(1)}$, i.e. to $B^{(1,1)}$.

In Section 7 we will provide nontrivial examples of functions in any level of the class $\mathcal{B}_n^{(t)}$. More precisely, we will prove that, for any t, $B^{(t,k)}$ is *strictly* contained in $B^{(t,k+1)}$ and that the levels $B^{(t,k)}$ are not empty, provided that $k > \frac{t-1}{t} n$. All these proofs will make use of the spectral characterization of these functions, which we derive in the following section.

5 Spectral Characterization of the Hierarchy of $\mathcal{B}_n^{(t)}$ Functions

We now derive a spectral characterization of the functions in any level of the class $\mathcal{B}_n^{(t)}$. We denote by f_0^* the zero-order Fourier coefficient.

Theorem 2 *A Boolean function $f : \{0,1\}^n \rightarrow \{0,1\}$ belongs to the k-th level of the class $\mathcal{B}_n^{(t)}$ if and only if the following two properties hold:*

(1) $f_0^* = 2^{-t}$;
(2) for any string w such that $0 < |w| \leq n - k, \;\; f^*(w) = 0$.

Proof.

- If $f \in \mathcal{B}_n^{(t)}$, then property (1) holds by definition, because f_0^* is equal to the probability that the function f takes the value 1. Thus, we only need to prove property (2).
  Let $\mu \equiv (\mu_1, \mu_2, \ldots, \mu_n)$ be a Boolean string such that $0 < |\mu| = n-\ell \leq n-k$. Moreover, let $U = \{i \mid \mu_i = 1\}$. For any string $u \in \{0,1\}^{n-\ell}$, let f_u denote the subfunction defined by the restriction that assigns to the variables x_i such that $i \in U$, the $(n-\ell)$ values taken from the string u, and leaves undetermined the other ℓ variables.
  Then, we have

$$f^*(\mu) = \frac{1}{2^n} \sum_w (-1)^{\mu^T w} f(w) = \frac{1}{2^n} \sum_w (-1)^{\sum_{i \in U} w_i} f(w)$$

$$= \frac{1}{2^n} \sum_{u \in \{0,1\}^{n-\ell}} \left[(-1)^{|u|} \sum_{v \in \{0,1\}^\ell} f_u(v) \right]$$

$$= \frac{1}{2^n} \sum_{u \in \{0,1\}^{n-\ell}} \left[(-1)^{|u|} \, |f_u| \right].$$

  For any $u \in \{0,1\}^{n-\ell}$, the subfunction f_u depends on $\ell \geq k$ variables and, since $f \in B^{(t,k)}$, and $B^{(t,k)} \subseteq B^{(t,\ell)}$ for any $\ell \geq k$, we have $f_u \in \mathcal{B}_\ell^{(t)}$ and $|f_u| = 2^{\ell-t}$. Thus, we get

$$f^*(\mu) = \frac{2^{\ell-t}}{2^n} \sum_{u \in \{0,1\}^{n-\ell}} (-1)^{|u|} = 0.$$

- We now prove that if properties (1) and (2) hold, then $f \in B^{(t,k)}$.
  Let us choose $(n-k)$ variables out of n, and let U be the set of the indices of these $(n-k)$ variables. For any $u \in \{0,1\}^{n-k}$, let f_u denote the subfunction obtained from f by assigning to the variables in the set U, the $(n-k)$ values taken from the string u, and leaving undetermined the other k variables.
  For any u, f_u depends on k variables. We show that any such subfunction accepts exactly 2^{k-t} inputs, i.e. for any string u, $|f_u| = 2^{k-t}$.
  Let $f_\#$ denote the vector of the cardinality of the 2^{n-k} subfunctions f_u, and let f_U^* denote the vector of the Fourier coefficients related to the 2^{n-k} strings $w \equiv (w_1, w_2, \ldots, w_n)$ such that $w_i = 0$ for any $i \notin U$. Note that all the 2^{n-k} coefficients in the vector f_U^* are of order less or equal to $n-k$. Because of the recursive definition of Hadamard matrices, it turns out that

$$f_U^* = \frac{1}{2^n} H_{n-k} \, f_\# .$$

  From properties (1) and (2), and from the fact that the zero order Fourier coefficient is equal to the probability that the function takes the value 1, it then follows

$$f_U^* = \frac{1}{2^t} \begin{pmatrix} 1 \\ 0 \\ \vdots \\ 0 \end{pmatrix},$$

from which

$$f_\# = 2^n \, H_{n-k}^{-1} \, f_U^* = \frac{2^n}{2^{n-k}} \, H_{n-k} \, f_U^* = 2^{k-t} \, H_{n-k} \begin{pmatrix} 1 \\ 0 \\ \vdots \\ 0 \end{pmatrix} = 2^{k-t} \begin{pmatrix} 1 \\ 1 \\ \vdots \\ 1 \end{pmatrix}.$$

Thus, the theorem follows by repeating the same argument for all the $\binom{n}{k}$ choices of the set U.

■

6 A Lower Bound on the Size Complexity of $\mathcal{B}_n^{(t)}$ Functions

We are now able to state and prove our main result, stating that AC^0-circuits cannot compute functions in the k-th level of $\mathcal{B}_n^{(t)}$, whenever $k = n - (\log n)^{\omega(1)}$ and $t = (\log n)^{O(1)}$.

We first make use of the spectral characterization derived in Theorem 2, together with Lemma 1, to determine a lower bound on the size required by a depth d circuit to compute functions in the k-th level of $\mathcal{B}_n^{(t)}$. Finally, an easy application of this bound will provide our thesis. Let $t > 0$.

Theorem 3 *Let $f \in B^{(t,k)}$ be a Boolean function depending on n variables, computable by a circuit of constant depth d and size M. Then*

$$M \geq 2^{\frac{(n-k)^{1/d}}{20} - t}.$$

Proof. An application of Lemma 1 yields the following inequality:

$$M \geq 2^{\frac{\theta^{1/d}}{20}} 2 \sum_{|w|>\theta} (f^*(w))^2.$$

Let us choose $\theta = n - k$. From the fact that $f^*(w) = 0$ for any $0 < |w| \leq n - k$ (see Theorem 2) it follows

$$\sum_{|w|>n-k} (f^*(w))^2 = \sum_{w:|w|\neq 0} (f^*(w))^2 = \sum_w (f^*(w))^2 - (f_0^*)^2,$$

where f_0^* denotes the zero-order Fourier coefficient. Then, by using the Parseval's identity $\sum_v (f^*(v))^2 = f_0^* = \frac{1}{2^t}$, we get

$$\sum_{|w|>n-k} (f^*(w))^2 = \frac{1}{2^t} - \frac{1}{2^{2t}} \geq 2^{-(t+1)},$$

and the thesis immediately follows:

$$M \geq 2^{\frac{(n-k)^{1/d}}{20}} 2 \sum_{|w|>n-k} (f^*(w))^2 = 2^{\frac{(n-k)^{1/d}}{20}-t}.$$

■

Notice how this result establishes a clear connection between complexity and combinatorial properties of Boolean functions.

Our main result, stating that AC^0-circuits cannot compute functions in the k-th level of $\mathcal{B}_n^{(t)}$, whenever $k = n - (\log n)^{\omega(1)}$, and $t = (\log n)^{O(1)}$, follows immediately as a corollary of Theorem 3.

Corollary 4 *Any function $f \in B^{((\log n)^{O(1)},\ n-(\log n)^{\omega(1)})}$ requires superpolynomial size to be computed by a constant depth circuit.*

Proof. Easily follows from Theorem 3. ■

Note how the lower bound to the size can become exponential:

Corollary 5 *Constant depth circuits require exponential size to compute functions in levels $B^{(t,k)}$ whenever k is s.t. $n - k = \Omega(n^\varepsilon)$, for any positive constant $\varepsilon < 1$, and $t = (\log n)^{O(1)}$.*

Proof. Immediate from Theorem 3. ■

7 Properties of the Hierarchy $\mathcal{B}_n^{(t)}$

In this section we provide nontrivial examples of functions in the levels of the hierarchy $\mathcal{B}_n^{(t)}$. More precisely, by applying the spectral characterization derived in Section 5, we prove that, for any t, $B^{(t,k)}$ is strictly contained in $B^{(t,k+1)}$ and that the sets $B^{(t,k)}$ are not empty, provided that $k > \frac{t-1}{t} n$.

Theorem 6 *For any n, $B^{(t,k)} \neq \emptyset$ if $k \geq \frac{t-1}{t} n + 1$.*

Proof. By induction on t.

Base

For any n, and for $t = 1$, the parity function and its complement belong to $B^{(1,1)}$ and $\frac{t-1}{t} n = 0$.

Induction step

Let us suppose that, for any n, $B^{(t,k)} \neq \emptyset$ if $k \geq \frac{t-1}{t} n + 1$.

Let g be a Boolean function, depending on $n - m$ variables, which belongs to the "deepest" level of $\mathcal{B}^{(t)}_{n-m}$. That is, $g \in B^{(t,k)}$ for $k = \frac{t-1}{t}(n-m) + 1$ (w.l.o.g. let us assume that t divides $n - m$).

We define f, depending on n variables, as follows:

$$f(\alpha\beta) = \begin{cases} 0 & \text{if } |\alpha| \text{ is odd} \\ g(\beta) & \text{if } |\alpha| \text{ is even}, \end{cases}$$

where $\alpha \in \{0,1\}^m$ and $\beta \in \{0,1\}^{n-m}$. First of all, note that $f \in \mathcal{B}^{(t+1)}_n$. Indeed, $|f| = 2^{m-1}|g| = 2^{m-1}2^{n-m-t} = 2^{n-(t+1)}$.

From the definition of f, and from the structure of Hadamard matrices, it turns out that the spectrum of f can be defined in terms of the spectrum of g and of the spectrum of the parity function, in the following way:

$$f^*(\alpha\beta) = parity^*(\alpha)g^*(\beta).$$

Thus, since the parity function has the following Fourier spectrum

$$p^*(\alpha) = \begin{cases} 1/2 & \text{if} \quad |\alpha| = 0 \\ 0 & \text{if} \quad 0 < |\alpha| < m \\ -1/2 & \text{if} \quad |\alpha| = m, \end{cases}$$

we get

$$f^*(\alpha\beta) = \begin{cases} \frac{1}{2}g^*(\beta) & \text{if} \quad |\alpha| = 0 \\ 0 & \text{if} \quad 0 < |\alpha| < m \\ -\frac{1}{2}g^*(\beta) & \text{if} \quad |\alpha| = m. \end{cases}$$

If we now use the fact that $g \in B^{(t, \frac{t-1}{t}(n-m)+1)}$, together with the spectral characterization of Theorem 2, we obtain that $f^*(w) = 0$, whenever $|w| < \min\{\frac{n-m}{t}, m\}$.

Therefore, the thesis follows if we choose $m = \lfloor\frac{n}{t+1}\rfloor$. Indeed, we have that $f^*(w) = 0$ whenever $|w| < \lfloor\frac{n}{t+1}\rfloor$, and from Theorem 2 it follows that $f \in B^{(t+1,k)}$ for $k > \frac{t}{t+1}n$, which completes our induction. ∎

Notice that, because of its construction, the function f defined in the proof of the above theorem is nondegenerated, i.e. it depends on all input variables.

By defining f in a more complicated way it is possible, in some cases, to decrease the bound on k, but only by a constant factor. Therefore, we conjecture that the bound on k given in Theorem 6 is not far from being asymptotically optimal.

Theorem 6 is an interesting result for the following two reasons. First of all, it allows us to verify that the classes of functions under investigation are not empty, at least for a significant number of levels. Moreover, since for constant values of t, the functions in the deepest levels of the hierarchy, can be regarded as "generalization" of the parity function, it is interesting to understand how "deep" we can go in such a generalization, i.e. how close the combinatorial structure of our functions is to the combinatorial structure of parity.

We now prove that, for any t, $B^{(t,k)}$ is strictly contained in $B^{(t,k+1)}$, provided that $k > \frac{t-1}{t} n$. In other words, we prove that nontrivial examples of functions do exist for any level of $B_n^{(t)}$.

Theorem 7 *If $k > \frac{t-1}{t} n$, then $B^{(t,k)}$ is strictly contained in $B^{(t,k+1)}$.*

Proof. The proof of the theorem is strictly related to that of Theorem 6. For $t = 1$, $B^{(1,k)}$ is strictly contained in $B^{(1,k+1)}$ for any $k \geq 1$ (see [1]).

For the induction step, we construct a function exactly as we did in the proof of Theorem 6. Then, by choosing for m any value between 1 and $\lfloor \frac{n}{t+1} \rfloor$, we get a function f such that $f \in B^{(t,n-m+1)}$ but $f \notin B^{(t,n-m)}$. ∎

8 Conclusion

Any attempt to find connections between mathematical properties and complexity has a strong relevance to the field of Complexity Theory. This is due to the lack of mathematical techniques to prove lower bounds for general models of computation. This work represents a step in this direction: we define combinatorial properties that makes Boolean functions "hard" to compute and show how the Fourier transform could be used as a mathematical tool for the analysis of Boolean functions complexity. Further work to be done includes a deeper analysis of the structure of the levels $B^{(t,k)}$, in order to get an optimal lower bound on k, and, more in general, a deeper analysis of the connections between combinatorial properties, spectral properties and complexity of Boolean functions.

References

1. A. Bernasconi. *On the Complexity of balanced Boolean Functions.* CIAC'97, Lecture Notes in Computer Science **1203** (1997) pp. 253-263.
2. M. Furst, J. Saxe, M. Sipser. *Parity, circuits, and the polynomial-time hierarchy.* Math. Syst. Theory, Vol. **17** (1984) pp. 13-27.
3. J. Håstad. *Computational limitations for small depth circuits.* Ph.D. Dissertation, MIT Press, Cambridge, Mass. (1986).
4. S.L. Hurst, D.M. Miller, J.C. Muzio. *Spectral Method of Boolean Function Complexity.* Electronics Letters, Vol. **18 (33)** (1982) pp. 572-574.
5. R. J. Lechner. *Harmonic Analysis of Switching Functions.* In *Recent Development in Switching Theory,* Academic Press (1971) pp. 122-229.
6. N. Linial, Y. Mansour, N. Nisan. *Constant Depth Circuits, Fourier Transform, and Learnability.* Journal of the ACM, Vol. **40 (3)** (1993) pp. 607-620.
7. H .U. Simon. *A tight $\Omega(\log \log n)$ bound on the time for parallel RAM's to compute nondegenerate Boolean functions.* FCT'83, Lecture Notes in Computer Science **158** (1983).
8. I. Wegener. *The complexity of Boolean functions.* Wiley-Teubner Series in Comp. Sci., New York - Stuttgart (1987).

Eulerian Secret Key Exchange
(Extended Abstract)

Takaaki Mizuki[1], Hiroki Shizuya[2] and Takao Nishizeki[1]

[1] Graduate School of Information Sciences, Tohoku University,
Sendai 980-8579, Japan
{mizuki, nishi}@ecei.tohoku.ac.jp

[2] Education Center for Information Processing, Tohoku University,
Sendai 980-8576, Japan
shizuya@ecip.tohoku.ac.jp

Abstract. Designing a protocol to exchange a secret key is one of the most fundamental subjects in cryptography. Using a random deal of cards, pairs of card players (agents) can share information-theoretically secure keys that are secret from an eavesdropper. In this paper we first introduce the notion of an Eulerian secret key exchange, in which the pairs of players sharing secret keys form an Eulerian circuit passing through all players. Along the Eulerian circuit any designated player can send a message to the rest of players and the message can be finally returned to the sender. Checking whether the returned message is the same as the original one, the sender can know whether the message circulation has been completed without any false alteration. We then give three efficient protocols to realize such an Eulerian secret key exchange. Each of the three protocols is optimal in a sense. The first protocol requires the minimum number of cards under a natural assumption that the same number of cards are dealt to each player. The second requires the minimum number of cards dealt to all players when one does not make the assumption. The third forms the shortest Eulerian circuit, and hence the time required to send the message to all players and acknowledge the secure receipt is minimum in this case.

1 Introduction

Suppose that there are k (≥ 2) players $P_1, P_2, \cdots, P_k$ and a passive eavesdropper, Eve, whose computational power is unlimited. All players wish to agree on a one-bit message that is secret from Eve. Since Eve is computationally unlimited, the secret message must be information-theoretically secure. Let C be a set of d distinct cards which are numbered from 1 to d. All cards in C are randomly dealt to players $P_1, P_2, \cdots, P_k$ and Eve. Let $C_i \subseteq C$ be P_i's hand, and let $C_e \subseteq C$ be Eve's hand. We denote this deal by $\mathcal{C} = (C_1, C_2, \cdots, C_k; C_e)$. Clearly $\mathcal{C}$ is a partition of set C. We write $c_p = |C_p|$ for each $p \in \{1, 2, \cdots, k\}$ and $c_e = |C_e|$. We say that $\gamma = (c_1, c_2, \cdots, c_k; c_e)$ is the *signature* of deal $\mathcal{C}$. The set C and the signature γ are public to all the players and even to Eve, but the hands C_i, $1 \leq i \leq k$, and C_e are held exclusively by P_i, $1 \leq i \leq k$, and Eve, respectively, as in the case of usual card games.

Fischer, Paterson and Rackoff [1] give a protocol for two players to exchange a one-bit key which is secret from Eve. A player encrypts a one-bit message by adding (modulo 2) the message and the key, and sends it to the other player. The decryption is done in the same way at the recipient. This cryptographic communication is information-theoretically secure. On the other hand, Fischer and Wright [2, 5] extends the protocol so that any number k of players can share a one-bit message. Their main idea is to form a spanning tree over the players. That is, they regard each player as a vertex of a graph G, and regard each pair of players sharing a one-bit key as an edge of G. Their protocol is designed so that G becomes a spanning tree whenever the protocol terminates normally. Along the tree a one-bit message can be spread over the players.

In this paper, we first introduce the notion of an Eulerian secret key exchange by which the graph G becomes an Eulerian circuit. See for example Figure 1(a) and Figure 2. Refer to [6, 7] for the graph-theoretic terminology. Along the Eulerian circuit any designated player can send a one-bit message to the rest of players using the keys shared among the players, and the message can be finally returned to the sender. Checking whether the returned message is the same as the original one, the sender can know whether the message circulation has been probably completed without any false alteration. Thus the secure receipt of the message to all players can be acknowledged. This acknowledgment is practically important when a computer network is unreliable, possibly due to traffic jam, disconnection, or error by some malfunction. We then give three efficient protocols to realize the Eulerian secret key exchange, using graph-theoretic techniques. Each of the three protocols is optimal in a sense among the class of all "key set protocols." The first protocol requires the minimum number of cards under a natural assumption that the same number of cards are dealt to each player, that is, $c_1 = c_2 = \cdots = c_k$. The second requires the minimum number of cards when $c_1 = c_2 = \cdots = c_k$ does not always hold. The third forms the shortest Eulerian circuit, that is, the Eulerian graph G has the fewest edges, and hence the time required to send the message to all players along the circuit and acknowledge the secure receipt is minimum.

2 Preliminaries

In this section we first briefly review the results and techniques given in [1, 2, 3, 4, 5, 8], and then define some terms.

The scenario is the same as before: there are k (≥ 2) players and Eve, and the players wish to exchange a one-bit key. The deal of cards is expressed as $\mathcal{C} = (C_1, C_2, \cdots, C_k; C_e)$. A *key set* $K = \{x, y\}$ consists of two cards x and y, one in C_i, the other in C_j with $i \neq j$, say $x \in C_i$ and $y \in C_j$. We say that a key set $K = \{x, y\}$ is *opaque* if $1 \leq i, j \leq k$ and Eve cannot determine whether $x \in C_i$ or $x \in C_j$ with probability greater than $\frac{1}{2}$. Note that both players P_i and P_j know that $x \in C_i$ and $y \in C_j$. If K is an opaque key set, P_i and P_j can share a secret key $r \in \{0, 1\}$, using the following rule agreed on before starting the protocol: $r = 0$ if $x > y$; $r = 1$, otherwise. Since Eve cannot determine whether $r = 0$

or $r = 1$ with probability greater than $\frac{1}{2}$, the key r is information-theoretically secure.

The players can obtain opaque key sets by executing the following protocol [1, 2, 5], where $C_1, C_2, \cdots, C_k, C_e$ denote the current hands of $P_1, P_2, \cdots, P_k$, Eve, respectively. Each player is allowed to discard a card x if necessary. *Discarding* x means all the k players agree that x has been removed from someone's hand, that is, $x \notin (\bigcup_{i=1}^{k} C_i) \cup C_e$.

1. Choose a proposer P_s such that $C_s \neq \emptyset$ and $1 \leq s \leq k$ by a certain procedure.
2. The proposer P_s determines in mind two cards x, y. The cards are randomly picked so that $x \in C_s$ and $y \notin C_s$. Then P_s proposes $K = \{x, y\}$ as a key set to all the players. (The key set is proposed just as a set. Actually it is sorted in some order, for example in ascending order, so Eve learns nothing about which card belongs to C_s unless Eve holds y.)
3. If there exists a player P_t holding y, P_t accepts K. (Since K is an opaque key set, P_s and P_t can exchange a one-bit key which is secret from Eve.) The players discard both x and y. Further, either P_s or P_t holding the fewer cards discards all her cards and drops out of the protocol. Return to Step 1.
4. If there exists no player holding y, that is, Eve holds y, then the players discard both x and y, and return to Step 1.

These steps 1–4 are repeated until at most one player remains in the protocol.

The protocol above makes the graph G form a spanning tree over the players under a certain condition. All the players are the vertices of G, and all the pairs of players sharing keys are the edges of G. A player P_i can encrypt a one-bit message for each P_j of the players adjacent to P_i in G by adding (modulo 2) the message to the key shared by P_i and P_j, and send the encrypted message to P_j. The receiver P_j decrypts it by adding it to the key (mod 2), and unless she is a leaf of the tree, she encrypts the message and sends it to each adjacent player except P_i in the same way using the key shared by her and her adjacent player. Thus any designated player can transmit a one-bit message to the rest of players along the tree with information-theoretic security. Fischer and Wright [2, 5] give the so-called SFP procedure for the step 1. They show that if $\gamma = (c_1, c_2, \cdots, c_k; c_e)$ satisfies $c_i \geq 1$ for every i, $1 \leq i \leq k$, and $\max c_i + \min c_i \geq k + c_e$, then graph G always becomes a spanning tree.

We then define a key exchange graph, an Eulerian key exchange, and a key set protocol, as follows.

Definition 1 *A* ***key exchange graph*** $G = (V, E)$ *is a multigraph with vertex set* V *and edge set* E *such that* $V = \{1, 2, \cdots, k\}$ *and, for any vertices* i *and* j *in* V*,* E *contains* m *multiple edges joining* i *and* j *if and only if players* P_i *and* P_j *share* m *one-bit keys, where* $m \geq 0$.

From now on we often identify player P_i, $1 \leq i \leq k$, with vertex $i \in V$ in the graph G.

Definition 2 *A key exchange is called an* **Eulerian key exchange** *if the key exchange graph G is an Eulerian graph, that is, G has an Eulerian circuit passing through every edge exactly once. (See for example Figure* 1(a).)

We consider the class of *key set protocols*, obtained by generalizing the protocol above.

Key Set Protocol

1. Choose a proposer P_s such that $C_s \neq \emptyset$ and $1 \leq s \leq k$ by a certain procedure (say "**Procedure A**").
2. The proposer P_s determines in mind two cards x and y. The cards are randomly picked so that $x \in C_s$ and $y \notin C_s$. Then P_s proposes $K = \{x, y\}$ as a key set to all the players.
3. If there exists a player P_t holding y, P_t accepts K. Both x and y are discarded.
4. If there exists no player holding y, that is, Eve holds y, then both x and y are discarded.
5. Determine a subset X of players by a certain procedure (say "**Procedure B**"). Each player in X discards her whole cards. Every player holding no card drops out of the protocol. Return to Step 1.

The steps 1–5 above are repeated until at most one player remains in the protocol.

At step 3 P_s and P_t succeed in sharing a one-bit key, an edge (s, t) joining vertices s and t is added to graph G, and the number of cards held by each of P_s and P_t decreases by one. At step 4 the numbers of cards held by P_s and Eve decrease by one, but no new edge is added to G.

The procedure A determines a player to propose a key set, and the procedure B determines the players to be forced to drop out of the protocol. By considering different procedures A and B, we obtain the class of *key set protocols*. The key set protocol that results from procedures A and B is called an (A,B)-*protocol*. We assume that procedures A and B can take only the signature, the sizes of current hands, and the current key exchange graph as the input data.

The protocol of [2, 5] mentioned above is a kind of key set protocol. Its procedure A, i.e. the SFP procedure, chooses a proposer s as follows: if there is a vertex $i \in V$ such that $|C_i| = 1$, $|C_e| = 0$ and $|C_j| \geq 2$ for all vertices $j \neq i$, then the SFP procedure chooses the vertex i as a proposer s; if there is no such i but there is a vertex j with $|C_j| \geq 2$, then the procedure chooses as s the vertex j having the minimum $|C_j|$ among all these vertices j; otherwise, the procedure chooses any vertex as s. Its procedure B determines a set X as follows:

$$X = \begin{cases} \{s\} & \text{if } y \in C_t \text{ and } |C_s| \leq |C_t|; \\ \{t\} & \text{if } y \in C_t \text{ and } |C_s| > |C_t|; \\ \emptyset & \text{if } y \in C_e. \end{cases}$$

Thus, when $y \in C_t$, either P_s or P_t holding the fewer cards drops out of the protocol.

Since we wish to make G an Eulerian circuit instead of a spanning tree, we need procedures A and B different from these procedures.

Definition 3 *A key set protocol* ***achieves*** *an Eulerian key exchange for a signature $\gamma = (c_1, c_2, \cdots, c_k; c_e)$ if the key exchange graph G always becomes an Eulerian graph for any deal C having the signature γ when the players follow the protocol.*

3 Protocol for Eulerian Key Exchange

In this section we give the first protocol, the (A_1,B_1)-protocol, of our three key set protocols. The (A_1,B_1)-protocol is optimal among the class of all key set protocols achieving an Eulerian key exchange for a signature $\gamma = (c_2, c_2, \cdots, c_k; c_e)$ with $c_1 = c_2 = \cdots = c_k$ in the sense that the number of required cards is minimum.

We give procedures A_1 and B_1 only for $k \geq 4$, but omit those for $k = 2, 3$ since they are similar to those for $k \geq 4$.

The smaller the degrees of vertices in a key exchange graph G are, the fewer the cards are required. We thus want to make the degree of each vertex of G as small as possible. However, we cannot always make every vertex have degree exactly two, that is, we cannot always form a Hamiltonian graph. We therefore make the degree of each vertex either two or four. This is our idea behind the (A_1,B_1)-protocol. Note that a graph is Eulerian if and only if the graph is connected and every vertex has even degree [6].

We denote by $d(i)$ the degree of vertex $i \in V$ in graph $G = (V, E)$. A vertex $i \in V$ having a non-empty current hand $C_i \neq \emptyset$ is called a *white vertex*, and a vertex $i \in V$ having an empty current hand $C_i = \emptyset$ is called a *black vertex*. White and black vertices are drawn respectively by white and black circles in all figures in the paper. A white vertex corresponds to a player remaining in a protocol, and a black vertex corresponds to a player who has dropped out of the protocol. Let W be the set of all white vertices. Figure 1 illustrates connected components of an intermediate graph G. During any execution of the protocol, the degree of each vertex of G is at most four; in particular, the degree of each black vertex is either two or four; each connected component of an intermediate graph G has one or two white vertices; if a component of an intermediate graph G has exactly one white vertex as illustrated in Figures 1(a) and (c), then the degree of this vertex is zero or two; a final Eulerian graph G has at most one white vertex, the degree of which is two or four; on the other hand, if a component of an intermediate graph G has two white vertices as in Figures 1(b), (d) and (e), then the degrees of these vertices are odd, namely one or three. Note that every connected component of any graph has an even number of vertices of odd degree [6].

Before the (A_1,B_1)-protocol terminates, the degree $d(i)$ of each white vertex i is either 0, 1, 2, or 3. We partition the set W of all white vertices i in an intermediate graph G into three subsets W_1, W_2 and W_3, depending on the

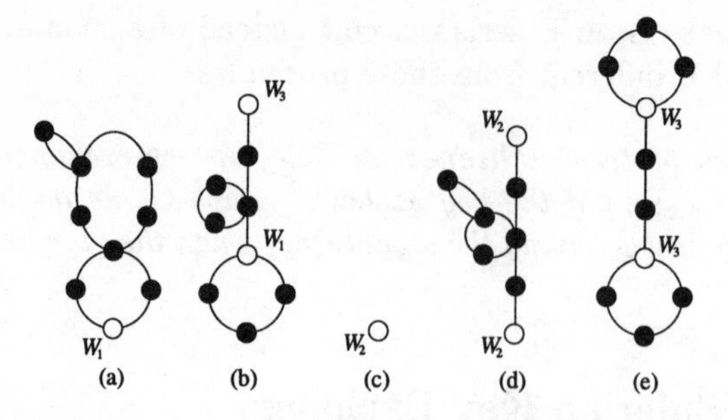

Fig. 1. Connected components of graph G during an execution of the protocol.

degrees $d(i)$ and $d(j)$ where j is the other white vertex, if any, in the connected component containing i, as follows:

$W_1 = \{i \in W \mid [d(i) = 2] \vee [d(i) = 3 \text{ and } d(j) = 1]\}$;

$W_2 = \{i \in W \mid [d(i) = 0] \vee [d(i) = 1 \text{ and } d(j) = 1]\}$; and

$W_3 = W - W_1 - W_2$.

Note that

$W_3 = \{i \in W \mid [d(i) = 1 \text{ and } d(j) = 3] \vee [d(i) = 3 \text{ and } d(j) = 3]\}$.

The white vertex of degree 2 in Figure 1(a) and the white vertex of degree 3 in Figure 1(b) are in W_1. The white vertices in Figures 1(c) and (d) are in W_2. The white vertex of degree 1 in Figure 1(b) and the white vertices in Figure 1(e) are in W_3.

Our procedure A_1 for step 1 chooses as a proposer s a vertex in W with the priority order of W_1, W_2 and W_3 in this order; we will explain the reason after giving procedures A_1 and B_1. Furthermore we keep every vertex $i \in W$ satisfying $|C_i| + d(i) \geq 4$. It should be noted that $|C_i| + d(i) \geq 4$ at the beginning of the protocol if $c_i \geq \lceil c_e/2 \rceil + 4$, and that if $|C_i| + d(i) \leq 3$ then $d(i)$ could not become four in the final graph G.

Procedure A_1: Choose a proposer $s \in W$ as follows.

Case A1: $C_e \neq \emptyset$ and there exists $i \in W_1 \cup W_2$ such that $|C_i| + d(i) \geq 5$.

If there exists a vertex $i \in W_1$ such that $|C_i| + d(i) \geq 5$ (Figures 1(a),(b)), choose any of these vertices as a proposer s. Otherwise, choose as s any $i \in W_2$ such that $|C_i| + d(i) \geq 5$ (Figures 1(c),(d)).

Case A2: $C_e = \emptyset$ and $W_1 \cup W_2 \neq \emptyset$.

If $W_1 \neq \emptyset$, choose any vertex in W_1 as a proposer s (Figures 1(a),(b)). If $W_1 = \emptyset$, choose as s any vertex in W_2 (Figures 1(c),(d)).

Case A3: Otherwise.

(We can show that in this case G is a connected graph with $|W| = 2$ and $W_3 \neq \emptyset$ like in Figure 1(b) or 1(e), and hence the protocol terminates when the two vertices i and j in W are joined by an edge.) Choose as s the vertex i in W such that $|C_i| \geq |C_j|$.

An edge in a graph is called a *bridge* if the graph has no cycle passing through the edge [6]. If vertices s and t were contained in different connected components

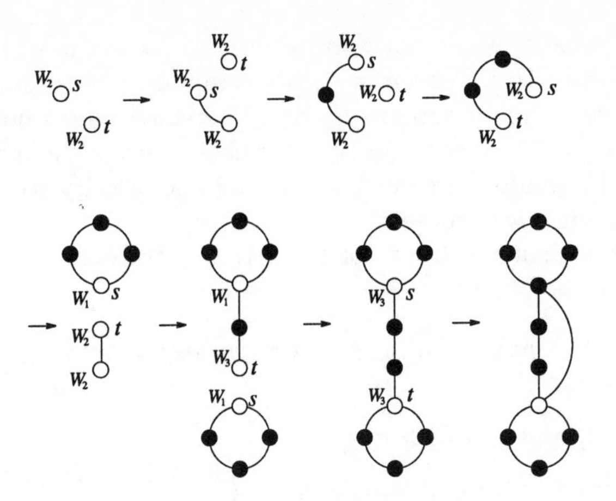

Fig. 2. A generating process of an Eulerian graph containing the graph in Figure 1(e).

before edge (s,t) is added, then the edge is a bridge in a new graph after (s,t) is added. On the other hand, if both s and t were in the same connected component, then the edge is not a bridge in a new graph. Our procedure B_1 for step 5 is simple, as follows.

Procedure B_1: Choose set X as follows.

$$X = \begin{cases} \{i \mid [i = s \text{ or } t] \wedge [d(i) \text{ is even}]\} & \text{if } [y \in C_t] \wedge [\text{edge } (s,t) \text{ is a bridge in } G]; \\ \{s\} & \text{if } [y \in C_t] \wedge [\text{edge } (s,t) \text{ is not a bridge in } G]; \\ \emptyset & \text{if } y \in C_e. \end{cases}$$

The degree of s or t changes only when $y \in C_t$. If the degree of s or t becomes an even positive number, then the vertex is forced to drop out of the protocol. If (s,t) is not a bridge in G, then the degrees of both s and t become even positive numbers in G but we make only s drop out of the protocol by setting $X = \{s\}$ as above, because, otherwise, the component containing s and t would have no white vertices and G could not become connected.

Thus we have given procedures A_1 and B_1. Figure 2 illustrates how the (A_1,B_1)-protocol generates an Eulerian graph.

We are now ready to explain the reason why we decide the procedure A_1 as above.

If s was contained in $W_1 \cup W_2$ and (s,t) is not a bridge in a new graph G, then $d(t) = 2$ in G. On the other hand, if s were contained in W_3 and (s,t) is not a bridge in G, then s would become a black vertex and t would remain a white vertex in G, due to procedure B_1, but the white vertex t would have degree 4 although the protocol may have not terminated. Thus we give priority to white vertices in $W_1 \cup W_2$ than those in W_3 when we choose s.

If G had two or more connected components having two white vertices of degree 3 like in Figure 1(e) and the two white vertices of one of the components were joined by an edge, then one of the two white vertices would become a white vertex of degree 4. Therefore we do not want to produce two or more connected components like in Figure 1(e). Whenever such a component is produced, there must have existed two or more vertices in W_1. Therefore we do not wish to increase the vertices in W_1. If a vertex in W_2 were chosen as s although $W_1 \neq \emptyset$, then the vertices in W_1 would increase. Thus we put the priority to vertices in W_1 than those in W_2 when we choose s.

We first have a sufficient condition for the (A$_1$,B$_1$)-protocol to achieve an Eulerian key exchange, as follows.

Theorem 1 *The* (A$_1$,B$_1$)*-protocol achieves an Eulerian key exchange for a signature* $\gamma = (c_1, c_2, \cdots, c_k; c_e)$ *if*

(a) $k = 2$, $c_1, c_2 \geq 2$, *and* $c_1 + c_2 \geq c_e + 4$;

(b) $k = 3$, *and every vertex* $i \in V$ *satisfies* $c_i \geq \begin{cases} c_e + 2 & if\ c_e \leq 1; \\ \lceil c_e/2 \rceil + 3 & if\ c_e \geq 2; \end{cases}$

or

(c) $k \geq 4$, *and every vertex* $i \in V$ *satisfies* $c_i \geq \lceil c_e/2 \rceil + 4$.

Proof. omitted in this extended abstract.

We next give a lower bound on the number of cards required for a key set protocol to achieve an Eulerian key exchange.

If a key set protocol achieves an Eulerian key exchange for signature $\gamma = (c_1, c_2, \cdots, c_k; c_e)$, then the key exchange graph G must become an Eulerian graph at the end of any run of the randomized protocol for any deal $\mathcal{C} = (C_1, C_2, \cdots, C_k; C_e)$ having signature γ. Hence, whoever has the card y in the proposed key set $K = \{x, y\}$, G should become an Eulerian graph. Considering a malicious **adversary** to choose y so that the protocol needs cards as many as possible, we have lower bounds on the number of cards as in the following Lemma 1.

Lemma 1 *If there exists a key set protocol achieving an Eulerian key exchange for* $\gamma = (c_1, c_2, \cdots, c_k; c_e)$, *then*

(a) $k = 2$, $c_1, c_2 \geq 2$, *and* $c_1 + c_2 \geq c_e + 4$;
(b-i) $k = 3$, $c_e \leq 1$, *and either* $c_i \geq c_e + 2$ *for every vertex* $i \in V$ *or there exists a pair of vertices* $i, j \in V$ *with* $c_i + c_j \geq c_e + 6$;
(b-ii) $k = 3$, $c_e \geq 2$, *and there exists a pair of vertices* $i, j \in V$ *with* $c_i + c_j \geq c_e + 6$;

or

(c) $k \geq 4$, *and either there exists a vertex* $i \in V$ *with* $c_i \geq c_e + 4$ *or there exists a pair of vertices* $i, j \in V$ *with* $c_i + c_j \geq c_e + 8$.

From Theorem 1 and Lemma 1 we immediately have Theorem 2 on a necessary and sufficient condition for a key set protocol to achieve an Eulerian key exchange when $c_1 = c_2 = \cdots = c_k$.

Theorem 2 *Let $\gamma = (c_1, c_2, \cdots, c_k; c_e)$ be any signature such that $c_1 = c_2 = \cdots = c_k$. Then there exists a key set protocol achieving an Eulerian key exchange for γ if and only if*

(a) $k = 2$, *and* $c_1 \geq \lceil c_e/2 \rceil + 2$;

(b) $k = 3$, *and* $c_1 \geq \begin{cases} c_e + 2 & if\ c_e \leq 1; \\ \lceil c_e/2 \rceil + 3 & if\ c_e \geq 2; \end{cases}$

or

(c) $k \geq 4$, *and* $c_1 \geq \lceil c_e/2 \rceil + 4$.

Thus the (A_1,B_1)-protocol is optimal among the class of all key set protocols achieving an Eulerian key exchange for a signature $\gamma = (c_2, c_2, \cdots, c_k; c_e)$ with $c_1 = c_2 = \cdots = c_k$ in the sense that the number kc_1 of required cards of players

$$kc_1 = \begin{cases} 2\lceil c_e/2 \rceil + 4 & \text{if } k = 2; \\ 3c_e + 6 & \text{if } k = 3 \text{ and } c_e \leq 1; \\ 3\lceil c_e/2 \rceil + 9 & \text{if } k = 3 \text{ and } c_e \geq 2; \\ k\lceil c_e/2 \rceil + 4k & \text{if } k \geq 4 \end{cases}$$

is minimum. Of course, the total number $d = kc_1 + c_e$ of cards dealt to the players and Eve is minimized by the (A_1,B_1)-protocol for any c_e, too.

4 Protocol Using the Minimum Number of Cards

In this section we give the second protocol, the (A_2,B_2)-protocol, which is optimal in the sense that the total number of cards dealt to the players, $\sum_{p=1}^{k} c_p$, is minimum when $c_1 = c_2 = \cdots = c_k$ does not always hold.

For the case $k = 2$, Theorem 1 and Lemma 1 imply that the (A_1,B_1)-protocol in the preceding section is optimal and the minimum number of cards is $c_1 + c_2 = c_e + 4$. We hereafter assume without loss of generality that $k \geq 3$ and $c_1 = \max_{p \in V} c_p$.

Our procedures A_2 and B_2 for the case $c_e \geq 1$ are very simple as follows.

Procedure A_2: Choose always vertex 1 as a proposer s.

Procedure B_2: Choose X as follows.

$$X = \begin{cases} \{t\} & \text{if } d(t) = 2; \\ \emptyset & \text{if } d(t) \leq 1. \end{cases}$$

For the case $c_e \geq 1$, an final Eulerian graph G has a pair of multiple edges joining vertex 1 and each of the other $k - 1$ vertices. Thus G is a so-called double star, and is an Eulerian graph. Although we omit the procedures A_2 and B_2 for the case $c_e = 0$, the (A_2,B_2)-protocol produces an Eulerian graph G in which vertex 1 and two other vertices induce a triangle, and vertex 1 is joined to each of the other $k - 3$ vertices by a pair of multiple edges.

Then the following theorem holds on the (A_2,B_2)-protocol.

Theorem 3 *Let $k \geq 3$. The* (A_2,B_2)*-protocol achieves an Eulerian key exchange for* $\gamma = (c_1, c_2, \cdots, c_k; c_e)$ *if*

$$\text{(a)} c_e \geq 1 \textit{ and } \begin{cases} c_1 \geq 2k + c_e - 2, \\ c_p \geq 2 \textit{ for each } p,\ 2 \leq p \leq k; \end{cases}$$

or

$$\text{(b)} c_e = 0 \textit{ and } \begin{cases} c_1 \geq 2k - 4, \\ c_p \geq 2 \textit{ for each } p,\ 2 \leq p \leq k. \end{cases}$$

Moreover, the (A_2,B_2)*-protocol is optimal among the class of key set protocols achieving an Eulerian key exchange for a signature in the sense that the number of required cards of players*

$$\sum_{p=1}^{k} c_p = \begin{cases} 4k + c_e - 4 \textit{ if } c_e \geq 1; \\ 4k - 6 \qquad \textit{if } c_e = 0 \end{cases}$$

is minimum.

Thus the number of cards required by the (A_2,B_2)-protocol is much smaller than that by the (A_1,B_1)-protocol especially when c_e is large.

One can easily observe that the (A_2,B_2)-protocol achieves an Eulerian key exchange for any signature $\gamma = (c_1, c_2, \cdots, c_k; c_e)$ satisfying (a) or (b) in Theorem 3. Thus it is easy to prove the first proposition in Theorem 3.

5 Protocol for Shortest Eulerian Circuit

In this section we give the third protocol, the (A_1',B_1)-protocol, which is optimal in the sense that the length of the Eulerian circuit is minimum. The length of the Eulerian circuit corresponds to the time needed for the circulation of a message with the acknowledgment of secure receipt. Denote the length by $l(G)$, then $l(G) = \frac{1}{2}\sum_{p=1}^{k} d(p)$ since $l(G)$ is equal to the number of edges in the Eulerian key exchange graph G. We define the *acknowledgment time* for a signature $\gamma = (c_1, c_2, \cdots, c_k; c_e)$ as follows.

Definition 4 *We call* $T(\mathrm{A}, \mathrm{B}; \gamma) = \max_{C} \max_{G_C} l(G_C)$ ***the acknowledgment time*** *of an* (A,B)*-protocol for a signature* γ*, where* C *runs over all deals having signature* γ*, and* G_C *runs over all key exchange graphs formed by any execution of the protocol for the deal* C*. If the protocol does not achieve an Eulerian key exchange for* γ*, we define* $T(\mathrm{A}, \mathrm{B}; \gamma) = \infty$.

We slightly modify the procedure A_1 in Section 3 so that, whenever a vertex in W_2 is chosen as a proposer s, an isolated vertex in W_2 is chosen as s with priority over non-isolated vertices in W_2. We denote this modified procedure by A_1'. The procedure A_1' is as follows.

Procedure A_1': Choose a proposer $s \in W$ as follows.

Case A1': $C_e \neq \emptyset$ *and there exists* $i \in W_1 \cup W_2$ *such that* $|C_i| + d(i) \geq 5$.

If there exists a vertex $i \in W_1$ such that $|C_i| + d(i) \geq 5$, choose any of these vertices as a proposer s. Otherwise, but if there exists an isolated vertex $i \in W_2$ such that $|C_i| + d(i) \geq 5$, then choose any of these vertices as s. Otherwise, choose as s any vertex $i \in W_2$ such that $|C_i| + d(i) \geq 5$.

Case A2': $C_e = \emptyset$ *and* $W_1 \cup W_2 \neq \emptyset$.

If $W_1 \neq \emptyset$, choose any vertex in W_1 as s. If $W_1 = \emptyset$ and there exists an isolated vertex $i \in W_2$, choose as s any of these vertices. Otherwise, choose as s any vertex in W_2.

Case A3': *Otherwise.*

In this case $W = \{i, j\}$. Choose as s the vertex i in W such that $|C_i| \geq |C_j|$.

We can prove the following theorem, showing that the (A_1',B_1)-protocol produces an Eulerian graph G in which at most half the vertices have degree four and all the remaining vertices have degree two.

Theorem 4 *Let* $k \geq 4$, *and let* $\gamma = (c_1, c_2, \cdots, c_k; c_e)$. *Then*

$$T(A_1', B_1; \gamma) \leq \begin{cases} \lfloor \frac{3}{2}k \rfloor & if\ \lceil c_e/2 \rceil + 4 \leq c_p\ for\ all\ p \in V; \\ \lfloor \frac{3}{2}k \rfloor - 1 & if\ c_e + 4 \leq c_p\ for\ all\ p \in V. \end{cases}$$

Considering a malicious adversary producing at least $\lfloor k/2 \rfloor$ edge-disjoint cycles in G, we have the following theorem on a lower bound on the acknowledgment time of a key set protocol.

Theorem 5 *Let* $k \geq 4$, *let an* (A,B)*-protocol be any key set protocol, and let* $\gamma = (c_1, c_2, \cdots, c_k; c_e)$ *be any signature. Then*

$$T(A, B; \gamma) \geq \left\lfloor \frac{3}{2}k \right\rfloor - 1.$$

In particular, if $c_p \leq c_e + 3$ *for all* $p \in V$, *then*

$$T(A, B; \gamma) \geq \left\lfloor \frac{3}{2}k \right\rfloor.$$

Theorems 1(c), 2(c), 4 and 5 immediately imply the following Theorem 6.

Theorem 6 *Let* $k \geq 4$. *For any signature* $\gamma = (c_1, c_2, \cdots, c_k; c_e)$ *satisfying* $c_1 = c_2 = \cdots = c_k$, *the* (A_1',B_1)*-protocol is optimal among the class of key set protocols in the sense that the acknowledgment time*

$$T(A_1', B_1; \gamma) = \begin{cases} \lfloor \frac{3}{2}k \rfloor - 1 & if\ c_e + 4 \leq c_1; \\ \lfloor \frac{3}{2}k \rfloor & if\ \lceil c_e/2 \rceil + 4 \leq c_1 \leq c_e + 3; \\ \infty & if\ c_1 \leq \lceil c_e/2 \rceil + 3 \end{cases}$$

is minimum. If γ *does not always satisfy* $c_1 = c_2 = \cdots = c_k$ *but satisfies* $\lceil c_e/2 \rceil + 4 \leq c_p$ *for every vertex* $p \in V$, *then the* (A_1',B_1)*-protocol is nearly optimal in the sense that* $T(A_1', B_1; \gamma)$ *does not exceed the minimum acknowledgment time more than one.*

Since procedure A_1' is a type of procedure A_1, the (A_1',B_1)-protocol not only forms the shortest Eulerian cycle, but also requires the minimum number of cards when $c_1 = c_2 = \cdots = c_k$.

One can observe that the acknowledgment time of the (A_1,B_1)-protocol satisfies $T(A_1, B_1; \gamma) = 2k - 2$ for a signature γ for which the protocol achieves an Eulerian key exchange, as follows. The protocol may produce a graph G of $2k-2$ edges in which exactly two vertices have degree 2 and all the other vertices have degree 4. Therefore $T(A_1, B_1; \gamma) \geq 2k - 2$. Any Eulerian graph G produced by the (A_1,B_1)-protocol has at most $2k-2$ edges, and hence $T(A_1, B_1; \gamma) \leq 2k-2$. Note that all vertices in G have degree 2 or 4, and that two or more vertices including the first black vertex have degree 2 in G.

On the other hand, the acknowledgment time of the (A_2,B_2)-protocol is

$$T(A_2, B_2; \gamma) = \begin{cases} 2k-2 \text{ if } c_e \geq 1; \\ 2k-3 \text{ if } c_e = 0. \end{cases}$$

Since $T(A_1', B_1; \gamma) \leq \lfloor \frac{3}{2}k \rfloor$, the ($A_1'$,$B_1$)-protocol produces an Eulerian circuit much shorter than those by the (A_1,B_1)- and (A_2,B_2)-protocols, and than an Eulerian circuit passing through every edge of a spanning tree twice.

Acknowledgment

We thank Dr. H. Suzuki of Ibaraki University for fruitful discussions and helpful comments.

References

1. M. J. Fischer, M. S. Paterson and C. Rackoff, "Secret bit transmission using a random deal of cards," DIMACS Series in Discrete Mathematics and Theoretical Computer Science, AMS, 2, pp. 173–181, 1991.
2. M. J. Fischer and R. N. Wright, "An application of game-theoretic techniques to cryptography," DIMACS Series in Discrete Mathematics and Theoretical Computer Science, AMS, 13, pp. 99–118, 1993.
3. M. J. Fischer and R. N. Wright, "An efficient protocol for unconditionally secure secret key exchange," Proceedings of the 4th Annual Symposium on Discrete Algorithms, pp. 475–483, 1993.
4. M. J. Fischer and R. N. Wright, "Bounds on secret key exchange using a random deal of cards," J. Cryptology, 9, pp. 71–99, 1996.
5. M. J. Fischer and R. N. Wright, "Multiparty secret key exchange using a random deal of cards," Proc. Crypto '91, Lecture Notes in Computer Science, 576, pp. 141–155, 1992.
6. F. Harary, "Graph Theory," Addison-Wesley, Reading, 1969.
7. T. Nishizeki and N. Chiba, "Planar Graphs: Theory and Algorithms," North-Holland, Amsterdam, 1988.
8. P. Winkler, "The advent of cryptology in the game of bridge," Cryptologia, 7, pp. 327–332, 1983.

One-Time Tables for Two-Party Computation

Donald Beaver[1]

IBM/Transarc; beaver@transarc.com

Abstract. In two-party secure computation, a pair of mutually-distrusting and potentially malicious parties attempt to evaluate a function $f(x, y)$ of private inputs x and y, held respectively by each, without revealing anything but $f(x, y)$ and without involving a trusted third party. This goal has been achieved with varying degrees of generality and efficiency using a variety of primitives, including combined oblivious transfer (OT) [GMW87], abstract oblivious transfer [K88], and committed oblivious transfer [CTG95].

This work introduces the concept of a *two-party one-time table* (OTT), a novel primitive that is theoretically equivalent to precomputed OT. The OTT is tailored to support field computations rather than single-bit logical operations, thereby streamlining higher-level computations, particularly where information-theoretic security is demanded.

The two-party one-time table is also motivated by the ease with which it can be constructed using simple resources provided by one or more partly-trusted external servers. This commodity-based approach strengthens overall security by ensuring that information flows strictly from servers to Alice and Bob, removing the need to trust third parties with the sensitive data itself.

1 Introduction

Two-party secure computation is a process by which two parties simulate a trusted but nonexistent mediator who helps them to compute a function $f(x, y)$ of private inputs held by each. This *virtual mediator* accepts x from Alice and y from Bob, and returns $f(x, y)$ to each, without revealing any intermediate results of the computation. Naturally, the goal is to achieve this end result with equal security, even when no such party is ready and willing to help.

Several successful designs for such protocols have been developed, including work by Yao [Yao82a, Yao82b], Goldreich, Micali, and Wigderson [GMW87], Chaum, Damgrd, and van de Graaf [CDG87], Kilian [K88], and Crépeau, Tapp and van de Graaf [CTG95], among others. These approaches span quite different fault models, ranging from computationally-bounded to infinitely-powerful attackers, and from static to adaptive security.

The work presented here addresses the information-theoretic domain, where perfect or statistical security is demanded, and adversaries have infinite computing resources. In this setting, Kilian's work is the pioneering achievement: two-party computation with statistical security against infinitely-powerful attackers is possible, using Oblivious Transfer as a primitive [K88]. (Oblivious

Transfer is a process by which Alice transmits bit b to Bob; it arrives with 50-50 probability; Alice does not know whether it arrived, but Bob does [Rab81].)

The protocol presented in [K88] certainly requires only polynomial time, but its complexity is high in practical terms. Recent results by Crépeau, Tapp and van de Graaf have reduced the complexity greatly [CTG95]. In both of these works, the fundamental steps are bitwise logical operations, built ultimately from oblivious transfers (including "committed" and "one-out-of-two" versions).

Our goal is to seek algorithmic speed-ups and simplifications by supporting field operations rather than AND's and EXOR's. Using general-purpose OT-based results, a direct multiplication of two m-bit elements involves a circuit of size m^2; each gate requires a subprotocol invoking $\Theta(k^3)$ OT's [CTG95]; thus communication complexity is on the order of $\Theta(m^6 k^3)$. The protocols presented here, however, require only $O(km)$ message bits. Moreover, a direct multiplication requires $\Omega(m)$ rounds of communication (or at least $\log m$ with more clever circuit design) using OT-based methods, whereas our protocol requires 1 round.

In part, we bring into question the centrality of OT and describe computations that are simpler to understand and to achieve using equally "arbitrary" (but certainly somewhat more complicated) primitives. These primitives are optimized for particular operations as well as for easy production by an underlying channel or third-party.

Which Tools to Use? The ubiquitous applicability of OT has led to an overwhelming amount of effort toward finding methods to implement it efficiently and securely.

But the engineer's central question of how to achieve OT overshadows a deeper one: if we employ significantly complicated mechanisms to achieve OT (in order to thereby enable cryptographic protocols), then can these mechanisms and their assumptions be used in a simpler or alternative fashion to power more suitable primitives?

We attempt to simplify the higher-level specification of a two-party task by supplying optimized arithmetic operations, rather than bitwise logical ones. Naturally, these results do not achieve anything more general than what OT already enables. But they achieve it faster and in a clearer, more direct manner, where field computations are involved.

Third-Party Assistance. Oblivious transfer is often treated as an underlying communication channel with specialized properties (such as guaranteed and measurable noise). A slightly different twist is to regard it as a service provided by a third-party, M.

From the view of trust management, it seems almost circular to use OT to help Alice and Bob compute $f(x, y)$: if M is trusted to implement OT on sensitive bits b, why doesn't M just accept x and y and compute $f(x, y)$ directly? In either case, M sees both secret inputs.

Through precomputed OT [B95], however, the distinction becomes clearer. By executing OT in advance on bits that are unrelated to their desired input bits, Alice and Bob can achieve the later computation of $f(x, y)$ without requiring on-line assistance from M. Equally importantly, they achieve their computation *without ever transmitting sensitive information to M.* Thus, the service provider (or "channel") M is trusted in a significantly less extensive manner.

One Time Tables. Assuming, then, that a trusted third party is willing to give or sell its assistance to Alice and Bob in an initial and relatively passive manner, is OT really the product they wish to purchase?

The *two-party one-time table* (OTT) is an appealing alternative. The OTT is theoretically equivalent to OT (in terms of privacy) but specially tailored for expressing computations in arithmetic terms. It simplifies Alice's and Bob's interaction by supporting direct, field-based computation, by avoiding extensive gate-based cut-and-choose operations, and by obviating intricate commitment protocols. The product is simple for the trusted party to generate, as well, requiring only random number generation and field-element computations.

In addition to simplifying secure computations, the OTT is designed to support off-line assistance – where, namely, the trusted service provider helps out initially but departs without learning (or needing to learn) Alice's and Bob's desired computation. This simplifies trust management for Alice and Bob, who need rely less extensively on the integrity of the server (or underlying channel).

Although we do not give details here, it is also possible to construct a robust OTT from resources provided by more than one server, even though one or more of the servers may be passively or actively faulty. Because Alice and Bob can seek out and choose among various, competing, cryptographic service providers, our solution matches well with the commodity-based approach for server-assisted cryptography proposed in [B97].

Results. Our main result is the following:

Theorem 1. *Let Alice and Bob have access to a* two-party one-time table. *There exists a statistically-secure two-party protocol for oblivious circuit evaluation that uses at most one message round per gate level and supports field operations including linear combinations, multiplication, and division.*

Properties and Comparisons. The goals sought and achieved in this paper differ from earlier approaches in several ways.

- *Computational* vs. *Information-Theoretic Security.* Speedups due to homomorphisms and various number-theoretic properties are inapplicable here, as they do not provide information theoretic security.
- *Lazy Verification.* Many earlier approaches use zero-knowledge proofs, cut-and-choose, and other methods to ensure correctness at each stage. Our OTT construction allows verification to be left until the final stage, thereby providing faster and simpler algorithms.
- *Commodity-Based Cryptography.* This work shares a basic motivation behind commodity-based cryptography [B97]: enlist third-party assistance to achieve a secure, joint computation. [B97] provided support for achieving OT using multiple servers. Here, we seek to replace OT and we address the case of only one server. While the methods presented here can be extended to rely on multiple servers, they are nonetheless distinct (whether or not extended to multiple servers).

2 Two-Party One-Time Tables

Let k be a security parameter, let $t = 2k+2$, let $N = 3t+1$, and let F be a field of size exceeding $N+1$. (If an arbitrary field is desired, an extension is suitable.) Let $R = \{\iota_1, \ldots, \iota_N\}$ be the roots of codewords for a BCH code over F [ML85].

We generally follow a notational convention that superscripts describe the party that holds a value; *e.g.* s^A is a string held by Alice.

Let $(\mathtt{CommA}^A(x), \mathtt{CommA}^B(x))$ be a committal by A to value x. That is, $\mathtt{CommA}^A(x)$ is a string held privately by A, and $\mathtt{CommA}^B(x)$ is a string held privately by B and which reveals nothing about x. A can *unveil* x by conveying $\mathtt{CommA}^A(x)$, while B can *verify* this unveiling by using $\mathtt{CommA}^B(x)$. (Further details are described later.)

2.1 Wires

A *wire* is a set of committed values known to each party, respectively. A wire exists in one of two states: *prepared*, and *dedicated.*

A prepared wire represents some random field element, a. Alice holds the values $\{\phi_i^A(a)\}_{i=1..N}$; Bob holds $\{\phi_i^B(a)\}_{i=1..N}$. (Unless otherwise noted, these sets will implicitly span $i = 1..N$.) These values are selected uniformly at random, constrained by $f(\iota_i) = \phi_i^A(a) + \phi_i^B(a)$ where $f(\cdot)$ is a (uniformly random) polynomial of degree $t = 2k+2$, satisfying $f(0) = a$. Although there are only two parties involved, this construction is similar to Shamir's secret sharing [Sha79] combined with sum-sharing of the pieces.

(As mentioned, all values are committed. That is, for each i, Alice holds $\mathtt{CommA}^A(\phi_i^A(a))$ while Bob holds $\mathtt{CommA}^B(\phi_i^A(a))$. Bob is likewise committed.)

A dedicated wire adjusts the represented value to some particular element. That is, a dedicated wire is the combination of the prepared wire with a public *correction* Δa to the (secret) represented value, a:

$$\psi^A(x) = (\Delta a, \{\phi_i^A(a))\}) \quad \text{(Alice)}$$
$$\psi^B(x) = (\Delta a, \{\phi_i^B(a))\}) \quad \text{(Bob).}$$

Here, the value represented by the wire is $x = a + \Delta a$. Again, although there are only two parties involved here, the construction is inspired by Beaver's circuit randomization technique [B91].

We refer to the value a on a dedicated wire w as the *core* value on w, and to Δa as the *correction* to w.

Input Wires. Prepared wires can also be constructed as *input* wires. An *input wire for Alice* is a prepared wire for which Alice is given *all* of the values in the representation, namely $\{\phi_i^B(a)\}$ as well. (One way to imagine this is as though Bob simply unveiled all of his values for that wire.) Input wires for Bob are defined similarly.

Virtual Wires. Another convenient concept is that of a *virtual* wire. A virtual wire is a wire (namely a collection of values) constructed from other wire values, and in this way distinguished from wires constructed before the outset of the protocol. Virtual wires are employed in the same way as original wires; the distinction is primarily for explanatory purposes.

2.2 Gates

A *multiplication gate* is a set of three prepared wires representing secret values r, s, and rs, where r and s are random field elements (*cf.* [B91]). Other gates (*e.g.* inversion) are described later.

Like wires, gates exist in two states, *prepared* and *dedicated*, depending on whether all of their wires are prepared or not. Gates are used in a one-time fashion: once dedicated, they are never rededicated. The evaluation of gates – including addition, multiplication, and multiplicative inversion gates – is described in §3.2.

2.3 Production and Supply

A trusted third party will supply some number N_I of input wires for Alice and N_I for Bob; some number N_P of prepared wires; some number N_M of multiplication gates, and some number N_D of inversion gates. The construction of these values requires a good random number generator and simple arithmetic. This collection of values comprises the two-party one-time table, with half destined for Alice and half destined for Bob.

The trusted party is thereafter not involved, although it (or another party) may later mediate key exchange to enable fair revelation of the final result.

2.4 Committal

The one-time table employs a statistically-secure committal scheme. Because the trusted party is supplying the one-time table, it builds the committal simultaneously, so that Alice and Bob need not execute a committal protocol. One arithmetic-based way to implement this is to commit Alice to a by selecting a random line $y = a + bx$ along with a random u, then supplying Alice with $\texttt{CommA}^A(a) = (a, b)$ and Bob with $\texttt{CommA}^B(a) = (u, v = a + bu)$ (*cf.* [RB89, TW86]). Alice unveils by revealing (a, b). Bob checks that $y = a + bx$. Her chance of successfully cheating is bounded by $1/(|F|-1)$. Other commitment schemes are also suitable.

2.5 Arithmetic: Moduli and Fields

Because there are only two clients, Alice and Bob, the protocol is not as immediately tied to particular moduli or fields as may be the case with multiparty computation, where field sizes must (for example) exceed the number of parties. Instead, the security parameter places a lower bound on the field size used for the OTT. Extension fields can be used to accommodate the case where a desired prime modulus (or given field) is smaller than the security parameter (more accurately, $2k+4$). Simultaneous execution using different prime moduli, using the Chinese Remainder Theorem at input and output, can accommodate factored, composite moduli. (Note that the multiplicative inverse protocol must still be restricted to elements that are guaranteed to be invertible, else information may be leaked.)

3 Unverified Computation

Input and circuit evaluation consist of turning prepared wires into dedicated ones, thereby propagating inputs from the input side of a circuit C_f for $f(x, y)$

to the output side. We shall assume that $f(x, y)$ has been described in terms of addition, multiplication, and multiplicative inversion operations over field F; we focus on evaluating individual operations.

3.1 Input

When Alice has input value x, she takes the next prepared input wire (for Alice) and announces $\Delta a = x - a$, without revealing x and a, of course. Bob provides inputs in a similar fashion. Clearly, Δa appears to Bob as a random element independent of x, since a is uniformly random (and not otherwise used).

3.2 Evaluation

We focus on evaluating addition and multiplication. Multiplicative inverses are a straightforward application of multiplication, using a method first described in [BB88], but require additional optional mechanisms.

Addition Let $(\psi^A(x), \psi^B(x))$ represent the value x on dedicated wire w_1, and $(\psi^A(y), \psi^B(y))$ represent the value y on dedicated wire w_2. We first show how to construct a virtual, dedicated wire w_3 representing the value $z = x + y$.

In more detail, let a be the core value on w_1 and b the core value on w_2; let Δa and Δb be the commonly-known corrections. To add the values held by two dedicated wires, Alice and Bob each locally add their shares as follows:

Wire Value	Alice	Bob
$x = a + \Delta a$	$\psi^A(x) = (\Delta a, \{\phi_i^A(a)\})$	$\psi^B(x) = (\Delta a, \{\phi_i^B(a)\})$
$y = b + \Delta b$	$\psi^A(y) = (\Delta b, \{\phi_i^A(b)\})$	$\psi^B(y) = (\Delta b, \{\phi_i^B(b)\})$
$z = x + y$	$\psi^A(z) = (\Delta a + \Delta b, \{\phi_i^A(a) + \phi_i^A(b)\})$	$\psi^B(z) = (\Delta a + \Delta b, \{\phi_i^B(a) + \phi_i^B(b)\})$

Remarks. (1) Even though Alice and Bob may have been committed to the values on the input wires, they are not directly committed to the value on virtual wire w_3. At the moment, there is nothing to prevent them from pretending that their values for w_3 are different, but this will be taken care of later. (2) Linear combinations are easily evaluated by replacing each sum above with the desired linear combination.

Multiplication The crux of the process is to multiply two values x and y to obtain a third, z, without revealing any of them. Let dedicated wires w_x and w_y represent the values x and y through core values a and b with corrections Δa and Δb, respectively. Alice and Bob select the next available undedicated multiplication gate g, which, as explained above, consists of three prepared wires w_1, w_2, and w_3, representing the values r, s and rs, respectively.

Let L interpolate a polynomial through n elements; namely:

$$L(\{(x_i, y_i)\}_{i=1..n}, x) = \sum_{i=1}^{n} (\prod_{j \neq i} \frac{x - x_j}{i - x_j}) y_i.$$

Let `Interpolate` denote a method that interpolates a set of elements (*i.e.* computes L) and returns the free term if the polynomial is of degree at most t. Otherwise, the method throws an exception, aborting the protocol.

The multiplication protocol is described in Fig. 1. For now, assume that

Multiply-Subroutine

1.1.	A:	$\alpha_{A,i} \leftarrow \phi_i^A(a) - \phi_i^A(r)$
		$\beta_{A,i} \leftarrow \phi_i^A(b) - \phi_i^A(s)$
1.2.	$A \to B$:	$\{\alpha_{A,i}, \beta_{A,i}\}$
1.3.	B:	$\alpha_{B,i} \leftarrow \phi_i^B(a) - \phi_i^B(r)$
		$\beta_{B,i} \leftarrow \phi_i^B(b) - \phi_i^B(s)$
1.4.	$B \to A$:	$\{\alpha_{B,i}, \beta_{B,i}\}$
2.1	A:	$\alpha \leftarrow$ **Interpolate**$(\{(\iota_i, \alpha_{A,i} + \alpha_{B,i})\})$
		$\beta \leftarrow$ **Interpolate**$(\{(\iota_i, \beta_{A,i} + \beta_{B,i})\})$
		$\phi_i^A(c) \leftarrow \phi_i^A(rs) + \phi_i^A(r)(\beta + \Delta b) + \phi_i^A(s)(\alpha + \Delta a)$
		$\Delta c \leftarrow (\alpha + \Delta a)(\beta + \Delta b)$
2.2	B:	$\alpha \leftarrow$ **Interpolate**$(\{(\iota_i, \alpha_{A,i} + \alpha_{B,i})\})$
		$\beta \leftarrow$ **Interpolate**$(\{(\iota_i, \beta_{A,i} + \beta_{B,i})\})$
		$\phi_i^B(c) \leftarrow \phi_i^B(rs) + \phi_i^B(r)(\beta + \Delta b) + \phi_i^B(s)(\alpha + \Delta a)$
		$\Delta c \leftarrow (\alpha + \Delta a)(\beta + \Delta b)$

Fig. 1. Two-pass protocol to evaluate a multiplication gate.

Alice and Bob are honest. Noting that L is linear (that is, $L(\{(\iota_i, u_i + v_i)\}, x) = L(\{(\iota_i, u_i)\}, x) + L(\{(\iota_i, v_i)\}, x)$ and $L(\{(\iota_i, cu_i)\}, x) = c \cdot L(\{(\iota_i, u_i)\}, x)$), it is easy to see that the values α and β evaluate to the following, by design:

$$\begin{aligned} \alpha &\equiv L(\{(\iota_i, \alpha_{A,i} + \alpha_{B,i})\}, 0) \\ &= L(\{(\iota_i, \phi_i^A(a) + \phi_i^B(a))\}, 0) + L(\{(\iota_i, -\phi_i^A(r) - \phi_i^B(r))\}, 0) = a - r \end{aligned}$$

and likewise for $\beta = b - s$. As a result,

$$\begin{aligned} c &= L(\{(\iota_i, \phi_i^A(c) + \phi_i^B(c))\}, 0) \\ &= L(\{(\iota_i, \phi_i^A(rs) + \phi_i^B(rs))\}, 0) + (\beta + \Delta b) L(\{(\iota_i, \phi_i^A(r) + \phi_i^B(r))\}, 0) \\ &\quad + (\alpha + \Delta a) L(\{(\iota_i, \phi_i^A(s) + \phi_i^B(s))\}, 0) \\ &= rs + (\beta + \Delta b) r + (\alpha + \Delta a) s, \end{aligned}$$

from which $c + \Delta c = xy$ can be verified through direct algebra.

3.3 Output

In §4, we describe how Alice and Bob verify that each has behaved sufficiently properly for them to continue with the final stage, namely the revelation of the output, $f(x, y)$. If that verification succeeds, they simply unveil their output wires and interpolate the results. That is, for each output z on a wire with core value a and correction Δa, Alice unveils $\{\phi_i^A(a)\}$ and Bob unveils $\{\phi_i^B(a)\}$. Using these values and the commonly-known value of Δa, they each use BCH decoding methods to correct up to t errors, computing (after error correction):

$$z = L(\{(\iota_i, \phi_i^A(a) + \phi_i^B(a))\}, 0) + \Delta a.$$

Fairness. Because this is a two-party setting, malicious faults enable one party to withold information after it has learned something from the other. The results of Cleve apply [Cle86]: with simple adjustments that cost extra rounds, the amount of advantage gained by witholding information can be limited to an optimal, inverse polynomial amount.

Alternatively, the help of a third party can be invoked. This exchange agent can be the original supplier or a new server. To prevent leaking the final value to the exchange agent, Alice and Bob modify f so that it accepts an extra random input from each and masks the output with that input:

$$\hat{f}(x \circ r, y \circ s) = (f(x,y) + r, f(x,y) + s).$$

After running the protocol to compute $\hat{f}$ instead, Alice and Bob provide the exchange agent with their complete set of information regarding the final wire, a. For Alice, this information comprises

$$\{(\phi_i^A(a), \texttt{CommA}^A(\phi_i^A(a)), \texttt{CommB}^A(\phi_i^B(a)))\}.$$

When the exchange agent receives the two messages containing this information, it performs the appropriate verification against each other. If no errors are found, it forwards each message to the other player. After receiving the forwarded message, each player then sends its message directly to the other (to avoid falsification on the part of the exchange agent). Each player then finishes the protocol as specified earlier (through error correction and interpolation), unmasking the result with r or s, respectively.

4 Lazy Verification

Notice that no mention of protection against malicious behavior dirtied the preceding description. Nor are we about to suggest the insertion of zero-knowledge proofs or cut-and-choose methods at each step. Instead, the verification occurs at the very end, before the output is revealed. All gates are verified simultaneously.

This means that Bob can trivially and maliciously add 1 to an intermediate wire value, for example, during the computation stage. All he need do is to increment each of his $\phi_i^B()$ values. Our goal is to enable Alice (and Bob) to detect such misbehavior just before the final output is revealed, and to abort if so.

Let us refer to the data associated with a given index $i \in [1..N]$ as a *channel.* For example, the values $\phi_i^A(x)$ and $\phi_i^B(x)$ over all x form part of channel i.

The verification process is simple: Alice randomly selects k channels $I = \{i_1, \ldots, i_k\} \subseteq [1..N]$ and announces them. Bob then randomly selects k remaining channels $J = \{j_1, \ldots, j_k\} \subseteq [1..N] \backslash I$ and announces them.

Alice and Bob then unveil all values indexed by $I \cup J$, throughout the entire circuit. For instance, where Alice holds $\{\phi_i^A(a)\}$, she unveils $\{\phi_i^A(a) : i \in I \cup J\}$.

Alice and Bob then individually check whether those values are consistent with the sums announced *on those channels* in multiplication steps. (This also requires propagating the values through virtual wires, *e.g.* where two intermediate results were added.) If a party finds any discrepancy whatsoever, it aborts.

5 Statistical Security

Because the core values are secret and uniformly random, the correction values are uniformly random values, regardless of the ultimate wire values. Because the secret wire values are used only once, the correction values reveal nothing about the actual values.

The verification procedure reveals some $2k$ channels. Because the representation on each wire employs a secret random polynomial of degree t, followed

Copy-Subroutine		
1.1.	A:	$\alpha_{A,i} \leftarrow \phi_i^A(a) - \phi_i^A(r)$
1.2.	$A \rightarrow B$:	$\{\alpha_{A,i}\}$
1.3.	B:	$\alpha_{B,i} \leftarrow \phi_i^B(a) - \phi_i^B(r)$
1.4.	$B \rightarrow A$:	$\{\alpha_{B,i}\}$
2.1	A:	$\alpha \leftarrow$ **Interpolate**$(\{(\iota_i, \alpha_{A,i} + \alpha_{B,i})\})$
		$\Delta r \leftarrow \Delta a + \alpha$
2.2	B:	$\alpha \leftarrow$ **Interpolate**$(\{(\iota_i, \alpha_{A,i} + \alpha_{B,i})\})$
		$\Delta r \leftarrow \Delta a + \alpha$

Fig. 2. Protocol to copy a wire value (fanout 1).

by a random pair summation, the set of values $\{(\phi_i^A(a), \phi_i^B(a)) : i \in I \cup J\}$ is uniformly distributed over $(F \times F)^{2k}$.

If Alice attempts to depart from the protocol, then Bob aborts the protocol unless Alice either (1) unveils at least one incorrect value successfully, or (2) announces at least $N - t$ incorrect values in some computation step. (If she makes fewer changes, then an α value will not interpolate correctly, and Bob will abort.)

Case (1) clearly occurs with probability $o(k^{-c})$ for any c. In case (2), each of Bob's k random index choices gives a chance of at least $k/N \geq 1/7$ to give detection. Alice's chance of success is at most $(6/7)^k$, which is $o(k^{-c})$ for any c.

6 Copying and Multiplicative Inverses

Because a fraction of the channels are revealed whenever a wire is verified, it is clear that wires cannot be reused indefinitely for multiple computations. This does *not* place a bound on the fan-out of a given gate or input during a single computation; the sister values "down the line" are verified simultaneously at the end. If wires are to be used in later computations, or revealed (for whatever reason) midstream, then further techniques are needed to realize unrestricted fan-out. These techniques are described below, for copying and for fast ($O(1)$-round), direct multiplicative inversion.

Copying. The one-time table is expanded to include copy gates. A *copy gate* is a set of two or three prepared wires representing the same secret random value r. The first wire is designated the input wire and the other one or two are outputs. A fanout-1 gate is essentially a renewal of the hidden randomness used to encode values, while a fanout-2 gate is useful for modelling multiple-fanout circuit gates.

The procedure described in Fig. 2 shows how to execute a copy fanout-1 operation on some dedicated wire w with core value a and correction Δa. Higher fanout is a simple generalization. It is straightforward to show that $\alpha = a - r$. Furthermore, $r + \Delta r = r + \Delta a + (a - r) = \Delta a + a = x$.

Premature Verification. To perform a verification before the final output has been reached, choose a slice across the circuit and duplicate each wire that crosses it, using fanout-1 *copy* gates. Perform the verification described in §4 on the portion of the circuit up to the copy gates. Later verifications operate on untouched wires after the copy gates and are fully independent of earlier verifications.

Multiplicative Inverses. Using methods pioneered in [BB88], Alice and Bob can take the multiplicative inverse of a nonzero intermediate value without revealing it. Rather than simulating some arithmetical or logical algorithm for multiplicative inverses, observe that for any nonzero x and u: $x^{-1} = (ux)^{-1}u$.

An *inverse* gate consists of two wires, each representing u for some random invertible u. It is employed to invert an arbitrary nonzero x as follows. First, x is copied using a fanout-2 copy gate; one copy is used for all later applications of x, while the other is used further in this subprotocol. Second, x is multiplied with the first copy of u using the multiplication protocol described above. Third, a premature verification is performed, where the slice includes the copy gate. Fourth, the product wire for ux is revealed, using the output protocol of §3.3. Finally, the value of $(ux)^{-1}$ is used as a constant multiplier to the second wire containing u: each of the values $\phi_i^A(u)$ and $\phi_i^B(u)$ is multiplied by $(ux)^{-1}$, and Δu is set to zero.

References

[BB88] J. Bar-Ilan, D. Beaver. "Non-Cryptographic Fault-Tolerant Computing in a Constant Expected Number of Rounds of Interaction." *Proceedings of PODC,* ACM, 1989, 201–209.

[B91] D. Beaver. "Efficient Multiparty Protocols Using Circuit Randomization." *Advances in Cryptology – Crypto '91 Proceedings,* Springer–Verlag LNCS 576, 1992, 420–432.

[B95] D. Beaver. "Precomputing Oblivious Transfer." *Advances in Cryptology – Crypto '95 Proceedings,* Springer–Verlag LNCS 963, 1995, 97–109.

[B97] D. Beaver. "Commodity-Based Cryptography." *Proceedings of the* 29th *STOC,* ACM, 1997, 446–455.

[Cle86] R. Cleve. "Limits on the Security of Coin Flips when Half the Processors are Faulty." *Proceedings of the* 18th *STOC,* ACM, 1986, 364–370.

[CDG87] D. Chaum, I. Damgrd, J. van de Graaf. "Multiparty Computations Ensuring Secrecy of Each Party's Input and Correctness of the Output." *Advances in Cryptology – Crypto '87 Proceedings,* Springer–Verlag LNCS 293, 1988.

[CTG95] C. Crépeau, A. Tapp, J. van de Graaf. "Committed Oblivious Transfer and Private Multi-Party Computations." *Advances in Cryptology – Crypto '95 Proceedings,* Springer–Verlag LNCS 963, 1995, 110–123.

[GMW87] O. Goldreich, S. Micali, A. Wigderson. "How to Play Any Mental Game, or A Completeness Theorem for Protocols with Honest Majority." *Proceedings of the* 19th *STOC,* ACM, 1987, 218–229.

[K88] J. Kilian. "Founding Cryptography on Oblivious Transfer." *Proceedings of the* 20th *STOC,* ACM, 1988, 20–29.

[ML85] A. Michelson, A. Levesque. *Error-Control Techniques for Digital Communication.* John Wiley and Sons, New York: 1985.

[Rab81] M.O. Rabin. "How to Exchange Secrets by Oblivious Transfer." TR-81, Harvard, 1981.

[RB89] T. Rabin, M. Ben-Or. "Verifiable Secret Sharing and Multiparty Protocols with Honest Majority." *Proceedings of the* 21st *STOC,* ACM, 1989, 73–85.

[Sha79] A. Shamir. "How to Share a Secret." *Communications of the ACM,* **22**, 1979, 612–613.

[TW86] M. Tompa, H. Woll. "How to Share a Secret with Cheaters." *Advances in Cryptology – Crypto '86 Proceedings,* Springer–Verlag LNCS 263, 1987, 261–265.

[Yao82a] A. Yao. "Protocols for Secure Computations." *Proceedings of the* 23rd *FOCS,* IEEE, 1982, 160–164.

[Yao82b] A. Yao. "Theory and Applications of Trapdoor Functions." *Proceedings of the* 23rd *FOCS,* IEEE, 1982, 80–91.

Author Index

Alber, J., 329
Atallah, M. J., 55

Baeza-Yates, R. A., 319
Beaver, D., 361
Bernasconi, A., 339
De Bonis, A., 127
Brattka, V., 184

Cai, J.-Y., 174
Chao, H. S., 309
Chao, K.-M., 299
Chen, D. Z., 45, 55
Chin, F. Y., 25

Daescu, O., 45
Dai, Y., 15
Du, D.-Z., 219

Eğecioğlu, Ö., 117

Fischer, M. J., 3
Fleischer, R., 65

Gargano, L., 127
Ge, Y., 289
Graf, T., 35
Graham, F. C., 2
Grebinski, V., 194

Hemaspaandra, L. A., 174
Hertrampf, U., 157
Ho, C-.W., 279
Hsu, F. R., 309
Hsu, L.-H., 269
Hsu, T.-s., 241, 279
Hwang, F. K., 204

Ibaraki, T., 105
Iwama, K., 229

Kamakoti, V., 35
Katoh, N., 15
Kim, D. S., 219
Kiwi, M., 137
Ko, M.-T., 279
Kosaraju, R. S., 4

Latha, N. S. J., 35
Lee, R. C. T., 309
Li, M., 85
Lin, M.-Y., 269
Lin, W.-D., 204
Lopez, D. R., 241
López-Ortiz, A., 75
Lu, C.-J., 167

Mans, B., 251
Miyano, E., 229
Mizuki, T., 349

Niedermeier, R., 329
Nishizeki, T., 349

Papadimitriou, C. H., 1
Pappalardi, F., 251
Peinado, M., 117
Peng, S.-L., 279
Pop, M., 4

Rangan, C. P., 35
Reith, S., 147

Schuierer, S., 75
Shizuya, H., 349
Shparlinski, I., 251
Spielman, D. A., 137
Sung, T.-Y., 269

Tajima, S., 229
Tamaki, H., 229
Tang, C. Y., 279, 299
Teng, S.-H., 137
Tsai, K.-H., 213

Vaccaro, U., 127

Wagner, K. W., 147
Wang, C. A., 25
Wang, D.-W., 213
Wang, J., 289
Wang, J.-J., 269
Wechsung, G., 174

Weihrauch, K., 184
Wu, B. Y., 299

Xue, G., 95

Yagiura, M., 105

Yang, B. T., 25

Zang, W., 261
Zhang, L., 85
Zheng, X., 184

Lecture Notes in Computer Science

For information about Vols. 1–1371

please contact your bookseller or Springer-Verlag

Vol. 1372: S. Vaudenay (Ed.), Fast Software Encryption. Proceedings, 1998. VIII, 297 pages. 1998.

Vol. 1373: M. Morvan, C. Meinel, D. Krob (Eds.), STACS 98. Proceedings, 1998. XV, 630 pages. 1998.

Vol. 1374: H. Bunt, R.-J. Beun, T. Borghuis (Eds.), Multimodal Human-Computer Communication. VIII, 345 pages. 1998. (Subseries LNAI).

Vol. 1375: R. D. Hersch, J. André, H. Brown (Eds.), Electronic Publishing, Artistic Imaging, and Digital Typography. Proceedings, 1998. XIII, 575 pages. 1998.

Vol. 1376: F. Parisi Presicce (Ed.), Recent Trends in Algebraic Development Techniques. Proceedings, 1997. VIII, 435 pages. 1998.

Vol. 1377: H.-J. Schek, F. Saltor, I. Ramos, G. Alonso (Eds.), Advances in Database Technology – EDBT'98. Proceedings, 1998. XII, 515 pages. 1998.

Vol. 1378: M. Nivat (Ed.), Foundations of Software Science and Computation Structures. Proceedings, 1998. X, 289 pages. 1998.

Vol. 1379: T. Nipkow (Ed.), Rewriting Techniques and Applications. Proceedings, 1998. X, 343 pages. 1998.

Vol. 1380: C.L. Lucchesi, A.V. Moura (Eds.), LATIN'98: Theoretical Informatics. Proceedings, 1998. XI, 391 pages. 1998.

Vol. 1381: C. Hankin (Ed.), Programming Languages and Systems. Proceedings, 1998. X, 283 pages. 1998.

Vol. 1382: E. Astesiano (Ed.), Fundamental Approaches to Software Engineering. Proceedings, 1998. XII, 331 pages. 1998.

Vol. 1383: K. Koskimies (Ed.), Compiler Construction. Proceedings, 1998. X, 309 pages. 1998.

Vol. 1384: B. Steffen (Ed.), Tools and Algorithms for the Construction and Analysis of Systems. Proceedings, 1998. XIII, 457 pages. 1998.

Vol. 1385: T. Margaria, B. Steffen, R. Rückert, J. Posegga (Eds.), Services and Visualization. Proceedings, 1997/1998. XII, 323 pages. 1998.

Vol. 1386: T.A. Henzinger, S. Sastry (Eds.), Hybrid Systems: Computation and Control. Proceedings, 1998. VIII, 417 pages. 1998.

Vol. 1387: C. Lee Giles, M. Gori (Eds.), Adaptive Processing of Sequences and Data Structures. Proceedings, 1997. XII, 434 pages. 1998. (Subseries LNAI).

Vol. 1388: J. Rolim (Ed.), Parallel and Distributed Processing. Proceedings, 1998. XVII, 1168 pages. 1998.

Vol. 1389: K. Tombre, A.K. Chhabra (Eds.), Graphics Recognition. Proceedings, 1997. XII, 421 pages. 1998.

Vol. 1390: C. Scheideler, Universal Routing Strategies for Interconnection Networks. XVII, 234 pages. 1998.

Vol. 1391: W. Banzhaf, R. Poli, M. Schoenauer, T.C. Fogarty (Eds.), Genetic Programming. Proceedings, 1998. X, 232 pages. 1998.

Vol. 1392: A. Barth, M. Breu, A. Endres, A. de Kemp (Eds.), Digital Libraries in Computer Science: The MeDoc Approach. VIII, 239 pages. 1998.

Vol. 1393: D. Bert (Ed.), B'98: Recent Advances in the Development and Use of the B Method. Proceedings, 1998. VIII, 313 pages. 1998.

Vol. 1394: X. Wu. R. Kotagiri, K.B. Korb (Eds.), Research and Development in Knowledge Discovery and Data Mining. Proceedings, 1998. XVI, 424 pages. 1998. (Subseries LNAI).

Vol. 1395: H. Kitano (Ed.), RoboCup-97: Robot Soccer World Cup I. XIV, 520 pages. 1998. (Subseries LNAI).

Vol. 1396: E. Okamoto, G. Davida, M. Mambo (Eds.), Information Security. Proceedings, 1997. XII, 357 pages. 1998.

Vol. 1397: H. de Swart (Ed.), Automated Reasoning with Analytic Tableaux and Related Methods. Proceedings, 1998. X, 325 pages. 1998. (Subseries LNAI).

Vol. 1398: C. Nédellec, C. Rouveirol (Eds.), Machine Learning: ECML-98. Proceedings, 1998. XII, 420 pages. 1998. (Subseries LNAI).

Vol. 1399: O. Etzion, S. Jajodia, S. Sripada (Eds.), Temporal Databases: Research and Practice. X, 429 pages. 1998.

Vol. 1400: M. Lenz, B. Bartsch-Spörl, H.-D. Burkhard, S. Wess (Eds.), Case-Based Reasoning Technology. XVIII, 405 pages. 1998. (Subseries LNAI).

Vol. 1401: P. Sloot, M. Bubak, B. Hertzberger (Eds.), High-Performance Computing and Networking. Proceedings, 1998. XX, 1309 pages. 1998.

Vol. 1402: W. Lamersdorf, M. Merz (Eds.), Trends in Distributed Systems for Electronic Commerce. Proceedings, 1998. XII, 255 pages. 1998.

Vol. 1403: K. Nyberg (Ed.), Advances in Cryptology – EUROCRYPT '98. Proceedings, 1998. X, 607 pages. 1998.

Vol. 1404: C. Freksa, C. Habel. K.F. Wender (Eds.), Spatial Cognition. VIII, 491 pages. 1998. (Subseries LNAI).

Vol. 1405: S.M. Embury, N.J. Fiddian, W.A. Gray, A.C. Jones (Eds.), Advances in Databases. Proceedings, 1998. XII, 183 pages. 1998.

Vol. 1406: H. Burkhardt, B. Neumann (Eds.), Computer Vision – ECCV'98. Vol. I. Proceedings, 1998. XVI, 927 pages. 1998.

Vol. 1407: H. Burkhardt, B. Neumann (Eds.), Computer Vision – ECCV'98. Vol. II. Proceedings, 1998. XVI, 881 pages. 1998.

Vol. 1409: T. Schaub, The Automation of Reasoning with Incomplete Information. XI, 159 pages. 1998. (Subseries LNAI).

Vol. 1411: L. Asplund (Ed.), Reliable Software Technologies – Ada-Europe. Proceedings, 1998. XI, 297 pages. 1998.

Vol. 1412: R.E. Bixby, E.A. Boyd, R.Z. Ri´´os-Mercado (Eds.), Integer Programming and Combinatorial Optimization. Proceedings, 1998. IX, 437 pages. 1998.

Vol. 1413: B. Pernici, C. Thanos (Eds.), Advanced Information Systems Engineering. Proceedings, 1998. X, 423 pages. 1998.

Vol. 1414: M. Nielsen, W. Thomas (Eds.), Computer Science Logic. Selected Papers, 1997. VIII, 511 pages. 1998.

Vol. 1415: J. Mira, A.P. del Pobil, M.Ali (Eds.), Methodology and Tools in Knowledge-Based Systems. Vol. I. Proceedings, 1998. XXIV, 887 pages. 1998. (Subseries LNAI).

Vol. 1416: A.P. del Pobil, J. Mira, M.Ali (Eds.), Tasks and Methods in Applied Artificial Intelligence. Vol.II. Proceedings, 1998. XXIII, 943 pages. 1998. (Subseries LNAI).

Vol. 1417: S. Yalamanchili, J. Duato (Eds.), Parallel Computer Routing and Communication. Proceedings, 1997. XII, 309 pages. 1998.

Vol. 1418: R. Mercer, E. Neufeld (Eds.), Advances in Artificial Intelligence. Proceedings, 1998. XII, 467 pages. 1998. (Subseries LNAI).

Vol. 1419: G. Vigna (Ed.), Mobile Agents and Security. XII, 257 pages. 1998.

Vol. 1420: J. Desel, M. Silva (Eds.), Application and Theory of Petri Nets 1998. Proceedings, 1998. VIII, 385 pages. 1998.

Vol. 1421: C. Kirchner, H. Kirchner (Eds.), Automated Deduction – CADE-15. Proceedings, 1998. XIV, 443 pages. 1998. (Subseries LNAI).

Vol. 1422: J. Jeuring (Ed.), Mathematics of Program Construction. Proceedings, 1998. X, 383 pages. 1998.

Vol. 1423: J.P. Buhler (Ed.), Algorithmic Number Theory. Proceedings, 1998. X, 640 pages. 1998.

Vol. 1424: L. Polkowski, A. Skowron (Eds.), Rough Sets and Current Trends in Computing. Proceedings, 1998. XIII, 626 pages. 1998. (Subseries LNAI).

Vol. 1425: D. Hutchison, R. Schäfer (Eds.), Multimedia Applications, Services and Techniques – ECMAST'98. Proceedings, 1998. XVI, 532 pages. 1998.

Vol. 1427: A.J. Hu, M.Y. Vardi (Eds.), Computer Aided Verification. Proceedings, 1998. IX, 552 pages. 1998.

Vol. 1430: S. Trigila, A. Mullery, M. Campolargo, H. Vanderstraeten, M. Mampaey (Eds.), Intelligence in Services and Networks: Technology for Ubiquitous Telecom Services. Proceedings, 1998. XII, 550 pages. 1998.

Vol. 1431: H. Imai, Y. Zheng (Eds.), Public Key Cryptography. Proceedings, 1998. XI, 263 pages. 1998.

Vol. 1432: S. Arnborg, L. Ivansson (Eds.), Algorithm Theory – SWAT '98. Proceedings, 1998. IX, 347 pages. 1998.

Vol. 1433: V. Honavar, G. Slutzki (Eds.), Grammatical Inference. Proceedings, 1998. X, 271 pages. 1998. (Subseries LNAI).

Vol. 1434: J.-C. Heudin (Ed.), Virtual Worlds. Proceedings, 1998. XII, 412 pages. 1998. (Subseries LNAI).

Vol. 1435: M. Klusch, G. Weiß (Eds.), Cooperative Information Agents II. Proceedings, 1998. IX, 307 pages. 1998. (Subseries LNAI).

Vol. 1436: D. Wood, S. Yu (Eds.), Automata Implementation. Proceedings, 1997. VIII, 253 pages. 1998.

Vol. 1437: S. Albayrak, F.J. Garijo (Eds.), Intelligent Agents for Telecommunication Applications. Proceedings, 1998. XII, 251 pages. 1998. (Subseries LNAI).

Vol. 1438: C. Boyd, E. Dawson (Eds.), Information Security and Privacy. Proceedings, 1998. XI, 423 pages. 1998.

Vol. 1439: B. Magnusson (Ed.), System Configuration Management. Proceedings, 1998. X, 207 pages. 1998.

Vol. 1441: W. Wobcke, M. Pagnucco, C. Zhang (Eds.), Agents and Multi-Agent Systems. Proceedings, 1997. XII, 241 pages. 1998. (Subseries LNAI).

Vol. 1443: K.G. Larsen, S. Skyum, G. Winskel (Eds.), Automata, Languages and Programming. Proceedings, 1998. XVI, 932 pages. 1998.

Vol. 1444: K. Jansen, J. Rolim (Eds.), Approximation Algorithms for Combinatorial Optimization. Proceedings, 1998. VIII, 201 pages. 1998.

Vol. 1445: E. Jul (Ed.), ECOOP'98 – Object-Oriented Programming. Proceedings, 1998. XII, 635 pages. 1998.

Vol. 1446: D. Page (Ed.), Inductive Logic Programming. Proceedings, 1998. VIII, 301 pages. 1998. (Subseries LNAI).

Vol. 1448: M. Farach-Colton (Ed.), Combinatorial Pattern Matching. Proceedings, 1998. VIII, 251 pages. 1998.

Vol. 1449: W.-L. Hsu, M.-Y. Kao (Eds.), Computing and Combinatorics. Proceedings, 1998. XII, 372 pages. 1998.

Vol. 1451: A. Amin, D. Dori, P. Pudil, H. Freeman (Eds.), Advances in Pattern Recognition. Proceedings, 1998. XXI, 1048 pages. 1998.

Vol. 1452: B.P. Goettl, H.M. Halff, C.L. Redfield, V.J. Shute (Eds.), Intelligent Tutoring Systems. Proceedings, 1998. XIX, 629 pages. 1998.

Vol. 1453: M.-L. Mugnier, M. Chein (Eds.), Conceptual Structures: Theory, Tools and Applications. Proceedings, 1998. XIII, 439 pages. (Subseries LNAI).

Vol. 1454: I. Smith (Ed.), Artificial Intelligence in Structural Engineering. XI, 497 pages. 1998. (Subseries LNAI).

Vol. 1456: A. Drogoul, M. Tambe, T. Fukuda (Eds.), Collective Robotics. Proceedings, 1998. VII, 161 pages. 1998. (Subseries LNAI).

Vol. 1457: A. Ferreira, J. Rolim, H. Simon, S.-H. Teng (Eds.), Solving Irregularly Structured Problems in Prallel. Proceedings, 1998. X, 408 pages. 1998.

Vol. 1458: V.O. Mittal, H.A. Yanco, J. Aronis, R-. Simpson (Eds.), Assistive Technology in Artificial Intelligence. X, 273 pages. 1998. (Subseries LNAI).

Vol. 1461: G. Bilardi, G.F. Italiano, A. Pietracaprina, G. Pucci (Eds.), Algorithms – ESA'98. Proceedings, 1998. XII, 516 pages. 1998.

Vol. 1464: H.H.S. Ip, A.W.M. Smeulders (Eds.), Multimedia Information Analysis and Retrieval. Proceedings, 1998. VIII, 264 pages. 1998.

Lecture Notes in Comp[uter Science] 840

Edited by G. Goos and J. Hartma[nis]

Advisory Board: W. Brauer D. Gries